Lecture Notes in Computer Science   14966

Founding Editor

Juris Hartmanis

The series Lecture Notes in Computer Science (LNCS), including its subseries Lecture Notes in Artificial Intelligence (LNAI) and Lecture Notes in Bioinformatics (LNBI), has established itself as a medium for the publication of new developments in computer science and information technology research, teaching, and education.

LNCS enjoys close cooperation with the computer science R & D community, the series counts many renowned academics among its volume editors and paper authors, and collaborates with prestigious societies. Its mission is to serve this international community by providing an invaluable service, mainly focused on the publication of conference and workshop proceedings and postproceedings. LNCS commenced publication in 1973.

Andrzej Dąbrowski · Josef Pieprzyk ·
Jacek Pomykała

Editors

Number-Theoretic Methods
in Cryptology

4th International Conference, NuTMiC 2024
Szczecin, Poland, June 24–26, 2024
Revised Selected Papers

 Springer

Editors
Andrzej Dąbrowski [iD]
University of Szczecin
Szczecin, Poland

Jacek Pomykała [iD]
University of Warsaw
Warsaw, Poland

Military University of Technology
Warsaw, Poland

Josef Pieprzyk [iD]
Polish Academy of Sciences
Warsaw, Poland

Data61, CSIRO
Sydney, Australia

ISSN 0302-9743   ISSN 1611-3349 (electronic)
Lecture Notes in Computer Science
ISBN 978-3-031-82379-4   ISBN 978-3-031-82380-0 (eBook)
https://doi.org/10.1007/978-3-031-82380-0

This Springer imprint is published by the registered company Springer Nature Switzerland AG
The registered company address is: Gewerbestrasse 11, 6330 Cham, Switzerland

If disposing of this product, please recycle the paper.

Preface

This volume contains a collection of papers presented at the NuTMiC 2024 conference. The conference was held at the Institute of Mathematics of the University of Szczecin, Szczecin, Poland, from June 24–26, 2024. It was the fourth edition of the series of conferences whose theme is Number-Theoretic Methods in Cryptology. The first edition was held at the University of Warsaw (Poland) in September 2017, the second edition was held at Paris-Sorbonne University (France) in June 2019, and the third edition was held at Adam Mickiewicz University in Poznań (Poland) in August 2022.

The aim of the NuTMiC conference series is to cross-fertilize Number Theory and Cryptology. On the one hand, the conference intends to explore current and future Number Theory challenges that arise from the rapidly evolving field of modern Cryptology. On the other hand, it targets investigating and applying number-theoretic methods in the design and analysis of cryptologic systems and protocols. Besides the well-established connections between the two domains, such as primality testing, factorization, elliptic and hyperelliptic curves, and lattices (to name a few), the conference endeavors to forge new ones that would encompass Number Theory structures and algorithms that have not yet been successfully used in Cryptology. It is expected that these new connections would lead to novel, more efficient, and more secure cryptographic systems and protocols. Conversely, addressing problems of cryptographic origins may help advance the state of the art on some problems and conjectures in Number Theory.

During the conference, 14 talks were delivered, including 5 invited talks and 9 contributed presentations. This book features 11 full papers (including 2 invited papers), carefully reviewed and selected from a total of 12 submissions. The papers explore a range of topics, including:

- Algorithms for elliptic isogeny computations over extensions of finite fields,
- Higher-dimensional isogeny representations,
- Elliptic-curve factorization using oracles,
- Methods for constructing families of pairing-friendly elliptic curves,
- Applications of learning with errors in cyclic division algebras,
- Average-case analysis of the strong Lucas test,
- The Okamoto-Uchiyama cryptosystem,
- Signature schemes and constructions,
- Message recovery attacks targeting the NTRU encryption algorithm.

We wish to thank the Program Committee members and the external reviewers for their time and effort. We also thank the local organizers who made the conference a success. We express our deep appreciation to the authors for their contributions. We would like to express our appreciation to Springer for their support and assistance in producing the conference proceedings.

We highly appreciate the support the conference received from the Rector of the University of Szczecin.

November 2024

Andrzej Dąbrowski
Josef Pieprzyk
Jacek Pomykała

Organization

General Chair

Andrzej Dąbrowski (Chair)   University of Szczecin, Poland

Organizing Committee

Ewa Ciechanowicz   University of Szczecin, Poland
Tomasz Jędrzejak   University of Szczecin, Poland
Dawid Kędzierski   University of Szczecin, Poland
Lucjan Szymaszkiewicz   University of Szczecin, Poland

| Daniel Waszkiewicz | National Institute of Telecommunications, Poland |
| Małgorzata Wieczorek | University of Szczecin, Poland |

Program Committee Chairs

Andrzej Dąbrowski	University of Szczecin, Poland
Josef Pieprzyk	IPI PAN, Poland and Data61, CSIRO, Australia
Jacek Pomykała	University of Warsaw and Military University of Technology, Poland

Program Committee

Grzegorz Banaszak	Adam Mickiewicz University, Poznań, Poland
Nicolas Courtois	University College London, UK
Andrej Dujella	University of Zagreb, Croatia
Christian Elsholtz	Technische Universität Graz, Austria
Faruk Gologlu	Charles University in Prague, Czechia
Louis Goubin	University of Versailles, France
Maciej Grześkowiak	Adam Mickiewicz University, Poznań, Poland
Tomasz Jędrzejak	University of Szczecin, Poland
Zbigniew Jelonek	IMPAN, Poland
Heiko Knospe	Technische Hochschule Köln, Germany
Przemysław Koprowski	University of Silesia, Poland
Piotr Krasoń	University of Szczecin, Poland
Mieczysław Kula	University of Silesia, Poland
Krystian Matusiewicz	Intel, Poland
Alexei A. Panchishkin	Université Grenoble Alpes, France
Andrzej Paszkiewicz	POL Cyber Command, Poland
Jerzy Pejaś	West Pomeranian University of Technology, Poland
Rene Peralta	NIST, USA
Olivier Ramaré	Aix-Marseille Université, France
Piotr Sapiecha	Intel, Poland
Mariusz Skałba	University of Warsaw, Poland
Gökhan Soydan	Bursa Uludağ University, Turkey
Jorge Urroz	Universitat Politècnica de Catalunya, Spain
Huaxiong Wang	Nanyang Technological University, Singapore
Michał Wroński	Military University of Technology, Poland
Chaoping Xing	Shanghai Jiao Tong University, China

Additional Revievers

Francois Arnault
Raul Falcon
Antonin Leroux
Carles Padro
Roope Vehkalahti

Abstracts of Invited Talks

Slow-Boiled Frogs

Daniel Bernstein

University of Illinois at Chicago, USA

In 2013, I introduced a name for cryptography that simply works, solidly resists attacks, and never needs any upgrades: I called it boring cryptography. This talk is about the opposite extreme, which is called lattice-based cryptography. I'll talk about some general context and some number-theoretic issues that appear in the area.

A Pot-Pourri on Equidistribution

Philippe Michel

École Polytechnique Fédérale de Lausanne, Switzerland

Since the seminal work of H. Weyl (who established that if α is irrational the sequence $(\alpha n)_{n \geq 1}$ is equidistributed modulo 1), equidistribution has played a central role in number theory and its applications. In this talk I will discuss recent examples of equidistribution for various sequences of arithmetic objects, the techniques used to establish these as well as some (number theoretic) applications with the hope that these will trigger further ideas from cryptographers.

Elliptic-Curve Factorization, Witnesses and Oracles

Jacek Pomykała

University of Warsaw and Military University of Technology, Warsaw, Poland

The survey talk concerns the selected approaches to integer factorization problem of positive integer N, with the aid of elliptic curves E over $\mathbb{Z}_N$. I will mainly focus on the investigation of various oracles related to EC-based factorization in one approach and on the investigation of so-called decomposition witnesses in the other. The second approach covers two aspects - first dealing with the extension of the classical Fermat factoring method for elliptic curves and the second inspired by familiar Lenstra's results based on the distribution of B-smooth local orders $\#E(\mathbb{F}_r)$ for prime $r|N$. Here the more detailed analysis concerns the investigation of parameters β and σ allowing to simultaneously control the B-smooth factor and the squarefree divisor of non B-smooth factor of $\#E(\mathbb{F}_r)$ in the suitable ranges. They are responsible for the choice the admissible elliptic curve E over $\mathbb{Z}_N$ and the deterministic time of factorization of N with the aid of witness $(u, v) \in \mathbb{Z}_N^2$.

On the Efficient Representation of Isogenies

Damien Robert

Université de Bordeaux, France

We survey different representations of isogenies between elliptic curves. Notably, we focus on the novel representation obtained via the techniques in the SIDH attacks, which showed that every isogeny admits an efficient representation by embedding it into a higher dimensional smooth degre isogeny. We explain how to work in practice with this representation (equality testing, division, push forwards and splittings), and discuss some applications, both in number theory (computation of endomorphism rings of ordinary elliptic curves, canonical lifts, deformations) and isogeny based cryptography (new signature schemes, class group action).

Integers of Prescribed Arithmethic Structure in Residue Classes

Igor E. Shparlinski

University of New South Wales, Sydney, Australia

We give an overview of recent results about the distribution of some special integers in residues classes modulo a large integer q. Questions of this type were introduced by Erdös, Odlyzko and Sarkozy (1987), who considered products of two primes as a relaxation of the classical question about the distribution of primes in residue classes. Since that time, numerous variations have appeared for different sequences of integers. The types of numbers we discuss include smooth, square-free, square-full and almost primes integers. We also expose, without going into technical details, the wealth of different techniques behind these results: sieve methods, bounds of short Kloosterman sums, bounds of short character sums and many others.

Contents

Invited Talks

On the Efficient Representation
of Isogenies
A Survey for NuTMiC 2024

Damien Robert[1,2]([✉])

[1] INRIA Bordeaux-Sud-Ouest, 200 avenue de la Vieille Tour,
33405 Talence Cedex, France
damien.robert@inria.fr
[2] Institut de Mathématiques de Bordeaux, 351 cours de la liberation,
33405 Talence Cedex, France

Abstract. We survey different (efficient or not) representations of isogenies, with a particular focus on the recent "higher dimensional" isogeny representation, and algorithms to manipulate them.

1 Introduction

The field of isogeny based cryptography changed drastically following the SIDH attacks [CD23, MMPPW23, Rob23b]. Indeed, the main byproduct of these attacks is a new efficient representation of isogenies, which we will call the *higher dimensional representation* or *HD representation*. This new representation quickly found cryptographic applications: SQIsignHD [DLRW24], FESTA [BMP23] and QFESTA [NO23], the Deuring VRF [Ler23b], an isogeny VDF [DMS23], SCALLOP-HD [CLP24], IS-CUBE [Mor23], LIT-SiGamal [Mor24], SILBE [DFV24], POKE [Bas24], SQIsign2d (West and East) [BDD+24, NO24], SQIPrime [DF24]... It gave rise to new methods to convert ideals to isogenies, both in the supersingular case [Ler23b, ON24, BDD+24] and in the oriented case [PR23b]. Apart from protocols, the HD representation was used to obtain new or better security reductions for isogeny based cryptography [MW23, ACD+23, PW24, ES24], and also better (classical) security reductions for the DLP between isogeneous elliptic curves [Gal24]. Finally, it also gave applications in number theory: computing the endomorphism ring of an ordinary elliptic curve E in polynomial time (if we are provided with the factorisation of the discriminant Δ_E), point counting for $E/\mathbb{F}_{p^n}$ in $\widetilde{O}(n^2 \log^{O(1)} p)$, canonical lift of an ordinary $E/\mathbb{F}_{p^n}$ to precision m in $\widetilde{O}(mn \log^{O(1)} p)$, and new algorithms to compute modular polynomials [Rob22b, KR24].

1.1 History

The groundbreaking idea to attack the SIDH supersingular elliptic curve cryptosystem [JD11, DJP14] using higher dimensional isogenies is due to Castryck and Decru, and independently to Maino and Martindale.

A. Dąbrowski et al. (Eds.): NuTMiC 2024, LNCS 14966, pp. 3–84, 2025.
https://doi.org/10.1007/978-3-031-82380-0_1

In 2022, Decru was working on building a VDF (Verifiable Delay Function) using isogenies in dimension 2. He realised that exploiting isogenies $E_1 \times E_2 \to E_1' \times E_2'$ between products of elliptic curves could potentially be used to attack the SIDH cryptosystem. He fully worked this attack out with Castryck, and in their preliminary article [CD22] (which was later published as [CD23]) they gave a working Magma implementation to break SIDH in (heuristic) polynomial time from the special starting curve E_0. In that attack, they made a crucial use of a technical result (now called Kani's lemma) due tu Kani [Kan97, § 2, Proof of Th. 2.3], and which will be key to derive the HD representation. The reason the attack works over the special curve E_0 is that it contains many known endomorphisms: notably endomorphisms of the form $a + bi$ which are of degrees $a^2 + b^2$. In the first version of their paper, Castryck and Decru sketched how their attack could be extended to an arbitrary starting supersingular curve E, but that it would be unlikely to be practical. De Feo[1] pointed out that this still gave an (heuristic) subexponential attack for an arbitrary starting curve E with unknown endomorphisms. And Wesolowski, in a note, also explained how Castryck and Decru's polynomial attack could extend to such an arbitrary E, provided that End(E) was known.

Independently, Maino, who had visited the Cosic group in Leuven earlier, during which he had discussed the dimension 2 VDF with Castryck and Decru, also had realised that dimension 2 isogenies could be used to attack SIDH. He first sketched with Martindale an effective attack path in [MM22], and then they gave a more fledged out attack, along with a Sage implementation, and with contributions by Panny, Pope and Wesolowski (and extra help by De Feo and Oudompheng) in [MMPPW23]. The original attack of [CD23], used a decisional version: they guess part of Alice's secret isogeny, then check using higher dimensional isogenies if that guess is correct. This allowed them to reconstruct Alice's isogeny step by step. By contrast the version of Maino and Martindale was direct: they used dimension 2 isogenies to directly recover Alice's secret isogeny. More precisely, they focused on an arbitrary starting curve E, and after a precomputation step to select appropriate parameters, they gave a direct algorithm to recover the secret isogenies[2]. The complexity of that direct key recovery depended on the parameters found, but the same heuristic complexity analysis as done by De Feo for Castryck and Decru's attack showed that it gave a subexponential attack. Combining the direct attack of Maino-Martindale with the exploitation of the special known endomorphisms of E_0 as in Castryck and Decru's version considerably improved the practical key recovery attack when the starting curve was E_0. In the SageMath reimplementation of the Magma code of Castryck and Decru, contributed by Oudompheng, Panny and Pope among others, incorporating this direct key recovery made the attack go from minutes or hours to seconds

---

[1] Private communication.

[2] After the first version of [CD23] was published, this direct key recovery improvement had also been independently found out by Oudompheng, Petit and Wesolowksi.

or minutes[3] (depending on the security parameters). This was the first hint that these attacks could potentially be applied for constructive use.

Once the idea of using dimension 2 isogenies to attack SIDH was introduced, it was natural to look at whether using even higher dimensions could improve these attacks: notably for a starting curve E with unknown endomorphisms (Wesolowski had the same idea (see footnote 1)). In [Rob23b], we explained how to combine Kani's lemma as used in [CD23, MMPPW23] with Zarhin's trick [Zar74] to attack SIDH in (proven) polynomial time even with a random starting curve E, by going to dimension up to $g = 8$. Indeed, the main obstacle preventing the polynomial time attack of [CD23] on the special curve E_0 to be applied to a random curve E is the lack of known endomorphisms (apart from integer multiplications) on E. Zarhin's trick solves this problem: we can always build many endomorphisms on E^2 and E^4 respecting the product principal polarisation by using suitable integer matrices. So we could apply the Castryck-Decru-Maino-Martindale attack replacing E by E^4, and using dimension 8 isogenies rather than dimension 2 isogenies. By luck, we had already used Zarhin's trick in our algorithm [CR15] to compute higher dimensional isogenies, so it was natural to apply it to the SIDH attack too. This also required to extend Kani's technical lemma from elliptic curves to abelian varieties, but that extension was completely straightforward[4].

In [Rob23b], the main algorithmic tool used for the attack was the following *embedding lemma*: for any $N > n$ coprime to n, an n-isogeny $\phi : E_1 \to E_2$ can be efficiently embedded into an N-isogeny $\Phi : A \to B$ in higher dimension $g = 2, 4, 8$, provided we know how ϕ acts on the N-torsion of E_1. More generally, the embedding lemma allows to embed an n-isogeny in dimension g into an N-isogeny in dimension $2g, 4g, 8g$. (For the embedding lemma with special curves like E_0, then g can often be smaller than for a generic curve, see Remark 13).

In the conclusion of [Rob23b], we asked the following question: "This tool allows one to break SIDH efficiently in all cases. Can it also be used to build new isogeny based cryptosystems?" One week after the first version of that article, we made in [Rob22a] the rather obvious remark that by taking N powersmooth above, the embedding lemma proves that any isogeny admitted an efficient representation, meaning a representation taking polynomial (in $\log n$ and $\log q$) space and time: this is the HD representation (although it was not given a name in that paper)! The conclusion of that article was: "The method presented above shows that the efficient computation of isogenies for higher dimensional abelian varieties has interesting algorithmic applications to elliptic curves. Hopefully, this is the start of many new results in this direction".

---

[3] With the low level C and Rust implementation we now have of the improved formulas for dimension 2 2^e-isogenies [DMPR24], what would take a few seconds in 2022 would now take only a few ms!.

[4] It also helped that we were already familiar with some of Kani's work which we had used to understand the denominators of Hilbert modular polynomials [MR19, Rob21, § 5.3.6].

And indeed, as illustrated by all the applications above, the HD representation quickly found lots of use. Still, these new applications are not immediate: the HD representation shows that if we know the evaluation of ϕ on sufficiently many nice points, we can efficiently evaluate ϕ everywhere. But it seems that we have a bootstrapping problem: how can we evaluate ϕ on these nice points to begin with? To answer this question, many new algorithms have been developed to work directly with these HD representations: divisions, duals, splittings, pushforwards...

1.2   A Survey

Almost two years after the SIDH attack, the NuTMiC (Number-Theoretic Methods in Cryptology) conference held in Szczecin seemed like a nice occasion to do a survey on the use of the HD representation and the algorithms developed to work with them.

Of course, the HD representation is not the only useful representation of an isogeny, so for this survey we will try to briefly explain the many ways we can represent an isogeny, and how we can convert between all these different representations.

In his invited talk for Eurocrypt 2024, Castryck gave a wonderful talk on: "An Attack Became a Tool: Isogeny-based Cryptography 2.0". The aim of this survey is to give an overview of the algorithms that have been developed for this renewal of isogeny based cryptography, so that hopefully they become accessible to a broader audience.

We apologize for the length of this survey, which is longer than initially expected.

1.3   Thanks

We thank Andrea Basso for several useful comments.

1.4   Outline

In Sect. 2, we give an overview of the different isogeny representations we will describe in this survey, what we mean by an efficient representation, and an overview of the algorithms we now have to work with efficient representations that may not be given (anymore!) by kernel generators of smooth order.

In Sect. 3, we survey the "standard" isogeny representations, meaning the ones that do not use higher dimension.

In Sect. 4, we survey the ideal representations of "horizontal" isogenies. Efficient ideal to isogeny algorithms have been key to develop efficient isogeny based cryptosystems (like SQIsign [DKLPW20, DLLW23]). The original version of SQIsign relied on the KLPT algorithm [KLPT14] for this conversion. But newer versions (SQIsignHD, SQIsign2d) have switched to an ideal to isogeny algorithm relying on the HD representation. We give a very brief overview of these different methods, both for the supersingular case and the oriented case.

In Sect. 5, we introduce in more details the HD representation. And then in Sect. 6, we describe algorithms to work with efficient representations of isogenies.

Finally in Sect. 7 we give a list of open questions.

In the appendices, we treat some technical subjects in more details. In Appendix A we look at the accessible torsion in an elliptic curve (or abelian variety). In Appendix B we explain how to relax the torsion requirement for the HD representation from $N > n$ to $N^2 > n$. In Appendix C, we give more details on the different generations of the ideal to isogeny algorithms in the supersingular case, and how they each gave rise to improvements to SQIsign. In Appendix D we give a geometric interpretation of the usual class group exact sequence for a non maximal quadratic order R, and explain the relationship between this geometric interpretation and the level structure encoded by going up isogenies. Finally in Appendix E we give some technical remarks on the Kodaira-Spencer isomorphism, which we need for the deformation representation and modular representation of isogenies in higher dimension.

2 Overview

In Sect. 2.1 we define what we mean by an efficient isogeny. Then in Sect. 2.2 we quickly review the "classical" representations of an isogeny, in Sect. 2.3 the ideal representations, and then we look at the HD representation in Sect. 2.4. We survey existing algorithms on the HD representation in Sect. 2.5.

2.1 Efficient Representation of an Isogeny

Let $\phi : E_1 \to E_2$ be an isogeny of degree n between elliptic curves defined over a field k (we will also say that ϕ is an n-isogeny). A representation of ϕ is any data that encodes the domain E_1, the codomain E_2, the degree n, along with a way (an algorithm) to evaluate the image by ϕ of a point $P \in E_1(k')$, where k'/k is a field extension. It can be useful to relax the condition that P is defined over a field, and allow for points over a k-algebra, notably to be able to work with formal points and their images.

In this survey paper, we will contend ourselves to work with this informal definition. For a more formal definition of the representation of an isogeny, we refer to [Ler22a].

For simplicity, we will assume that k is a finite field $k = \mathbb{F}_q$, $q = p^d$, and that the degree n of ϕ is prime to p. We will also stick to isogenies between elliptic curves, although many of the constructions we will introduce generalise to principally polarised abelian varieties (and some of our constructions for elliptic curves will actually use abelian varieties of dimension $g > 1$ as explained in Sect. 1). In practice, we will also always work with Weierstrass equations for E_1, E_2. For most applications, one can also assume that ϕ is cyclic, that is that its kernel is isomorphic as a group to $\mathbb{Z}/n\mathbb{Z}$, although for some applications it will be convenient to treat all cases.

We will say that an isogeny representation is *compact*, or *space efficient*, if the data to encode it is polynomial in $\log n$ and $\log q$. We will say that it is *efficient* if, from the encoded data, it can evaluate the image of a point $P \in E_1(\mathbb{F}_{q'})$ in time polynomial in $\log n$ (so polylogarithmic in n) and $\log(q')$.

Remark 1 (Efficient versus practical representations). With our definitions, an isogeny representation that allows to compute the image of a point in $O(\log^C n)$ arithmetic operations will be an efficient representation, even if C is a big constant. For *practical* applications of isogeny representations, rather than theoretical results, we will want our representations to be computable in *practice*, and as *fast* as possible.

For instance, the HD representation embeds an n-isogeny ϕ in dimension 1 into an N isogeny Φ in dimension $g = 2, 4, 8$. Taking N smooth and the N-torsion accessible (see Appendix A for this notion), we can decompose Φ into a product of small ℓ-isogenies, with $\ell = O(\log n)$. So for this choice of N, the HD representation is efficient according to our definitions.

But there will be a huge timing difference, depending on whether we can take $g = 2$ and $N = 2^e$ and work with rational points of 2^e-torsion, or if we need to work in dimension $g = 8$, with N having prime factors up to $O(\log n)$, and needing to work with torsion points defined over a field extension of degree up to $O(\log^2 n)$. For cryptographic applications, we want to find parameters allowing the former situation rather than the later. We will come back to this at several points during this survey.

We will explore several representations.

2.2 The Classical Representations

The *function representation* encodes E_1, E_2 by their short Weierstrass equations $E_i : y^2 = x^3 + a_i x + b_i$ (assume $p > 3$ here), and ϕ as a rational function $\phi(x,y) = \left(\frac{g(x)}{h(x)}, cy \left(\frac{g(x)}{h(x)} \right)' \right)$, where g, h are polynomials of degree $\leqslant n$ in x. This function representation takes linear space $O(n \log q)$ to encode, and linear time $O(n \log q')$ to evaluate.

The *kernel representation* encodes ϕ by the tuple $(E_1, \operatorname{Ker} \phi)$, this takes linear space. Vélu's formula can be used to convert from the kernel representation to the rational function representation (this includes computing a Weierstrass equation for E_2) in linear time. There is a slight ambiguity here since $\operatorname{Ker} \phi$ only determines ϕ up to post-composition by an automorphism of E_2. Since, unless $j(E_2) = 0, 1728$, the only automorphisms are multiplication by ± 1, we will ignore this subtlety.

A useful variant is to encode $\operatorname{Ker} \phi$ by a generator T, we call this the *generator representation*. In general, T may live in a field extension of $\mathbb{F}_q$, but when the generator is rational, this gives a compact (i.e., space efficient) representation of ϕ: $O(\log q)$ bits. Furthermore, in that case the isogeny image can be evaluated in $\tilde{O}(\sqrt{n} \log q')$ by the `sqrtVelu` algorithm of [BDLS20]. A variant, if a generator T lives in a too large extension, is to use a *multigenerator representation*: $\operatorname{Ker} \phi = \langle T_1, \ldots, T_m \rangle$. For instance, if $n = \prod \ell_i^{e_i}$, we can take for T_i a generator of $\operatorname{Ker} \phi[\ell_i^{e_i}]$, by the CRT the T_i generate the full kernel, and may live in smaller field extensions than T.

Given a subgroup G of E_1, the question of whether we can find generators of G living in a small enough extension of $\mathbb{F}_q$ (by small we mean polynomial in $\log G$), will appear frequently in this paper. We say that the G-torsion is *accessible* when this is the case. For example, if $E[n]$ is accessible (for instance, we have $E[\ell_i^{e_i}] \subset E(\mathbb{F}_{q_i})$ with $\mathbb{F}_{q_i}$ a small field extension of $\mathbb{F}_q$ for each i), then all n-isogenies $\phi : E_1 \to E_2$ have a space efficient (multi) generator representation. We refer to Appendix A for more details.

An isogeny ϕ of degree $\leqslant n$ is completely determined by its image on $4n + 1$ distinct points [UJ18]. Let $G \subset E(\mathbb{F}_q)$ be a (rational) subgroup of order $N > 4n$. By additivity of ϕ, it suffices to know the images $\phi(P_i)$ of ϕ on generators $\langle P_i \rangle$ of G to know ϕ on G. We call this the *interpolation representation*. Indeed, one can use a standard rational function reconstruction to reconstruct the rational function $\frac{g(x)}{h(x)}$ from this interpolation data. If these generators are rational, this rational reconstruction can be done in quasi-linear time, using standard algorithms [BCG+17]. The space needed for this representation depends on the fields of definition of the P_i, similarly to the generator representation of the kernel.

An alternative (when p is large enough compared to n), which is always space efficient, is to represent ϕ via the image $(\mathbf{P}, \phi(\mathbf{P}))$ of a formal point $\mathbf{P}$ at precision 2, and reconstruct ϕ (in quasi linear time) by solving a differential equation. We call this the *deformation representation*, because it encodes how ϕ deforms: see Sect. 3.3.

We can also use modular polynomials to represent an isogeny (this is the *modular representation*): if $(j(E_1), j(E_2))$ is a non singular point of the modular polynomial $\Phi_n(X, Y)$, this tuple is enough to reconstruct the deformation representation: one can use Φ_n to express the derivative $j'(E_2, dx/y)$ in term of $j'(E_1, dx/y)$ (for normalised differentials), which is enough to obtain the deformation of E_2 induced by the deformation of E_1 via the isogeny ϕ. Since $j(E_2)$ can be described as a specific root of $\Phi_n(j(E_1), Y)$ (via a deterministic ordering of roots), this representation is the only one, out of those presented in this paper, which, given E_1, only needs $O(\log n)$ extra bits of information to encode ϕ. However, the best algorithms currently know to evaluate $\Phi_n(j(E_1), Y)$ cost $\tilde{O}(n^2 \log p + n \log q)$, so are not linear in n.

If $n = \prod \ell_i^{e_i}$, so that ϕ is an isogeny of smooth degree (say with smoothness bound $\ell_i \leqslant B = O(\log^C n)$), we can decompose $\phi : E_1 \to E_2$ as a product of small degree isogenies ϕ_i. Any reasonable representation of the small isogenies ϕ_i (say linear in their degree l_i) then gives an efficient representation of ϕ. We call this the *decomposition representation*.

Note however that given a representation of ϕ, computing its decomposition representation may be expensive; we will come back to this in Sect. 3.4. In the particular case when n is smooth and the n-torsion is accessible (this is always the case if n is powersmooth), then we can efficiently convert a (multi)generator representation into a decomposed representation, so the (multi)generator representation is efficient.

2.3   The Ideal Representations

Another type of representation, which is quite different from the ones above, is to represent the isogeny ϕ by an ideal I_ϕ.

There are two different cases here. When E_1, E_2 are supersingular curves defined over $\mathbb{F}_{p^2}$ (and such that all their geometric endomorphisms are already defined over $\mathbb{F}_{p^2}$), by Deuring's correspondence the isogeny ϕ is always represented by an ideal I_ϕ in a quaternion algebra. We call this the *supersingular ideal representation*.

The other case concerns oriented isogenies between oriented elliptic curves. Here, an orientation is an embedding of a quadratic imaginary order R inside $\mathrm{End}_k(E_1)$ and $\mathrm{End}_k(E_2)$. The isogeny ϕ is said to be oriented if $\phi \circ \gamma = \gamma \circ \phi$ for all $\gamma \in R$. The embedding of R inside $\mathrm{End}_k(E_1)$ is required to be primitive (or saturated), meaning that $\mathrm{End}_k(E_1) \cap (R \otimes_{\mathbb{Z}} \mathbb{Q}) = R$, but not the embedding of R inside $\mathrm{End}_k(E_2)$. If R is primitive in $\mathrm{End}_k(E_2)$, we say that ϕ is an horizontal oriented isogeny, otherwise we say that ϕ is an ascending oriented isogeny. In both cases, ϕ can also be represented by an ideal I_ϕ (this would not necessarily be true if we allowed isogenies of degree divisible by p), we call this the *oriented ideal representation*.

Whenever the Frobenius π_k is not trivial (not given by a multiplication by some integers), we can consider the natural $\mathbb{Z}[\pi_k]$ orientation on E_1, E_2; since ϕ is rational it will also be oriented. This case includes ordinary elliptic curves, or supersingular elliptic curves over $k = \mathbb{F}_p$. (Note however that $\mathbb{Z}[\pi_k]$ may not be saturated in $\mathrm{End}_k(E_1)$, so we may need to replace it by its saturation R to find an ideal representation of ϕ; such an ideal representation exists only if ϕ is horizontal or ascending).

The ideal representation is always space efficient; it is even efficient in the supersingular or horizontal cases (provided we have an efficient representation of the orientation), but this is much harder to prove (see Sect. 4). To convert an ideal I of norm $N(I)$ (we define $N(I)$ to be the reduced norm in the supersingular case) into an isogeny, one can compute $E[I] \subset E[N(I)]$, the intersection of the kernel of all endomorphisms $\alpha \in I$. The isogeny ϕ_I corresponding to I is the isogeny with kernel $E[I]$, so this allows to go from the ideal representation to the kernel (or generator) representation in polynomial time in $n = N(I)$ (see Sect. 4 and Example 13 for more details). A strategy to find an evaluation algorithm polylogarithmic in n is to find an equivalent ideal J of (power)smooth norm, and compute a decomposed representation of the isogeny ϕ_J (from which we can derive an algorithm to evaluate ϕ_I efficiently if we have an efficient representation of the endomorphism α linking J to I). Such a *smoothening* algorithm in the supersingular case is given by the KLPT algorithm [KLPT14] (a non heuristic version is proved under GRH in [Wes22]). For the oriented case, there is no known polynomial time smoothening algorithm. Hence the question of efficiently converting an arbitrary ideal I into an isogeny ϕ_I had remained an important open problem in isogeny based cryptography. Thus an efficient evaluation algorithm for an (horizontal) ideal representation, was only recently given in [PR23b], and makes essential use of the HD representation which will be the main topic of this survey.

The ideal representation is very convenient, but it has a crucial drawback: in the supersingular case, publishing I leaks the endomorphism ring of E_1 and E_2, which is the one thing that is supposed to stay secret for isogeny based cryptography. In [Ler22a], Leroux introduced the suborder representation which leaked less informations; this representation is now essentially superseded by the HD representation.

2.4   The HD Representation

As explained in Sect. 1, the main byproduct of the SIDH attacks is that for any $N \geqslant n$ (coprime to p), by combining a lemma due to Kani in [Kan97] and Zahrin's trick, it is always possible to embed ϕ into a higher N-dimensional isogeny Φ between abelian varieties A, B of dimension $g = 2$ or 4 or 8. Taking N to be (power)smooth, and computing a decomposition representation of Φ, this allows to give an efficient representation of ϕ.

In practice, if N is prime to n, it is enough to know the action of ϕ on $E_1[N]$ to be able to reconstruct $\mathrm{Ker}\,\Phi$. From this kernel, one can use a higher dimensional version of Vélu's formula [Rob21, LR22] to compute the decomposition of Φ. We remark that, in some cases, there is an optimisation where we can relax the condition $N > n$ to $N^2 > n$, see [Rob23b, § 6.4], [DLRW24, Appendix C.2] or Appendix B. There is also a more recent version when we just need to know ϕ on a large enough subgroup rather than on the full N-torsion [CDM+24]. To simplify the exposition in this survey, we will mainly stick to the case $N > n$.

To describe the action of ϕ on $E_1[N]$, it suffices to give its action on a basis (P_1, P_2). It can be convenient to replace this basis by generators, like we did for the generator representation of the kernel: if $N = \prod l_i^{e_i}$ we give the action of ϕ on a basis of each $E_1[\ell_i^{e_i}]$, this can allow to work with smaller field extensions.

Technically, there are two subvariants of the HD representation, depending on how we represent Φ (see Definition 3): the one where we give the action of ϕ on (generators of) $E_1[N]$, which is essentially the kernel generator representation of Φ; and the one where we represent Φ by its decomposition into a product of smaller isogenies. The key point is that if N is smooth, and the basis of each $E_1[\ell_i^{e_i}]$ live in a small enough extension (which is always the case if N is powersmooth), then going from the kernel representation of Φ to its decomposition representation can be done efficiently, and then evaluating Φ (hence ϕ) amount to evaluating small ℓ_i-isogenies (in higher dimension), which is efficient.

We remark that the HD representation (in its kernel version) is very close to the interpolation representation of ϕ: we represent ϕ by its images $(P_{1,i}, P_{2,i}, \phi(P_{1,i}), \phi(P_{2,i}))$ on the basis $(P_{1,i}, P_{2,i})$ of the $E_1[\ell_i^{e_i}]$ torsion (say with ℓ_i prime to n to simplify), such that $\prod \ell_i^{e_i} > n$. The key point is that, to evaluate ϕ, rather than reconstruct the function representation via a generic interpolation algorithm, like we used in the interpolation representation, there is a much more clever algorithm that uses the higher dimensional isogeny Φ to evaluate $\phi(P)$ in time polynomial in $\log n$, ℓ_i, $\log q'$ and the $\log q_i$, where $P_{1,i}, P_{2,i} \in E(\mathbb{F}_{q_i})$. (So not all HD representations are efficient, but we can always find an efficient HD representation by selecting the $\ell_i^{e_i}$ appropriately!) This is probably one of the most complicated structured[5] univariate interpolation algorithm in the world, but it works very well in practice!

We can summarize the HD representation as follows: if we know how to evaluate ϕ on enough nice points, we know how to efficiently evaluate it everywhere. Restated like that, it seems we have a bootstrapping problem: how do we find the evaluation of ϕ on these nice points in the first place? This was the main problem stated in [Rob22a] when this representation was introduced.

One answer is when we have a supersingular ideal representation I_ϕ of ϕ: we want to publish an efficient representation of ϕ without publishing I_ϕ which would leak the endomorphism ring. We can use the HD representation for that. This is the main idea behind SQIsignHD [DLRW24].

2.5   Algorithms for the HD Representation

More generally, since the publication of [Rob22a], many algorithmic tools have been found to work directly with an efficient representation of an isogeny rather than the more customary kernel (or ideal) representation. See for instance [Rob22b, Ler23b, CLP24, NO23, PR23b]. The main goal of this survey is to give a convenient reference for these algorithms.

In the following, by efficient we still mean polynomial in $\log n$ and $\log q$.

1. **Universality:** The HD representation is universal[6]: from any efficient representation of ϕ, we can efficiently extract an (efficient) HD representation. This means that an algorithm available for an HD representation applies to any efficient representation of ϕ, by converting it to an HD representation first if necessary. Notably, since the HD representation can be used to evaluate ϕ on a point P defined over a k-algebra, but we only need to be able to evaluate ϕ on points defined over finite field extensions of k to derive an HD representation, we see that any representation able to evaluate ϕ over fields k'/k can be converted to a representation able to evaluate it over any k-algebra!
2. **Equality:** Given efficient representations of two isogenies $\phi_1, \phi_2 : E_1 \to E_2$, one can efficiently test if $\phi_1 = \phi_2$.
3. **Composition and addition:** Given efficient representations of $\phi_1 : E_1 \to E_2$ and $\phi_2 : E_2 \to E_3$ (resp. $\phi_2 : E_1 \to E_2$), one can find[7] an efficient representation of $\phi_2 \circ \phi_1 : E_1 \to E_3$ (resp. $\phi_1 + \phi_2$).
4. **Dual isogeny:** Given an efficient representation of $\phi : E_1 \to E_2$, one can find(see footnote 7) an efficient representation of the dual isogeny $\tilde{\phi} : E_2 \to E_1$.

---

[5] Here the structure that is being exploited is that ϕ commutes with the addition law, and that the interpolation points have smooth order.

[6] This term was coined by Benjamin Wesolowski.

[7] Efficiently!.

5. **Division:** Given efficient representations of $\phi : E_1 \to E_2$ and $\psi : E_1 \to E_1'$, one can find(see footnote 7) an efficient representation of an isogeny $\phi' : E_1' \to E_2$ such that $\phi = \phi' \circ \psi$ if it exists, or answer $\perp$ if such an isogeny ϕ' does not exist. Likewise when $\psi : E_2' \to E_2$ and we try to write $\phi = \psi \circ \phi'$. As a particular case, when $\psi = [m]$, we can efficiently test if ϕ is divisible by m (equivalently: if $\operatorname{Ker} \phi \supset E[m]$), and if so compute an efficient representation of ϕ/m.

6. **Lifts and deformations:** Given an efficient representation of $\phi : E_1 \to E_2$ over a finite field $\mathbb{F}_q$, and given $R = \mathbb{F}_q[\epsilon]/\epsilon^m$ or $R = \mathbb{Z}_q/p^m \mathbb{Z}_q$, one can find(see footnote 7) an efficient representation of the isogeny ϕ deformed/lifted to R.

7. **Splitting isogenies:** If ϕ is an isogeny of degree $n = n_1 n_2$ with $n_1 \wedge n_2 = 1$, ϕ decomposes uniquely as $\phi = \phi_2 \circ \phi_1$ where ϕ_i is of degree n_i. If we have an efficient representation of ϕ, we can find(see footnote 7) an efficient representation of ϕ_2 and ϕ_1.

8. **Pushforward of isogenies:** if $\phi_1 : E_1 \to E_2$ is of degree n_1, $\phi_2 : E_1 \to E_2'$ is of degree n_2, with $n_1 \wedge n_2 = 1$, and we have an efficient representation of ϕ_1, ϕ_2, then we can find(see footnote 7) an efficient representation of the pushforward isogeny $\phi : E_1 \to E_3$ (this is the isogeny with kernel $\operatorname{Ker} \phi = \operatorname{Ker} \phi_1 + \operatorname{Ker} \phi_2$). We can write $\phi = \phi_2' \circ \phi_1 = \phi_1' \circ \phi_2$ where ϕ_2' (resp. ϕ_1') is the pushforward of ϕ_2 (resp. ϕ_1) by ϕ_1 (resp. ϕ_2), and, by the splitting algorithm above, we can find(see footnote 7) an efficient representation of ϕ_1', ϕ_2'.

9. **Kernel:** Given an efficient representation of an n-isogeny $\phi : E_1 \to E_2$, we can recover an equation for its kernel in quasi-linear time $\tilde{O}(n)$ arithmetic operations. (Note that the output is of size $O(n \log q)$ so we cannot do better than linear time). If $E_2[n]$ is accessible, we can also find a generator representation of $\operatorname{Ker} \phi$ in polylogarithmic time.

We note the following remarkable feature of the splitting and pushforward algorithms. By assumption, an isogeny representation of $\phi : E_1 \to E_2$ contains both the domain and codomain data. But in some of the representations we have discussed, only the domain is needed; the codomain can be recovered from the data. This is the case for the kernel representations (via the kernel equation or generators), and also for the ideal representations. We remark that, as long as we know how to evaluate ϕ (say in Weierstrass coordinates), we can evaluate ϕ on some points and recover the Weierstrass equation of E_2 via linear algebra, or we can evaluate ϕ on the formal group at low precision (see Sect. 3.2).

The HD representation of $\phi : E_1 \to E_2$ uses interpolation data $(P, Q, \phi(P), \phi(Q))$ for a basis (P, Q) of the N-torsion, hence in particular require to know the codomain E_2 already. However, when $n = n_1 n_2$ with n_1 coprime to n_2, then the splitting algorithm recovers the middle curve $E_1 \to^{\phi_1} E_{12} \to^{\phi_2} E_2$ where $\phi = \phi_2 \circ \phi_1$ and ϕ_i is a n_i-isogeny. This is a powerful tool to construct new elliptic curves (see Examples 7 and 8), and it is amazing that we can extract E_{12} efficiently just from the interpolation data above, which seems on first sight to be completely unrelated to this curve!

In particular, if ϕ admits an efficient representation, then (ϕ, n_1) gives an example of an efficient representation of ϕ_1 where the codomain E_{12} of ϕ_1 is not specified in advance. For instance to construct such a ϕ, we can try to find $\phi_2 : E_{12} \to E_1$, of degree coprime to $\deg \phi_1$, so that ϕ is an endomorphism (we often have many convenient ways to represent endomorphisms, see Example 5).

3   The Standard Representations

In this section, we survey the classical representations: the function representation in Sect. 3.1, the kernel representations in Sect. 3.2, the interpolation representations and its variants in Sect. 3.3. An important case is when an isogeny can be decomposed into a product of isogenies of smaller degree, this is treated in Sect. 3.4.

3.1   The Function Representation

Let $\phi : E_1 \to E_2$ be a cyclic isogeny of degree n over $k = \mathbb{F}_q$, with $E_i : y_i^2 = x_i^3 + a_i x_i + b_i$ given by short Weierstrass equations, and n prime to p. We can write ϕ as $\phi(x_1, y_1) = (R_1(x_1, y_1), R_2(x_1, y_1))$ for some rational functions $R_1, R_2 \in k(E_1)$: we have a morphism $\phi^* : k(E_2) \to k(E_1)$ and R_1 (resp. R_2) is the image of x_2 (resp. y_2) by this morphism.

Since the divisor (0_{E_1}) has self intersection 1, and $\deg x_i = 2$, $\deg y_i = 3$, the rational functions R_1, R_2 have total degree $2n$ and $3n$ respectively (taking into account the point at infinity).

Since $\phi(-P) = \phi(P)$, we see that we can rewrite $\phi(x_1, y_1) = (R_1(x_1), y_1 R_2(x_1))$ where now $R_1(x_1), R_2(x_1) \in k(x_1) = k(\mathbb{P}^1)$.

Let $\omega_i = dx_i / y_i$ be the "canonical" basis of global differentials on E_i. We have $\phi^* \omega_2 = \frac{1}{c} \omega_1$ for some constant $c \neq 0$, because the space of global differentials is of dimension 1 (an elliptic curve has genus 1), and we only deal with separable isogenies so the action on differentials is non trivial. Plugging the formulas for ϕ, we obtain $\frac{dR_1}{y_1 R_2} = \frac{1}{c} \frac{dx_1}{y_1}$, so we can further simplify the formulas for ϕ to: $\phi(x, y) = (R(x), cyR'(x))$.

We say that ϕ is *normalised* if $c = 1$. Using the change of variable $x_2' = \frac{1}{c^2} x_2$ and $y_2' = \frac{1}{c^3} y_2$, we have $dx_2' / y_2' = c dx_2 / y_2$, so $\phi : E_1 \to E_2'$ becomes normalised where $E_2' : y_2'^2 = x_2'^3 + a_2 c^4 x_2' + c^6 b_2$.

The kernel of ϕ is given, along with 0_{E_1}, by all points that are sent to infinity by ϕ, hence by the denominator $h(x) = 0$ of $R(x) = \frac{g(x)}{h(x)}$. Let $K = \operatorname{Ker} \phi$, we have (since ϕ is separable the geometric points in the kernel have multiplicity one): $h(x) = \prod_{T \in K \setminus \{0\}} (x - x(T))$. We remark that if n is odd, $h(x)$ is a square because $T, -T$ are two distinct points with the same x coordinate.

The function representation takes space $O(n \log q)$, and $O(n \log q')$ to evaluate on a point $P \in E(\mathbb{F}_{q'})$.

3.2   The Kernel Representations

Kernel Equation. Conversely, given the equation $h(x) = 0$ representing a (cyclic) kernel K, we can recover the function representation. Kohel in [Koh96] gives these explicit formulas, adapted from Vélu [Vél71]: for $f(x) = x^3 + a_1 x + b_1$, and $\sigma_1 = \sum_{T \in K \setminus \{0\}} x(T)$ the trace of h, we have that

$$\frac{g}{h}(x) = nx - \sigma_1 - f'(x)\frac{h'}{h}(x) - 2f(x)\left(\frac{h'}{h}\right)'(x), \tag{1}$$

is the function representation for a normalised isogeny $\phi : E_1 \to E_2, (x, y) \mapsto (\frac{g}{h}(x), y(\frac{g}{h})'(x))$ with kernel K. Here $E_2 : y^2 = x^3 + a_2 x + b_2$ is given by (see for instance [Feo10, § 8.2]):

$$a_2 = a_1 - 5t, \quad b_2 = b_1 - 7w, \quad t = \sum_{T \in K \setminus \{0\}} f'(T), \quad u = \sum_{T \in K \setminus \{0\}} 2f(T) + x(T)f'(T).$$

The kernel representation takes space $O(n \log q)$. More precisely, it is given by a polynomial $h(x)$ in $\mathbb{F}_q[x]$ of degree $n - 1$, and if n is odd we can write $h(x) = h_1(x)^2$, where h_1 is a polynomial of degree $(n - 1)/2$. Converting to the function representation takes $O(n)$ arithmetic operations using Eq. (1).

Vélu's formula from [Vél71] were originally given in term of the generator representation: if $K = \langle T \rangle$, Vélu argues that the functions

$$
\begin{aligned}
R_1(P) &:= \left(x_1(P) + \sum_{T \in K \setminus \{0\}} x_1(P + T) - x_1(T) \right), \\
R_2(P) &:= \left(y_1(P) + \sum_{T \in K \setminus \{0\}} y_1(P + T) - y_1(T) \right)
\end{aligned}
\tag{2}
$$

are invariant by translation by $T \in K$ and have the correct polar divisors to be the pullback by ϕ of Weierstrass coordinates x_2, y_2 on $E_2 = E_1/K$.

To recover the equation of E_2, Vélu use the formal group law of the elliptic curve. More precisely, he fixes the uniformiser $z_1 = -x_1/y_1$ on E_1, and look at the development of x_1, y_1 along z_1. He then plugs the formula of R_1, R_2 to obtain their development along z_1, hence the development of $z_2 = -R_1/R_2$ in term of z_1. Inversing this to get z_1 in term of z_2, he recover the development of R_1, R_2 in term of z_2, from which he obtain the equation of E_2. Weierstrass coordinates are not unique; Vélu explains that he made the choice of normalisations on R_1, R_2 such that z_2 and z_1 coincide up to order 5: $z_2 = z_1 + O(z_1^5)$. In particular, the resulting isogeny is normalised.

Kernel Generator. To evaluate a cyclic isogeny ϕ on a point P given a generator T of the kernel $K = \langle T \rangle$, one can use the formulas from Eq. (2), this costs $O(n)$ arithmetic operations in the compositum field of the fields of definition of P and T. An alternative is to first convert to the kernel equation:

$h(x) = \prod_{i=1}^{n-1}(x - x(i \cdot T))$, this uses $O(n)$ arithmetic operations in the field of definition $\mathbb{F}_{q^e}$ of T, and then to use Eq. (1) to evaluate on P in $O(n)$ arithmetic operation in $\mathbb{F}_{q'}$.

When the field of definition $\mathbb{F}_{q^e}$ of T is small, the generator representation gives our first compact isogeny representation: representing T takes space $O(e \log q)$. In the worst case we have $e = \Theta(n)$, so we do not gain anything compared to the $O(n \log q)$ size of the kernel representation, but in the best case $e = 1$ and the isogeny takes $O(\log q)$ to represent. (Note that in this case, since T is rational we have $n \mid \#E(\mathbb{F}_q)$, so necessarily $n = O(q)$).

Another advantage of the generator representation is that we can use the sqrtVelu algorithm of [BDLS20], which requires $\widetilde{O}(\sqrt{n})$ arithmetic operations in the compositum field of T and P to evaluate $\phi(P)$.

As mentioned in Sect. 2, we can also use a multigenerator representation $K = \langle T_1, \ldots, T_m \rangle$. This is particularly useful when $n = \prod \ell_i^{e_i}$ is powersmooth, in that case we can take T_i to be a generator of $K[\ell_i^{e_i}]$, which will live in an extension of degree at most $\ell_i^{e_i}$. If the powersmooth bound B is $B = O(\log^{O(1)} n)$, this gives a compact representation. By contrast, the generator $T = T_1 + \cdots + T_m$ live in an extension of degree the compositum of all these fields, which can only be bounded by n in the generic case.

More precisely, since $K = \langle T \rangle$ is cyclic of degree n, π_q acts by multiplication by $\lambda \in (\mathbb{Z}/n\mathbb{Z})^*$ on T, and the order e of λ gives the degree of the extension where T is defined, so in particular $e \mid \phi(n)$.

Recall that we have defined a torsion subgroup G of cardinal N to be accessible whenever we can find generators that live in a small (polynomial in $\log N$) extension of $\mathbb{F}_q$. From this definition, we see that the multigenerator representation of the kernel K is space efficient whenever the K-torsion is accessible.

Remark 2 (Radical isogenies). Sometimes for applications, like for the CGL hash function [CLG09], we wish to iterate ℓ-isogenies: we want to compute a cyclic ℓ^n-isogeny. As we will see in Sect. 3.4, the best case happens when we have a rational generator of the big ℓ^n-isogeny, because we can then evaluate it in $\widetilde{O}(n \log \ell)$. However, it often happens that only the ℓ-torsion is rational (or lives in a small extension), and that the ℓ^n-torsion would live in a too big extension. The standard solution is then to compute this ℓ-torsion, extract the first generator of order ℓ from it, compute the first isogeny $E_0 \to E_1$ of degree ℓ, and iterate, computing the ℓ-torsion on E_1 again (typically by sampling points and multiplying by the cofactor).

The idea of radical isogenies is to start with the first kernel $K_0 = \langle P_0 \rangle$ too, but to compute the next kernel $K_1 = \langle P_1 \rangle \subset E_1$ directly, and to iterate this construction. This idea was introduced in [CDV20], where the authors showed that P_1 could be computed from a choice of ℓ-th root of the self Tate pairing $e_{T,\ell}(P_0, P_0)$. The original motivation of the radical isogeny formulas were for the CSIDH cryptosystem, and they have been improved in [OM22, CDHV22, Pri24, Dec24]. The theory of multiradical isogenies (radical isogenies in higher dimension) was developed in [CD21, Rob23c].

3.3 Interpolation Representations

Standard Lagrange Interpolation. By the Cauchy-Schwarz inequality,

$$\deg(\phi_1 + \phi_2) \leqslant \left(\sqrt{\deg \phi_1} + \sqrt{\deg \phi_2} \right)^2 .$$

Hence if ϕ_1, ϕ_2 are two n-isogenies such that $\phi_1 \neq -\phi_2$, $\deg(\phi_1 - \phi_2) \leqslant 4n$. It follows that if ϕ_1, ϕ_2 coincide on at least $4n + 1$ points, they have to be equal.

The interpolation representation of ϕ is given by $(E_1, E_2, (P_i, \phi(P_i)))$ for a list of $N > 4n$ points P_i. By the above argument, this completely determines ϕ.

In practice, we can recover ϕ as follows from the interpolation data:

- We first recover interpolation data for $R(x) = \frac{g}{h}(x)$, with the notations from Sect. 3.1. Each interpolation data $(P_i, \phi(P_i))$ gives an interpolation data $(x(P_i), R(x(P_i)))$. Since the points P_i are distinct, and only $\pm P_i$ have the same x-coordinate, we obtain at least $2n + 1$ different interpolation points for R.
- The function $R(x)$ has total degree $2n$ seen as a function on E (taking into account the point at infinity), and since $\deg(x) = 2$ we get that $\deg_x R \leqslant n$. Hence our $2n + 1$ points of interpolation are enough to recover $R(x)$ in quasi-linear time $\widetilde{O}(n)$ by the standard rational function reconstruction algorithms [BCG+17].
- It remains to recover the normalisation constant c, which we can obtain from any point P_i such that $\phi(P_i)$ is not of 2-torsion. There are at most $4n$ such points, and we have $4n + 1$ points, so we can always find one.

The interpolation representation takes space $O(n \log q)$, assuming that all the P_i are defined in $\mathbb{F}_q$. On the other hand, if $P_i \in E_1(\mathbb{F}_{q^e})$, we can act by the Galois action (i.e., the Frobenius) on the interpolation data $(P_i, \phi(P_i))$ to obtain "for free" e different interpolation points.

But in fact, we can do better: by additivity of ϕ, knowing the interpolation data on some points P_i allows to recover the action of ϕ on the full subgroup $G = \langle P_i \rangle$. So it suffices to give interpolation data for generators P_i of a rational (for simplicity) subgroup $G \subset E_1$ to recover ϕ by interpolation, as long as $\#G > 4n$. Assume that $\#G = N$, with $N = \prod \ell_i^{e_i}$. Like in the multi generator representation for the kernel, it can be helpful to diminish the size of the representation to take generators of each of the CRT subgroups $G[\ell_i^{e_i}]$. (However, the complexity of the rational function reconstruction will depend on the degree of the field of definition of all points of G).

Note that here N is decoupled from n. This can be used to find a space efficient representation of ϕ even if its kernel K has its torsion not accessible: we only need to find a large enough accessible torsion subgroup $G \subset E$.

For instance, if $P \in E_1(\mathbb{F}_q)$ is a rational point of N-torsion, we can use $(P, \phi(P))$ (along with (E_1, E_2, n) as usual) as a compact interpolation representation of any n-isogeny ϕ with $4n < N$. We remark that in this case, $N = O(q)$, hence $n = O(q)$.

The Deformation Representation as an Hermite-Padé interpolation.
There is a convenient way to find such a rational torsion "point" of large order.
Namely, let $\mathbf{P}$ be fat point above 0_{E_1}, that is a point $\operatorname{Spec} k[\epsilon]/\epsilon^2 \to E$ such
that the natural composition $\operatorname{Spec} k \to \operatorname{Spec} k[\epsilon]/\epsilon^2 \to E$ gives 0_{E_1}. One can
see $\mathbf{P}$ as the choice of a tangent vector $v \in T_{O_E}(E)$ of E at 0_E. Then $\mathbf{P}$ is a
rational "point", that we can also view as a formal point in the formal group of E
truncated to precision 2, which is of order[8] p, because $[n]$ acts by multiplication
by n on the tangent space at 0_{E_1}.

Concretely, since ϕ maps the neutral point to the neutral point, to give the
interpolation data $(\mathbf{P}, \phi(\mathbf{P}))$ (to precision 2) is equivalent to giving the action
of ϕ on the tangent space of E_1 and E_2 at their neutral points, or equivalently
to give the normalisation constant c such that $\phi^* \omega_2 = \frac{1}{c}\omega_1$. Indeed, if we use
$z_1 = x_1/y_1$ as a uniformiser on E_1, then a formal point at precision m is given by
$\mathbf{P} = a_1 z_1 + a_2 z_1^2 + a_3 z_1^3 + O(z_1^{m+1})$. We have explicit formulas for $x_1 = z_1^{-2} + \cdots$,
$y_1 = z_1^{-3} + \cdots$, from the curve equation of E_1, so we have the Weierstrass
formal coordinates of $\mathbf{P}$, to which we can apply ϕ to obtain the Weierstrass
formal coordinates x_2, y_2 of $\phi(\mathbf{P})$. Letting $z_2 = x_2/y_2$, we can thus recover
$z_2(\phi(\mathbf{P}))$ to precision m: $z_2(\phi(\mathbf{P})) = \frac{a_1}{c} z_1 + a_2' z_1^2 + \ldots + O(z_1^{m+1})$. In particular,
$dz_2 \circ d\phi = \frac{1}{c} dz_1$. Since $dz_i = dx_i/y_i$, we see that working at precision $m = 2$ is
enough to recover c such that $\phi^* \omega_2 = \frac{1}{c}\omega_1$.

By the same argument as in the usual interpolation reconstruction, the
isogeny ϕ is completely determined from this Hermite-Padé interpolation data c
as long as p is large enough compared to n: $p > 4n$. To reconstruct ϕ in practice,
i.e. recover the rational function R, one can solve the differential equation

$$c^2(x^3 + a_1 x + b_1) R'(x)^2 = R(x)^3 + a_2 R(x) + b_2; \tag{3}$$

derived from the equation $\phi^* \omega_2 = \frac{1}{c}\omega_1$.

This differential equation can be solved in quasi-linear time $\widetilde{O}(n)$ arithmetic
operations [BMSS08] using fast methods on power series, to first reconstruct R
as a power series and then as a rational function, as long as $p > 8n - 5$.

When n is too large compared to p, one can instead take a lift $\widetilde{\phi} : \widetilde{E_1} \to \widetilde{E_2}$ of
ϕ to $\mathbb{Z}_q/p^m \mathbb{Z}_q$ (where $\mathbb{Z}_q$ is the ring of Witt vectors over $\mathbb{F}_q$), at p-adic precision
m large enough: $n = O(p^m)$. Such a technique is used in [LS08] to reconstruct an
isogeny from roots of the modular polynomial ϕ_n even in cases where p is small
compared to n.

We call this representation the *deformation representation*; the advantage of
this representation is that it takes space $O(m \log q) = O(\log q + d \log n)$ for any
n-isogeny, where $q = p^d$. The rational function reconstruction takes quasi-linear

---

[8] This is not a coincidence, as we will see in Remark 3, by [Oda69], the Dieudonné module $\mathbb{D}(E[p])$ of $E[p]$ is canonically isomorphic to the de Rham cohomology $H^1_{DR}(E)$,
and the Frobenius filtration on $E[p]$ corresponds to the Hodge filtration on $H^1_{DR}(E)$
[Oda69, Corollary 5.11], so in particular $D(E[\pi_p]) \simeq H^0(E, \Omega_E)$ (up to a Frobenius twist). In other words, differentials on E corresponds via the Dieudonné antiequivalence of category to infinitesimal points of p-torsion in $E[\pi_p]$.

time. This representation will (one day) be explained in more details in [KR22], in the meanwhile see [Rob21, § 4.7.3].

The reason for the name *deformation representation* comes from the following useful reformulation. The Kodaira-Spencer isomorphism gives a canonical isomorphism between the Sym^2 of the tangent space of an elliptic curve E at 0_E and the tangent space of the moduli stack $\mathcal{A}_1$ parametrizing elliptic curves at E (see Appendix E). So, reformulating in terms of differential, the choice of a differential ω_E on E is essentially the same as the choice of a deformation $\widetilde{E}/\text{Spec}(k[\epsilon]/\epsilon^2)$ of E (essentially because of the Sym^2, which means that $\pm\omega_E$ gives the same deformation). In terms of modular functions, this is just a reformulation of the fact that the modular function $j'(\tau)$ is of weight 2, where $j'(\tau) = dj(\tau)/2\pi i d\tau$. From the algebraic interpretation of modular forms as sections of suitable powers of the Hodge line bundle, a modular function of weight 2 such as j' means that j' depends (in a functorial way) on both the choice of E and a differential ω_E on E, and that $j'(E, \lambda\omega_E) = \lambda^2 j'(E, \omega_E)$.

All of this can be made very explicit: if $\omega_E = dx/y$ is the canonical differential form on $E : y^2 = x^3 + ax + b$ we have $j'(E, \omega_E) = 18j(E)b/a$ by [Sch95, § 7]. Conversely, fixing $j'(E, \omega_E)$ for some differentials ω_E on E fixes ω_E up to a sign because j' is of weight 2, hence in particular fixes a, b such that $\omega_E = dx/y$ on $E : y^2 = x^3 + ax + b$ because a is of weight 4 and b is of weight 6.

And the deformation $\widetilde{E}$ associated to (E, ω_E) is given by $\widetilde{E} : y^2 = x^3 + a(\epsilon)x + b(\epsilon)$ such that $j(\widetilde{E}) = j(E) + j'(E, \omega_E)\epsilon$ modulo ϵ^2, which gives a linear equation between the coefficients a', b' of $a(\epsilon) = a + a'\epsilon$ and $b(\epsilon) = b + b'\epsilon$. We cannot do better because multiplying $a(\epsilon)$ and $b(\epsilon)$ by $u(\epsilon)^4$ and $u(\epsilon)^6$ respectively with $u(\epsilon) = 1 + \gamma\epsilon$ does not change the isomorphism class of $\widetilde{E}$ and the curve still reduces to E.

In other words, the following data are essentially equivalent for an elliptic curve E:

- A short Weierstrass equation for E, and its associated differential $\omega_E = dx/y$
- The j-invariant $j(E)$ and a choice of differential ω_E up to a sign
- $j(E)$ and $j'(E, \omega_E)$
- The j-invariant $j(\widetilde{E})$ of the deformation $\widetilde{E}$ of E to $k[\epsilon]/\epsilon^2$ associated to $\text{Sym}^2 \omega_E$

Now, given a deformation $\widetilde{E_1}$ of E_1, the isogeny $\phi : E_1 \to E_2$ deforms uniquely to an isogeny $\widetilde{E_1} \to \widetilde{E_2}$. If $\widetilde{E_1}$ is the deformation associated to a differential $\text{Sym}^2 \omega_{E_1}$, then $\widetilde{E_2}$ is the deformation associated to the differential $\text{Sym}^2 \omega_{E_2}/n$ where $\phi^*\omega_{E_2} = \omega_{E_1}$ (see [KPR24]). Given two short Weierstrass equations $y^2 = x^3 + a_i x + b_i$ for E_1, E_2, the normalisation constant $\phi^* dx_2/y_2 = \frac{1}{c}dx_1/y_1$ then shows that if $\widetilde{E_1}$ is the deformation associated to $\omega_{E_1} = dx_1/y_1$, then $\widetilde{E_2}$ is the deformation associated to $\omega_{E_2} = cdx_2/y_2$.

In summary, given E_1, E_2, the deformation representation consists of any of these essentially equivalent data (meaning that we can easily switch between them):

- The pullback action of ϕ on differentials;
- The datum $j'(E_1, \omega_{E_1})$, $j'(E_2, \omega_{E_2})$ for the modular function j', where $\omega_{E_1} = \phi^* \omega_{E_2}$;
- The isogeny normalisation constant;
- The deformation $\widetilde{\phi} : \widetilde{E_1} \to \widetilde{E_2}$ to any non trivial deformation $\widetilde{E_1}/(k[\epsilon]/\epsilon^2)$ of E_1;
- The action of ϕ on the formal group to precision $m = 2$;
- The differential equation Eq. (3), from which we can recover ϕ in $\widetilde{O}(n)$ (if p is large enough).

Remark 3 (Dieudonné modules, points of p-torsion and differentials). From the point of view of Dieudonné modules, the Dieudonné module $\mathbb{D}(E[p^n])$ (in the contravariant Dieudonné theory) is precisely given by the de Rham cohomology $H^1_{DR}(\widetilde{E})$ of a lift $\widetilde{E}$ of E to p-adic precision n. This is a consequence of the fact that the Dieudonné module $\mathbb{D}(E(p))$ of the p-divisible group is isomorphic to the crystal $H^1_{\mathrm{crys}}(E, \mathbb{Z}_p)$ associated to E, which itself can be computed as the hypercohomology of the de Rham Witt complex [Ill79]. Furthermore, by the Serre-Tate theorem lifts $\widetilde{E}$ of E corresponds to lifts of the p-divisible group $E(p)$, which themselves correspond by the Grothendieck-Mazur theorem to lifts of their associated crystals (which over a field is the Dieudonné module), which is encoded by the linear data of a choice of lift of their Hodge filtration from modulo p to modulo p^n. The filtration on $\mathbb{D}(E[p^n])$ modulo p^n associated to a choice of lift $\widetilde{E}$ is precisely given by the Hodge filtration on $H^1_{dR}(\widetilde{E}, \mathbb{Z}/p^n\mathbb{Z})$ via the isomorphism above.

This is really a beautiful theory, and unfortunately it is hard to find a concise reference for all these facts. At the core, Grothendieck's crystalline topology should be seen as a far reaching generalisation that an infinitesimal point can be associated to some differential data.

Also, a very useful fact proved by Tate in [Tat67], is that over a complete Noetherian local ring R with residue field of characteristic p, connected p-divisible groups correspond to formal Lie groups (when seen as fppf sheafs; and the Dieudonné module of the group is the same as the Cartier module of the Lie group). And as expected, the formal Lie group corresponding to the connected part of the p-divisible group $E(p)$ is precisely the formal group of E.

All these results extend to an abelian variety A. For the isogeny practitioner, what the Dieudonné theory means is that the p-torsion of A corresponds to $H^1_{DR}(A)$ (and a point of p^n-torsion to a De Rham differential of a lift of A modulo p^n). As mentioned earlier, by [Oda69, Corollary 5.11], this is refined by the Hodge decomposition $0 \to H^0(A, \Omega^1_{A/k}) \to H^1_{dR}(A) \to H^1(A, O_A) \to 0$ which corresponds via the Dieudonné functor to the Frobenius filtration $0 \to A[\hat{\pi}] = \Im\pi \to A[p] \to A[\pi] = \mathrm{Ker}\,\pi \to 0$; more precisely $\mathbb{D}(A[\hat{\pi}]) \simeq H^1(A, O_A) \simeq H^0(A^\vee, \Omega^1_{A^\vee/k})^\vee$ while $\mathbb{D}(A[\pi])$ is a Frobenius twist of $H^0(A, \Omega^1_{A/k})$. In particular, the choice of an infinitesimal point in $A[\pi]$ (note that this is always a connected infinitesimal proper group scheme) corresponds to a choice of global differential on A. And by the discussion above, the formal group law encodes the connected part $A^\circ(p)$ of the p-divisible group.

In some sense, since the global differentials form a vector space of dimension g for an abelian variety of dimension g (because it is smooth), we always have a "free" basis of rank g of points of p-torsion. Furthermore, to any algorithm which is expressed in term of usual geometric points on A, as long as it is functorial enough to work as well with points with multiplicities, so in particular p-torsion points, then there should exist a traduction of this algorithm (via Dieudonné's antiequivalence of category) which can work in terms of differentials. (This does not mean that the resulting algorithm is still as efficient though). This is precisely what we have done above: we had an algorithm to interpolate an n-isogeny from its value at a point of ℓ-torsion, with $\ell \gg n$, and we adapted it to an algorithm that reconstitutes the isogeny from how it acts on the global differential dx/y, as long as $p \gg n$. The main advantage, as we have noted, is that the global differential is always there, while the point of ℓ-torsion may live in a large extension. This point of view can often be helpful in isogeny based cryptography when one need extra points of ℓ-torsion and how isogenies act on them: using $\ell = p$ allows to work with rational "points".

The last remark is that, implicitly, many existing isogeny algorithms in the literature track these points of p-torsion (i.e. track the global differentials). It is well known that Vélu formula gives normalised isogenies; and the condition $\phi^*\omega_2 = \omega_1$ for differentials/p-torsion "points" is the pendant of the equation $\phi(P_1) = P_2$ for normal (étale) points. It is a bit less known, but still true, that the usual isogeny formulas in Mongomery coordinates or theta coordinates still track differentials. For Vélu formulas, the differential was encoded through a choice of short Weierstrass equation $y^2 = x^3 + ax + b$, and indeed the Weierstrass coefficients a, b are modular forms of weight 4 and 6 respectively. In the Montgomery model $y^2 = x^3 + \mathcal{A}x^2 + x$, the coefficient $\mathcal{A}$ is only a modular function (of level $\Gamma^0(4)$), so does not keep track of differentials. Likewise, the level 2 projective theta null point $(\vartheta_0(E) : \vartheta_1(E)) = \vartheta_1(E)/\vartheta_0(E)$ is given by a modular function (of level $\Gamma(2,4)$). But in practice, during the isogeny algorithms we work with the numerator A and denominator C of $\mathcal{A} = A/C$ separately to avoid divisions, and likewise for the numerator $\vartheta_1(E)$, and denominator $\vartheta_0(E)$ of the theta null point. This time we do have modular forms. We refer to [KNRR21, § 4.3] for more details on the modular interpretation of the theta isogeny formulas, to [KNRR21, § 4.5] for applications on how to compute Siegel modular forms algebraically, and to [Rob21, § 6.4] for applications to point counting via canonical lifts.

Modular Polynomials. This point of view allows us to make the link with the *modular representation*. Recall that the modular polynomial $\Phi_n(X, Y)$ is a symmetric polynomial in $\mathbb{Z}[X, Y]$, such that $\Phi_n(j(E_1), Y) = \prod(Y - j(E_{1,i}))$ where $E_{1,i}$ are all the n-isogeneous codomain curves from E_1. When this evaluated polynomial has no multiplicity, a root $y = j(E_2)$ completely determines the n-isogeny $\phi : E_1 \to E_2$.

Elkies algorithm [Elk92, Elk97] gives a way to reconstruct ϕ in practice from $\Phi_n, j(E_1), j(E_2)$. We can reformulate it as follows: start with a short Weierstrass equation for E_1, and look at the deformation $\widetilde{E_1}$, with $j(\widetilde{E_1}) = j(E_1) + \epsilon j'(E_1, \omega_{E_1})$ associated with $\text{Sym}^2 \omega_{E_1}$ for the differential $\omega_{E_1} = dx_1/y_1$. Then ϕ deforms to $\widetilde{\phi} : \widetilde{E_1} \to \widetilde{E_2}$, where $\widetilde{E_2}$ is associated to the differential $\text{Sym}^2 \omega_{E_2}/n$, where $\phi^* \omega_{E_2} = \omega_{E_1}$. In other words, $j(\widetilde{E_2}) = j(E_2) + \epsilon j'(E_2, \omega_{E_2}/\sqrt{n}) = j(E_2) + \epsilon j'(E_2, \omega_{E_2})/n$. Plugging the equation $\Phi_n(j(\widetilde{E_1}), j(\widetilde{E_2})) = 0$ amount to differentiating Φ_n, and gives the equation $j'(E_1, \omega_{E_1}) \partial \Phi_n / \partial X(j(E_1), j(E_2)) + j'(E_2, \omega_{E_2})/n \partial \Phi_n / \partial Y(j(E_1), j(E_2)) = 0$. This allow us to recover $j'(E_2, \omega_{E_2})$ and also $j(\widetilde{E_2}) = j(E_2) + \epsilon j'(E_2, \omega_{E_2})/n$. From $j'(E_2, \omega_{E_2})$, we recover the short Weierstrass equation of E_2 such that the isogeny ϕ is normalised with respect to ω_{E_1} and $\omega_{E_2} = dx_2/y_2$. In other word, we recover the deformation representation of ϕ (normalised in this case to have $c = 1$), from which we can recover ϕ by using [BMSS08] in quasi-linear time when $n = O(p)$. See [Rob21, § 5.4.1] for more details, and [KPR24] for a generalisation to abelian surfaces.

Given E_1, in the case where the codomains of all the n-isogenies from E_1 are different, ϕ is completely determined from $(E_1, j(E_2), n)$ In fact, we could even replace $j(E_2)$ by using a deterministic numerotation of the roots of $\Phi_n(j(E_1), Y)$, and giving which root number corresponds to $j(E_2)$; this representation takes $O(\log n)$ space rather than $O(\log p)$.

On the other hand, to reconstruct the isogeny ϕ from this data, we need to compute Φ_n, which takes quasi-linear time in its size $\widetilde{O}(n^3)$ [Eng09, BLS12, KR24]: this time the reconstruction is not quasi-linear in n. But for the reconstruction we just need the evaluations $\Phi_n(j(E_1), Y)$ and $\partial \Phi_n / \partial X(j(E_1), Y)$ which can be computed faster [Sut13b, Kie20, Ler23a, Rob22b]. For instance, we have algorithms in $\widetilde{O}(n^2 \log p)$ to recover Φ_n modulo p directly. However, we have no algorithms so far able to compute $\Phi_n(j(E_1), Y)$ in quasi-linear time $\widetilde{O}(n \log q)$.

An alternative approach to reconstruct ϕ given (E_1, E_2, n), without using Φ_n, is given in [DHPS16].

3.4   Decomposing an Isogeny

So far, we have seen some representations that were space efficient, but none able to evaluate ϕ on a point P in better than linear time. When $n = \prod \ell_i^{e_i}$ is smooth, the decomposition representation will be our first efficient representation. The key point is that we can decompose ϕ as a product of ℓ_i-isogenies: $\phi = \phi_1 \circ \phi_2 \circ \dots$.

To represent ϕ, it then suffices to give a representation of each of the intermediate ϕ_i; and any decent representation of ϕ_i (for instance the kernel representation or even the function representation), where by decent we mean a representation which takes linear space and time in the degree, will then give a representation of ϕ which takes space $O(B \log n \log q)$ and time $O(B \log n \log q')$ for evaluation, where B is a smoothness bound on n. More precisely, the *decomposed representation* takes space $O((\sum e_i \ell_i) \log q)$ and can be evaluated in

$O((\sum e_i \ell_i) \log q')$. If n is smooth (B is polynomial in $\log n$), we obtain our fist representation that is both compact and efficient.

One can also look at the cost of converting from one of the above representations to the decomposed representation. The main case of interest is the generator representation: $K = \langle T \rangle$. For simplicity, we will focus on the particular case where $n = 2^e$ and $T \in E_1(\mathbb{F}_q)$ is a point of 2e-torsion.

The naive way to decompose ϕ as a product of 2-isogenies is to start with T, compute $T_1 = 2^{e-1}T$ of order 2 which generates the kernel of our first 2-isogeny ϕ_1, compute the equations for ϕ_1 using Vélu's formulas from Sect. 3.2, and then pushforward T via ϕ_1 to get $\phi_1(T)$ which generates a kernel of order 2^{e-1} on $E_1/\langle T_1 \rangle$, and iterate this construction. This costs $O(e^2)$ arithmetic operations because we need $O(e^2)$ doublings.

A clever solution to achieve a better complexity was introduced in [DJP14]: the authors show how one can push more (carefully chosen) points along our intermediate isogenies ϕ_i to obtain a complexity of $O(e \log e)$ arithmetic operations to obtain the decomposition of ϕ.

The same method holds in the general case: provided that n is smooth and the kernel torsion is accessible (e.g., n is powersmooth, or $n = 2^e$ and T lives in a small extension), one can convert a multigenerator representation of the kernel into a decomposed representation in polylogarithmic time.

More precisely (restricting to ϕ cyclic for simplicity, but the same bounds hold in the general case):

Proposition 1. *Let* $n = \prod_{i=1}^{m} \ell_i^{e_i}$, $T_1, \ldots, T_i, \ldots, T_m$ *generators of* $K[\ell_i^{e_i}]$, *where* $T_i \in E(\mathbb{F}_{q^{d_i}})$. *Then one can compute the decomposed representation of the isogeny* ϕ *with kernel* K *in time* $\widetilde{O}(\sum_i e_i \ell_i \log q (d_i \log e_i + \sum_{j>i} d_j)) = \widetilde{O}(m^2 de\ell \log q)$ *where* d *(resp. e, ℓ) is a bound on the* e_i *(resp. d_i, ℓ_i).*

An alternative method, using `sqrtVelu`, *cost* $\widetilde{O}(\sum_i e_i \sqrt{\ell_i} \log q (d_i + \sum_{j>i}(d_i \vee d_j)) = \widetilde{O}(m^2 d'e\sqrt{\ell} \log q)$ *where this time* d' *is the maximum of the degree of the compositum field extension of the fields of definitions of* T_i, T_j *for* i, j.

Proof. We refer to Remark 17 for our assumptions about the lattice of field extensions used to represent our accessible CRT basis $\langle T_i \rangle$.

One first decomposes the isogeny with kernel $K[\ell_1^{e_1}]$ using its generator T_1 in $O(e_1 \log e_1 \ell_1)$ arithmetic operations over $\mathbb{F}_{q^{d_1}}$; this costs $\widetilde{O}(e_1 \log e_1 \ell_1 d_1 \log q)$. Then we push the other generators $T_2, \ldots, T_m$ and we iterate. Pushing T_i through this $\ell_1^{e_1}$ isogeny takes $O(e_1 \ell_1)$ arithmetic operations over $\mathbb{F}_{q^{d_i}}$, so time $\widetilde{O}(e_1 \ell_1 d_i \log q)$, using the kernel representation.

An alternative to using the kernel representation to represent the intermediate ℓ_i-isogenies $\phi_{i,u}$ is to keep the generator representation and use the `sqrtVelu` formulas. This replaces the $O(\ell_i)$ part by $\widetilde{O}(\sqrt{\ell_i})$, but on the other hand one has to work with the compositum field of the field of definition of the generator T_i' of the kernel of ϕ_i (which is included in the field of definition of T_i), and the field of definition of the current image T_j' of T_j (which is included in the field of definition of T_j). This compositum is thus of degree at most $d_i \vee d_j$. □

In isogeny based cryptography, for *practical* reasons we prefer to work with isogenies of smooth degree defined by rational generators, this is in practice much faster than working with general accessible torsion, which requires (small) field extensions. Still, it can be convenient to take small field extensions for some applications, and the case of non rational generators has been quite optimised too, see [EPSV, BGDS23].

Example 1 (Decomposing an isogeny from its kernel equation). If $\phi : E_1 \to E_2$ is given by its kernel equation $K = \mathrm{Ker}\, \phi \colon H(x) = 0$, we can apply Proposition 1 by treating the point $\mathbf{P} : x \mod H$ as a "formal" generator of K (working in x-only coordinates for simplicity).

Let's assume that $n = n_1 n_2$. We can first compute $n_2 \mathbf{P}$, then compute its characteristic polynomial (for instance via a resultant). To avoid divisions by potentially non invertible elements modulo H, one can work with the homogenisation $H(X, Z)$ of $H(x)$, and work in $(X : Z)$ coordinates; the homogenised characteristic polynomial of $n_2 \mathbf{P}$ is given by $H_1(X, Z) = \prod_{P \in K}(Z(n_2 P)X - X(n_1 P)Z) = \prod_{Q \in K[n_1]}(Z(Q)X - X(Q)Z)^{n_2}$. Taking the gcd with H, we recover $G_1(X, Z) = \prod_{Q \in K[n_1]}(Z(Q)X - X(Q)Z)$, where G_1 gives the equation of the kernel $K_1 = K[n_1]$.

We remark that non invertible elements modulo H are actually useful since taking a gcd with H allows to partially factorize it. For instance we could recover G_1 directly (without a resultant) by computing $n_1 \mathbf{P} = (P(X, Z), Q(X, Z))$ where Q is the division polynomial $\phi_n(X, Z)$ computed modulo H, and then taking $G_1 = Q \wedge H$. This is essentially the same as computing $\phi_n \mod H$ via the recurrence formula for ϕ_n where we reduce modulo H at each step, and this costs $O(\log n)$ arithmetic operations in $k[x]/H(x)$.

From this equation for K_1, we can compute the isogeny equation for the first step $\phi_1 : E_1 \to E_1'$ of the isogeny: $\phi = \phi_2 \circ \phi_1$. We can then compute $\mathbf{P}_2 = \phi_1(\mathbf{P})$, and take a resultant (or better use power projection [Sho99]) to get the kernel equation $H_2(X, Z) = 0$ of $\mathrm{Ker}\, \phi_2$. Using [KU11] for the power projection over a finite algebra, this step can be done in pseudo-linear time in n.

We refer to [EPSV, BGDS23] for optimisations, like using minimal polynomials rather than characteristic polynomials or using the Frobenius action.

4   The Ideal Representations

In this section we survey the ideal representation. We explain the link between ideals and isogenies in Sect. 4.1, then we treat the case of supersingular curves in Sect. 4.2 and the case of oriented curves (in particular ordinary curves) in Sect. 4.3.

4.1   Isogenies Represented by Ideals

Ordinary Isogeny Graph: Volcanoes. It is well known that if $E/\mathbb{F}_q$ is an ordinary elliptic curve, its ℓ-isogeny graph form the structure of a volcano, see [Koh96, FM02, Sut13a].

Let $\phi : E_1 \to E_2$ be an n-isogeny between elliptic curves defined over a finite field $\mathbb{F}_q$. In the ordinary case, we say that ϕ is horizontal if $\mathrm{End}(E_1) = \mathrm{End}(E_2)$ (in the ordinary case $\mathrm{End}_{\overline{\mathbb{F}}_q}(E) = \mathrm{End}_{\mathbb{F}_q}(E)$ is a quadratic imaginary field, so the notation is unambiguous), ascending if $\mathrm{End}(E_1) \subset \mathrm{End}(E_2)$ (and it is not horizontal), and descending if $\mathrm{End}(E_1) \supset \mathrm{End}(E_2)$ (and it is not horizontal). If n is not prime we can also have incomparable orders.

An horizontal or ascending isogeny ϕ of kernel K can always be represented by an ideal $I \subset \mathrm{End}(E_1)$: if we let $I = I(\phi) := \{\alpha \in \mathrm{End}(E_1), \alpha(K) = 0\}$ to be the set of all endomorphisms that are 0 on $K = \mathrm{Ker}\,\phi$, then we have $E[I] = K$, where $E[I] := \{P \in E(\overline{\mathbb{F}}_q) \mid \alpha(P) = 0 \forall \alpha \in I\}$ (we have the obvious inclusion $K \subset E[I]$ by definition, and one can prove the converse when ϕ is not descending). Note that $E[I] \subset E[N(I)]$. In fact, I can always be represented as $I = (\alpha, N(I))$, and $E[I] = E[\alpha] \cap E[N(I)]$. The isogeny ϕ is horizontal if and only if I is invertible in $\mathrm{End}(E_1)$, otherwise ϕ is ascending and $\mathrm{End}(E_2) = O(I)$ is the order associated to I. Note that if ϕ is descending, it cannot be represented by an ideal in $\mathrm{End}(E_1)$, but its dual $\tilde{\phi} : E_2 \to E_1$ can be represented by a non invertible ideal in $\mathrm{End}(E_2)$. For more details, see [Wat69, Koh96, Kan11].

In particular, if $R = \mathrm{End}(E_1)$, we have a commutative group action (which is free and transitive) from the Picard group $\mathrm{Pic}(R)$ of R (the class group of invertible ideals, this is the same as the usual class group when R is a maximal quadratic order) on the set of elliptic curves E' horizontal isogeneous to E_1 (meaning that $\mathrm{End}(E') = R$). To an invertible ideal class $[I]$ with representative I, we associate the codomain E'_I of $\phi_I : E' \to E'_I$. This group action was first used in the context of cryptography by Couveignes in 1997 (but published much later in [Cou06]), and revisited in [RS06, DKS18]. The first practical version was CSIDH [CLMPR18]. Having a commutative group action allows to translate many cryptographic constructions coming from the DLP problem. For instance the CSIDH key exchange is very similar to the Diffie-Helman's key exchange. However, while Shor's quantum polynomial time algorithm for the DLP does not apply to group actions, there exists a quantum subexponential time algorithm due to Kuperberg [Kup05, Pei20].

Supersingular Isogeny Graphs. If E_1, E_2 are supersingular elliptic curves defined over $\mathbb{F}_{p^2}$, with all their geometric endomorphisms defined over $\mathbb{F}_{p^2}$ (equivalently $\mathrm{End}_{\mathbb{F}_{p^2}}(E_i)$ is of rank 4), then ϕ is "almost horizontal", in the sense that $\mathrm{End}(E_i)$ are both maximal quaternionic order. (By assumption, $\mathrm{End}_{\overline{\mathbb{F}}_p}(E_i) = \mathrm{End}_{\mathbb{F}_{p^2}}(E_i)$ so the notation is also unambiguous). By the "supersingular case", we will always refer to this situation, i.e. when considering a supersingular elliptic curve $E/\mathbb{F}_{p^2}$ we will always assume it is a maximal or minimal curve.

In that case, ϕ is also always represented by the ideal $I = I(\phi)$, i.e. we also have $\mathrm{Ker}\,\phi = E[I]$: this is the well known Deuring correspondence (see [Ler22b] for a nice overview, and [Voi21] for a complete reference). Furthermore, $\mathrm{End}(E_1)$ is the left order associated to I and $\mathrm{End}(E_2)$ its right order.

In both the oriented and supersingular cases, the dual/contragredient isogeny $\widetilde{\phi_I} : E_2 \to E_1$ is represented by the ideal $\overline{I}$, and we let $N(I)$ be the integer such that $I\overline{I} = \overline{I}I = N(I)$, so that ϕ_I is a $N(I)$-isogeny. The number $N(I)$ is the norm of I in the quadratic case, and its reduced norm in the quaternionic case.

It is well known that supersingular isogeny graphs are expander (and even Ramanujan in dimension 1). This means that following a sufficiently long path of small degree isogenies, the codomain elliptic curve is close to statically uniform among all supersingular curves. Furthermore, since the endomorphism rings are not commutative, there is no commutative group action (we only have a groupoid of ideal classes), so Kuperberg's attack does not apply in the supersingular setting. However, this absence of commutativity makes the key exchange more difficult: given two isogenies $\phi_1 : E_0 \to E_1$ and $\phi_2 : E_0 \to E_2$ where the isogenies ϕ_i are secret and the domains and codomains curve are public, we could still define a common secret via a pushforward square $\phi : E_0 \to E_{12}$, but the curve E_{12} depends on the choice of ϕ_1, ϕ_2: using other isogenies $\phi_1' : E_0 \to E_1, \phi_2' : E_0 \to E_2$ with the same codomains E_1, E_2 could give another result E_{12}'. We will go back to this when we look at Eichler orders in Appendix C.2. The SIDH protocol used the clever solution to go around this problem by publishing extra "torsion information" on ϕ_1, ϕ_2 to allow for Alice and Bob to compute the pushforward isogenies $E_1 \to E_{12}, E_2 \to E_{12}$. We refer to [De 17] for a nice introduction. Unfortunately (but fortunately for the HD representation!), this extra information turned out to reveal too much information.

Orientations. In [CK20], Colo and Kohel introduce the notion of orientations on the supersingular isogeny graph. With E_1, E_2 are supersingular curves over $\mathbb{F}_{p^2}$ as above, and given a quadratic imaginary order R, an R-orientation on E_i is an embedding $R \hookrightarrow \mathrm{End}(E_i)$. The isogeny ϕ is said to be R-oriented if it commutes with the orientations on E_1 and E_2. If R is saturated in $\mathrm{End}(E_1)$ (the orientation is also said to be primitive), we obtain an action by the class group of R on the oriented horizontal isogeny graph from E_1 (like we had for the horizontal isogeny graph of ordinary curves with $R = \mathrm{End}(E_1)$). This is convenient construction to build a group action on elliptic curves with a controlled number of points: $E_i(\mathbb{F}_{p^2}) = \mathbb{Z}/(p \pm 1)\mathbb{Z}^2$ (depending on the quadratic twist used). More generally, allowing going up R-oriented isogenies we obtain a volcano structure.

An important example is when E_1 is defined over $\mathbb{F}_p$, in that case $\mathrm{End}_{\mathbb{F}_p}(E_1)$ is only a quadratic imaginary order containing $\mathbb{Z}[\pi_p]$ (which is either maximal or of index 2 in its maximal order O_p), and we have a natural orientation given by the Frobenius endomorphism π_p. This is the orientation at play for the CSIDH cryptosystem, and we have a free transitive group action on $\mathrm{Cl}(\mathbb{Z}[\pi_p])$ or $\mathrm{Cl}(O_p)$ (depending on whether $\mathrm{End}_{\mathbb{F}_p}(E_1) = \mathbb{Z}[\pi]$ or O_p) on the supersingular curves $E/\mathbb{F}_p$ at the same 2-volcano level as E_1. This situation is thus very similar to the ordinary case, where the Frobenius also provides a natural orientation. More generally, looking at the graph of supersingular elliptic curves E d-isogeneous to their Galois conjugate: $\phi_d : E \to E^{(p)}$ (with isogenies respecting this morphism) is essentially equivalent to looking at an orientation by $\sqrt{-dp}$. The case $d = 1$ corresponds to the case supersingular curves defined over $\mathbb{F}_p$ [CS21].

More recently, other orientations on supersingular curves have been considered, for SCALLOP [FFK+23]. Here the orientation is given by a specific endomorphism $\alpha \in \text{End}(E_1)$ rather than the Frobenius, so one needs to provide an efficient representation of α.

One can see orientations as a way of providing volcano like graph structure on top of supersingular isogeny graphs (an orientation is an extra structure in the terminology of categories). The notion of orientation allows to treat in a unified setting both ordinary curves and supersingular curves over $\mathbb{F}_p$ (which are both naturally oriented by the Frobenius), and even more general cases like SCALLOP.

A word of warning: for general R-oriented curves, if p is inert in R (this does not happen in the ordinary case or for supersingular curves over $\mathbb{F}_p$), then not all horizontal isogenies come from an ideal: the Frobenius isogeny π_p is horizontal and has degree p, and cannot come from an ideal of R since there are none of norm p. This is essentially the only obstruction [Onu21]. In particular, the class group action $\text{Cl}(R)$ is not quite transitive in that case: there are two orbits.

Remark 4. The functor from ideal to isogenies extends to R-modules, where R is a primitive orientation on E. More precisely, one can build a (contravariant) functor from modules to group schemes and module maps to group morphisms, see [JKP+18] for ordinary and supersingular elliptic curves, and its extension to the oriented case in [PR23a]. This general module setting has several advantages. First, this functor, applied to a projective R-module of rank g produces an abelian variety of dimension g isogeneous to E^g; so using modules allow to understand the higher dimensional isogeny graph of E^g. This was used for instance to find the formulas for Clapotis [PR23b].

Secondly, non projective modules encode level structure informations. This can be used to handle level structure at the module level, we will see an example of this in Appendix D.

Finally, for supersingular curves, it can be used to understand the "forgetting the orientation" functor; from the module point of view if O is the full endomorphism ring and R an orientation, forgetting the orientation amount to sending M to $M \otimes_R O$. This is, implicitly, the approach used in [ACL+23, ACL+22] to study the interaction between the R-oriented isogeny graph and the full supersingular graph.

4.2   The Supersingular Case

Recall that if E_1, E_2 are supersingular elliptic curves defined over $\mathbb{F}_{p^2}$, with all their geometric endomorphisms defined over $\mathbb{F}_{p^2}$ then any isogeny $\phi : E_1 \to E_2$ is represented by an ideal I.

We will assume that we have an efficient representation of the endomorphism ring $\text{End}(E_1)$. This means that we both know its representation as an abstract maximal $\mathcal{O}_1$ order in $B_{p,\infty}$, the quaternion algebra ramified at p and infinity, and that we are also able to evaluate endomorphisms $\alpha \in \mathcal{O}_1$ on points of E_1. Typically this is done by giving an efficient representation of a basis $(\alpha_1, \alpha_2, \alpha_3, \alpha_4)$ of the $\mathbb{Z}$-lattice $\mathcal{O}_1$, basis given by endomorphisms of reasonable degree (polynomial in p). (In practice, we take $\alpha_1 = 1$).

We remark that given an effective representation of a basis α_i, it is possible to recover the abstract structure of $\text{End}(E_1)$ by computing Weil pairings. Conversely, given an abstract representation of $\mathcal{O}_1$, it is possible to recover an effective representation for the endomorphisms: this is done by starting with an elliptic curve E_0 with an effective representation, computing a suitable effective isogeny between E_0 and E_1, and transporting the effective endomorphisms representation from E_0 to E_1. We refer to [EHLMP18, Wes22] for more details.

Given this effective representation of $\text{End}(E_1)$, the ideal I provides a compact representation of the isogeny ϕ. Furthermore, $\text{Ker}\,\phi = E[I]$ so, by Sect. 3.2, this representation is effective if $n = N(I)$ (where N denote the reduced norm) is smooth and the N-torsion is accessible. See Example 13 for more details on how to compute generators for $E[I]$.

In fact, the ideal representation is always effective, even when $N(I)$ is a large prime. Indeed, for isogeny based cryptography, many algorithms of the form `IdealToIsogeny` have been proposed, to make the Deuring correspondence between ideals and isogenies not only effective but practical.

We can broadly distinguish between four (minor) generations of these algorithms (we refer to Appendix C for more details):

1. The KLPT [KLPT14] algorithm can be (heuristically) used to smoothen in polynomial time an arbitrary ideal I into a smooth equivalent ideal J of large norm. A proven version (under GRH) is given in [Wes22]. The first generation then takes J to be powersmooth (or smooth with accessible $N(J)$-torsion). We can then reconstruct the associated isogeny ϕ_J from its kernel $E[J]$. From ϕ_J, we can recover ϕ_I.

2. The idea for the second generation is to take J to be of (reduced) norm 2^e rather than powersmooth, and to select a prime p such that a large 2^f-torsion is rational on supersingular elliptic curves $E/\mathbb{F}_{p^2}$. This gives the first version of the Deuring correspondence that is practical enough to be used in a signature protocol: SQIsign [DKLPW20].

   A problem that has to be solved is that since $2^f \mid p \pm 1$, we have $2^f < p$ while the KLPT algorithm gives (generically) an ideal of norm $p^{4.5} \gg p$. The solution is to cut the isogeny represented by J into chunks of 2^f-isogenies, so it gets decomposed as $\phi_J : \phi_m \circ \cdots \circ \phi_1$. The problem is then to find the kernel of the intermediate isogenies $\phi_i : E_i \to E_{i+1}$, this is also known as "refreshing the 2^f-torsion".

   The solution used by the second generation is to construct isogenies $E_0 \to E_i$ from a special elliptic curve E_0 to refresh the torsion; this is combined with an improved version of KLPT which use the theory of Eichler orders to decrease the degree produced by the KLPT smoothening algorithm from $\approx p^6$ to $\approx p^{4.5}$.

A further improvement to this method is described in [DLLW23], where a suitable endomorphism is constructed on each intermediate curve E_i to refresh the torsion, rather than via a special isogeny $E_0 \rightarrow E_I$

We refer to [Ler22b] for more details on this second generation.

3. In the third generation, one still compute an equivalent ideal J of large reduced norm 2^e, which is then split into chunks of 2^f-isogenies, but the ideas of the HD representation are used there to refresh the torsion.

   In [Ler23b], Leroux explain how to use a dimension 2 representation of arbitrary endomorphisms (similar to the techniques of SCALLOP-HD) to refresh the torsion on E_i, while in [ON24], Onuki and Nakagawa, inspired by SQIsignHD, use isogenies $E_0 \rightarrow E_i$ as in the original SQIsign version, but represented in dimension 2.

4. In the fourth generation, the smoothening KLPT step is completely bypassed, and the ideal I is directly converted into an isogeny, using higher dimensional isogenies like for the third generation.

   This algorithm (Clapotis, see Example 8) was introduced in [PR23b] to convert an ideal in the oriented case (see Sect. 4.3), and adapted to the supersingular setting for SQIsign2d in [BDD+24].

In the algorithms above, rather than computing $\phi_I : E_1 \rightarrow E_2$ directly, we often compute another equivalent but easier isogeny $\phi_J : E_1 \rightarrow E_2$. To recover ϕ_I from ϕ_J, one can use a double path algorithm, see Appendix C.1.

Conceiving improved ideal to isogeny algorithms that are not only effective in theory, but really practical, has thus been a subject of intense research that has made tremendous progress since the HD representation was introduced.

We have seen that in the fourth generation, used in the higher dimensional versions of SQIsign, like SQIsignHD [DLRW24] or the even more recent SQIsign2d variants [BDD+24, NO24, DF24], the KLPT "smoothening" algorithm is no longer needed. A hope for a fifth generation would be to find a way to improve KLPT to generate smooth ideals of norm $\approx p$ rather than $\approx p^{4.5}$, see Appendix C.5.

Remark 5. The main drawback of the ideal representation is that it requires the knowledge of $\mathrm{End}(E)$. Furthermore, giving the ideal I leaks the endomorphism ring (which is the left order of I), so one cannot share this ideal representation in settings where $\mathrm{End}(E)$ is supposed to remain secret, like in isogeny based cryptography. For that reason, Leroux introduced in [Ler22a] *the suborder* representation, which is an isogeny representation which is still efficient while only leaking part of the endomorphism ring. This representation has now been essentially superseded by the HD representation which leaks no informations on the endomorphism ring.

4.3   The Oriented Case

Let's assume that we have an efficient primitive orientation on E_1 by a quadratic imaginary order R. We recall that this includes the case of ordinary elliptic curves and supersingular curves over $\mathbb{F}_p$, where in both cases R is given by the saturation of $\mathbb{Z}[\pi]$ in $\mathrm{End}_{\overline{\mathbb{F}}_q}(E)$.

Like in the supersingular case, converting an horizontal (i.e. invertible) ideal $I \subset R$ into an horizontal isogeny $\phi_I : E_1 \to E_2$, requires to compute the kernel $E[I]$ and then use the algorithms from Sect. 3.2. This is only effective if $N(I)$ is smooth and the $N(I)$-torsion is accessible.

In the supersingular case, we saw that it was always possible, thanks to the KLPT algorithm, to efficiently *smoothen* I, i.e., find in polynomial time an equivalent ideal J of smooth norm. In the ordinary case, this is much harder. We know that (under GRH), $\mathrm{Cl}(R)$ is generated by ideals of norm $O(\log^2 \Delta_R)$. So one can definitively decompose I into an equivalent smooth ideal J of smoothness bound $O(\log^2 \Delta_R)$, and heuristically this holds for a powersmooth decomposition too (increasing the smoothness bound if needed). But as we will explain below, we currently only have subexponential algorithms to smoothen an ideal in the oriented case.

If we can find J, then once we evaluate ϕ_J, we can evaluate ϕ_I if needed as we did in the supersingular cases, by using the effective R-orientation on E_1 (like in Appendix C.1, this may involve constructing a double path or using a division algorithm if we want to evaluate ϕ_I on points of torsion non coprime to $N(J)$).

If R is not given by the natural Frobenius orientation, we also need to push the R-action into an effective representation on E_2. There are two ways to do that: if we have an effective orientation from some special curve E_0, then we can build a double path from E_0 to E_1, E_2 in order to propagate the orientation to each. The problem is that this way of representing the orientation leaks an isogeny path from E_0 to E_i, which we want to avoid in cryptographic applications of isogenies. The other solution, if $R = \mathbb{Z} \oplus \mathbb{Z}\alpha$ is to simply give an HD representation of α on each E_i.

Like for the supersingular ideal to isogeny algorithms, we can distinguish four minor generations in the oriented case:

1. The first generation simply bypass the smoothening problem by considering only ideals of the form $I = \prod \mathfrak{l}_i^{e_i}$ where the $\mathfrak{l}_i$ are prime ideals of small norm and the exponents e_i are small. In other words: we consider a *restricted group action* [ADMP20, DHK+23]. This is sufficient for a key exchange like CSIDH [CLMPR18], and even for signatures via rejection sampling [DG19].
2. The second generation is to do a huge class group computation, using the standard classical subexponential algorithms. This is the approach of CSI-FiSh [BKV19].
3. The third generation is to take an orientation by a non maximal conductor, in order to simplify the class group computation. A question is then how to represent the endomorphism α giving the orientation; in SCALLOP α is taken to be of smooth-norm, while in SCALLOP-HD α is arbitrary and represented via a dimension 2 HD isogeny.
4. The fourth generation, Clapoti(s) [PR23b] completely bypass the smoothening step (this is the same method as the fourth generation in the supersingular case).

As we will explain below, even when the class group is easy to compute as in SCALLOP-HD, the smoothening step is still subexponential. By contrast, Clapoti(s) allows to compute the action of an invertible ideal I in polynomial time (in $\log p$ and $\log \Delta_R$) through higher dimensional isogenies. So once again, the HD representation proves crucial.

Like in the supersingular case, we now have a polynomial time algorithm to translate I into an isogeny, when I is *invertible*. In summary: the ideal representation is both compact and efficient, both in the supersingular case, or in the oriented *horizontal* case. It is still unknown how to ascend or descend a large ℓ-isogeny volcano in polynomial time in $\log \ell$, see Sect. 7.

Remark 6. Let R be a primitive orientation on E_0, and $I \subset R$ an invertible ideal. Although we do not know how to smoothen in polynomial time an ideal $I \subset R$ in the oriented case, it is possible to smoothen in polynomial time the module map $I \oplus \overline{I} \to R \oplus R$. Technically, to handle the polarisations, we work with Hermitian modules and similitudes. Here the Hermitian form on R is the one induced by the norm,, the one on $R \oplus R$ is the product Hermitian form and the Hermitian form on $I \oplus \overline{I}$ comes from pullback. And for any N large enough (heuristically[9] $N \gg \min(N(I), \Delta_R^{1/2})^2 \Delta_R^2$), we can find a module map ϕ which is a N-similitude for these Hermitian form. Taking $N \gg p^3$ smooth gives our smoothening. Using the module equivalence of category Remark 4, we can convert this smoothened module map into an N-isogeny $E_0 \times E_0 \to E_I \times E_{\overline{I}}$ with respect to the product principal polarisations. More details will be given in [PR23a].

Smoothening an Ideal. We give more details on the smoothening step used by the second and third generations. We first want to find $I \sim J = \prod \mathfrak{l}_i^{e_i}$ for small prime ideals $\mathfrak{l}_i$ of norm ℓ_i; this is already hard classically.

Furthermore, if the e_i are too large, so that the $\ell_i^{e_i}$ torsion is not accessible, we need to split the $\mathfrak{l}_i^{e_i}$ isogeny into chunks of $\mathfrak{l}_i^{f_i}$ isogenies, that we compute one by one. The problem is that if the exponent e_i is too large, we might need superpolynomially many chunks. We can reformulate this problem by saying that, even when $\mathfrak{l}$ is small, the action of $\mathfrak{l}^n$ takes time $O(n)$ to compute rather than $O(\log n)$. So we really want to find a powersmooth decomposition, i.e. such that the e_i are small. This is hard even with a quantum computer.

The current solution is as follows:

1. Find the group structure of $\mathrm{Cl}(R)$ with respect to small generators $\mathfrak{l}_i$, and in particular the lattice of relation L.
2. Find a decomposition $I \sim \prod \mathfrak{l}_i^{e_i}$, with potentially very large e_i. In practice, this step can be bypassed when I is directly given as $I = \prod \mathfrak{l}_i^{e_i}$ with large e_i (like for some signatures schemes).

---

[9] The similitude with the KLPT algorithm from Appendix C.2 is not a coincidence, in both case the lattice of endomorphisms (of the special curve E_0 for KLPT and of R-oriented endomorphisms on $E_0 \times E_0$ in our situation) is of rank 4 and has elements of small norms.

3. Solve (an approximation of) the close vector problem on the lattice of relations L, to obtain $I \sim \prod \mathfrak{l}_i^{e'_i}$ where the vector of exponents (e'_i) is small. To get a good approximation of CVP requires to find a good short basis of the lattice of relations L (the better the basis, the better the CVP solution).

Class Group Computation. For a generic quadratic order, the best classical algorithms to compute the class group of R are in subexponential time $L_{1/2}(\Delta_R)$.

The classical algorithm first finds relations between smooth ideals with a $L_{1/2}(\Delta_R)$ smoothness bound. Then a linear algebra step constructs the lattice of relations with respect to generators of norm $O(\log^C \Delta_R)$ for some appropriate constant C. We also have polynomial time quantum algorithms.

In practice, as shown by [BKV19] which gave a record class group computation for the CSI-Fish signature, this class group computation can be done for Δ_R of around 512 bits.

To go further (classically), the idea of SCALLOP [FFK+23] is to use special quadratic orders R, with conductor f carefully chosen inside a maximal order R_0 of small discriminant Δ_{R_0}.

From the conductor square (see Appendix D), we obtain the exact sequence from Eq. (4), which shows that computing $\mathrm{Cl}(R)$ amounts to computing the group structure of $\mathrm{Cl}(R_0)$ and of $(R_0/fR_0)^*/(R/fR)^*$. Since R_0 is chosen to have small discriminant, computing $\mathrm{Cl}(R_0)$ is easy. If $f = \ell$ is a prime, the second part amount to computing discrete logarithms in $\mathbb{F}_\ell^*$ if ℓ splits in R_0, and in $\mathbb{F}_\ell^{2,*}/\mathbb{F}_\ell$ if ℓ is inert in R_0. Carefully choosing f (typically f a large prime that splits in R_0 and such that $f - 1$ is smooth so that DLPs in $\mathbb{F}_\ell$ are easy) allows to compute the large class group $\mathrm{Cl}(R)$ in practice.

In SCALLOP [FFK+23], due to the constraints of having to give an efficient representation of the orientation, f could be chosen smooth, but with a quite large smoothness bound. In SCALLOP-HD [CLP24], the authors show that switching to a dimension 2 HD representation for the orientation allows to take $f = \ell = 2^m u + 1$ with a small u, which allows for very fast DLPs in $\mathbb{F}_\ell^*$.

Decomposing an Ideal. Decomposing an ideal I into an equivalent ideal $I \sim \prod \mathfrak{l}_i^{e_i}$ (where $\mathfrak{l}_i$ are our chosen small generators for $\mathrm{Cl}(R)$) is very similar to the class group computation, except it needs to be done online, this is not a precomputation. The classical algorithm is in $L_{1/2}(\Delta_R)$ for a generic quadratic order R, while quantum algorithms are polynomial time. If R is a special order of conductor f inside a maximal order R_0 of small discriminant, and such that $f = \ell$ is a large prime that splits in R_0 and such that $\ell - 1$ is smooth, like in SCALLOP or SCALLOP-HD, the decomposition amounts to a multi-DLP in $\mathbb{F}_\ell^*$, which is easy.

Closest Vector Problem. Since we have constructed the lattice of relations L between the $\mathfrak{l}_i$, we can use this lattice to shrink the exponents e_i: this is a closest vector problem. We can solve (an approximation of) the close vector problem on L by first finding a good short basis of L.

Heuristically, as explained in https://yx7.cc/blah/2023-04-14.html, using a lattice reduction algorithm of complexity $\approx L_\alpha(\Delta_R)$ allows to find exponents bounded by $\approx L_\beta(\Delta_R)$, where $\beta = \frac{1}{2} - \frac{1}{2}\alpha$ (and neglecting polynomial factors in $\log \Delta_R$), using $d \approx \log^{1-\beta} \Delta_R$ many small ideals (so working with a lattice of dimension d). We find out that, using a lattice reduction which takes polynomial time like LLL, we get exponents of size $L_{1/2}(\Delta_R)$. This is because we are in the special case of class groups of quadratic imaginary field, so we can expect the class group to be almost cyclic, and we heuristically expect it to be generated by very small elements of norm up to $\log^{1/2}(\Delta_R)$, so even through the LLL algorithm gives an exponential approximation factor in the dimension of the lattice, this dimension is small enough that the resulting approximation factor is subexponential in Δ_R. With a lattice reduction taking time $L_{1/3}(\Delta_R)$, we get exponents of size $L_{1/3}(\Delta_R)$. With a lattice reduction taking time $L_{1/2}(\Delta_R)$, we get exponents of size $L_{1/4}(\Delta_R)$, and with a lattice reduction taking exponential time we get exponents of size $O(1)$. And quantum algorithms do not seem to help for this step, because we do not know how to exploit that the lattices relations come from a class group.

This means that asymptotically, when using the generation 2 algorithm to compute the full class group of $\mathbb{Z}[\sqrt{-p}]$ like in CSI-FiSh, since this step takes $L_{1/2}(\Delta_R)$ already we can allow the precomputation phase to be of $L_{1/2}(\Delta_R)$ and expect exponents of size $L_{1/4}(\Delta_R)$. However, in a setting like SCALLOP or SCALLOP-HD where the class group computation is taylored to be easy (polynomial time), we still need a $L_{1/3}(\Delta_R)$ precomputation to get exponents of size $L_{1/3}(\Delta_R)$ (or we could spend more precomputation time to get smaller exponents).

In practice, for the specific examples of class group action as computed in CSI-FiSh, SCALLOP, or SCALLOP-HD, the lattice reduction phase was not the bottleneck (for CSI-Fish it was computing the lattice itself, step bypassed in SCALLOP or SCALLOP-HD, at the cost of no longer having a natural orientation). But this asymptotic bottleneck prevents an asymptotic instantiation of the class group action in polynomial time, even with a quantum computer.

Only the fourth generation, which bypass the smoothening step, can give a fully unrestricted group action.

5   The HD Representation

The ideal representation is our first isogeny representation that gives an effective representation for horizontal isogenies, provided that we know an effective representation of the endomorphism ring or orientation of the domain.

We have seen in Sect. 3.3 that the interpolation representation can give a compact representation of *any* isogeny. However, recovering the isogeny from the interpolation data is at least in $\widetilde{O}(n)$.

Recently, as a side product of the SIDH attacks [CD23, MMPPW23, Rob23b], a much more sophisticated algorithm has been found to reconstruct the isogeny ϕ from suitable interpolation data, via higher dimensional isogenies.

The most flexible version is the following [CDM+24]:

Theorem 1. *Let* $\phi : E_1 \rightarrow E_2$ *be an* n-*isogeny (with as usual* n *prime to* p*). Let* G *be a subgroup of* E_1 *of order at least* $4n + 1$*, and* $(P_1, \ldots, P_r)$ *generators of* G*. Given a point* $P \in E_1(\mathbb{F}_{q'})$ *and the interpolation data* $(P_i, f(P_i))$*, there is an algorithm to evaluate* $\phi(P)$ *in time polynomial in* $\log n$*,* r*,* $\log q$*, the largest prime factor of* n*, the extension degrees of the field of definition of* P *and the points* P_i*, and the extension degree of the points of* $\ell^{\lfloor e/2 \rfloor}$ *torsion for each prime power* $\ell^e \mid \#G$.

For our applications, we will only need a simplified version where we assume that we have interpolation data on the full N-torsion, version contained in [Rob23b] and which was exploited in [Rob22a] for isogeny representations. In that setting, Theorem 1 can be rephrased as follows: assume that we have interpolation data on the N-torsion with N large enough ($N > n$, but see Appendix B for a relaxation to $N^2 > n$), smooth and the N-torsion accessible on E_1. Then we have an efficient representation of the isogeny ϕ. This is done by embedding ϕ into a N-isogeny Φ in higher dimension; we call this the *HD representation*.

Since the HD representation use isogenies of higher dimension, we first review isogenies between abelian varieties and algorithms to compute them in Sect. 5.1. We then explain Kani's lemma in Sect. 5.2, and how it can be used to embed isogenies to obtain the HD representation in Sect. 5.3.

Remark 7 (Level structure). Theorem 1 shows that, in the context of isogeny based cryptography, revealing too much torsion information is insecure. Namely, a secret isogeny $\phi : E_1 \rightarrow E_2$ can be reconstructed efficiently if the adversary know its degree n, and its action on the N-torsion (provided that N is smooth and the N-torsion is accessible), as long as $N^2 \approx n$ (or, by Theorem 1, even just the action on a subgroup $G \subset E_1[N]$ of size $\approx n$). Still, revealing some torsion information (i.e., suitable image points under ϕ), as was done in SIDH, can be very useful to build isogeny based cryptosystems.

We discuss several solutions:

– Mask the torsion revealed. For instance in M-SIDH [FMP23] the torsion information $(\phi(P_1), \phi(P_2))$ is masked by some common scalar λ (one needs to be careful that the Weil pairing $e_{W,N}$ will reveal λ^2 modulo N). In FESTA [BMP23] the authors hide the torsion by a diagonal matrix. The correct notion under which to study these maskings is the concept of level structure, we refer to [DFP24] for an overview. An alternative, as in [Bas24] is to also hide the degree.
– Only reveal torsion information for a subgroup G of size much smaller than the degree n. This is for instance the approach for IS-CUBE and LIT-SIGAMAL [Mor23, Mor24].

We recall that supersingular isogeny graphs are expander. This property remains true even when adding level structure information [BCC+23, CL23, PW24], at the condition of increasing the degree of the path relative to the degree of the revealed level structure. This means that we can heuristically expect that it is not easier to recover a degree n'-isogeny for which the action on the N-torsion is revealed than to recover a degree n-isogeny for which no torsion information is revealed, provided that $n' \approx nN^2$.

– Reveal the torsion information on the N-torsion for N large with respect to n, but prime rather than smooth. Indeed, in that case, while the embedding lemma still allows to embed ϕ into a higher dimensional N-isogeny Φ, without the smoothness condition we do not know how to evaluate Φ efficiently. This is an open question in dimension 1 already, see also Sect. 7. Still, this is quite a strong cryptographic assumption to make (that we don't know how to evaluate an isogeny from kernel generators of non smooth order), and we would recommend to use the other solutions if possible.

5.1   Isogenies Between Abelian Varieties

The key idea behind Theorem 1 is to embed the n-isogeny ϕ into a higher dimensional N-isogeny Φ given by the interpolation data. We first briefly describe the type of isogenies we will work with, and existing algorithms to compute them.

If $\phi : A \to B$ is an isogeny, we denote by $\hat{\phi} : \widehat{B} \to \widehat{A}$ its dual isogeny.

Definition 1. *An isogeny $\phi : (A, \lambda_A) \to (B, \lambda_B)$ between polarised abelian varieties is called an n-isogeny if the polarisation $\phi^* \lambda_B := \hat{\phi} \circ \lambda_b \circ \phi = n\lambda_A$.*

We recall the following standard results, see [Rob23b].

Lemma 1. *Let $\phi : (A, \lambda_A) \to (B, \lambda_B)$ be an isogeny between principally polarised abelian varieties, and let $\widetilde{\phi} = \lambda_A^{-1} \circ \hat{\phi} \circ \lambda_B : B \to A$. Then ϕ is an n-isogeny if and only if $\widetilde{\phi} \circ \phi = n$ (or equivalently, $\phi \circ \widetilde{\phi} = n$).*

Furthermore, if $\Phi = \begin{pmatrix} \phi_{11} & \phi_{12} \\ \phi_{21} & \phi_{22} \end{pmatrix} : (A_1 \times A_2, \lambda_{A_1} \times \lambda_{A_2}) \to (B_1 \times B_2, \lambda_{A_2} \times \lambda_{B_2})$ is an isogeny between the product abelian varieties endowed with their product principal polarisations, then $\widetilde{\Phi} = \begin{pmatrix} \widetilde{\phi_{11}} & \widetilde{\phi_{21}} \\ \widetilde{\phi_{12}} & \widetilde{\phi_{22}} \end{pmatrix}$.

Remark 8. If $\phi : (A, \lambda_A) \to (B, \lambda_B)$ is an n-isogeny between principally polarised abelian varieties, the kernel $K = \operatorname{Ker} \phi \subset A[n]$ is maximal isotropic for the (polarised) Weil pairing $e_{W,n\lambda_A}$. And if n is prime (and $n \neq p$), $K(\overline{k}) \simeq (\mathbb{Z}/n\mathbb{Z})^g$. Conversely, if $K \subset A[n]$ is maximal isotropic for $e_{W,n\lambda_A}$, then the polarisation $n\lambda_A$ descends to a principal polarisation λ_B on $B = A/K$. Note however, that unlike the dimension one case, there can be several non equivalent principal polarisations on an abelian variety, so being an n-isogeny (with respect to specific principal polarisations) is a stronger condition than just the kernel being maximal isotropic for $e_{W,n\lambda_A}$.

We have partial generalisations of the isogeny representations from Sect. 3 for elliptic curves to abelian varieties.

In low dimension, $g \leqslant 3$, any principally polarised abelian variety (over an algebraically closed field) is a Jacobian of a curve or a product of Jacobians. For instance in dimension 2 we have either a product of two elliptic curves or a Jacobian of an hyperelliptic curve of genus 2. In dimension 3, if (A, λ_A) is a principally polarised abelian variety with an indecomposable polarisation λ_A, then A is a Jacobian $A = \text{Jac}(C)$, with C either hyperelliptic of genus 3 or a quartic curve.

The polarisation is important here: for instance every superspecial abelian variety $A/\mathbb{F}_q$ of dimension $g > 1$ is isomorphic (over the algebraic closure) as an unpolarised abelian variety to E_0^g, where $E_0/\overline{\mathbb{F}}_q$ is any supersingular curve, but in general a principal polarisation on A won't be given by the product principal polarisation. In dimension 2 over $\mathbb{F}_{p^2}$, among superspecial principally polarised abelian surfaces, there are roughly $\approx p^3/2880$ superspecial Jacobians, and $\approx p^2/288$ product of supersingular elliptic curves.

Starting in dimension 4, a generic abelian variety won't be a Jacobian. We can instead use the theory of algebraic theta functions, as developed by Mumford in [Mum66, Mum67a, Mum67b], with useful extensions by Kempf [Kem88, Kem89a, Kem89b, Kem90, Kem92] (see also [Mum83, Mum84, Mum91] for the analytic theory of theta functions).

Vélu's Like Formulas. We have an extension of Vélu's formula which allows to evaluate an n-isogeny $\phi : (A, \lambda_A) \to (B, \lambda_B)$ between principally polarised abelian varieties, given a generator representation $K = \langle T_1, \ldots, T_g \rangle$ of its kernel, which is isotropic for the Weil pairing $e_{W, n\lambda_A}$, in the algebraic theta model induced by a symmetric theta structure of level m on A (this algebraic theta model is completely determined by the theta null point of A if $m \geqslant 4$; and if $m = 2$ we obtain a model for the Kummer variety $A/\pm 1$ if the polarisation is indecomposable). We note that the rank condition on the kernel K is similar to having a cyclic isogeny in dimension 1. Namely, by [LR12, CR15, LR22], in the case that n is prime to the level m and the characteristic p is coprime[10] to mn, [LR22] gives an algorithm in $O((mn)^g)$ arithmetic operations over the field of definition of the generators T_i to compute the theta null point of B, and also to compute the image of a point P (this time over the field of definition of P and the T_i). A sketch of an adaptation of this algorithm to the case when m is not coprime to n (this includes the important case where $n = 2$ is even) is given in [Rob21]. The formulas of [CR15] have been implemented in Magma [BCR10]; in the repository there is also a branch which contains the fork ThetaAV written by Anna Somoza and David Lubicz in Sage. (The faster formulas from [LR22]

---

[10] A symmetric theta structure has even level, so this condition imposes that $p \neq 2$. But we can use the same lift and reduce technique as in [GL09] to handle the case $p = 2$. For instance [BCR10] contains code to compute dimension 2 isogenies in characteristic two, derived by adapting the formulas for characteristic 0 in such a way that they have good reduction modulo 2.

are currently only in the private development version). These isogeny formulas are part of a research effort [LR12, CR15, LR15b, LR22, LR16, LR10, LR15a] to adapt the usual algorithms on elliptic curves (arithmetic, pairings, isogenies) to arbitrary abelian varieties represented by algebraic theta functions; see [Rob21] for an overview.

Remark 9 (Level 2 vs level 4). In [Mum67a, § 6], Mumford constructs the universal abelian scheme of level m over $\mathbb{Z}[1/m]$ via Riemann's relations, when the level m is divisible by 4. But level m requires m^g theta coordinates, so for efficiency we prefer to work in level $m = 2$. This adds some technical difficulties, notably for gluing and splitting, see [DMPR24, § 4.1].

For ℓ-isogenies between Jacobians, we also have an algorithm in $O(\ell^g)$ by [CE14], formulas which were optimised for $g = 2, 3$ in [Mil20] (see also [Tia24] for gluing and splitting isogeny formulas).

For small ℓ and in the Jacobian model in dimension 2, formulas for 2-isogenies were already known by Richelot [Ric36, Ric37], and specific formulas for the case $\ell = 3, 5$ were given in [BFT14, Fly15]. A general method to find ℓ-isogeny formulas in the generic Kummer model (rather than the theta model; this model was described by Cassel and Flynn [CF+96, Fly93]), are given in [Nic18, CCS24]. Smith also gave 2-isogeny formulas in dimension 3 in [Smi08] when the domain is a Jacobian of an hyperelliptic curve (this allows to transfer the DLP problem to the Jacobian of a non hyperelliptic curve when the codomain is not a Jacobian of an hyperelliptic curve), see also [FK11]. Smith's approach to transfer the DLP from a Jacobian of a genus 3 hyperelliptic curve to a genus 3 quartic was generalized using the ℓ-isogeny formulas of [CE14, Mil20] in [Tia20].

Decomposing a Smooth Isogeny. We can then apply the same quasi-optimal strategies to decompose a smooth n-isogeny into a product of small isogenies as in dimension 1, when the n-torsion is accessible.

We give the statement for the theta model, since it handles all abelian varieties, but the same strategy works for isogenies between Jacobian models.

Lemma 2. *Let $(A, \lambda_A)/\mathbb{F}_q$ be a principally polarised abelian variety represented by a symmetric theta structure of level m. Let $K \subset A[n]$ be a maximal isotropic subgroup of rank g. Let $n = \prod_{i=1}^{a} \ell_i^{e_i}$, and $T_{1,i}, \ldots, T_{g,i}$ generators of $K[\ell_i^{e_i}]$, which live in $\mathbb{F}_{q^{d_i}}$.*

If n is not coprime to m, we let $n' = n \vee m$ if n is odd and $n' = 2(n \vee m)$ if n is even, and we also need to assume that we are given points $T_1', \ldots, T_g'$ of n'-torsion above the generators of K, that are compatible with our theta structure. Then the T_i' induce a uniquely determined theta structure of level m on $B = A/K$, and we can compute the theta null point of B in time $\widetilde{O}(\sum_i e_i m^g \ell_i^g (d_i \log e_i + \sum_{j>i} d_i \vee d_j) \log q) = \widetilde{O}(a^2 d' e m^g \ell^g \log q)$, where e is a bound on the e_i, d' is a bound on the $d_i \vee d_j$, and ℓ is a bound on the ℓ_i.

Given a point $P \in A(\mathbb{F}_{q^{d_P}})$, represented by its theta coordinates, we can compute the theta coordinates of $\Phi(P)$ in $\widetilde{O}(\sum_i e_i m^g \ell_i^g (d_i \vee d_P) \log q) = \widetilde{O}(a e m^g \ell^g d'' \log q)$ where d'' is a bound on the $d_i \vee d_P$.

Proof. This follows from the general isogeny formulas from [Rob21], and the decomposition process is explained in more details in [DLRW24, Appendix F]. See Remark 17 for our assumptions about the lattice of field extensions used to represent the CRT basis.   □

Remark 10. The condition on having explicit points of n'-torsion in Lemma 2 is to ensure that the theta structure of $B = A/K$ is uniquely determined. We can relax this condition at the cost of taking some roots. For instance in level $m = 2$, to compute 2^e-isogenies in dimension 2, in [DMPR24, Rob23a] we describe an algorithm to compute them from only their kernel, by using appropriate square roots at the last two steps where we don't have enough information to determine the theta structure uniquely. These square roots are not needed if we have points of 2^{e+2}-torsion above the kernel.

Fast Formulas. Due to the importance of the HD representation for dimension 1 isogenies, the specific case of 2^e-isogenies (or ℓ^e-isogenies with small ℓ) in dimension 2 and 4 has recently been the focus of optimisations. In dimension 2, [Kun22] Kunzweiler gives efficient 2^e-isogeny formulas in the Jacobian model, which she has extended to the Kummer Jacobian model (private communication). In [DK23], Decru and Kunzweiler give efficient 3^e-isogeny formulas in the Kummer model. Recently, in [CCS24], Corte-Real Santos, Costello, Smith gave optimised 3-isogeny formula in dimension 2 in the theta squared Kummer model.

Fast 2^e-isogeny formulas in dimension 2 in the level 2 theta model (hence with theta Kummer surfaces) were given in [DMPR24]. These formulas achieve a factor 10 speed up for the codomain computation compared to Richelot isogenies, and a factor 30 for the image of points. These fast formulas are the basis of all the recent isogeny based cryptosystems which use a 2^e-HD representation in dimension 2; they are notably used for the verification of the SQIsign2d variants [BDD+24, NO24, DF24], which currently give the record verification time for the SQIsign family of signature schemes.

Fast formulas for 2^e-isogenies in the level 2 theta model in any dimension were sketched in the notes [Rob23a]; these have been fully worked out by Dartois in dimension 4 in [Dar24], along with explicit base change formulas. The main application of Dartois' work is for the verification of the SQIsignHD signature [DLRW24], which is done in dimension 4.

The reason we prefer to use as small a dimension as possible for the HD representation is that the level 2 theta model in dimension g requires 2^g coordinates, so we expect an exponential slow down with respect to the dimension. As a rule of thumb: in dimension g we work with 2^g theta coordinates, and to decompose a 2^e-isogeny we need to push g points in the kernel at each step, so we expect a time of $Cg2^g$ for the decomposition by step. The formulas are very similar (a combination of Hadamard transforms, squarings, and multiplication by suitable constants) across all dimensions, so we expect C to be roughly independent of g. So as a very rough approximation, we can expect dimension 2 2^e-isogenies to be 4 times slower than in dimension 1, dimension 4 isogenies to be 8 times slower than dimension 2 and 32 times slower than dimension 1, and dimension 8 isogenies to be 32 times slower than dimension 4, 256 times slower than dimension 2 and 1024 times slower than dimension 1.

In practice, a low level Rust and C implementations of the formulas of [DMPR24] need around 2–3 ms to compute a 2^{128}-isogeny in dimension 2 over a field of 256 bits (the C implementation was written for SQIsign2d-West and will soon be available, the Rust implementation is already available). Compared in dimension 1, this is roughly a 4.5 slow down, this is expected with respect to our factor 4 slow down we argued previously, because in dimension 1 the best 2^e-isogeny formulas use Montgomery coordinates which are slightly faster than theta coordinates, and decompose the 2^e-isogeny into chunks of 4-isogenies, which is also slightly faster than decomposing into chunks of 2-isogenies.

Remark 11 (Fast Kummer surfaces versus generic Kummer surfaces). As mentioned, the fast 2^n and 3^n isogeny formulas on Kummer surfaces from [DMPR24, CCS24] use Mumford's algebraic theta model (or twisted variants of this model, like the squared theta model of a Kummer surface). It is quite amazing that this theta model is so fast, since it is also universal and give models for abelian varieties in any dimension: Dartois Sage's implementation [Dar24] for 2^n-isogenies in dimension 4 is already very promising. One reason is that the algebraic theta model is taylored to have very efficient formulas for the map $(P, Q) \rightarrow (P + Q, P - Q)$; this is the duplication formula.

This efficient duplication formula (in level 2, hence on the Kummer variety) gives fast arithmetic: doubling and differential additions. This fast arithmetic has been exploited since a long time, notably for classical DLP-based cryptography in dimension 1 and 2 [CC86, Gau07, RSSB16, HR19]. This theta model (or its twisted variant given by squared thetas), is often referred as the fast Kummer model in the literature. It is thus not surprising that the theta model is also used for fast isogenies in higher dimension. In fact, it was used even before the HD representation! In supersingular isogeny based cryptography, an isogeny $\phi : E_1 \rightarrow E_2$ over $\mathbb{F}_{p^2}$ induces via the Weil restriction of scalar functor an isogeny $W_{\mathbb{F}_{p^2}/\mathbb{F}_p}\phi : A_1 := W_{\mathbb{F}_{p^2}/\mathbb{F}_p}E_1 \rightarrow A_2 := W_{\mathbb{F}_{p^2}/\mathbb{F}_p}E_2$ defined over $\mathbb{F}_p$, where A_1, A_2 are principally polarised abelian surfaces (which are neither Jacobians nor product of elliptic curves over $\mathbb{F}_p$!). Furthermore, A_1, A_2 are 2-isogeneous over $\mathbb{F}_p$ to Jacobians $\mathrm{Jac}(C_1), \mathrm{Jac}(C_2)$ of hyperelliptic curves. In [Cos18], Costello explains that, using the squared theta model, computing the induced isogeny $\mathrm{Jac}(C_1) \rightarrow \mathrm{Jac}(C_2)$ in dimension 2 over $\mathbb{F}_p$ is potentially faster than computing the original isogeny ϕ in dimension 1 but over $\mathbb{F}_{p^2}$. This idea was recently revisited in the context of SQIsign in [CR24].

The main drawback of the theta model (of level 2) is that the fast duplication formula is obtained by splitting the map $(P, Q) \rightarrow (P+Q, P-Q)$ into two. This requires a specific Galois structure on the 2-Tate module, more precisely an abelian variety $A/\mathbb{F}_q$ has a rational theta model of level 2 if and only if there is a symplectic basis of $A[2]$ for the Weil pairing with trivial self Tate pairings [Rob21, § 2.11]. For general abelian varieties, the theta model is thus only defined over a field extension. In that case, at least for abelian surfaces, it can be useful to work with the Jacobian Kummer model or the generic Kummer model of Cassels and Flynn [CF+96].

Other Isogeny Representations for Abelian Varieties. For the other "classical" isogeny representations, the main difficulty is that while in dimension 1 we can (almost always) reduce to question about univariate polynomials or rational functions, in higher dimensions we need to work with multivariate rational functions, for which it is harder to obtain quasi-linear algorithms.

For instance, in theory, one could find a function representation of an isogeny ϕ in higher dimension by using any evaluation algorithm on the generic point η_A of A, but in practice equations for A are quite involved in higher dimension, so the function representation is not really used. Still, in dimension 2, when we have an isogeny $\phi : \mathrm{Jac}(C_1) \to \mathrm{Jac}(C_2)$ between Jacobian of hyperelliptic curves, it is convenient to express ϕ in terms of the Weierstrass coordinates of C_1, C_2. One method to do that is to evaluate ϕ on a formal point at precision 2, deduce a differential equation for ϕ, and to solve the differential system [CE14, § 6.2], [KPR24, CMSV19], so at least this case is tractable.

We have the same problem with kernel equations: unlike the dimension 1 case where the kernel is described by a univariate polynomial, kernel equations in higher dimension are given by multivariate polynomials, which make them harder to work with. An algorithm to compute ϕ when we have a univariate parametrisation of the kernel K is given in [LR15b].

And of course, for the interpolation representation, we need multivariate rational function reconstruction. Likewise for adapting the `sqrtVelu` algorithm in higher dimension: the obstacles were informally worked out in a Ciao workshop meeting in Bordeaux in December 2022. The `sqrtVelu` algorithm is described by a resultant in [BDLS20], but can be reformulated as evaluating a degree $O(\sqrt{n})$ univariate polynomial P on $O(\sqrt{n})$ points, which can be done in $\tilde{O}(\sqrt{n})$ by fast multipoint evaluation algorithms. We can extend this approach to higher dimension using the known isogeny formulas, but then we need fast *multivariate* multipoint evaluations, which exist in theory over finite fields by [KU11], but without practical implementations.

Finally, we also mention that reconstructing an isogeny from a root of the modular polynomial Φ_ℓ has been worked out in dimension 2 in [KPR24] (see Example 16).

5.2   Kani's Lemma and Its Applications

Definition 2. *A (n_1, n_2)-isogeny diamond is a commutative diagram of isogenies between polarised abelian varieties:*

$$
\begin{array}{ccc}
A_0 & \xrightarrow{\ \phi_1\ } & A_1 \\
\downarrow{\scriptstyle \phi_2} & & \downarrow{\scriptstyle \phi_2'} \\
A_2 & \xrightarrow{\ \phi_1'\ } & A_{12}
\end{array}
$$

where $\phi_1 : A_0 \to A_1$ and $\phi'_1 : A_2 \to A_{12}$ are n_1-isogenies, $\phi_2 : A_0 \to A_2$ and $\phi'_2 : A_1 \to A_{12}$ are n_2-isogenies.

Remark 12. If n_1 is coprime to n_2, then an isogeny diamond as above is the same thing as a pushforward square from ϕ_1, ϕ_2 or a pullback square from ϕ'_1, ϕ'_2.

We can now state Kani's lemma, which is contained in [Kan97, § 2, Proof of Th. 2.3].

Theorem 2 (Kani's lemma). *Let n_1 and n_2 be two integers. Given a (n_1, n_2)-isogeny diamond, the isogeny $\Phi : A_0 \times A_{12} \to A_1 \times A_2$ given matricially by*

$$\Phi = \begin{pmatrix} \phi_1 & \widetilde{\phi}'_2 \\ -\phi_2 & \widetilde{\phi}'_1 \end{pmatrix}$$

is a (n_1+n_2)-isogeny between these product of abelian varieties with their product polarisations. If n_1 is coprime to n_2, the kernel of Φ is given by

$$\mathrm{Ker}\, \Phi = \{(\widetilde{\phi_1}(P), \phi'_2(P)) \mid P \in A_1[n_1 + n_2]\}$$
$$= \{(-\widetilde{\phi_2}(P), \phi'_1(P)) \mid P \in A_2[n_1 + n_2]\}$$
$$= \{(n_1 P, \phi'_2 \circ \phi_1(P)) \mid P \in A_0[n_1 + n_2]\}.$$

Proof. We compute $\widetilde{\Phi} \circ \Phi = \begin{pmatrix} \widetilde{\phi_1} & -\widetilde{\phi_2} \\ \phi'_2 & \phi'_1 \end{pmatrix} \begin{pmatrix} \phi_1 & \widetilde{\phi'_2} \\ -\phi_2 & \phi'_1 \end{pmatrix} = \begin{pmatrix} n_1 + n_2 & 0 \\ 0 & n_1 + n_2 \end{pmatrix}$ which

shows that Φ is a $(n_1 + n_2)$-isogeny for the product polarisations.

The kernel of Φ, of cardinal $(n_1 + n_2)^{2g}$, is given by the image of $\widetilde{\Phi}$ on $(A_1 \times A_2)[n_1 + n_2]$. If n_1 is coprime to n_2, the restriction of $\widetilde{\Phi}$ to $A_1[n_1] \times 0_{A_2}$ is injective so its image already spans the full kernel: $\mathrm{Ker}\, \Phi = \{(\widetilde{\phi_1}(P), \phi'_2(P)) \mid P \in A_1[n_1 + n_2]\}$. The second equality follows by symmetry, and the third by plugging $P = \phi_1(P_0)$ with $P_0 \in A_0[n_1 + n_2]$.

We refer to [MMPPW23, Theorem 1] or [Rob23b] for more details. □

Corollary 1 (Embedding an isogeny in dimension 2). *Let $\phi : E_1 \to E_2$ be an n-isogeny. Let $N > n$ be coprime to n, $n' = N - n$, and assume that we know an efficient representation of an n'-isogeny $\psi : E_1 \to E'_1$. Then we can embed ϕ into an N-isogeny $\Phi = \begin{pmatrix} \phi & \widetilde{\psi}' \\ -\psi & \widetilde{\phi}' \end{pmatrix} : E_1 \times E'_2 \to E_2 \times E'_1$. Furthermore, if N is smooth, the N-torsion is accessible on E_1, and we know how ϕ acts on $E_1[N]$, then Φ can be efficiently computed.*

We have $\phi = p \circ \Phi \circ i : E_1 \to E_1 \times E'_2 \to E_2 \times E'_1 \to E_2$ where $i(P) = (P, 0)$ and $p(P, Q) = P$, so we can recover $\phi(P)$ via $\Phi((P, 0)) = (\phi(P), -\psi(P))$.

Similar constructions hold when we have $\widetilde{\psi} : E'_1 \to E_1$, $\psi' : E_2 \to E'_2$ or $\widetilde{\psi}' : E'_2 \to E_2$.

Proof. Given $\psi : E_1 \to E_1'$ of degree n' coprime to n, we can look at the push-forward square, which form a (n, n')-isogeny diamond:

$$
\begin{array}{ccc}
E_1 & \xrightarrow{\phi} & E_2 \\
\downarrow{\psi} & & \downarrow{\psi'} \\
E_1' & \xrightarrow{\phi'} & E_2'
\end{array}
$$

We then have $\Phi : E_1 \times E_2' \to E_2 \times E_1'$ a N-isogeny. Furthermore, the kernel of $\widetilde{\Phi}$ is given by $\operatorname{Ker}\widetilde{\Phi} = \{(\phi(P), -\psi(P)) \mid P \in E_1[N]]\}$. From our assumption on ϕ, and since ψ is supposed to be efficient, we can compute this kernel, then compute $\widetilde{\Phi}$, and then compute Φ whose kernel is $\widetilde{\Phi}((E_1' \times E_2')[N])$.

If we now assume that we know an efficient representation of $\widetilde{\psi} : E_1' \to E_1$, then we can recover how ψ acts on $E_1[N]$, which is enough to get $\operatorname{Ker}\widetilde{\Phi}$.

If we are given $\psi' : E_2 \to E_2'$ (or simply its action on the N-torsion), we can directly recover $\operatorname{Ker}\Phi = \{(\hat{\phi}(P), \psi'(P)) \mid P \in E_2[N]]\} = \{(n_1 P, \psi' \circ \phi(P)) \mid P \in E_1[N]\}$. (The first equality requires to extract the action of $\hat{\phi}$ on $E_2[N]$ from the action of ϕ on $E_1[N]$, but not the second). Finally if we are given $\widetilde{\psi'}$, we can just extract the action of ψ' on $E_2[N]$. $\qquad\square$

Example 2. Let $p \equiv 3 \mod 4$ and $E_0 : y^2 = x^3 + x/\mathbb{F}_p$, this is a supersingular elliptic curve whose endomorphism ring over $\mathbb{F}_{p^2}$ contains $\mathbb{Z}[i, \pi_p]$ with $i^2 = -1$ and $\pi_p^2 = -p$. In particular, we can build endomorphisms $x + yi + u\pi_p + vi\pi_p$ of norm $x^2 + y^2 + p(u^2 + v^2)$. So on this special curve, if n' is a sum of two squares or n' is large enough ($n' \gg p$), we can build dimension 1 endomorphisms of norm n'. More generally, a similar strategy works for a supersingular curve which contains a quadratic order of small discriminant. We will see in Example 7 how to leverage this construction of endomorphisms of large enough degree to construct isogenies from E_0 of any degree.

It remains to find an efficient representation of some isogeny $\psi : E_1 \to E_1'$ of some fixed degree. The main insight is that this is always possible, provided we go in higher dimension (this is Zahrin's trick).

Proposition 2. *Let n' be an integer. If $n' = a_1^2$, then $[m] : E_1 \to E_1$ is an n'-isogeny.*

If $n' = a_1^2 + a_2^2$, then $\begin{pmatrix} a_1 & a_2 \\ -a_2 & a_1 \end{pmatrix} : E_1^2 \to E_1^2$ is an n'-isogeny.

If $n' = a_1^2 + a_2^2 + a_3^2 + a_4^2$ (which is always the case), then $\begin{pmatrix} a_1 & -a_2 & -a_3 & -a_4 \\ a_2 & a_1 & a_4 & -a_3 \\ a_3 & -a_4 & a_1 & a_2 \\ a_4 & a_3 & -a_2 & a_1 \end{pmatrix} :$

$E_1^4 \to E_1^4$ is an n'-isogeny.

Proof. If we let M be the matrix of endomorphism appearing in the proposition, then the contragredient isogeny $\widetilde{M}$ for the product polarisation is given by $\widetilde{M} = \overline{M}^T$ by Lemma 1, and we check that $\widetilde{M}M = n' \operatorname{Id}$. $\qquad\square$

Example 3 (Embedding an isogeny in higher dimension). Let α be the matrix from Proposition 2, which induces an n'-endomorphism on E_1^u and E_2^u, where $u = 1, 2, 4$ according to whether n' is a sum of $1, 2, 4$ squares. Then as in Corollary 1, we can embed ϕ into the dimension $g = 2u$ endomorphism $\varPhi = \begin{pmatrix} \phi\, \mathrm{id}_u & \widetilde{\alpha} \\ -\alpha & \widetilde{\phi}\, \mathrm{id}_u \end{pmatrix}$: $E_1^u \times E_2^u \to E_2^u \times E_1^u$.

Then $\varPhi$ embeds ϕ (and its dual): $\phi = p \circ \varPhi \circ i : E_1 \to E_1^u \times E_2^u \to E_1^u \times E_2^u \to E_2$, and has kernel $\mathrm{Ker}\, \varPhi = \{(-\widetilde{\alpha}(P), \phi\, \mathrm{id}_u(P)) \mid P \in E_1^u[N]\} = \{(nP, \alpha\phi\, \mathrm{id}_u(P)) \mid P \in E_1^u[N]\}$.

Furthermore, in that case we can split $\varPhi$ in two, see Appendix B, which is not possible in Corollary 1 (unless we already know all four curves E_1, E_2, E_1', E_2').

Remark 13 (Choice of dimension). Due to the cost of computing isogenies in high dimension, for efficiency we would like to use Example 3 with g as small as possible. In practice, there are sufficiently many integers which are sum of two squares that it is very rare that we need to go all the way to $g = 8$, usually $g = 4$ is sufficient.

Furthermore, in the case where E_1 (or E_2) is the special curve from Example 2, then we can use endomorphisms γ of the form $a_1 + a_2 i$ on E_1 to relax the dimension u in Proposition 2: we can use $u = 1$ when n' is a sum of two squares and $u = 2$ in the general case when n' is a sum of four squares. This means that in that case we have $g = 2$ or 4 rather than 4 or 8. However, the pushforward of γ by ϕ will in general be an isogeny rather than an endomorphism, so this means that we cannot split $\varPhi$ in two in that case if we use endomorphisms like γ rather than integers in our matrix α from Example 3. A similar strategy works for a curve containing a quadratic order with small discriminant.

Remark 14. It is not always easy to check if an integer n' is a sum of two squares (and if so find the decomposition). This usually requires to know the factorisation of n'. Once the factorisation is known, one can then use Cornacchia's algorithm [Cor07].

When we have flexibility on the choice of n' (we will see in Theorem 3 that we can take any n' such that $N = n + n'$ is smooth with N-torsion accessible, e.g., N is powersmooth), the fastest way is to try several n' until we find n' a prime congruent to 1 modulo 4 (or possibly a product of small primes congruent to 1 modulo 4 time a prime congruent to 1 modulo 4).

There is a randomized algorithm in time $O(\log^2 n')$ to decompose n' as a sum of four squares [RS86, PT18].

5.3   The HD Representation

We have seen in Example 3 how we can now use Theorem 2 to obtain a special form of Theorem 1. We can give a more precise complexity statement:

Theorem 3. *Let $\phi : E_1 \to E_2$ be an n-isogeny, $(P_1, Q_1, \ldots, P_r, Q_r)$ a CRT basis of the N-torsion, with $N = \prod_{i=1}^m \ell_i^{e_i} > n$ coprime to n. Let $u = 1, 2, 4$ according to whether $N - n$ is a sum of $1, 2, 4$ squares. Let $(P_i, \phi(P_i), Q_i, \phi(Q_i))$*

be interpolation data of ϕ on $E_1[N]$. Then ϕ can be efficiently embedded into a 2u-dimensional N-isogeny Φ, which can be decomposed as a product of ℓ_i-isogenies in time $O(\sum_i e_i \ell_i^{2u}(d_i \log e_i + \sum_{j>i(d_i \vee d_j)})) = \tilde{O}(m^2 d' e \ell^{2u})$ arithmetic operations over $\mathbb{F}_q$.

Given a point $P \in E_1(\mathbb{F}_{q'})$ with $q' = q^{d_P}$, and the decomposed representation of Φ, evaluating P requires $O(\sum_i e_i \ell_i^{2u}(d_i \vee d_P)) = O(me\ell^{2u}d'')$ arithmetic operations over $\mathbb{F}_q$.

Proof. We use Proposition 2 to build an n'-endomorphism $\psi : E_1^u \to E_1^u$. We can then use Corollary 1 to ψ and $\phi\,\mathrm{Id} : E_1^u \to E_2^u$ to embed ϕ into a N-isogeny of dimension $2u$ $\Phi : E_1^u \times E_2^u \to E_1^u \times E_2^u$. (We remark that since ψ is a matrix of integers, it commutes with $\phi\,\mathrm{Id}$ and so $\psi' : E_2^u \to E_2^u$ is given by the same matrix as ψ). Although Corollary 1 is stated for elliptic curves, by Theorem 2 which was stated for principally polarised abelian varieties, it extends to product of elliptic cuves (and their product principal polarisation).

Once we have Φ we can use the higher dimensional Vélu's like formula to decompose it. The complexity then follows from Lemma 2. See also [Rob22a, Rob22b] for more details. ☐

Remark 15.

- The complexity of the HD representation depends on the dimension $g = 2u$ of Φ, the isogeny we embed ϕ into, but also the largest prime divisor of N and the field of definition of the points of $E_1[\ell_i^{e_i}]$ torsion. In the best case, we have $g = 2$, $N = 2^e$, and $E_1[2^e]$ has rational points. We note that to achieve $g = 2$ we need to find a $n' = N - n$ dimension 1 isogeny $\psi : E_1 \to E_1'$, which can be hard (unless n' is a square). However, if we know $\mathrm{End}(E_1)$, it is much easier to build isogenies of appropriate degree. That's the main idea behind the algorithms of SQIsign2d [BDD+24, NO24, DF24], which build an efficient representation of the response isogeny by embedding it into a 2^e-isogeny in dimension 2.

- If N is smooth and the N-torsion on E_1 is accessible, but N is not coprime to n, we can write $N = dN_1$, $n = dn_1$ with $d = N \wedge n$. Then ϕ splits as $\phi = \phi_1 \circ \phi_0$ where $\phi_0 : E_1 \to E_{11}$ is a d-isogeny whose kernel is efficiently computed since $d \mid N$ and we know how ϕ acts on the N-torsion, and $\phi_1 : E_{11} \to E_2$ is a n_1-isogeny. We can then apply Theorem 3 to ϕ_1 to embed it into an N_1-isogeny.

Definition 3. *The HD representation of ϕ can use any efficient representation of an higher dimensional isogeny Φ such that ϕ embeds into Φ. In practice, following Theorem 3, the HD representation of ϕ can consist:*

1. *The interpolation data: $(P_i, \phi(P_i), Q_i, \phi(Q_i))$, with N smooth and $E_1[N]$-accessible. By Theorem 2, this is essentially equivalent to a multigenerator representation of $\mathrm{Ker}\,\Phi$. We call this the torsion HD-representation.*

2. *The decomposition of Φ as a product of small isogenies. This only requires N-smooth, but of course to convert from the torsion HD-representation into a decompose HD-representation we also need N accessible.*

Corollary 2. *The HD representation is universal. Any other efficient representation of ϕ can be efficiently converted (meaning in polynomial time in $\log n$ and $\log q$) into an HD representation.*

Proof. It suffices to select a smooth integer N with accessible N-torsion (for instance N powersmooth), construct a CRT basis of the N-torsion, and use our efficient representation to evaluate ϕ on this CRT basis. We thus obtain a torsion HD representation. (The HD representation needs the degree of ϕ, but we have assumed that this is always part of the data to represent ϕ).   □

As mentioned in the introduction, we can use Corollary 2 to show that we just need to be able to evaluate ϕ efficiently on enough nice points $P \in E(\mathbb{F}_{q'})$ to be able to efficiently evaluate ϕ on all points P, even in k-algebras.

Remark 16. Given an HD like data, say a CRT basis (P_i, Q_i) of $E_1[N]$, and points $(R_i, S_i) \in E_2[N]$ that are the putative images of P_i, Q_i by some isogeny ϕ, one can ask whether it really encodes some isogeny.

The HD representation takes the torsion information above, and convert it into the kernel of an N-isogeny $\Phi : A \rightarrow B$, where A, B are supposed to split as product of elliptic curves: $A = \prod E_i$, $B = \prod E'_j$. (The expected decomposition depends on the dimension used, and whether we used auxiliary endomorphisms or isogenies).

If the codomain B of Φ, does not split as a product of elliptic curves, we know that Φ does not encodes a dimension one isogeny. When B splits, then Φ consists of a matrix of n_{ij}-isogenies $\phi_{ij} : E_i \rightarrow E'_j$, with $n_{ij} \leqslant N$ (we can recover n_{ij} using pairings, see Lemma 3). In that case, Φ does encode (several!) dimension one isogenies. To check whether it encodes a particular candidate, i.e. whether a ϕ_{ij} is equal to some specific isogeny $\phi : E_1 \rightarrow E_2$, we can first look for a E'_j isomorphic to E_2, and then (if we know how ϕ is supposed to act on sufficiently many points) use the equality algorithm of Proposition 3 to check for equality.

6 Algorithms on Efficient Representation of Isogenies

Using the universality of the HD representation (Corollary 2), we can build many algorithms on efficiently represented isogenies.

6.1 Equality and Sum

We start with standard algorithms.

Proposition 3. *Given two efficient representations of two n-isogenies $\phi_1, \phi_2 : E_1 \rightarrow E_2$, we can efficiently test if ϕ_1 is equal to ϕ_2.*

Proof. Here we assume that the degree of ϕ_1, ϕ_2 are the same, otherwise they are clearly not equal (see Lemma 3 on how to recover the degrees if we only have a bound on it). We also assume that the codomain is the same; it can be useful to test equality up to postcomposition by an automorphism, but it suffices to apply Proposition 3 to all automorphisms.

To test equality, we simply construct a CRT basis of the N-torsion on E_1 for any $N > 2\sqrt{n}$ where the N-torsion is accessible. Then $\phi_1 = \phi_2$ if and only if they agree on this CRT basis. □

If we have an efficient representation of a n_1-isogeny $\phi_1 : E_1 \to E_2$ and of a n_2-isogeny $\phi_2 : E_2 \to E_3$ then we have an efficient representation of the $n_1 n_2$-isogeny $\phi_2 \circ \phi_1 : E_1 \to E_3$ (say by keeping it decomposed as ϕ_1 followed by ϕ_2).

If we have a n_1-isogeny $\phi_1 : E_1 \to E_2$ and a n_2-isogeny $\phi_2 : E_1 \to E_2$ that have an efficient representation, then of course we can also evaluate the sum $\phi_1 + \phi_2$ efficiently on any point. But a catch is that in our representation data, we require to know the degree of the isogeny; and Cauchy-Schwarz only gives us a bound on the degree of $\phi_1 + \phi_2$. This is in fact enough, thanks to pairings:

Lemma 3. *Let $\phi : E_1 \to E_2$ be an isogeny of degree at most n. Assume that we know the evaluation of ϕ on a CRT basis of the N-torsion, with N smooth and the N-torsion accessible, and $N > n$. Then we can efficiently recover the degree of ϕ.*

Proof. If $\deg \phi = d$, for $P, Q \in E[N]$, we have $e_N(\phi(P), \phi(Q)) = e_N(P, Q)^d$. We can thus recover d by computing Weil pairings and discrete logarithms (working separately on each prime power of N). □

Corollary 3. *If we have an efficient representation of two isogenies $\phi_1, \phi_2 : E_1 \to E_2$, we have an efficient representation of their sum.*

6.2   Duals and Divisions

Now, we describe algorithms that need the HD representation. These are algorithms that are relatively straightforward, simply using the idea that the HD representation only needs to know the evaluation of ϕ on sufficiently many nice points to extract from it an HD representation.

Proposition 4. *If we have an efficient representation of an n-isogeny $\phi : E_1 \to E_2$, then we also have an efficient representation of the dual n-isogeny $\tilde{\phi} : E_2 \to E_1$.*

Proof. Take a CRT basis (P_i, Q_i) of $E_1[N]$, with N large enough and prime to n, smooth and the N-torsion accessible. Then $(P_i' = \phi(P_i), Q_i' = \phi(Q_i))$ is a CRT basis of $E_2[N]$ (because N is prime to n), which can be efficiently evaluated by our assumption on ϕ. Now, by definition of the contragredient isogeny, we have $\tilde{\phi}(P_i') = nP_i, \tilde{\phi}(Q_i') = nQ_i$. Thus we have an HD representation of $\tilde{\phi}$.

(We remark that[11], if we are given a basis (P, Q) of $E_1[N]$ and the action of ϕ on this basis, and we are given a basis (P', Q') of $E_2[N]$, then using that

---

[11] This argument was communicated by Benjamin Wesolowski.

$e_{W,N}(\phi(R), S) = e_{W,N}(R, \tilde{\phi}S)$ where $R \in E_1[N]$ and $S \in E_2[N]$, we can recover how $\tilde{\phi}$ acts on P', Q' via Weil pairings and dlp computations in μ_N, even if N is not coprime to n. This can be useful to reconstruct the action of $\tilde{\phi}$ on some N-torsion without going all the way to an HD representation).   □

Example 4. If $E/\mathbb{F}_q$ is an elliptic curve, the Frobenius π_p can be efficiently computed: $\pi_p(x(P), y(P)) = (x^p(P), y^p(P))$. Hence its dual, the Verschiebung $\hat{\pi}_p$ can also be efficiently computed in $O(\log^{O(1)}(p))$. Computing the Verschiebung via its kernel (which consist of the étale points of p-torsion if E is ordinary) would take $O(p)$. If E is ordinary, evaluating the Verschiebung on differentials, we can recover the invertible eigenvalue of the Frobenius modulo p, hence its trace modulo p. This gives a polynomial point counting algorithm if $q = p$.

Let $\phi : E_1 \to E_2$ be an isogeny, and m an integer. The isogeny ϕ is divisible by m if and only if $\phi(E_1[m]) = 0$. If we have an efficient representation of ϕ and the m-torsion of E_1 is accessible, we can thus test divisibility by m efficiently. Using the HD representation, we can handle the general case:

Proposition 5. *If we have an efficient representation of an n-isogeny $\phi : E_1 \to E_2$, then we can test if ϕ is divisible by some integer m, and if so obtain an efficient representation of ϕ/m.*

Proof. Take a CRT basis (P_i, Q_i) of $E_1[N]$, with N large enough and prime to n, smooth and the N-torsion accessible. Then $(P_i, \phi(P_i)/m, Q_i, \phi(Q_i)/m)$ gives a torsion HD-representation of ϕ/m, if it exists.

To test if this putative HD data is valid for ϕ/m, we can use Remark 16, because we know how ϕ/m is supposed to act if it exists.   □

Example 5. Assume that we have an efficient representation of an endomorphism $\alpha \in \mathrm{End}(E)$, and let $R = \mathbb{Q}[\alpha] \cap \mathrm{End}(E)$ be the saturation of $\mathbb{Z}[\alpha]$ in $\mathrm{End}(E)$. Assume that we know the factorisation of the conductor f_α of $\mathbb{Z}[\alpha]$ in its maximal order R_0. Then we can apply the ℓ-division algorithm to suitable endomorphisms on $\mathbb{Z}[\alpha]$ for $\ell \mid f_\alpha$ to recover the conductor f_R of R in R_0. In other words, we can recover in polynomial time (in $\log \Delta_\alpha$) the saturation of a quadratic order $\mathbb{Z}[\alpha]$ in $\mathrm{End}(E)$, provided we have a factorisation of the conductor f_α.

When E is ordinary, $\mathrm{End}(E)$ is the saturation of $\mathbb{Z}[\pi_q]$, and we can apply the above algorithm since π_q has an efficient representation. This gives a polynomial time algorithm to compute $\mathrm{End}(E)$, provided we have a factorisation of the conductor of $\mathbb{Z}[\pi_q]$. See [Rob22b] for more details, and [MW23, PW24, ES24] for other applications of the saturation algorithm.

Similarly, for any elliptic curve, if R is the saturation of $\mathbb{Z}[\pi]$ in $\mathrm{End}(E)$ (this saturation is the same whether we take $\mathrm{End}_{\mathbb{F}_q}(E)$ or $\mathrm{End}_{\overline{\mathbb{F}}_q}(E)$ for $\mathrm{End}(E)$), then each endomorphism $\alpha \in R$ admits an efficient representation, namely polynomial in $\log \Delta_\alpha$ and $\log q$. Indeed, we can write $\alpha = (a + b\pi)/f$, with f dividing the conductor of $\mathbb{Z}[\pi]$ (hence at most $O(q)$), and $\log|a|, \log|b|$ are in $O(\Delta_\alpha + \log q)$. We can evaluate $a + b\pi$ efficiently, hence α too by the division algorithm.

Corollary 4. *Given efficient representations of $\phi : E_1 \to E_2$ and of $\phi_r : E_1' \to E_2$ and $\phi_l : E_1 \to E_2'$, one can efficiently test if $\phi = \phi_r' \circ \phi_l$ and if $\phi = \phi_r \circ \phi_l'$, and if so output an efficient representation of ϕ_r' and ϕ_l'.*

Proof. We treat the case of ϕ_l, the other case being symmetric. We have $\phi = \phi_r' \circ \phi_l$ if and only if $\phi \circ \widetilde{\phi_l} = \phi_r' \circ [\deg \phi_l]$, so, since we know an efficient representation of $\widetilde{\phi_l}$ by Proposition 4, the question reduces to the question of division by an integer, which is handled by Proposition 5. □

6.3   Advanced Algorithms on Efficient Representations

In this section, we describe some algorithms that need to delve into how the HD representation works rather than treating it simply as a black box.

Proposition 6. *Given an efficient representation of $\phi : E_1 \to E_2$ over a finite field $\mathbb{F}_q$, and given $R = \mathbb{F}_q[\epsilon]/\epsilon^m$ or $R = \mathbb{Z}_q/p^m\mathbb{Z}_q$ and a deformation/lift $\widetilde{E_1}/R$ of E_1 to R, one can find an efficient representation of the isogeny ϕ deformed/lifted to R.*

Proof. The idea is as follows: we compute a decomposed HD representation $\Phi : A \to B$ of ϕ, where Φ is split into a product of small higher dimensional isogenies Φ_i. We lift ϕ by lifting Φ, which amount to lifting the kernel of each Φ_i.

There is however one technical difficulty with this approach. Let's say we use a 8-dimensional representation $\Phi : E_1^4 \times E_2^4 \to E_1^4 \to E_2^4$ of $\phi : E_1 \to E_2$. Then when we want to compute the deformation $\widetilde{\phi}$ of ϕ to $\widetilde{E_1}/R$, we don't yet know the codomain $\widetilde{E_2}$ of $\widetilde{\phi}$. So if $A = E_1^4 \times E_2^4$, we don't know which is the correct deformation $\widetilde{A}/R$ of A we need to embed $\widetilde{\phi}$!

The solution is to make an arbitrary choice $\widetilde{E_2}'$ for $\widetilde{E_2}$. If our choice is incorrect, we will get a $B = \widetilde{E_1}^4 \times (\widetilde{E_2}'')^4$ with $\widetilde{E_2}'' \neq \widetilde{E_2}'$. In other words, to get the correct deformation $\widetilde{\Phi}$, we want to find $\widetilde{E_2}'$ such that $\widetilde{E_2}''$ is equal to $\widetilde{E_2}'$. We can solve this by a Newton algorithm, where each steps reduces to a linear algebra problem on the deformation spaces of E_1 and E_2. The corresponding matrix can be computed by computing the $\widetilde{E_2}''$ associated to the $\widetilde{E_2}'$ for a basis of the deformation space. We refer to [Rob22b, KR24] for more details. □

Example 6. If $E/\mathbb{F}_q$ is ordinary, one can use Proposition 6 to compute the canonical lift $\widetilde{E}/\mathbb{Z}_q$ to p-adic precision m in polynomial time in $\log p$. Combined with Example 4, this gives a point counting algorithm in $O(n^2 \log^C p)$ when $q = p^n$ [Rob22b].

One can also use deformations of isogenies to $\mathbb{F}_q[\epsilon]/\epsilon^m$ to compute modular polynomials efficiently [KR24].

Proposition 7. *Assume that we have an efficient representation of an isogeny $\phi : E_1 \to E_2$ of degree $n = n_1 n_2$ with $n_1 \wedge n_2 = 1$. Then ϕ splits uniquely as $\phi = \phi_2' \circ \phi_1 = \phi_1' \circ \phi_2$ where ϕ_1, ϕ_1' are of degree n_1 and ϕ_2, ϕ_2' are of degree n_2. Furthermore, we can efficiently find an efficient representation of $\phi_1, \phi_2, \phi_1', \phi_2'$.*

7777777777777

Proof. If $K = \mathrm{Ker}\,\phi$, we can define ϕ_1 (resp. ϕ_2) to be the isogeny with kernel $K[n_1]$ (resp. $K[n_2]$) and ϕ_2' (resp. ϕ_1') to be the isogeny with kernel $\phi_1(K)$ (resp. $\phi_2(K)$). Then $\phi_1, \phi_2, \phi_1', \phi_2'$ form a (n_1, n_2)-isogeny diamond, hence embed into a N-isogeny Φ in dimension 2 with $N = n_1 + n_2$. Furthermore, we can recover the kernel of Φ from the action of ϕ on $E_1[N]$. This solves the problem if N is smooth and E_1, E_2 have accessible N-torsion: computing Φ gives all four isogenies $\phi_1, \phi_2, \phi_1', \phi_2'$ at once (we can distinguish ϕ_1 from ϕ_2 using pairings since they don't have the same degree).

For the general case, we pad ϕ with extra isogenies. Namely we consider $\phi' = \psi_2 \circ \phi \circ \psi_1$ where ψ_1 (resp. ψ_2) is an efficient isogeny of degree u (resp. v). The goal is to find such isogenies ψ_1, ψ_2 such that $N = un_1 + vn_2$, with N smooth and the N-torsion accessible, so that we can apply the splitting algorithm above to split ϕ' into $\psi_2 \circ \phi_2'$ and $\phi_1 \circ \psi_1$. Then we can apply the division algorithm to recover ϕ_2', ϕ_1. If we want to recover ϕ_1', ϕ_2 instead, we need to search for u, v such that $N = un_2 + vn_1$.

A simple pigeonhole argument shows that for any $N > n_1 n_2$, we can find $u, v > 0$ such that $N = un_1 + vn_2$. We remark that if N is coprime to $n_1 n_2$ then automatically un_1 will be coprime to vn_2.

Once we have fixed u, v, we can always find ψ_1, ψ_2 by going to higher dimension if needed: namely we work with E_1^r, E_2^r with $r = 1, 2, 4$, and letting by abuse of notations $\phi : E_1^r \to E_2^r$ to be the diagonal isogeny of $\phi : E_1 \to E_2$.

For instance, if $u = u_1^2 + u_2^2 + u_3^2 + u_4^2$ we build a quaternion matrix for ψ_1, and then using the algorithm above in dimension 4 (so using a Φ in dimension 8), we recover $\phi_1 \circ \psi_1$. This means that we obtain ϕu_i for $i = 1, 2, 3, 4$, hence we can directly recover $\phi_1 \gcd(u_i)$. Thus we don't even need to apply the division algorithm to recover ϕ_1 in the cases where we can find u_i coprime. This includes all odd integers. However if u has 2-adic valuation e, then 2^f has to divide all the u_i for $f = \lfloor (e-1)/2 \rfloor$ (and conversely we can find a solution with $\gcd(u_i) = 2^f$). $\square$

Example 7 (QFESTA). The first application of splitting isogenies was given in QFESTA [NO23]. Namely we have seen in Example 2 that on $E_0 : y^2 = x^3 + x/\mathbb{F}_{p^2}$, with $p \equiv 3 \mod 4$, we can efficiently build endomorphisms of norm $D \gg p$. Suppose that we want to build an efficient isogeny of degree n. We first look for an endomorphism γ of norm $D = n(N-n)$ where N is chosen to be prime to n, and smooth with accessible N-torsion on E_0. Then a direct application of Proposition 7 allows to split γ into $\gamma = \gamma_2 \circ \gamma_1$ where γ_1 is a n-isogeny. Our D is chosen so that the splitting can be done directly in dimension 2 without any padding required for γ. To find γ, we require that D is large enough ($D \gg p$); but on the other hand for efficiency we also want to take $N = 2^e$ where $2^e \mid p \pm 1$ so that the 2^e-torsion is rational over $\mathbb{F}_{p^2}$, this already imposes $N < p$. For some applications, we might have only have available 2^e-torsion with 2^e significantly smaller than p, making the condition $D \gg p$ hard to satisfy. A solution is to simply iterate the splitting: we take $D_1 = n(N_1 - n)$, $D_2 = D_1(N_2 - D_1)$, $D_3 = D_2(N_3 - D_2)$ for appropriately chosen divisors $N_1, N_2, N_3 \ldots$ of N, until we can find an endomorphism γ_i of norm D_i. Then we successively split γ_i to obtain an isogeny γ_{i-1} of degree D_{i-1}, then γ_{i-2} of degree D_{i-2} until we find γ_1 of degree n.

In [NO24], Nakagawa and Onuki extend this construction to build an isogeny of degree n from an arbitrary supersingular elliptic curve E, provided that the connecting ideal I between E_0 and E is known. The idea is to sample γ in the Eichler order of E_0 and E, using the tools introduced for the original SQIsign scheme (see [Ler22b] for a nice overview of these tools).

Example 8 (Clapotis). The full power of the splitting algorithm of Proposition 7 was introduced for the Clapotis group action in [PR23b]. We want to compute the isogeny $\phi_I : E_1 \to E_2$ associated to an *invertible* class group ideal I, provided that we know an efficient representation of $\text{End}(E_1)$. It suffices to find two equivalent ideals J_1, J_2 to I of coprime degree. Then $\phi = \widetilde{\phi_{J_2}} \circ \phi_{J_1}$ is an endomorphism of E_1, which we know how to efficiently evaluate. We can thus apply Proposition 7 to recover an efficient representation of ϕ_{J_1}, ϕ_{J_2}; this build a double path from $E_1 \to E_2$ from which it is easy to get an efficient representation of ϕ_I too (but for the class group action usually we only need to recover the codomain E_2 rather than the full ϕ_I).

The same idea apply to convert an ideal to isogeny in the supersingular setting, and is the basis of [BDD+24] (see Appendix C for more details).

Example 9. Let $E_1/\mathbb{F}_q$ be an elliptic curve of cardinal divisible by a prime ℓ, and assume that $\#E(\mathbb{F}_q)$ is not divisible by ℓ^2. Let $K = \langle T \rangle$ be the kernel generated by a point $T \in E_1[\ell](\mathbb{F}_q)$. Then we can efficiently compute $\phi : E_1 \to E_2 = E_1/K$. Indeed, $\pi_q - 1$ is efficient (see Example 5) of degree $\#E(\mathbb{F}_q)$, and we can recover ϕ by splitting it.

Similarly, assume that $\chi_\pi \mod \ell$ splits and is separable: $\chi_\pi(X) = (X - \lambda_1)(X - \lambda_2)$ modulo ℓ with $\lambda_1 \neq \lambda_2$; in other words the Frobenius has two distincts eigenvalues. Then ℓ split in $\mathbb{Z}[\pi]$, hence splits in R as $\ell = \mathfrak{l}_1 \mathfrak{l}_2$, the saturation of $\mathbb{Z}[\pi]$ in $\text{End}_{\mathbb{F}_q}(E_1)$, because from our assumptions ℓ does not divide the conductor of $\mathbb{Z}[\pi]$. In that case there are only two rational kernels K in $E_1[\ell]$ (given by the eigenvectors of λ_i), and we can apply Example 8 to compute the corresponding isogenies efficiently. The preceding example is a special case of this when $\lambda_1 = 1$ and $\lambda_2 = q \neq 1 \mod \ell$.

When $\lambda_1 = \lambda_2 = \lambda$, but π_q is not diagonal, then $\pi_q = \begin{pmatrix} \lambda & 1 \\ 0 & \lambda \end{pmatrix}$, and there is a unique rational kernel K in $E_1[\ell]$. Either E_1 is at the top of the volcano and $\ell = \mathfrak{l}^2$ is ramified in R, or we are at the bottom of the volcano: $R = \mathbb{Z}[\pi_q]$ and K is the kernel of the unique ascending isogeny (see Example 12). In the first case we can apply Example 8 to compute efficiently the associated isogeny, but not in the second case because the ideal associated to K is not invertible.

Proposition 8. *Assume that we have an efficient representation of a n_1-isogeny $\phi_1 : E_0 \to E_1$ of degree n_1 and of a n_2-isogeny $\phi_2 : E_0 \to E_2$ with $n_1 \wedge n_2 = 1$. Then we have a pushforward square, with $\phi_2' : E_1 \to E_{12}$ the pushforward of ϕ_2 by ϕ_1, and $\phi_1' : E_2 \to E_{12}$ the pushforward of ϕ_1 by ϕ_2. And we can efficiently find an efficient representation of ϕ_1', ϕ_2'.*

Proof. Let $\phi = \phi_2 \circ \widetilde{\phi_1} : E_1 \to E_2$, which admits an efficient representation by Proposition 4. Then we can split ϕ as $\phi = \phi_1'' \circ \phi_2''$ by Proposition 7, and we have $\phi_2' = \phi_2''$, $\phi_1' = \widetilde{\phi_1''}$. □

Example 10. In the setting of Proposition 8, if ϕ_2 has smooth accessible kernel $K_2 = \mathrm{Ker}\,\phi_2$, then the pushforward ϕ_2' of ϕ_2 by ϕ_1 is given by the kernel $K_2' = \phi_1(K_2)$ (even if n_1 is not coprime to n_2). So we can directly compute ϕ_2' from its kernel K_2'.

Furthermore, since $\phi_1' \circ \phi_2 = \phi_2' \circ \phi_1$, we can find the evaluation of ϕ_1' on nice points, hence we can easily build an HD representation of ϕ_1'.

Thus, pushforwards in an hybrid setting, where ϕ_1 is given by an HD representation and ϕ_2 by a smooth accessible kernel allows for easier pushfowards; this is exploited to great effect in the POKE framework of [Bas24].

Example 11. Proposition 8 allows to generalise the "SIDH proof of knowledge" of isogenies from [DDGZ22, BCC+23, GPV24] from smooth isogenies to arbitrary efficient isogeny representations.

Namely, to prove the knowledge of an efficient n-isogeny $\phi : E_1 \to E_2$, provided that we have an algorithm to sample efficient isogenies of large degrees (coprime to $\deg \phi$) from E_1 (for instance because we know its endomorphism ring), one could build a pushforward square:

$$
\begin{array}{ccc}
E_1 & \xrightarrow{\ \phi\ } & E_2 \\
\downarrow{\scriptstyle\psi} & & \downarrow{\scriptstyle\psi'} \\
E_1' & \xrightarrow{\ \phi'\ } & E_2'
\end{array}
$$

and commit the efficient representations of ψ, ϕ', ψ', where the pushforwards ψ', ϕ' are computed by Proposition 8. Then the verifier ask to reveal one out of the three of ψ, ϕ', ψ', verify it encodes a valid isogeny representation, and this protocol is repeated sufficiently many times.

The soundness associated to this protocol is that the Prover knows some isogeny $E_1 \to E_2$. The Zero-Knowledge property is harder, since revealing ϕ' could leak informations on ϕ, that's why ψ is required to be of large enough degree.

A subtlety appear when the prover wants to prove that he knows an isogeny $\phi : E_1 \to E_2$ of explicit degree $n = n_1$. Even if he publicly commit to the degrees n_2 of ψ, ψ' and n_1 of ϕ' (which the verifier can check if the corresponding isogeny is revealed), he can only prove that he knows an isogeny $\widetilde{\psi'} \circ \phi' \circ \psi$ of degree $n_1 n_2^2$.

Indeed, the verification process does not allow to prove that ψ' is the pushforward of ψ by ϕ (or equivalently, that ϕ' is the pushforward of ϕ through ψ, or equivalently that $\psi' \circ \phi = \phi' \circ \psi$, or equivalently that $\widetilde{\psi'} \circ \phi' \circ \psi$ is divisible by $[n_2]$).

In the SIDH proof of isogeny knowledge, to prove the degree of ϕ, the authors of [DDGZ22] explain how the prover can commit informations about the kernels of ψ, ψ' (or more precisely their duals) to convince the verifier that they form a pushforward square. See also the survey [BDGP23] and [DFP24, § 5.5]. However, their solution requires that if n_2 is the degree of ϕ, then n_2 should be smooth and the n_2-torsion accessible on E_1, E_2, E_1', E_2'. So we cannot use an arbitrary efficient isogeny for ψ anymore.

Still, the method of [DDGZ22] only require to be able to construct the pushforward ϕ' and to evaluate it on points, so by Proposition 8 it still works for any efficient representation of ϕ. In fact, since this proof of knowledge with explicit degree requires taking ψ to be smooth with accessible torsion, we could simply use Example 10 directly instead of Proposition 8. We remark also that, using Proposition 8, we can generalize the method of [DDGZ22] to any ψ such that the n_2-torsion is accessible on E_1', E_2', provided that we can find an efficient representation of ψ (since we are relaxing the constraint n_2-smooth, we cannot use the decomposed representation anymore). Although this constraint on ψ is not too big in practice, as far as we know it remains an open question to handle the case of an arbitrary efficient representation for ψ, where the n_2-torsion might not be accessible. Maybe a solution would be to use many distinct degrees n_2 for ψ: if the prover can convince the verifier that he knows isogenies of degree $n_1 n_2^2$ between E_1 and E_2 for sufficiently many distinct degrees n_2, then maybe this is enough to prove that he knows an isogeny of degree n_2, because the quadratic lattice $(\mathrm{Hom}(E_1, E_2), \deg)$ is of rank at most four.

6.4 Kernel

Given an efficient representation of a cyclic n-isogeny $\phi : E_1 \to E_2$, one can ask whether we can recover its kernel. If the n-torsion is accessible on E_1, we can evaluate ϕ on a basis of the n-torsion, and solve a DLP in E_2 (or in μ_n if the n-torsion is also accessible on E_2 so that we use the Weil pairing) to recover $\mathrm{Ker}\,\phi$. This is efficient if n is smooth.

If the n-torsion is accessible on E_2, we can compute a basis (R, S) of $E_2[n]$, and compute $(P, Q) = (\widetilde{\phi}(R), \widetilde{\phi}(S))$ by Proposition 4. Then (P, Q) is a multi-generator representation of $\mathrm{Ker}\,\phi$. If we know the factorisation of n, we can even compute the orders of P, Q and from that extract a generator. This can allow to relax the condition that n is smooth.

In all cases, we can at least recover the kernel in time $\widetilde{O}(n)$:

Lemma 4. *If we have an efficient representation of an n-isogeny $\phi : E_1 \to E_2$, we can recover its function representation and an equation for its kernel in $\widetilde{O}(n)$ arithmetic operations.*

Proof. Since we know how to evaluate ϕ on every point, in principle it would be enough to evaluate it on the generic point $\mathbf{P} \in E_1(k(E_1))$ of E_1. The problem is that $k(E_1)$ is a rational function field over k, and the cost on the arithmetic on $k(E_1)$ depends on the degrees of the intermediate functions in $k(E_1)$ computed during the evaluation. To bound these intermediate degrees, a solution is to work with the formal group to some precision $m = \Theta(n)$, and do a rational reconstruction at the end. But we have seen in Sect. 3.3 that it is enough to work in precision $m = 2$ to obtain the deformation representation, from which we can reconstruct the isogeny. (An alternative strategy would be to try to interpolate from the evaluation on several rational points).

Let us detail this formal group strategy. We evaluate ϕ on a fat point $\mathbf{P} \in E_1(k[\epsilon]/\epsilon^2)$ to recover the action of ϕ on differentials, hence recover the deformation representation. We then solve a differential equation in $\widetilde{O}(n)$ to recover the function representation, from which we can extract the kernel. This assumes that $p > 8n - 5$. For large n, we just need to lift E_1 arbitrarily to $\mathbb{Z}_q/p^m\mathbb{Z}_q$ with $m = O(\log n)$ and then lift ϕ; we know how to do that efficiently thanks to Proposition 6.   □

7   Open Questions

Although the toolbox to manipulate efficient representations of isogenies has considerably expended, there are still many open questions.

We discuss some of these here:

1. The most important one is to find in polynomial time in $\log n$ an efficient representation of an n-isogeny $\phi : E_1 \to E_2$ represented by a rational generator $\operatorname{Ker} \phi = \langle T \rangle$, $T \in E_1[n](\mathbb{F}_q)$. The best available algorithm (if n is not smooth, e.g. when n is a large prime), sqrtVelu, takes "exponential" time $\widetilde{O}(n^{1/2})$.
2. If $\phi : E_1 \to E_1$ is a cyclic ℓ^2-isogeny, then it decomposes uniquely as $\phi = \phi'_2 \circ \phi_1$ with ϕ_1, ϕ'_2 ℓ-isogenies. But Proposition 7 needs the coprimality condition to split ϕ efficiently. Thus, assuming that we have an efficient representation of ϕ, computing (efficient representations of) ϕ'_2, ϕ_1 efficiently remains an open question.
3. We have the same question about computing the pushforward of isogenies ϕ_1, ϕ_2 whose degree are not coprime.
   In certain cases, we can give some answers. If $\phi_1 : E_0 \to E_1$ is an ℓ-isogeny and $\phi_2 : E_0 \to E_2$ is also an ℓ-isogeny, with ℓ prime, then either they have the same kernel in which case their pushforwards are given by an isomorphism $E_1 \simeq E_2$, or they have disjoint kernel in which case their pushforwards are $\widetilde{\phi_1}$ and $\widetilde{\phi_2}$. Thus, when ϕ_1 is a n_1 isogeny and ϕ_2 is a n_2-isogeny, we can still compute their pushfoward in some cases even where n_1 is not coprime to n_2. Namely we split them into isogenies of coprimary degree, and use pushout squares to reduce to the case $n_1 = \ell^{e_1}$ and $n_2 = \ell^{e_2}$. We have seen how to handle the case $e_1 \leqslant 1$ and $e_2 \leqslant 1$. We can assume $e_1 \leqslant e_2$. We can use Corollary 4 to test if ϕ_2 is divisible by ϕ_1: $\phi_2 = \phi'_2 \circ \phi_1$, in which case the pushforward of ϕ_1 is the identity and the pushforward of ϕ_2 is ϕ'_2. But to handle the general case, we would need to know how to split ϕ_1, ϕ_2 into chunks of ℓ-isogenies. Thus the pushforward question reduces to the splitting question.
4. Given a supersingular elliptic curve $E/\mathbb{F}_{p^2}$, it admits rational isogenies of any degree. Can we efficiently construct one of fixed arbitrary degree n (or even

random large degree)? Currently we only know how to do that for n smooth (by taking a product of small isogenies) or when we know $\text{End}(E)$ (by using an IdealToIsogeny algorithm).

5. Can we climb up in a R-oriented ℓ-isogeny volcano when ℓ is large, i.e., can we find an efficient way to compute an ascending isogeny? We do not know how to compute the climbing up isogeny efficiently, even if we already know the codomain (but see [Gal24] for a speed up when the codomain is known). Solving the climbing up problem with known codomain would have important implications on the post-quantum security of SCALLOP and SCALLOP-HD, due to [CII+23].

   For the climbing up isogeny, we do have a compact representation given by a non invertible ideal. Indeed, the conductor ideal $\mathfrak{f} = \mathbb{Z} + fR_0 \subset R$ of $R = \text{End}(E)$ encodes the isogeny ascending all the way to the top, so $(\ell, \mathfrak{f})$ is the ideal representing one step up. In that case, due to the non invertibility, Clapoti(s) does not apply and we do not now how to convert this ideal to an isogeny, other than by computing it from its kernel, which takes time $\widetilde{O}(\ell^2 \log^2 q)$. But see [Gal24] again for some improvements to this naive method in some cases.

   Solving the ℓ^2-splitting problem would give an algorithm to climb: climbing up from E_1 to E_2 and then descending back down to some E_1' corresponds to an invertible ideal $\mathfrak{a}$ of norm ℓ^2 in $\text{Cl}(\text{End}(E_1))$, and thanks to Clapotis we know how to convert $\mathfrak{a}$ into an ℓ^2-isogeny ϕ. Splitting ϕ would then give us E_2.

6. The same question holds for a descending isogeny, but in that case we do not even have an ideal representation.

7. Can we find an improvement to the KLPT algorithm that produces powersmooth ideals of norm $\approx p$ rather than $\approx p^{4.5}$ (see Appendix C.5)?

Finally, as we have illustrated several times, having a theoretical polynomial time algorithm is often not good enough: for isogeny based cryptography we want fast algorithms. In practice, this means finding elliptic curves with large accessible 2^e-torsion, and embedding isogenies into higher dimensional 2^e-isogenies of dimension r, with r as small as possible (and ideally $r = 2$). So even through the algorithms in Sect. 6 are polynomial time, there are probably many tricks still open to speed them up.

A   On the Accessible N-Torsion of an Abelian Variety

In this section, we define the notion of accessible torsion.

Definition 4. *The N-torsion of an elliptic curve $E/\mathbb{F}_q$ is said to be* accessible *if we can work with points of N-torsion by using field extensions of small degree (meaning polynomial in $\log N$).*

   In practice, this means that if $N = \prod \ell_i^{e_i}$, we want to find CRT generators $(P_i, Q_i) \in E[\ell_i^{e_i}]$ which live in extensions of (uniform) degree $O(\log^{O(1)} N)$.

This definition extends to an abelian variety (and to subgroups G): the N-torsion (resp. G-torsion) on $A/\mathbb{F}_q$ is accessible if we can find CRT generators $(P_1, \ldots, P_g) \in A[\ell_i^{e_i}]$ (resp. $G[\ell_i^\infty]$) which live in extensions of (uniform) degree $O(\log^{O(1)} N)$ (resp. $O(\log^{O(1)} \#G)$).

Remark 17. We remark that while the (P_i, Q_i) are assumed to live in small extensions, the basis of N-torsion $P = \sum P_i, Q = \sum Q_i$ could be defined over an extension of large degree. So to work with accessible N-torsion, we need to work with the CRT basis directly.

Sometimes it will be useful to be able to work with linear combinations of the points (P_i, Q_i, P_j, Q_j). Let d_i (resp. d_j) be the degree of the field extension where (P_i, Q_i) (resp. (P_j, Q_j)) are defined.

If d_i is coprime to d_j, then one can construct the compositum field of degree $d_i d_j$ in quasi-linear time $\widetilde{O}(d_i d_j)$ along with embeddings $\mathbb{F}_{q^{d_i}} \to \mathbb{F}_{q^{d_i d_j}}$ and $\mathbb{F}_{q^{d_j}} \to \mathbb{F}_{q^{d_i d_j}}$ in quasi-linear time $\widetilde{O}(d_i d_j)$ too [DDS14]. Switching between the bivariate representation $\mathbb{F}_{q^{d_i d_j}} = \mathbb{F}_q[X, Y]/\langle \Phi_{d_i}(X), \Phi_{d_j}(Y) \rangle$ and the univariate representation of $\mathbb{F}_{q^{d_i d_j}}$, where $\Phi_{d_i}(X)$ (resp. $\Phi_{d_j}(Y)$) are irreducible defining polynomials for $\mathbb{F}_{q^{d_i}}$ (resp. $\mathbb{F}_{q^{d_j}}$), can be done by modular composition, which can be computed in pseudo-linear time in the bit model by [KU11].

In the general case, if $d' = d_i \wedge d_j$, we can construct the subfield $\mathbb{F}_{q^{d'}} \subset \mathbb{F}_{q^{d_i}}$ by taking a random element $\alpha \in \mathbb{F}_{q^{d_i}}$, applying the trace $\beta = \mathrm{Tr}_{\mathbb{F}_{q^{d_i}}/\mathbb{F}_{q^{d'}}}(\alpha)$, and computing the minimal polynomial of β by power projection [Sho99] which can be done in pseudo-linear time by the transposition principle. Doing the same to construct $\mathbb{F}'_{q^{d'}} \subset \mathbb{F}_{q^{d_j}}$, we can find an isomorphism $\mathbb{F}'_{q^{d'}} \simeq \mathbb{F}_{q^{d'}}$ by computing the factorisation of the degree d' polynomial defining $\mathbb{F}'_{q^{d'}}$ over $\mathbb{F}_{q^{d'}}$; this costs $\widetilde{O}(d'^3 \log^2 q)$. For better algorithms for the isomorphism problem, we refer to [BDDFS19]. Evaluating the isomorphism is then again done by modular composition.

To simplify the complexity statements when decomposing an N-isogeny into product of ℓ_i-isogenies using accessible N-torsion, we will always suppose that we have precomputed the embedding of P_i, Q_i, P_j, Q_j into their common overfield $\mathbb{F}_{q^{d_i \vee d_j}}$. And similarly to push a point P defined over $\mathbb{F}_{q^{d_P}}$, we will assume that we have precomputed the embedding of P, P_i, Q_i into $\mathbb{F}_{q^{d \vee d_i}}$. (This last point is only needed if we represent the small isogenies by the generator of their kernel rather than by a kernel equation). This step can be done in pseudo-linear time using fast modular composition.

To know whether the N-torsion is accessible, we can look at the smallest degree where the points of $\ell_i^{e_i}$ torsion on an elliptic curve E are defined. This is well known, and we summarize this in the following lemma (which we state for abelian varieties since this case is not harder than for elliptic curves).

Lemma 5. *Let $A/\mathbb{F}_q$ be an abelian variety, and $\ell \neq p$ a prime. The minimal degree d_0 such that all points of ℓ-torsion in A are defined satisfy $d_0 \leqslant \ell^{2g} - 1$. Let $e > 0$ be the highest integer such that all points of ℓ^e-torsion of A are defined* over $\mathbb{F}_{q^{d_0}}$. Then the minimal degree d such that all points of ℓ^{e+f}-torsion of A are defined is $d = d_0 \ell^f$.

Proof. Let $\chi_\pi(X)$ be the characteristic polynomial of the Frobenius, this is a monic polynomial of degree $2g$. The degree d_0 is the minimal integer such that $\pi_q^{d_0} - 1 \equiv 0 \mod \ell$. The minimal polynomial $\Phi_{\pi, \ell}$ of the Frobenius acting on the ℓ-torsion $A[\ell]$ divides $\chi_\pi \mod \ell$, hence the order d_0 of π_q modulo ℓ is at most $\ell^{2g} - 1$.

Now we look at the action of π_q on the Tate module $T_\ell A$. By assumption, we have that $\pi_q^{d_0} - 1$ is of ℓ-adic valuation e: $\pi_q^{d_0} = 1 + \ell^e P + O(\ell^{e+1})$. Then $\pi_q^{d_0 r} = 1 + r\ell^e P + O(\ell^{e+1})$, so $\pi_q^{d_0 r} - 1$ is of ℓ-adic valuation at least $e+1$ if and only if $\ell \mid r$. It follows that the ℓ^{e+1}-torsion is defined over $\mathbb{F}_{q^{d_0 \ell}}$, and not over a smaller subfield. We conclude by recurrence. □

Example 12. Let $\chi_\pi = X^2 - tX + q$ be the characteristic polynomial of the Frobenius π_q acting on an elliptic curve $E/\mathbb{F}_q$, then $\chi_\pi \mod \ell$ is the characteristic polynomial of π_q acting on $E[\ell]$ ($\ell \neq p$).

– If this polynomial is irreducible modulo ℓ, then we have two distinct eigenvalues λ_1, λ_2 of π_q defined over $\mathbb{F}_{\ell}^2$. We let d_i be the order of λ_i, and d_q the order of q modulo ℓ (this is called the embedding degree in pairing based cryptography). There are no rational cyclic kernel in $E[\ell]$, and the ℓ-torsion becomes defined over an extension of degree $d_1 \vee d_q = d_2 \vee d_q$. In that case the ℓ-isogeny volcano is of height 1: the conductor of $\mathbb{Z}[\pi]$ is not divisible by ℓ, and in particular if R is the saturation of $\mathbb{Z}[\pi]$ in $\text{End}(E)$, the index $[R : \mathbb{Z}[\pi]]$ is not divisible by ℓ. Furthermore, ℓ is inert in $\mathbb{Z}[\pi]$ and R.

– If $\pi_q = (X - \lambda_1)(X - \lambda_2) \mod \ell$ splits with no multiplicity, we have two distinct eigenvalues $\lambda_1, \lambda_2 \in \mathbb{F}_\ell$, we have two rational kernels (formed by the eigenvectors for λ_i), whose points are defined over an extension of degree d_1 and d_2 respectively. The full ℓ-torsion is defined over $d_1 \vee d_q = d_2 \vee d_q$. Like in the above example the ℓ-isogeny volcano is of height 1, but in this case ℓ splits in $\mathbb{Z}[\pi]$ and R, and the two kernels corresponds to the two ideals $\mathfrak{l}_i$ where $\ell = \mathfrak{l}_1 \mathfrak{l}_2$ in R.

– If $\pi_q = (X - \lambda)^2 \mod \ell$, then either $\pi_q = \begin{pmatrix} \lambda & 0 \\ 0 & \lambda \end{pmatrix}$ or $\pi_q = \begin{pmatrix} \lambda & 1 \\ 0 & \lambda \end{pmatrix}$. We can

distinguish these two cases by testing whether $\pi - \lambda$ is divisible by ℓ, this is a special case of the endomorphism ring algorithm of Example 5.

In the first case all kernels are rational, and in the second case only one is rational. If d is the order of λ, then since $\lambda^2 = q$ we have that $d = d_q$ or $d = 2d_q$. In the first case the full ℓ-torsion is defined over the extension of degree d, while in the second case over the extension of degree ℓd.

In the first case, we have descending isogenies so we are not at the bottom of the volcano. In the second case, either the isogeny volcano is of height 1, $\ell = \mathfrak{l}^2$ is ramified in R (and $\mathbb{Z}[\pi]$), and the unique kernel is $K = E[\mathfrak{l}]$, with $\mathfrak{l}$ invertible. Or the isogeny volcano is of height > 1, we are on the bottom of the ℓ-isogeny volcano, and $K = E[\mathfrak{l}]$ where $\mathfrak{l}$ is the unique ideal of norm ℓ in R (necessarily non invertible in R whose conductor is divisible by ℓ). Climbing up via K, the Frobenius become diagonal on the codomain.

To generate a basis of the N-torsion of an elliptic curve, there are several methods.

- The simplest is to factorize the division polynomial ψ_N; this is a polynomial of degree $O(N^2)$. Using [KU11], this costs $\widetilde{O}(N^3)$.
- If the N-torsion lives in a small extension d, a more efficient way is to sample points in $E(\mathbb{F}_{q^d})$ and then multiply by appropriate cofactors to get points of N-torsion.

  This can be more tricky than it looks if we want a basis rather than simply points of N-torsion: imagine that $N = \ell$ is a large prime and that $E(\mathbb{F}_{q^d})[\ell^\infty] = \langle P_1, P_2 \rangle$ with P_1 of order ℓ^2 and P_2 of order ℓ. Then sampling a random point in $E(\mathbb{F}_{q^d})$ and multiplying by the cofactor, we get a random linear combination $P = aP_1 + bP_2$ which is of order ℓ^2 (unless by luck $\ell \mid a$), and so to get a point of ℓ torsion we need to look at ℓP which is a multiple of ℓP_1. So the naive algorithm only produce points in $\langle \ell P_1 \rangle$ with overwhelming probability.

  A general solution to this problem is given in [Sut11], for a specific solution on elliptic curves using the Weil or Tate pairing to speed up this computation, see [Rob21, § 5.6.2]. Using the Tate pairing and Pollard's rho to solve the DLP in $\mu_N \subset \mathbb{F}_{q^d}$, the complexity of this method is bounded by $\widetilde{O}(d^2 \log^2 q + \sqrt{N}d \log q)$.

  Rather than multiplying by a cofactor (an integer), it can be helpful to multiply our random point by a suitable endomorphism, see [Cou09, BGDS23].
- In supersingular isogeny based cryptography, we work with elliptic curves $E/\mathbb{F}_{p^2}$ such that $E(\mathbb{F}_{p^2}) = (\mathbb{Z}/p \pm 1)^2$ (depending on which quadratic twist we take).

  So it is very easy to find p such that the N-torsion is accessible (and even rational over $\mathbb{F}_{p^2}$): simply take N such that $N \mid p+1$ or $N \mid p-1$. One can even split N as $N = N_1 N_2$ where $N_1 \mid p+1$ and $N_2 \mid p+2$, and work on E or its quadratic twist E' whenever appropriate (in practice we work with their Kummer line, which does not depend on the twist, over $\mathbb{F}_{p^2}$). This is enough as long as we can treat points of N_1-torsion independently from the points of N_2-torsion (otherwise we need to go to $\mathbb{F}_{p^4}$).

  Furthermore, to generate a basis of the N-torsion it suffice to sample random points and multiply by the appropriate cofactor: the problem mentioned above does not appear in that case. We can use the Weil pairing to test when we have found generators. Thus we do not need DLPs, and the complexity is bounded by $\widetilde{O}(d^2 \log^2 q)$.

  For the special case when $N = 2^e$, efficient algorithms to sample a deterministic basis of the 2^e-torsion have been developed using the Tate pairing (this is often called an entangled basis in the literature), see for instance [CJL+17, ZSPDB18, CEMR24, § 5.1].

Example 13 (The kernel of an isogeny represented by an ideal I). If a cyclic n-isogeny ϕ is represented by an ideal I, then both in the oriented and supersingular cases we have $\operatorname{Ker}\phi = E[I] \subset E[N(I)]$. We will assume that we have an efficient representation of the order associated to I. To find a multigenerator representation of the kernel, we have two methods. First we factorize $N(I) = \prod \ell_i^{e_i}$, then a CRT basis of $E[I]$ consists of finding a generator P_i of each $E[I, \ell_i^{e_i}]$, so we can reduce to the case $n = N(I)$ a prime power. We can write $I = (N(I), \alpha)$.

- The first idea is to work over the field $\mathbb{F}_{q^d}$ where the points of $E[I]$ are defined: $E[I](\overline{\mathbb{F}}_q) \subset E[N(I)](\mathbb{F}_{q^d})$. We sample generators of $E[N(I)](\mathbb{F}_{q^d})$ using the methods above, and then eventually solve a DLP (combined with pairings) to find the kernel of α.
- The second idea, used in [EPSV, § 4.1], relies on selecting α such that $N(\alpha) \wedge N(I)^2 = N(I)$, then $E[I] = \overline{\alpha}(E[N(I)])$. So sampling a basis of $E[N(I)]$ and applying $\overline{\alpha}$ to this basis, we recover generators for $E[I]$.

B   Relaxing the Torsion Requirement for the HD Representation

In this section, we explain how splitting the HD isogeny Φ in two can allow to reduce the torsion information we need on ϕ to recover Φ.

B.1   Splitting a Smooth Cyclic Isogeny in Two

Let $\phi : E_1 \to E_2$ be a cyclic n-isogeny with kernel K, and assume that $n = n_1 n_2$. Then we can split ϕ as $\phi = E_1 \xrightarrow{\phi_1} E_{12} \xrightarrow{\phi_2} E_2$ where $K_1 = \operatorname{Ker} \phi_1 = K[n_1]$, and $\operatorname{Ker} \phi_2 = \phi_1(K)$.

If ϕ is smooth and the kernel K is accessible, we can efficiently find these kernels (see Proposition 7 for the general case, but the general case needs n_1 coprime to n_2)). It is convenient to work with $\widetilde{\phi_2}$ instead, whose kernel is given by $K_2 = \phi(E_1[n])$.

From the kernels K_1, K_2 of $\phi_1, \widetilde{\phi_2}$, we can recover ϕ: we compute a decomposed representation of $\phi_1 : E_1 \to E_{12}$ from K_1, $\widetilde{\phi_2} : E_2 \to E'_{12}$ from K_2, and we glue in the middle, i.e., compute an isomorphism $E_{12} \simeq E'_{12}$. Then we just need to compute the dual of the decomposed representation of $\widetilde{\phi_2}$ to obtain a way to evaluate ϕ. In other words: $(K_1 \subset E_1, K_2 \subset E_2)$ gives a representation of ϕ (the *split representation*), which is efficient if n is smooth, K_1 is accessible in E_1, and K_2 accessible in E_2.

For a supersingular curve over $\mathbb{F}_{p^2}$, since the Frobenius is $\pi_q = \pm[p]$ acts like a scalar on the n-torsion, and furthermore all supersingular curves have the same torsion structure, asking for K to be accessible is the same as asking the n-torsion to be accessible. In particular, the split representation relax the conditions on accessibility: if the n-torsion is accessible then certainly the n_1 and n_2-torsion are, but not conversely.

We will apply the same strategy to Φ in higher dimension for the HD representation.

B.2   Splitting the HD Representation in Two

In Sect. 5.3, we embed the n-isogeny $\phi : E_1 \to E_2$ into a higher dimensional N-isogeny $\Phi : A = E_1^u \times E_2^u \to B = E_1^u \times E_2^u$ of dimension $g = 2u$ by using an auxiliary $N - n$-endomorphism α on E_1^u given by a suitable matrix of integers, where $u = 1, 2, 4$ depending on whether $N - n$ is a sum of $1, 2, 4$ squares. As usual, we assume N smooth and the N-torsion accessible.

To recover the full kernel K of Φ, we need to know how ϕ acts on the N-torsion. But in fact, we need less information, because K is of rank g. First we remark that if we know how ϕ acts on the N-torsion, then we also have the full kernel of $K' = \widetilde{\Phi} = \Phi(A[N])$. If we write $N = N_1 N_2$, and we only require to know how ϕ acts on the $N_1 \vee N_2$-torsion, we can still recover $K[N_1]$ and $K'[N_2] = \Phi(A[N_2])$. This is enough to compute Φ: we decompose Φ as $\Phi = \Phi_2 \circ \Phi_1$, a N_1-isogeny followed by a N_2-isogeny. The kernel of Φ_1 is given by $K[N_1]$, which is indeed of degree N_1^g since K is of rank g, while the kernel of $\widetilde{\Phi_2}$ is given by $K'[N_2]$. So we can compute $\Phi_1 : A \to C$ and $\widetilde{\Phi_2} : C \to A$, and glue together in the middle at C to reconstruct Φ.

In summary, we just need to know how ϕ acts on the N-torsion, with $N^2 > n$, to be able to efficiently represent it via an N^2-isogeny Φ, because we can split Φ in two: $\Phi = \Phi_2 \circ \Phi_1$ with Φ_i a N-isogeny, and glue in the middle. We refer to [Rob23b, DLRW24] for more details.

We warn the reader that this strategy works only if the auxiliary isogeny α is given by a matrix of integers. That's because in that case we know the pushforward of α by $\phi \operatorname{Id}$ (this is simply the same matrix because it commutes with $\phi \operatorname{Id}$), and we know the codomain of Φ already. Using a general auxiliary isogeny α in order to embed ϕ in Φ requires (in general) to know ϕ on the $N > n$ torsion.

Remark 18. The group $E[N]$ is of cardinality N^2, so the condition $N^2 > n$ is not enough to uniquelty determine ϕ. By Sect. 3.3 we would need $N^2 > 4n$ instead. The reason for this discrepancy is as follows.

First, we describe Φ by its kernel, but that only determines Φ up to post-composition by an automorphism of its codomain. When a principally polarised abelian variety (A, λ_A) splits as $A = A_1^{e_1} \times A_2^{e_2} \times \dots$ as a product of principally polarised abelian varieties (A_i, λ_{A_i}) with λ_A the product polarisation of the λ_{A_i} and (A_i, λ_{A_i}) not isomorphic to (A_j, λ_{A_j}), then by the polarised Poincare's decomposition theorem, polarised automorphisms of (A, λ_A) are generated by the polarised automorphisms of the (A_i, λ_{A_i}) along with the permutations of the e_i factors of each A_i. In our setting, the codomain of Φ is a product of elliptic curve, whose automorphisms are ± 1 (unless we are in the special case of $j = 0, 1728$), so there is only a sign ambiguity for the individual isogenies composing Φ, hence in particular for ϕ. This is the same situation as if we only had the kernel of ϕ rather than ϕ.

Secondly, Φ embed several isogenies at once (in particular it embeds ϕ and its dual), and sometimes we require the extra information to identify ϕ among these isogenies. For instance, take a curve E with complex multiplication by

$\sqrt{-2}$, and take the endomorphism $\phi = 3 + \sqrt{-2}$. Then ϕ and $-\widetilde{\phi} = -3 + \sqrt{-2}$ have norm 11 and act the same on the 6-torsion (remark that $11 < 6^2 < 4 \times 11$). So while we can embed both of them at the same time into a higher dimensional 6-isogeny, to distinguish between them we need to evaluate them on more points.

We have seen in Sect. 6 ways to embed an isogeny $\phi : E_1 \to E_2$ into a higher dimensional isogeny without even knowing its codomain. For instance, if we have an efficient representation of an $n_1 n_2$-isogeny ϕ with $n_1 \wedge n_2 = 1$, we can split ϕ in $\phi = \phi_2 \circ \phi_1 = \phi_1' \circ \phi_2'$ and we can embed $\phi_1, \phi_2', \widetilde{\phi_1'}, \widetilde{\phi_2}$ at the same time into some higher dimensional isogeny Φ. One solution to identify which isogeny is which is to use pairings, see Lemma 3.

B.3   The Codomain Product Theta Structure

When we embed ϕ into an higher dimensional smooth N-isogeny Φ, and we compute Φ from its kernel using the theta isogeny algorithm described in Sect. 5.1, we have the following technical difficulty: by construction of the HD representation, the domain A and codomain B of Φ are a product of elliptic curves with their product polarisations. So we start with the product theta structure on A: we can efficiently convert between Weierstrass coordinates on E and level 2 or 4 theta coordinates on E (taking a small extension if needed to make the theta structure rational), and then we take the product theta structure to work on A: the resulting theta embedding on A is simply the Segre embedding.

However, when applying Lemma 2 to compute B, there is no reason that the resulting theta structure Θ_B on B is the product theta structure. Since we need to project back to elliptic curves to recover ϕ from Φ, as a first step we first need to convert Θ_B to a product theta structure Θ_B' (this is a linear change of variable). For this, we could simply test all the automorphisms of theta structure until we find a product theta structure. But, under the conditions of Lemma 2, we can actually predict which theta structure we will obtain on B, and know in advance which automorphism to apply to get a product theta structure in the end, see [DLRW24, Appendix F] for the technical details.

In dimension 2, the situation is simpler, because we can detect when we are on a product by the annulation of some even level 4 theta constant ϑ_i, and we are on a product theta structure precisely when $i = i_0 = (11; 11)$, and so it suffices to take any automorphism sending i to i_0 to recover a product theta structure (see [DMPR24, § 4]).

We note also that while Φ is completely determined by its kernel, to decompose Φ via the theta isogeny algorithm as in Lemma 2, we might need a bit more information. Typically, we embed ϕ into a 2^e-isogeny Φ in dimension g, and we work in level $m = 2$. To get a well defined theta structure on the codomain of Φ, we need points of 2^{e+2}-torsion above its kernel, and for that the most convenient way is to require to know how ϕ acts on the 2^{e+2}-torsion, not only the 2^e-torsion (of course we could guess it from the information we have).

B.4   Gluing in the Middle

The same remark applies when we can split Φ in two: $\Phi = \Phi_2 \circ \Phi_1$, and we compute $\Phi_1 : A \to C$ and $\widetilde{\Phi}_2 : B \to C$. For instance, if Φ is a 2^e-isogeny, we only need to know the action of ϕ on the $2^{e/2}$-torsion to recover the kernel of Φ_1 and $\widetilde{\Phi}_2$. But to be sure we recover the same theta structure on C from A and from B, we need to know (or guess) the action of ϕ on the $2^{e/2+2}$-torsion. We refer again to [DLRW24, Appendix F] and [Dar24, Appendices A, B] for more details. The advantage of imposing a theta structure of level m on C is that it kills the automorphisms of C (all automorphisms if $m \geqslant 3$, but if $m = 2$ there remain the ± 1 automorphisms), which solves the problem of gluing up to automorphisms we would have had if we had not rigidified C thusly.

C   Ideal to Isogeny Algorithms in the Supersingular Case and Applications to the SQIsign Family

In this section we give more details on the ideal to isogeny algorithms in the supersingular case. Among its many cryptographic applications, we can mention the SQIsign family of signature scheme, which we describe in Appendix C.6.

C.1   Double Paths

In Sect. 4.2, we gave an overview of different algorithms to convert an ideal I to an isogeny $\phi_I : E_1 \to E_2$, assuming that we know an efficient representation of the endomorphism ring $\mathrm{End}(E_1)$ of E_1.

Actually, what we explained is only how to convert an equivalent ideal J into an isogeny $\phi_J : E_1 \to E_2$; where J is a smoothening of I in the first generations, and simply the smallest equivalent ideal in the fourth generation.

There remains two questions: how do we compute ϕ_I once we know how to compute ϕ_J? And how can we compute an efficient representation of $\mathrm{End}(E_2)$, which would be needed if we want to convert a new ideal I_2 from E_2?

For the first question, since we know I and J, we know the endomorphism $\alpha = \widetilde{\phi_J} \circ \phi_I$; in terms of ideal this is a generator of the principal ideal $\overline{J}I$. Furthermore, by assumption we know how to evaluate it on E_1. And if $P \in E_1$, we have $\phi_J(\alpha(P)) = N(J)\phi_I(P)$. We can thus recover the image of P by ϕ_I up to the factor $N(J)$. This at least allow us to evaluate ϕ_I on all points P of torsion prime to $N(J)$. We have seen in Proposition 5 that we have a division algorithm which allows to evaluate ϕ_I efficiently when we know how to evaluate $N(J)\phi_I$ efficiently.

Another solution is to simply compute another "nice" ideal J' equivalent to I, of norm prime to $N(J)$. In other words: we build a *double path* of isogenies $\phi_J, \phi_{J'} : E_1 \to E_2$ where both isogenies have an efficient representation and are of coprime degrees. In fact, the Clapotis algorithm used in the fourth generation already directly constructs a double path. From this double path, we can translate any other ideal I into an isogeny ϕ_I efficiently, by evaluating a suitable endomorphism on E_1 first and then applying ϕ_J or $\phi_{J'}$.

The same solution works to evaluate endomorphisms on E_2. First we know that abstractly $\mathrm{End}(E_2)$ is the right order of I. Next, given $\beta \in \mathrm{End}(E_2)$, we can consider the endomorphism $\alpha = \widetilde{\phi_J} \circ \beta \circ \phi_J$, which we know how to evaluate efficiently on E_1 by assumption. Since we know an efficient representation of ϕ_J, then by Proposition 4 we can build an efficient representation of $\widetilde{\phi_J}$. But in fact, in our setting, when J is taken to be a smoothening of I, then it decomposes as a product of small degree isogenies so it is easy to compute its dual directly; and in the fourth generation where we use the Clapotis algorithm to embed a double path $\phi_{J_1}, \phi_{J_2} : E_1 \to E_2$ into a smooth HD isogeny Φ, then Φ is also decomposed into a product of small HD isogenies; hence we can also directly compute $\widetilde{\Phi}$, and so $\widetilde{\phi_{J_1}}, \widetilde{\phi_{J_2}}$ too. In any case we can compute $\phi_J \circ \alpha \circ \widetilde{\phi_J}(Q) = N(J)^2 \beta(Q)$, hence recover $\beta(Q)$ at points of order prime to $N(J)$. The general case to evaluate β proceed by division or via a double path like for ϕ_I above.

In fact, it is often more practical to build everything from a special nice curve E_0, with particularly nice endomorphism evaluation (typically E_0 will be a supersingular curve defined over $\mathbb{F}_p$ and with small discriminant). Assume that we have built a isogenies $\phi_{J_1} : E_0 \to E_1$ and $\phi_{J_2} : E_0 \to E_2$ represented by ideals J_1, J_2. Then if $P \in E_1$, we have $N(J_1 J_2) \phi_I(P) = \phi_{J_2} \circ \vartheta \circ \widetilde{\phi_{J_1}}$, where $\vartheta = \widetilde{\phi_{J_2}} \circ \phi_I \circ \phi_{J_1}$ is an endomorphism on E_0. This allows to evaluate any isogeny ϕ_I on points of order coprime to $N(J_1 J_2)$. The same technique holds to evaluate endomorphisms on E_1 and E_2 at points of order coprime to $N(J_1)$ and $N(J_2)$ respectively. We can then invoke the division algorithm for the general case. Or, simpler, we only need to compute a double "double path", i.e. an efficient double path $\phi_{J_1}, \phi_{J_1'}$ between E_0 and E_1, and another ϕ_{J_2}, ϕ_{J_2}' between E_0 and E_2 (with $N(J_1 J_2)$ coprime to $N(J_1' J_2')$).

C.2   The KLPT Algorithm and Eichler Orders

KLPT. The first generation of an ideal to isogeny algorithm is the KLPT *smoothening* algorithm from [KLPT14]. The idea is to find an ideal J equivalent to I, but of smooth norm.

If J is equivalent to I, ϕ_I, ϕ_J are both isogenies from E_1 to E_2, so $\widetilde{\phi_J} \circ \phi_I$ is an endomorphism $\alpha \in \mathrm{End}(E_1)$, which is in I because the endomorphisms factorizes through ϕ_I by construction. It follows that $J = I \frac{\overline{\alpha}}{N(I)}$, which is of norm $N(\alpha)/N(I)$. So the goal is to find $\alpha \in I$ such that $N(\alpha)/N(I)$ is of smooth norm.

The KLPT algorithm shows that, if $\mathrm{End}(E_1)$ is a special quadratic order (called special extremal order), it is heuristically always possible to find (in polynomial time!) $\alpha \in I$ such that $N(\alpha)/N(I) = N$ for any N large enough: $N \gg p^3$. A proven algorithm (under GRH) was given in [Wes22], for $N = p^C$ but with no explicit bound on C. An example of a special extremal order is the elliptic curve $E_0 : y^2 = x^3 - x$ when $p \equiv 3 \mod 4$, whose endomorphism ring contains $\mathbb{Z}[i]$.

For the general case where E_1 is not special extremal, one can instead build a smooth connecting ideal J_1 between E_0 and E_1, and another J_2 between E_0 and E_2. Then $J = J_2\overline{J_1}$ is a smooth connecting ideal between E_1 and E_2. But we can only expect to find such a J of norm $N \gg p^6$, because $N(J_i) \gg p^3$.

Taking N powersmooth, this gives an algorithm which is efficient in theory, but not in practice (but see [EPSV] for several tricks). Indeed, the size of N means that we cannot expect to find rational N-torsion, we need to take field extensions.

Eichler Orders. The first practical algorithm to convert an arbitrary ideal to an isogeny was given in [DKLPW20]: using the theory of Eichler order, it is proven that one can (heuristically) always find a smooth ideal J connecting E_1 and E_2 of norm $N \gg p^{4.5}$. More precisely, let I_0 be an ideal connecting E_0 to E_1. Then one can (heuristically) find an ideal J equivalent to I and of norm N as long as $N \gg p^{1.5}N(I)^3N(I_0)^3$. By Minkowski's bound, the (reduced) norm of the smallest ideal connecting two supersingular curves is in $O(\sqrt{p})$. Replacing I and I_0 by small equivalent ideals, we obtain the $p^{4.5}$ bound mentioned above. And the closer E_1 is to E_0 (compared to a uniform supersingular curve), the better the bound, up to p^3 when $E_1 = E_0$.

Let us quickly explain how Eichler orders are used. Let E_0 be a special extremal curve, and I_0 the ideal connecting E_0 to E_1, and assume that it is coprime to I for simplicity. Let $I' = I_0^*I$ be the pullback ideal of I (in term of isogeny this corresponds to the isogeny $\phi_I : E_1 \to E_2$ pullbacked to $\phi_{I'} : E_0 \to E_0'$ via $\phi_{I_0} : E_0 \to E_1$). Since E_0 is special extremal, by KLPT we can find a nice equivalent ideal J' to I' of norm N for $N \approx p^3$. We thus have two isogenies: $\phi_{I'}, \phi_{J'} : E_0 \to E'$. But while $\phi_I : E_1 \to E_2$ is the pushforward of $\phi_{I'}$ to E_1, the pushforward ϕ_J of $\phi_{J'}$ to E_1 may give an isogeny $E_1 \to E_2'$ with a different codomain than E_2!

Let $K = E_0[I_0]$ be the kernel of $\phi_{I_0} : E_0 \to E_1$. We have an endomorphism $\alpha = \widetilde{\phi_{J'}} \circ \phi_{I'}$, and in term of ideals $J' = I'\overline{\alpha}/N(I')$. One can show that the codomain of ϕ_J is still the same E_2 if and only if $\alpha(K) = K$ [Ler22b, Proposition 2.3.11]. Thus, for the above construction to work, J' cannot be taken to be any ideal equivalent to I', it has to come from some α such that $\alpha(K) = K$.

The endomorphisms of E_0 such that $\alpha(K) = K$ form precisely the Eichler order $O = O_0 \cap O_1$ where $O_0 = \mathrm{End}(E_0)$ and $O_1 = \mathrm{End}(E_1)$. We refer to [Ler22b, Arp22] for more details on Eichler order and their relationship with kernel of isogenies.

Thus the KLPT algorithm has to be adapted to sample elements α that are in the Eichler order O, in order for the pushfoward strategy to work. We refer to [Ler22b, DKLPW20] for these algorithms, which give the bound described above.

C.3   Ideal to Isogeny: Splitting a Large Dimension One 2^e-Isogeny

For practical instanciation of KLPT, we would like to:

1. Use smooth ideals with a very small smoothness bound, ideally take $N = 2^e$

2. Work over $\mathbb{F}_{p^2}$ and avoid taking any field extension.

Unfortunately, this cannot happen because, even with a carefully chosen p, $E(\mathbb{F}_{p^2}) = (\mathbb{Z}/(p \pm 1)\mathbb{Z})^2$ can fit at most the 2^f torsion with $2^f \mid p \pm 1$. While the KLPT algorithm (the improved version with Eichler orders) can only generate 2^e-isogenies with $2^e \approx p^{4.5}$.

Using E_0 to Refresh the Torsion. As mentioned in Sect. 4.2, the first SQIsign paper [DKLPW20] introduced the idea of splitting the large 2^e-isogeny into chunks of 2^f-isogenies. First, select a p with large available rational (over $\mathbb{F}_{p^2}$) 2^f-torsion, and search for J equivalent to I of norm 2^e, $2^e \approx p^{15/4}$ (this is better than the bound of $p^{4.5}$ above, because in the context of SQIsign E_1 is only at distance $p^{1/4}$ from E_0 and $p^3 p^{3/4} = p^{15/4} = p^{3.75}$).

If we split the isogeny ϕ_J into chunks of 2^f-isogenies, then $J_1 = J + (2^f)$ is the ideal corresponding to the first chunk: $\phi_{J_1} : E_1 \to E_{1,1}$. Let $J' = J_{1,*}J$ the pushforward of J by J_1, we now need to split J' into chunks to evaluate the next step $\phi_{J_2} : E_{1,1} \to E_{1,2}$ from $E_{1,1}$, where $J_2 = J' + (2^f)$. The problem is that to compute the next kernel, $E_{1,1}[J_2] = E_{1,1}[2^f, J']$, we need to evaluate J' on the 2^f-torsion, and we cannot go back to E_1 because J_1 is of reduced norm 2^f, so we have "consumed" all our available torsion when we evaluated ϕ_{J_1}.

We need to "refresh" the 2^f-torsion on $E_{1,1}$. The solution in SQIsign is to build a T-isogeny $\phi_T : E_0 \to E_{1,1}$ with T smooth and odd. We can then use this isogeny from E_0 to evaluate the action of J' on the 2^f-torsion of $E_{1,1}$, since it is of degree coprime to 2^f. Since E_0 is special extremal we can build an isogeny ϕ_T as long as $T \gg p^3$. This is still bigger than the available torsion, but we are saved by two things.

The first one, is that in that case we already know the codomain $E_{1,1}$, we don't need to construct it. This means that we can split ϕ_T in two: write $T = T_1 T_2$, ϕ_T as $\phi_{T_2} \circ \phi_{T_1} : E_0 \to E' \to E_{1,1}$ for some random intermediate curve E'. We can compute ϕ_{T_1} from its kernel in $E_0[T_1]$, and $\widetilde{\phi_{T_2}}$ from its kernel in $E_{1,1}[T_2]$. In that case, since T_2 is odd we can use ϕ_{J_1} to go back to E_1 to evaluate suitable endomorphisms on $E_{1,1}[T_2]$ to obtain $\mathrm{Ker}\, \widetilde{\phi_{T_2}}$ (in practice, by also using a suitable isogeny connecting E_0 to E_1). We then just need to glue things in the middle at E'. This trick of splitting an isogeny in two, when both the domain and codomain are already know is very useful, and allows to get a square factor in the amount of effective available torsion (see Appendix B for the same trick for the HD representation).

So we can expect T_1, T_2 to be of size roughly $p^{3/2}$, which is still bigger than p. The next trick is to work both with the curves $E_0, E_{1,1}$ and their quadratic twists. This allows to have both the $p-1$-torsion and the $p+1$-torsion available, while still working over $\mathbb{F}_{p^2}$ (the only condition is to not use them at the same time, which would involve working over the quadratic extension $\mathbb{F}_{p^4}$).

Example 14. In the original SQIsign, one use available 2^f-torsion with $2^f \approx p^{1/8}$ (so that the response is split into ≈ 30 blocks of 2^f-isogenies), and T_1^2-torsion (i.e., $T_2 = T_1$) with $T_1 \approx p^{3/2}$, split into one part dividing $p-1$ and the other part dividing $p + 1$. There remain approximatively $\approx p^{1/4}$ available torsion not used, because it is very hard to find such smooth primes with smoothness conditions both on $p - 1$ and $p + 1$ already, see [BSC+23].

For a security parameter λ, p has size 2λ, and the signature has size around $23/2 \cdot \lambda$.

Using an Endomorphism to Refresh the Torsion. In [DLLW23], a novel idea is to use an endomorphism γ of reduced norm T on $E_{1,1}$ instead to refresh the 2^f-torsion. This endomorphism γ (which need to satisfy some technical conditions) can be evaluated from the 2^f-isogeny $\phi_{J_1} : E_1 \to E_{1,1}$ because T is odd. The advantage compared to building a T-isogeny $E_0 \to E_{1,1}$ is that one can find a suitable endomorphism γ of smooth norm T for $T \approx p^{2.5}$, compared to p^3 for the isogeny; splitting it in two we only need accessible torsion of size $\approx p^{1.25}$ compared to $\approx p^{1.5}$. This leaves more room for the rest of the torsion, makes it easier to find good primes p, and allows to increase the available 2^f-torsion.

We briefly explain how the algorithm works. The idea is that we know the kernel K_1 of the contragredient isogeny $E_{1,1} \to E_1$, it is given by $\phi_{J_1}(E_1[2^f])$. What we want is to find the kernel $K = E_{1,1}[J_2]$ of the next isogeny $E_{1,1} \to E_{1,2}$. If we can find an effective representation of any endomorphism $\gamma \in \mathrm{End}(E_{1,1})$ such that $\gamma(K_1) \cap K_1 = 0$ (this condition can be translated into the fact that γ should not be in some Eichler order), then by linear algebra we can find an endomorphism $\gamma' = a\gamma + b$ such that $\gamma'(K_1) = K$. Evaluating γ on K_1 thus allow us to find our next kernel K.

Example 15. For SQIsign, using an endomorphism rather than an isogeny from E_0 drops the requirement on the smooth accessible torsion T from $T \approx p^{3/2}$ to $T \approx p^{5/4}$. The relaxed torsion requirement makes finding suitable p easier and improves the signing time by enabling the use of primes p with $2^f \approx p^{1/4}$ accessible torsion, which increases the size of each block of 2^f-isogenies and reduces the number of steps to ≈ 15. Furthermore, the signature is compressed further and has size around $17/2 \cdot \lambda$. See Appendix C.6 for more details.

Using an Endomorphism Represented by a Dimension 2 Isogeny. In [Ler23b], using tools from [CLP24], Leroux introduces the first version of an ideal to isogeny algorithm using the newer HD representation. Namely, rather than searching for γ of smooth norm in order to be able to evaluate it efficiently, he uses a dimension two 2^f-representation of γ. As we have seen in Sect. 5, an HD representation allows to efficiently represent any γ of odd norm $T < 2^f$ (even non smooth), as long as we know how γ acts on the 2^f torsion. In fact, by the same splitting trick as above, we can work up to γ of norm $T < 2^{2f}$. Since we need to be able to evaluate γ on the 2^f-torsion to obtain this representation, the only change is that J is taken to be a 3^h-isogeny rather than a 2^e-isogeny as previously, and it is split into chunks of 3^g-isogenies rather than chunks of 2^f-isogeny. For instance, one can take $p = 2^f 3^g u - 1$ for a small cofactor u. Since we relax the smoothness bound on γ, one can find a suitable endomorphism γ of reduced norm $T \approx p^{2/3}$. We thus only require $2^f \approx p^{1/3}$ to represent γ, leaving ample room for the chunks of 3^g-isogenies.

Rather than computing chunks of 3^g-isogenies in dimension 1, and using T-endomorphisms embedded in 2^f-isogenies in dimension 2, we could reverse the role of 2 and 3, but going from 2-isogenies to 3-isogenies is less expensive in dimension 1 than dimension 2.

Using an HD Representation for Isogenies from E_0. A recent alternative, in [ON24], is to go back to the original SQIsign protocol, using isogenies from E_0 rather than endomorphisms, but like in [Ler23b] to use an HD representation to relax the smoothness requirement on these isogenies, as was pioneered in SQIsignHD [DLRW24]. In that version, one take $p = 2^f u - 1$ for a small cofactor u, so that the full $2^f \approx p$ is available, and we write $f = f_1 + f_2$. The isogeny ϕ_J is split into chunks of 2^{f_1}-isogenies, and the refreshing isogenies $\phi_T : E_0 \to E_{1,1}$ of degree T are represented by an HD 2^{f_2}-representation, which requires to be able to evaluate ϕ_T on the 2^{f_2}-torsion.

Using a previously built isogeny $\phi_T' : E_0 \to E_1$, and the isogeny $\phi_{J_1} : E_1 \to E_{1,1}$, we can indeed recover the action of ϕ_T on the 2^{f_2}-torsion: ϕ_{J_1} is a 2^{f_1}-isogeny, so only consumes 2^{f_1} out of our available $2^{f_1+f_2}$-torsion!

One can use a dimension 4 HD representation, like in SQIsignHD. This allows to split ϕ_T in two and allows to use any T such that $T < 2^{2f_2}$. Or we can use the fact that E_0 contains $\mathbb{Z}[i]$ to build a dimension two representation, which will be more efficient that the dimension four representation (see Remark 13); but we cannot split in two in this case, so we need $T < 2^{f_2}$. We already mentioned that Minkowski's bound show that we can always find an isogeny $\phi_T : E_0 \to E_{1,1}$ of (non smooth) norm T for $T = O(\sqrt{p})$. So even the dimension 2 representation leaves ample room to split ϕ_J into chunks of 2^{f_1}-isogenies.

C.4   The Clapotis Algorithm

Another recent alternative is given in [BDD+24], inspired by the Clapoti(s) algorithm from [PR23b], see Example 8. It allows to directly convert any isogeny $\phi_I : E_1 \to E_2$ in one go by using a dimension 4 HD-representation embedding at the same time two isogenies $\phi_{J_1}, \phi_{J_2} : E_1 \to E_2$, with $N(J_1)$ coprime to $N(J_2)$. In other words this algorithm builds a double path from E_1 to E_2 in one go by using a dimension 4 2^e-isogeny. The constraint on torsion is $2^e \approx N(J_1)N(J_2)$, and taking small equivalent ideals $J_1, J_2 \sim I$ we can have $N(J_1)N(J_2) \approx p$, which is barely enough to fit into the available 2^f-torsion (we cannot split in two in that case because we don't know E_2 yet). Then ϕ_I can be recovered from this double path and the effective endomorphism representation $\mathrm{End}(E_1)$ on E_1.

In practice, it is better, as explained in [BDD+24], to build a double path between E_0 and E_1, and another between E_0 and E_2. Because $\mathrm{End}(E_0)$ contains $\mathbb{Z}[i]$, which gives a lot of small endomorphisms, this allows to find a dimension 2 representation of the double paths (again, see Remark 13). This replaces the need to compute   one dimension 4 isogeny by two dimension 2 isogeny, which is much faster in practice.

Of course, the parameters given in the discussions above, like splitting a big 2^e-isogeny into chunks of 2^f-isogeny, or using HD representations to embed T-isogenies into higher dimensional 2^f-isogeny could be generalised to any $B \mid B'$ with B, B' smooth (in our examples, $B = 2^f$ and $B' = 2^e$). This replaces the constraint of having accessible 2^f-torsion to having accessible B-torsion. But 2^e-isogenies are the one we prefer to use in practice, because they are the most convenient, especially in higher dimension. When we have the choice of the prime p, so when building protocols, we might as well choose p to have large accessible 2^f-torsion.

C.5 Improving KLPT?

From the diameter of the supersingular ℓ-isogeny graph, we know that there exist isogenies of norm $\ell^e \approx p$ between any two elliptic curves E_1, E_2. But currently the KLPT algorithm, even using the improved variant via Eichler orders, can only generate isogenies of norm $\ell^e \approx p^{4.5}$ for generic elliptic curves (for SQIsign the parameters allow to reach $\ell^e \approx p^{15/4}$). This is too big to fit into the available 2^f-rational torsion, even taking $p = u2^f - 1$ a pseudo-Mersenne prime. So for the SQIsign signature, this large 2e-isogeny has to be split into chunks of 2^f-isogenies as we have seen. And even for the verification, taking f as large as possible like in AprèsSQI [CEMR24] (which impacts the signature time), the verifier still has to compute several 2^f-isogenies in dimension 1 with $2^f \approx p$.

By contrast, the Clapotis algorithm can directly convert the smallest ideal between E_1 and E_2, which is of norm $\approx \sqrt{p}$ (or less) by Minkowski's theorem, and this ideal can be efficiently found by the LLL algorithm. Hence the ideal can be converted into an isogeny in one step, which drastically speeds up the signature; and the verification only needs a 2^e-isogeny in dimension 2 with $2^e \approx \sqrt{p}$. From the current state of the art, it seems that computing this smaller $2^e \approx \sqrt{p}$-isogeny for verification in dimension 2 is faster than computing a degree $2^e \approx p^{15/4}$ isogeny in dimension 1. Still, [CEMR24] gives several tricks using uncompressed signatures and hints to speed up verification in dimension 1, so SQIsign might still stay competitive with SQIsign2d for verification, thanks to AprèsSQI (but in any case the signature will be much faster in SQIsign2d, see Appendix C.6).

However, being able to generate directly an equivalent smooth ideal J of norm $2^e \approx p$ would be enough to fit into the available 2^f-torsion, and allow to convert it into an isogeny in one step, as is also currently done in the fourth generation, all the while staying in dimension 1 rather than going in dimension 2. The verification would also be faster: since we expect a factor $\approx \times 4$ (at least) slowdown between dimension 1 and 2 as explained in Sect. 5.1, we expect an isogeny of dimension 1 of norm $\approx p$ to be at least twice as fast to compute as a $\approx \sqrt{p}$-isogeny in dimension 2.

C.6 The SQIsign Family

We conclude this section by comparing different versions of SQIsign (which use different generations of an ideal to isogeny algorithm), along with their concrete instantiation for NIST level 1 (which means that the security parameter is $\lambda = 128$ bits). In all cases, for a security parameter λ, we work with supersingular curves over $\mathbb{F}_{p^2}$ with p of size 2λ.

We recall the general structure of the SQIsign identification protocol: the Prover wants to prove the knowledge of a secret isogeny $\varphi_{\mathsf{sk}} : E_0 \to E_{\mathsf{pk}}$; where only the codomain E_{pk} is public. He computes a secret commitment isogeny $\varphi_{\mathsf{com}} : E_0 \to E_{\mathsf{com}}$, and publishes E_{com}. The verifier challenges by an isogeny $\varphi_{\mathsf{chl}} : E_{\mathsf{pk}} \to E_{\mathsf{chl}}$. The prover responds by an isogeny $\varphi_{\mathsf{rsp}} : E_{\mathsf{chl}} \to E_{\mathsf{com}}$ connecting the challenge curve to the commitment curve. See for instance [BDD+24] for more details.

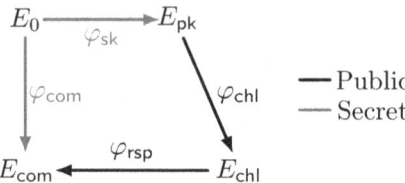

SQIsign. This is the version in [DKLPW20] with the improvements of [DLLW23]. This version uses dimension 1 both for the signature and the verification, and the signature takes $\approx 17/2 \cdot \lambda$ bits. The verification is reasonably fast but the signature is very slow, and is very hard to implement in constant time due to the many operations on the quaternion side required. Furthermore the security assumptions rely on many ad-hoc heuristics.

The public key is given by an isogeny of degree $\approx p^{1/4}$, which is not enough to reach the whole supersingular keyspace (this would require a degree $\approx p^{1/2}$). In particular, it is not statically uniform (a provable bound to reach the uniform distribution up to $2^{-\lambda}$ would even require a degree $\approx p^2$). The commitment is a smooth isogeny of degree $\approx p$. The response is an isogeny of degree $2^e \approx p^{3.75}$ which is split into 15 isogenies of degree $2^f \approx \sqrt{p}$. The torsion is refreshed using 15 endomorphisms of smooth degree $T \approx p^{5/4}$.

For the NIST submission, at level 1, we have p with accessible 2^f-torsion and T-torsion with $f = 75$ and $T = 3^{36} \cdot 7^4 \cdot 11 \cdot 13 \cdot 23^2 \cdot 37 \cdot 59^2 \cdot 89 \cdot 97 \cdot 101^2 \cdot 107 \cdot 109^2 \cdot 131 \cdot 137 \cdot 197^2 \cdot 223 \cdot 239 \cdot 383 \cdot 389 \cdot 491^2 \cdot 499 \cdot 607 \cdot 743^2 \cdot 1033 \cdot 1049 \cdot 1193 \cdot 1913^2 \cdot 1973$. The commitment is of degree $T/3^g$ ($g = 36$), and the challenge is of degree $2^f 3^g$. The available 2^f torsion is enough to split the response in 14 rather than 15, so the signature needs to compute 14 2^f-isogenies and 13 endomorphisms of degree T. The verification needs to compute the challenge again along with the 14 2^f-isogenies in dimension 1. The signature takes 177 Bytes.

SQIsignHD. This is the version in [DLRW24], which actually describes two variants.

FastSQIsignHD: this version uses dimension 1 for the signature and dimension 4 for the verification, and the signature takes $\approx 13/2 \cdot \lambda$ bits. The prime p

is a SIDH prime of the form $p = 2^f 3^g - 1$, with $2^f \approx 3^g \approx \sqrt{p}$. The public key and commitment are generated by a double path procedure, which generates a double path of degree 2^{2f} and $3^{2g} \approx p$ to the same codomain. In practice, the double path is computed via three 2^f-isogenies and three 3^g-isogenies in dimension 1. This double path procedure is assumed to be computationally uniform. The challenge is generated by an isogeny of degree 3^g. The response is an isogeny of degree $q \leqslant 2^{2e}$ such that $2^{2e} - q$ is a sum of two squares, this allows to find a dimension 4 HD representation of this response. In practice, we take $2e \approx \log_2(p) + 16$.

The signature only needs to compute the commitment, the challenge is represented by its kernel, and the response by its action on the 2^f-torsion. The verification computes the challenge from its kernel, and compute the response via its dimension 4 representation, namely it needs to compute two 2^e-isogenies in dimension 4 with $e \approx \log_2(p)/2 + 8 \leqslant f$.

For NIST level 1, we have $p = 13 \cdot 2^{126} 3^{78} - 1$, and the signature takes 109 Bytes. The commitment needs three 2^{126}-isogenies and three 3^{78}-isogenies in dimension 1. The verification computes the challenge of degree 3^{78} and checks the response via two 2^{73}-isogenies in dimension 4.

RigorousSQIsignHD: this version uses dimension 1 for the signature and dimension 8 for verification. This version is optimised for security: the public key, and commitment are provably statically uniform, and the zero knowledge property has a clean proof which relies on a RADIO oracle. However, while this protocol is polynomial time with respect to the security parameter λ, it is much too slow to be used in practice and no serious instantiation has been proposed.

Indeed, the public key, challenge and commitments are given by isogenies of degree $\approx p^3$, this requires accessible T-torsion with $T \approx p^3$, hence imposes to work with field extensions. The response is an isogeny of degree $q \approx p^2 \leqslant 2^{2e}$, which will be computed during the verification via two 2^e-isogenies in dimension 8, with $2^e \approx p$. This imposes to take $p = c2^e - 1$, and T to be powersmooth. The signature thus needs to split large degree isogenies into block of T-isogenies with the T-torsion defined over extensions, so is much slower than in SQIsign. The verification needs to compute two $2^e \approx p$-isogenies in dimension 8, compared to FastSQIsignHD which only needs two $2^e \approx \sqrt{p}$-isogenies in dimension 4. A 2^n-isogeny in dimension 8 is expected to be roughly 32 times slower than a 2^n-isogeny in dimension 4, so the verification time of RigorousSQIsignHD would also be much slower than in FastSQIsignHD.

SQIsign2d. In this section, we will focus on the variant SQIsign2d-West of [BDD+24], but see also [NO24, DF24]. This version has faster signature and verification times than SQIsign, a shorter signature and a cleaner security proof.

The prime p is chosen to be a Montgomery friendly prime of the form $p = c2^e - 1$ so that the 2^e-torsion is accessible, with $2^e \approx p$. The signature takes $\approx 8 \cdot \lambda$ bits. The public key is provably statically uniform and computed via the Clapotis version of the ideal to isogeny algorithm applied a uniformly random ideal class. The Clapotis algorithm requires one 2^e-isogeny in dimension 2, and between zero and two 2^f-isogenies in dimension 2, with and $2^f \approx \sqrt{p}$, depending on the chosen trade off between work on the quaternion side and on the isogeny side. The commitment is also statically uniform and use the same algorithm. Like in RigorousSQIsignHD, there is a clean zero knowledge proof which relies on a slightly different oracle.

Like in FastSQIsignHD the signature only generates the kernel of the challenge, which is computed during the verification. This challenges is a 2^e-isogeny in dimension 1. The verification also checks the response, via a 2^f-isogeny in dimension 2. Compared to FastSQIsignHD, the signature step also needs to compute an auxiliary isogeny $\beta : E_{\text{chl}} \to E_{\text{aux}}$ of appropriate degree to embed the response into a 2^f-isogeny $\Phi : E_{\text{chl}} \times E_{\text{aux}} \to E_{\text{com}} \times E'_{\text{aux}}$ in dimension 2, this is also done via the Clapotis version of the ideal to isogeny algorithm. More precisely, it is $\widetilde{\Phi}$ which is computed first from the commitment curve and $\beta' : E_{\text{com}} \to E'_{aux}$, and then $\widetilde{\Phi}$ is evaluated to recover Φ. The reason we want Φ rather than $\widetilde{\Phi}$ to embed the response is to compress the signature by using commitment recoverability. See [BDD+24, Figure 2] for the full diagram.

There is a slightly more heuristic version where the commitment is computed using the QFESTA algorithm (Example 7) rather than Clapotis (Example 8), which is faster but lack the provable statical uniform property. The challenge is also taken to be of degree 2^f, this allows to compute β hence Φ directly from E_{chl}.

For NIST level 1, we have $p = 5 \cdot 2^{248} - 1$ and the signature takes 148 Bytes using hints to speed up the verification. Removing hints could reduce the signature size up to 128 Bytes, at the cost of a slight increase (5 to 10%) to the verification time.

The signature computes the commitment via Clapotis, which uses a 2^{248}-isogeny and two 2^{126}-isogenies in dimension two. The auxiliary isogeny β' is also generated by Clapotis, this gives a 2^{126}-isogeny $\widetilde{\Phi}$ in dimension 2, and computing it allows to recover Φ. The verification computes the challenge, which is a 2^{248}-isogeny in dimension one, and the response is checked via a 2^{126}-isogeny in dimension two.

In the heuristic version the commitment only needs one 2^{248}-isogeny in dimension 2. The challenge is a 2^{122}-isogeny in dimension 1 which is computed during the verification. This allows to build Φ directly from β, computed via Clapotis. This heuristic version gives a signature roughly 60% faster than the cleaner version. The response is checked via a 2^{126}-isogeny in dimension two too, so the verification time does not change much (only the challenge is smaller).

The verification time of SQIsign2d is the fastest out of all the variants described in this section, and the signature is only slightly slower and less compact than FastSQIsignHD and much faster than SQIsign. But as explained in [BDD+24, Remark 24], we can adapt FastSQIsignHD to the prime used in SQIsign2d, this gives an even faster signature time than the original FastSQIsignHD. That version has a signature time roughly 3× faster than the heuristic version of SQIsign2d. However, this comes at the cost of the verification time, which needs to compute an isogeny in dimension 4 rather than 2, for an expected slow down of roughly $\times 8$.

D   The Class Group of a Non Maximal Order

Let R be an order in some number field K, R_0 be the maximal order of K, and $\mathfrak{f}$ the conductor ideal of R: $\mathfrak{f} = (R : R_0)$. This is both an ideal in R and in R_0.

The conductor square

$$
\begin{array}{ccc}
R & \longrightarrow & R_0 \\
\downarrow & & \downarrow \\
R/\mathfrak{f} & \longrightarrow & R_0/\mathfrak{f}
\end{array}
$$

is a Milnor square (also called an excision datum): it is both a pullback and a pushforward.

In particular, $\operatorname{Spec} R = \operatorname{Spec} R_0 \coprod_{\operatorname{Spec} R_0/\mathfrak{f}} \operatorname{Spec} R/\mathfrak{f}$ is given by the gluing of $\operatorname{Spec} R_0$ and $\operatorname{Spec} R/\mathfrak{f}$ over $\operatorname{Spec} R_0/\mathfrak{f}$. This is a pushout in the category of schemes, and gives an explicit description of $\operatorname{Spec} R$ as a blow down of the regular scheme $\operatorname{Spec} R_0$.

As shown by Milnor, finitely presented vector bundles (so in particular line bundles) satisfy excision, which means that to specify a line bundle $\mathcal{L}$ on $\operatorname{Spec} R$ is the same thing as specifying a line bundle $\mathcal{L}_0$ on $\operatorname{Spec} R_0$ and $\mathcal{L}_\mathfrak{f}$ on $\operatorname{Spec} R/\mathfrak{f}$, along with an isomorphism $\mathcal{L}_0 \simeq \mathcal{L}_\mathfrak{f}$ over $\operatorname{Spec} R_0/\mathfrak{f}$. For vast generalisations of this result, see [BM21, EHIK21, AHHR24].

From this we obtain a Mayer-Vietoris exact sequence:

$$1 \to U(R) \to U(R/\mathfrak{f}) \oplus U(R_0) \to U(R_0/\mathfrak{f}) \to \operatorname{Pic}(R) \to \operatorname{Pic}(R_0) \oplus \operatorname{Pic}(R_0/\mathfrak{f}) \to 0,$$

which gives the usual exact sequence between $\operatorname{Pic}(R)$ and $\operatorname{Cl}(R_0)$:

$$1 \to R_0^*/R^* \to (R_0/\mathfrak{f})^*/(R/\mathfrak{f})^* \to \operatorname{Pic}(R) \to \operatorname{Cl}(R_0) \to 0. \tag{4}$$

This geometric derivation of Eq. (4) might seem overkill, but it has the following nice application for isogenies between elliptic curves. Let $\phi : E_1 \to E_2$ be an horizontal isogeny between R-oriented elliptic curves, so that $\phi = \phi_I$ for some invertible ideal $I \in \operatorname{Pic}(R)$. The conductor ideal $\mathfrak{f}$ of $R \subset R_0$ gives ascending isogenies $\phi_{\mathfrak{f},1} : E_1 \to E_1'$ and $\phi_{\mathfrak{f},2} : E_2 \to E_2'$. Looking at the image $I' = IR_0 \in \operatorname{Cl}(R_0)$ of I under Eq. (4), we also have an horizontal isogeny $\phi_{I'} : E_1' \to E_2'$, which commutes with ϕ_I under the ascending isogenies (and in fact is a pushforward: $\operatorname{Ker} \phi_{\mathfrak{f},2} = \phi_I(\operatorname{Ker} \phi_{\mathfrak{f},1})$).

Conversely, take some ideal class $[I'] \in \operatorname{Cl}(R_0)$ represented by I' giving some isogeny $\phi_{I'} : E_1' \to E_2'$. Let K_1 be the kernel of the descending isogeny $\widetilde{\phi_{\mathfrak{f},1}} : E_1' \to E_1$. If we take I' to be of prime norm to the conductor of R, the image $K_2 = \phi_{I'}(K_1)$ of K_1 by $\phi_{I'}$ is the kernel of a descending isogeny $\widetilde{\phi_{\mathfrak{f},2}} : E_2' \to E_2$, and we obtain an isogeny $\phi_I : E_1 \to E_2$ completing the pushforward square. Changing the class of I' amount to postcomposing by an endomorphism α :

$E_2' \to E_2'$; in general we obtain a different isogeny $E_1 \to E_2$ unless $\alpha(K_2) = K_2$. We can thus reinterpret isogenies $E_1 \to E_2$ as isogenies $E_1' \to E_2'$ with some extra level structure information (namely the image of the kernel K_1). In other words: E_2' is determined from E_2 and K_2, and it is also determined by the class of I. From the conductor square, we know that the class of I is determined by the class of some invertible ideal I' in R_0, along with some invertible ideal in $R/\mathfrak{f}$ and some gluing data. But we know that E_2 is given by the class of I', so this means that the data of K_2 should correspond to some module over $R/\mathfrak{f}$. One can make this statement rigourous by using the fact that the usual ideal to isogeny functor extends to modules, see Remark 4.

More generally, there are two ways to extend the usual relationship between ideals and isogenies to include level structure information. The first is to consider ideals with respect to a suborder whose conductor is related to the level structure. For instance, extending the Deuring's correspondence to keep track of the kernel of a specific isogeny $E_0 \to E_1$ corresponds to looking at ideals in the Eichler order $O_0 \cap O_1$. This is the point of view adapted in [Ler22b, Arp22].

Another approach is to use again that the ideal to isogeny functor extend to modules. For instance, if $R = \mathrm{End}(E)$, the map associated to the module $R \hookrightarrow R/I$ is the inclusion $E[I] \subset E$; in particular $E[n]$ corresponds to the module R/nR. Keeping track of level structure along isogenies corresponds to looking at modules over R along with modules over the endomorphism data of the level structure, with some compatible gluing conditions. Again, the conductor square allows to go back and forth between the two approaches (for sufficiently nice modules).

In other words, as argued in [PR23a], the module point of view, which extends the usual relationship between ideals and isogenies (as described in Sect. 4), is very fruitful since it can handle in a unified framework level structures and the higher dimensional isogeny graph starting from E^g. And the geometric description of $\mathrm{Spec}\, R$ as an excision datum of $\mathrm{Spec}\, R_0$ and $\mathrm{Spec}\, R/\mathfrak{f}$ over $\mathrm{Spec}\, R_0/\mathfrak{f}$ allows to better understand modules on R.

E   The Kodaira-Spencer Isomorphism for Abelian Varieties

In this section, we briefly recall the Kodaira-Spencer isomorphism. IF A/k is an abelian variety, the Kodaira-Spencer map is a canonial isomorphism between $\mathrm{Sym}^2(T_0(A))$ and the tangent $T_A \mathcal{A}_g$ to A on the moduli space of principally polarised abelian varieties $\mathcal{A}_g$. In other words, deformations of A are controlled by the Sym^2 of differentials on A.

Let $p \colon A \to S$ be an abelian scheme over a smooth base. Recall that we have a canonical flat connection on the De Rham (hyper)cohomology, the Gauss-Manin connection:

$$\nabla \colon R^1 p_* \Omega_{A/S} \to R^1 p_* \Omega_{A/S} \otimes \Omega_S^1.$$

Combining the Gauss-Manin connection with the Hodge filtration, one can define the *Kodaira–Spencer map* (see [And17, § 1.4], [FC90, § III.9]):

$$\kappa \colon T_S \to R^1 p_* T_{A/S},$$

where $T_{A/S}$ denotes the dual of $\Omega^1_{A/S}$. Since $\mathrm{Lie}_S A = p_* T_{A/S} = s^* T_{A_S}$ where $s \colon S \to A$ is the zero section [EGM12, Prop. 3.15], by the projection formula [Stacks, Tag 0943], we have

$$R^1 p_* T_{A/S} = \mathrm{Lie}_S(A) \otimes_{\mathcal{O}_S} R^1 p_* \mathcal{O}_A.$$

Moreover, $R^1 p_* \mathcal{O}_A$ is naturally isomorphic to $\mathrm{Lie}_S(A^\vee)$, where $A^\vee \to S$ denotes the dual of A. Therefore, we can also write the Kodaira–Spencer map as

$$\kappa \colon T_S \to R^1 p_* T_{A/S} \simeq \mathrm{Lie}_S(A) \otimes_{\mathcal{O}_S} \mathrm{Lie}_S(A^\vee).$$

The Kodaira-Spencer map κ is invariant by duality. A polarization $A \to A^\vee$ induces another version of the Kodaira–Spencer map:

$$\kappa \colon T_S \to \mathrm{Sym}^2 \mathrm{Lie}_S(A) = \mathrm{Hom}_{\mathrm{Sym}}(\Omega^1_{A/S}, \Omega^{1 \vee}_{A^\vee/S}) = \mathrm{Hom}_{\mathrm{Sym}}(\mathrm{Lie}_S(A)^\vee, \mathrm{Lie}_S(A^\vee)).$$

If we apply this construction to the universal abelian scheme $\mathcal{X}_g \to \mathcal{A}_g$, the Kodaira–Spencer map is an isomorphism [And17, § 2.1.1]. In particular, if $x \colon \mathrm{Spec}\, k \to \mathcal{A}_g$ is a point represented by a principally polarised abelian variety A/k, we have a canonical isomorphism $T_x \mathcal{A}_g \simeq \mathrm{Sym}^2(T_0(A))$. Moreover, if j is a modular invariant (i.e. a rational map $\mathcal{A}_g \to \mathbb{A}^1$), then via the Kodaira–Spencer isomorphism, its differential dj becomes a Siegel modular function of weight Sym^2.

Example 16. In [KPR24], we use the Kodaire-Spencer isomorphism to work with *deformation representations* of isogenies in dimension 2.

Namely, if $H : y^2 = f(x)$ is an hyperelliptic curve, with $\deg f = 6$, we can associate a canonical basis of differentials $(dx/y, x\,dx/y)$ on $\mathrm{Jac}(H)$, via the canonical isomorphism $\Omega^1(\mathrm{Jac}(C)) \simeq H^1(C)$. Then the curve equation gives a universal vectorial modular $(\mathrm{Jac}(H), dx/y, x\,dx/y) \mapsto f(x)$ form of weight Sym^6. If (j_1, j_2, j_3) are the Igusa invariants, we can thus express the vectorial modular functions dj_1, dj_2, dj_3 of weight Sym^2 in terms of the coefficients of the curve equation $f(x)$: explicit formulas are in [KPR24].

We can then proceed as in Sect. 3.3 to reconstruct an isogeny from the modular polynomial: letting $J = (j_1, j_2, j_3)$, and Φ_ℓ the modular polynomial(s) in dimension 2, if $\Phi_\ell(J(A), J(B)) = 0$, differentiating Φ_ℓ gives the relationship between $dJ(A, \omega_A)$ and $dJ(B, \omega_B)$ where ω_A, ω_B are normalised basis of differentials on A, B, i.e., if $\phi : A \to B$ is the associated ℓ-isogeny, we have $\phi^* \omega_B = \omega_A$. Using the formulas expressing dJ in terms of the curve equation, we can find equations for H_A, H_B such that the isogeny is normalised with respect to the canonical basis $(dx/y, x\,dx/y)$. We can then solve a differential equation to recover ϕ in Mumford coordinates. See [Rob21, § 5.4.2] for a summary, and [KPR24] for all details.

In theory, the heat equation

$$2\pi i(1 + \delta_{jk})\frac{\partial \vartheta_i}{\partial \tau_{jk}} = \frac{\partial^2 \vartheta_i}{\partial z_j \partial z_k}.$$

gives the Kodaira-Spencer isomorphism for theta functions. So we could extend the approach of [KPR24] to the theta model in higher dimension. The main problem is that the size of the modular polynomial explodes in higher dimension (see [Kie22b] for bounds on the degree and height). For instance, the Siegel modular polynomial in dimension g is of size $\widetilde{O}(\ell^{N(N+2)})$ where $N = g(g+1)/2$ is the dimension of the moduli space $\mathcal{A}_g$; and the evaluated modular polynomial $\Phi_\ell(J(A), Y)$ for $A/\mathbb{F}_q$ is of size $\widetilde{O}(\ell^N \log q)$. When $g = 1$, we recover that ϕ_ℓ is of size $\widetilde{O}(\ell^3)$, and the evaluated modular polynomial of size $O(\ell \log q)$. In dimension 2, the Siegel modular polynomial Φ_ℓ is of size $\widetilde{O}(\ell^{15})$, and the evaluated modular polynomial is of size $O(\ell^3 \log q)$, but the best algorithm we have to evaluate it [Kie20] uses analytic method and cost time $O(\ell^6 \log q)$, if $q = p$ (see also [Rob22b] for CRT and p-adic variants which use the HD representation). In dimension 3, the Siegel modular polynomial Φ_ℓ is of a staggering size $\widetilde{O}(\ell^{48})$, and even the evaluated modular polynomial of size $O(\ell^6 \log q)$.

In higher dimension, using Hilbert modular polynomials to parametrize β-isogenies, with $N(\beta) = \ell$, is more reasonable: they take space $\widetilde{O}(\ell^{g+2})$ and their evaluation take space $\widetilde{O}(\ell \log q)$. We refer to [Kie22a, Kie21] for applications to point counting in dimension 2.

References

[ADMP20] Alamati, N., De Feo, L., Montgomery, H., Patranabis, S.: Cryptographic group actions and applications. In: Advances in Cryptology–ASIACRYPT 2020: 26th International Conference on the Theory and Application of Cryptology and Information Security, Daejeon, South Korea, 7–11 December 2020, Part II, pp. 411–439. Springer (2020)

[AHHR24] Alper, J., Hall, J., Halpern-Leistner, D., Rydh, D.: Artin algebraization for pairs with applications to the local structure of stacks and Ferrand pushouts. In: Forum of Mathematics, Sigma, vol. 12, p. e20. Cambridge University Press (2024)

[And17] André, Y.: On the Kodaira–Spencer map of abelian schemes. Ann. Sc. Norm. Super. Pisa Cl. Sci. (5) **17**(4), 1397–1416 (2017)

[Arp22] Arpin, S.: Adding level structure to supersingular elliptic curve isogeny graphs. arXiv preprint arXiv:2203.03531 (2022)

[ACL+23] Arpin, S., Camacho-Navarro, C., Lauter, K., Lim, J., Nelson, K., Scholl, T., Sotáková, J.: Adventures in supersingularland. Exp. Math. **32**(2), 241–268 (2023)

[ACL+22] Arpin, S., Chen, M., Lauter, K.E., Scheidler, R., Stange, K.E., Tran, H.T.: Orientations and cycles in supersingular isogeny graphs. In: Proceedings of Women in Number Theory, vol. 5 (2022)

[ACD+23] Arpin, S., Clements, J., Dartois, P., Eriksen, J.K., Kutas, P., Wesolowski, B.: Finding orientations of supersingular elliptic curves and quaternion orders. arXiv preprint arXiv:2308.11539 (2023)

[BGDS23] Banegas, G., Gilchrist, V., Dévéhat, A.L., Smith, B.: Fast and frobenius: rational isogeny evaluation over finite fields. In: International Conference on Cryptology and Information Security in Latin America, pp. 129–148. Springer (2023)

[Bas24] Basso, A.: POKE: a framework for efficient PKEs, Split KEMs, and OPRFs from higher-dimensional isogenies. In: Cryptology ePrint Archive (2024)

[BCC+23] Basso, A., et al.: Supersingular curves you can trust. In: Annual International Conference on the Theory and Applications of Cryptographic Techniques, pp. 405–437. Springer (2023)

[BDD+24] Basso, A., et al.: SQIsign2D-West: The Fast, the Small, and the Safer. In: Asiacrypt 2024 (2024). https://asiacrypt.iacr.org/2024/

[BMP23] Basso, A., Maino, L., Pope, G.: FESTA: fast encryption from super-singular torsion attacks. In: International Conference on the Theory and Application of Cryptology and Information Security, pp. 98–126. Springer (2023)

[BDLS20] Bernstein, D., De Feo, L., Leroux, A., Smith, B.: Faster computation of isogenies of large prime degree. In: Algorithmic Number Theory Symposium (ANTS XIV), vol. 4, pp. 39–55. Mathematical Sciences Publishers (2020). arXiv: 2003.10118, https://msp.org/obs/2020/4/p04.xhtml

[BDGP23] Beullens, W., De Feo, L., Galbraith, S.D., Petit, C.: Proving knowledge of isogenies: a survey. Designs Codes Cryptogr. **91**(11), 3425–3456 (2023)

[BKV19] Beullens, W., Kleinjung, T., Vercauteren, F.: CSI-FiSh: efficient isogeny based signatures through class group computations. In: International Conference on the Theory and Application of Cryptology and Information Security), pp. 227–247. Springer (2019)

[BM21] Bhatt, B., Mathew, A.: The arc-topology. Duke Math. J. **170**(9), 1899–1988 (2021)

[BCR10] Bisson, G., Cosset, R., Robert, D.: AVIsogenies. Magma package devoted to the computation of isogenies between abelian varieties (2010). https://www.math.u-bordeaux.fr/~damienrobert/avisogenies/. Free software (LGPLv2+), registered to APP (reference IDDN.FR.001.440011.000.R.P.2010.-000.10000). Latest version 0.7, released on 13 Mar 2021

[BMSS08] Bostan, A., Morain, F., Salvy, B., Schost, E.: Fast algorithms for computing isogenies between elliptic curves. Math. Comput. **77**(263), 1755–1778 (2008)

[BCG+17] Bostan, A., et al.: Algorithmes efficaces en calcul formel (2017). https://hal.inria.fr/hal-01431717/document

[BDDFS19] Brieulle, L., De Feo, L., Doliskani, J., Flori, J.-P., Schost, É.: Computing isomorphisms and embeddings of finite fields. Math. Comput. **88**(317), 1391–1426 (2019)

[BLS12] Bröker, R., Lauter, K., Sutherland, A.: Modular polynomials via isogeny volcanoes. Math. Comput. **81**(278), 1201–1231 (2012). arXiv: 1001.0402

[BFT14] Bruin, N., Flynn, E.V., Testa, D.: Descent via (3, 3)-isogeny on Jacobians of genus 2 curves. Acta Arith **165**(3), 201–223 (2014)

[BSC+23]   Bruno, G., et al.: Cryptographic smooth neighbors. In: International Conference on the Theory and Application of Cryptology and Information Security, pp. 190–221. Springer (2023)

[CF+96]   Cassels, J.W.S., Flynn, E.V., et al.: Prolegomena to a Middlebrow Arithmetic of Curves of Genus 2, vol. 230. Cambridge University Press (1996)

[CD21]   Castryck, W., Decru, T.: Multiradical isogenies. Arithmetic Geom. Cryptogr. Coding Theory **779**, 57–89 (2021)

[CD22]   Castryck, W., Decru, T.: An efficient key recovery attack on SIDH (preliminary version). Cryptology ePrint Archive, Paper 2022/975 (2022). https://eprint.iacr.org/2022/975

[CD23]   Castryck, W., Decru, T.: An efficient key recovery attack on SIDH. In: Hazay, C., Stam, M. (eds) EUROCRYPT 2023. LNCS, vol. 14008, pp. 423–447. Springer, Cham (2023). https://doi.org/10.1007/978-3-031-30589-4_15

[CDHV22]   Castryck, W., Decru, T., Houben, M., Vercauteren, F.: Horizontal racewalking using radical isogenies. In: International Conference on the Theory and Application of Cryptology and Information Security (Asiacrypt), pp. 67–96. Springer (2022)

[CDM+24]   Castryck, W., et al.: Interpolating isogenies and breaking the SIDH cryptosystem (2024)

[CDV20]   Castryck, W., Decru, T., Vercauteren, F.: Radical isogenies. In: International Conference on the Theory and Application of Cryptology and Information Security (Asiacrypt). Lecture Notes in Computer Science, vol. 12492, pp. 493–519. Springer (2020)

[CLMPR18] Castryck, W., Lange, T., Martindale, C., Panny, L., Renes, J.: CSIDH: an efficient post-quantum commutative group action. In: International Conference on the Theory and Application of Cryptology and Information Security (Asiacrypt 2018), pp. 395–427. Springer (2018)

[CLG09]   Charles, D., Lauter, K., Goren, E.: Cryptographic hash functions from expander graphs. J. Cryptol. **22**(1), 93–113 (2009). ISSN 0933-2790

[CII+23]   Chen, M., Imran, M., Ivanyos, G., Kutas, P., Leroux, A., Petit, C.: Hidden stabilizers, the isogeny to endomorphism ring problem and the cryptanalysis of pSIDH. In: International Conference on the Theory and Application of Cryptology and Information Security, pp. 99–130. Springer (2023)

[CLP24]   Chen, M., Leroux, A., Panny, L.: SCALLOP-HD: group action from 2-dimensional isogenies. In: IACR International Conference on Public-Key Cryptography, pp. 190–216. Springer (2024)

[CS21]   Chenu, M., Smith, B.: Higher-degree supersingular group actions. arXiv preprint arXiv:2107.08832 (2021)

[CC86]   Chudnovsky, D., Chudnovsky, G.: Sequences of numbers generated by addition in formal groups and new primality and factorization tests. Adv. Appl. Math. **7**(4), 385–434 (1986). https://doi.org/10.1016/0196-8858(86)90023-0. ISSN 0196-8858

[CL23]   Codogni, G., Lido, G.: Spectral theory of isogeny graphs. arXiv preprint arXiv:2308.13913 (2023)

[CK20]   Colo, L., Kohel, D.: Orienting supersingular isogeny graphs. J. Math. Cryptol. **14**(1), 414–437 (2020)

[Cor07]   Cornacchia, G.: Su di un metodo per la risoluzione in numeri interi dell'equazione $\sum_{h=0}^{n} C_h x^{n-h} y^h = P$, vol. 46, pp. 33–90 (1907)

[CCS24]   Corte-Real Santos, M., Costello, C., Smith, B.: Efficient (3, 3)-isogenies on fast Kummer surfaces. arXiv preprint arXiv:2402.01223 (2024)

[CEMR24] Corte-Real Santos, M., Eriksen, J.K., Meyer, M., Reijnders, K.: AprésSQI: extra fast verification for SQIsign using extensionfield signing. In: Annual International Conference on the Theory and Applications of Cryptographic Techniques, pp. 63–93. Springer (2024)

[CR24] Corte-Real Santos, M., Reijnders, K.: Return of the Kummer: a toolbox for genus 2 cryptography. Cryptology ePrint Archive, Paper 2024/948 (2024). https://eprint.iacr.org/2024/948

[CR15] Cosset, R., Robert, D.: An algorithm for computing (ℓ, ℓ)-isogenies in polynomial time on Jacobians of hyperelliptic curves of genus 2. Math. Comput. **84**(294), 1953–1975 (2015). https://doi.org/10.1090/S0025-5718-2014-02899-8

[CMSV19] Costa, E., Mascot, N., Sijsling, J., Voight, J.: Rigorous computation of the endomorphism ring of a Jacobian. Math. Comput. **88**(317), 1303–1339 (2019)

[Cos18] Costello, C.: Computing supersingular isogenies on Kummer surfaces. In: Advances in Cryptology–ASIACRYPT 2018: 24th International Conference on the Theory and Application of Cryptology and Information Security, Brisbane, QLD, Australia, 2–6 December 2018, Part III, pp. 428–456. Springer (2018)

[CJL+17] Costello, C., Jao, D., Longa, P., Naehrig, M., Renes, J., Urbanik, D.: Efficient compression of SIDH public keys. In: Annual International Conference on the Theory and Applications of Cryptographic Techniques, pp. 679–706. Springer (2017)

[Cou09] Couveignes, J.: Linearizing torsion classes in the Picard group of algebraic curves over finite fields. J. Algebra **321**(8), 2085–2118 (2009). ISSN 0021-8693

[Cou06] Couveignes, J.M.: Hard Homogeneous Spaces. IACR Cryptology ePrint Archive, vol. 2006, p. 291 (2006)

[CE14] Couveignes, J.-M., Ezome, T.: Computing functions on Jacobians and their quotients. LMS J. Comput. Math. **18**(1), 555–577 (2014). arXiv: 1409.0481

[Dar24] Dartois, P.: Fast computation of 2-isogenies in dimension 4 with the theta model and cryptographic applications (2024)

[DLRW24] Dartois, P., Leroux, A., Robert, D., Wesolowski, B.: SQIsignHD: new dimensions in cryptography. In: Joye, M., Leander, G. (eds.) EUROCRYPT 2024. LNCS, vol. 14651, pp. 3–32. Springer, Cham (2024). https://doi.org/10.1007/978-3-031-58716-0_1

[DMPR24] Dartois, P., Maino, L., Pope, G., Robert, D.: An algorithmic approach to $(2, 2)$-isogenies in the theta model and applications to isogeny-based cryptography. In: Asiacrypt 2024 (2024). https://asiacrypt.iacr.org/2024/

[De 17] De Feo, L.: Mathematics of isogeny based cryptography. arXiv: 1711.04062 (2017)

[DDGZ22] De Feo, L., Dobson, S., Galbraith, S.D., Zobernig, L.: SIDH proof of knowledge. In: International Conference on the Theory and Application of Cryptology and Information Security, pp. 310–339. Springer (2022)

[DDS14] De Feo, L., Doliskani, J., Schost, É.: Fast arithmetic for the algebraic closure of finite fields. In: Proceedings of the 39th International Symposium on Symbolic and Algebraic Computation, pp. 122–129 (2014)

[DFP24] De Feo, L., Fouotsa, T.B., Panny, L.: Isogeny problems with level structure. In: Annual International Conference on the Theory and Applications of Cryptographic Techniques, pp. 181–204. Springer (2024)

[DG19]   De Feo, L., Galbraith, S.D.: SeaSign: compact isogeny signatures from class group actions. In: Advances in Cryptology–EUROCRYPT 2019: 38th Annual International Conference on the Theory and Applications of Cryptographic Techniques, Darmstadt, Germany, 19–23 May 2019, Part III, pp. 759–789. Springer (2019)

[DHPS16] De Feo, L., Hugounenq, C., Plût, J., Schost, É.: Explicit isogenies in quadratic time in any characteristic. LMS J. Comput. Math. **19**(A), 267–282 (2016)

[DJP14]   De Feo, L., Jao, D., Plût, J.: Towards quantum-resistant cryptosystems from supersingular elliptic curve isogenies. J. Math. Cryptol. **8**(3), 209–247 (2014)

[DKS18]   De Feo, L., Kieffer, J., Smith, B.: Towards practical key exchange from ordinary isogeny graphs. In: International Conference on the Theory and Application of Cryptology and Information Security, pp. 365–394. Springer (2018). arXiv: 1809.07543

[DKLPW20] De Feo, L., Kohel, D., Leroux, A., Petit, C., Wesolowski, B.: SQISign: compact post-quantum signatures from quaternions and isogenies. In: International Conference on the Theory and Application of Cryptology and Information Security (Asiacrypt 2020), pp. 64–93. Springer (2020)

[DLLW23] De Feo, L., Leroux, A., Longa, P., Wesolowski, B.: New algorithms for the Deuring correspondence: towards practical and secure SQISign signatures. In: Annual International Conference on the Theory and Applications of Cryptographic Techniques, pp. 659–690. Springer (2023)

[Dec24]   Decru, T.: Radical Vélu N-isogeny formulae. In: Annual International Cryptology Conference (Eurocrypt), pp. 107–128. Springer (2024)

[DK23]   Decru, T., Kunzweiler, S.: Efficient computation of (3 n, 3 n)-isogenies. In: International Conference on Cryptology in Africa, pp. 53–78. Springer (2023)

[DMS23]   Decru, T., Maino, L., Sanso, A.: Towards a quantum-resistant weak verifiable delay function. In: International Conference on Cryptology and Information Security in Latin America, pp. 149–168. Springer (2023)

[DHK+23] Duman, J., Hartmann, D., Kiltz, E., Kunzweiler, S., Lehmann, J., Riepel, D.: Generic models for group actions. In: IACR International Conference on Public-Key Cryptography, pp. 406–435. Springer (2023)

[DF24]   Duparc, M., Fouotsa, T.B.: SQIPrime: a dimension 2 variant of SQISignHD with non-smooth challenge isogenies. Cryptology ePrint Archive (2024)

[DFV24]   Duparc, M., Fouotsa, T.B., Vaudenay, S.: Silbe: an updatable public key encryption scheme from lollipop attacks. Cryptology ePrint Archive (2024)

[EGM12]   Edixhoven, B., van der Geer, G., Moonen, B.: Abelian varieties. Book project (2012). http://van-der-geer.nl/~gerard/AV.pdf

[ES24]   Eisentraeger, K., Scullard, G.: Connecting Kani's Lemma and pathfinding in the Bruhat-Tits tree to compute supersingular endomorphism rings. arXiv: 2402.05059 (2024)

[EHLMP18] Eisenträger, K., Hallgren, S., Lauter, K., Morrison, T., Petit, C.: Supersingular isogeny graphs and endomorphism rings: reductions and solutions. In: Advances in Cryptology–EUROCRYPT 2018: 37th Annual International Conference on the Theory and Applications of Cryptographic Techniques, Tel Aviv, Israel, 29 April–3 May 2018, Part III, pp. 329–368. Springer (2018)

[Elk92] Elkies, N.: Explicit isogenies. In: Manuscript, Boston, MA (1992)

[Elk97] Elkies, N.: Elliptic and modular curves over finite fields and related computational issues. In: Computational Perspectives on Number Theory: Proceedings of a Conference in Honor of AOL Atkin, September 1995, University of Illinois at Chicago, vol. 7, p. 21. American Mathematical Society (1997)

[EHIK21] Elmanto, E., Hoyois, M., Iwasa, R., Kelly, S.: Cdh descent, cdarc descent, and Milnor excision. Math. Ann. **379**(3), 1011–1045 (2021)

[Eng09] Enge, A.: Computing modular polynomials in quasi-linear time. Math. Comput. **78**(267), 1809–1824 (2009)

[EPSV] Eriksen, J.K., Panny, L., Sotáková, J., Veroni, M.: Deuring for the people: supersingular elliptic curves with prescribed endomorphism ring in general characteristic. Contemp. Math. **796**, 339–373 (2023). LuCaNT: LMFDB, Computation, and Number Theory (Providence), Proceedings

[FC90] Faltings, G., Chai, C.-L.: Degeneration of abelian varieties. Ergebnisse der Mathematik und ihrer Grenzgebiete (3) 22. Springer, Berlin (1990)

[Feo10] de Feo, L.: Algorithmes Rapides pour les Tours de Corps Finis et les Isogénies. Ph.D. thesis. Ecole Polytechnique X (2010). http://hal.inria.fr/tel-00547034/en

[FFK+23] Feo, L.D., et al.: SCALLOP: scaling the CSI-FiSh. In: IACR International Conference on Public-Key Cryptography, pp. 345–375. Springer (2023)

[Fly15] Flynn, E.V.: Descent via (5, 5)-isogeny on Jacobians of genus 2 curves. J. Number Theory **153**, 270–282 (2015)

[Fly93] Flynn, E.V.: The group law on the Jacobian of a curve of genus 2. J. für die reine angewandte Math. **439**, 45–69 (1993)

[FMP23] Fouotsa, T.B., Moriya, T., Petit, C.: M-SIDH and MDSIDH: countering SIDH attacks by masking information. In: Annual International Conference on the Theory and Applications of Cryptographic Techniques, pp. 282–309. Springer (2023)

[FM02] Fouquet, M., Morain, F.: Isogeny volcanoes and the SEA algorithm. In: Fieker, C., Kohel, D.R. (eds.) ANTS 2002. LNCS, vol. 2369, pp. 276–291. Springer, Heidelberg (2002). https://doi.org/10.1007/3-540-45455-1_23

[FK11] Frey, G., Kani, E.: Correspondences on hyperelliptic curves and applications to the discrete logarithm. In: International Joint Conferences on Security and Intelligent Information Systems, pp. 1–19. Springer (2011)

[Gal24] Galbraith, S.: Climbing and descending tall volcanoes. Cryptology ePrint Archive, Paper 2024/924 (2024). https://eprint.iacr.org/2024/924

[Gau07] Gaudry, P.: Fast genus 2 arithmetic based on Theta functions. J. Math. Cryptol. **1**(3), 243–265 (2007)

[GL09] Gaudry, P., Lubicz, D.: The arithmetic of characteristic 2 Kummer surfaces and of elliptic Kummer lines. Finite Fields Their Appl. **15**(2), 246–260 (2009)

[GPV24] Ghantous, W., Pintore, F., Veroni, M.: Efficiency of SIDHbased signatures (yes, SIDH). J. Math. Cryptol. **18**(1), 20230023 (2024)

[HR19] Hisil, H., Renes, J.: On kummer lines with full rational 2-torsion and their usage in cryptography. ACM Trans. Math. Softw. (TOMS) **45**(4), 1–17 (2019)

[Ill79] Illusie, L.: Complexe de de Rham-Witt et cohomologie cristalline. Ann. Sci. l'École Normale Supérieure **12**(4), 501–661 (1979)

[JD11] Jao, D., De Feo, L.: Towards quantum-resistant cryptosystems from supersingular elliptic curve isogenies. In: International Workshop on Post-Quantum Cryptography (PQCrypto 2011), pp. 19–34. Springer (2011)

[JKP+18] Jordan, B.W., Keeton, A.G., Poonen, B., Rains, E.M., Shepherd-Barron, N., Tate, J.T.: Abelian varieties isogenous to a power of an elliptic curve. Compos. Math. **154**(5), 934–959 (2018)

[Kan97] Kani, E.: The number of curves of genus two with elliptic differentials. J. reine angewandte Math. **485**, 93–122 (1997)

[Kan11] Kani, E.: Products of CM elliptic curves. Collectanea Math. **62**(3), 297–339 (2011)

[KU11] Kedlaya, K.S., Umans, C.: Fast polynomial factorization and modular composition. SIAM J. Comput. **40**(6), 1767–1802 (2011)

[Kem88] Kempf, G.: Multiplication over abelian varieties. Am. J. Math. **110**(4), 765–773 (1988)

[Kem89a] Kempf, G.: Linear systems on abelian varieties. Am. J. Math. **111**(1), 65–94 (1989)

[Kem92] Kempf, G.: Equations of Kümmer Varieties. Am. J. Math. **114**(1), 229–232 (1992)

[Kem89b] Kempf, G.: Projective coordinate rings of abelian varieties. In: Algebraic Analysis, Geometry and Number Theory, pp. 225–236 (1989)

[Kem90] Kempf, G.R.: Some wonderful rings in algebraic geometry. J. Algebra **134**(1), 222–224 (1990)

[Kie20] Kieffer, J.: Evaluating modular polynomials in genus 2. arXiv: 2010.10094 [math.NT] (2020). hal-02971326

[Kie21] Kieffer, J.: Higher-dimensional modular equations, applications to isogeny computations and point counting. Thèse de doctorat dirigée par Damien Robert, Mathématiques Pures, Université de Bordeaux. Ph.D. thesis (2021). http://www.theses.fr/2021BORD0188

[Kie22a] Kieffer, J.: Counting points on abelian surfaces over finite fields with Elkies's method (2022). arXiv: 2203.02009 [math.NT]

[Kie22b] Kieffer, J.: Degree and height estimates for modular equations on PEL Shimura varieties (2022)

[KPR24] Kieffer, J., Page, A., Robert, D.: Computing isogenies from modular equations between Jacobians of genus 2 curves. J. Algebra (2024). arXiv: 2001.04137 [math.AG]

[KR22] Kieffer, J., Robert, D.: Fast evaluation of modular polynomials and compact representation of isogenies between elliptic curves (2022)

[KNRR21] Kirschmer, M., Narbonne, F., Ritzenthaler, C., Robert, D.: Spanning the isogeny class of a power of an elliptic curve. Math. Comput. **91**(333), 401–449 (2021). https://doi.org/10.1090/mcom/3672, arXiv: 2004.08315

[Koh96] Kohel, D.: Endomorphism rings of elliptic curves over finite fields. Ph.D. thesis. University of California (1996)

[KLPT14] Kohel, D., Lauter, K., Petit, C., Tignol, J.-P.: On the quaternionisogeny path problem. LMS J. Comput. Math. **17**(A), 418–432 (2014)

[Kun22] Kunzweiler, S.: Efficient Computation of $(2^n, 2^n)$-Isogenies. Cryptology ePrint Archive (2022)

[KR24] Kunzweiler, S., Robert, D.: Computing modular polynomials by deformation. In: ANTS XVI Conference (2024). https://antsmath.org/ANTSXVI/

[Kup05] Kuperberg, G.: A subexponential-time quantum algorithm for the dihedral hidden subgroup problem. SIAM J. Comput. **35**(1), 170–188 (2005)

[LS08]   Lercier, R., Sirvent, T.: On Elkies subgroups of ℓ-torsion points in elliptic curves defined over a finite field. J. théorie nombres Bordeaux **20**(3), 783–797 (2008)

[Ler22a]   Leroux, A.: A new isogeny representation and applications to cryptography. In: International Conference on the Theory and Application of Cryptology and Information Security, pp. 3–35. Springer (2022)

[Ler22b]   Leroux, A.: Quaternion algebras and isogeny-based cryptography. Ph.D. thesis. LIX (2022)

[Ler23a]   Leroux, A.: Computation of Hilbert class polynomials and modular polynomials from supersingular elliptic curves. Cryptology ePrint Archive (2023)

[Ler23b]   Leroux, A.: Verifiable random function from the Deuring correspondence and higher dimensional isogenies (2023)

[LR10]   Lubicz, D., Robert, D.: Efficient pairing computation with theta functions. In: Hanrot, G., Morain, F., Thomé, E. (eds.) ANTS 2010. LNCS, vol. 6197, pp. 251–269. Springer, Heidelberg (2010). https://doi.org/10.1007/978-3-642-14518-6_21

[LR12]   Lubicz, D., Robert, D.: Computing isogenies between abelian varieties. Compos. Math. **148**(5), 1483–1515 (2012). https://doi.org/10.1112/S0010437X12000243, arXiv:1001.2016 [math.AG]

[LR15a]   Lubicz, D., Robert, D.: A generalisation of Miller's algorithm and applications to pairing computations on abelian varieties. J. Symb. Comput. **67**, 68–92 (2015). https://doi.org/10.1016/j.jsc.2014.08.001

[LR15b]   Lubicz, D., Robert, D.: Computing separable isogenies in quasioptimal time. LMS J. Comput. Math. **18**, 198–216 (2015). https://doi.org/10.1112/S146115701400045X, arXiv:1402.3628

[LR16]   Lubicz, D., Robert, D.: Arithmetic on Abelian and Kummer Varieties. Finite Fields Appl. **39**, 130–158 (2016). https://doi.org/10.1016/j.ffa.2016.01.009

[LR22]   Lubicz, D., Robert, D.: Fast change of level and applications to isogenies. In: Research in Number Theory (ANTS XV Conference), vol. 9, no. 1 (2022). https://doi.org/10.1007/s40993-022-00407-9

[MM22]   Maino, L., Martindale, C.: An attack on SIDH with arbitrary starting curve. Cryptology ePrint Archive, Paper 2022/1026 (2022). https://eprint.iacr.org/2022/1026

[MMPPW23] Maino, L., Martindale, C., Panny, L., Pope, G., Wesolowski, B.: A direct key recovery attack on SIDH. In: Annual International Conference on the Theory and Applications of Cryptographic Techniques, pp. 448–471. Springer (2023)

[MW23]   Merdy, A.H.L., Wesolowski, B.: The supersingular endomorphism ring problem given one endomorphism. arXiv preprint arXiv:2309.11912 (2023)

[Mil20]   Milio, E.: Computing isogenies between Jacobians of curves of genus 2 and 3. Math. Comput. **89**(323), 1331–1364 (2020). arXiv:1709.06063

[MR19]   Milio, E., Robert, D.: Denominators of modular polynomials on Hilbert surfaces (2019)

[Mor23]   Moriya, T.: IS-CUBE: an isogeny-based compact KEM using a boxed SIDH diagram. Cryptology ePrint Archive (2023)

[Mor24]   Moriya, T.: LIT-SiGamal: an efficient isogeny-based PKE based on a LIT diagram. Cryptology ePrint Archive (2024)

[Mum66] Mumford, D.: On the equations defining abelian varieties. I. Invent. Math. **1**, 287–354 (1966)

[Mum67a] Mumford, D.: On the equations defining abelian varieties. II. Invent. Math. **3**, 75–135 (1967)

[Mum67b] Mumford, D.: On the equations defining abelian varieties. III. Invent. Math. **3**, 215–244 (1967)

[Mum83] Mumford, D.: Tata Lectures On Theta I. Progress in Mathematics, vol. 28, pp. xiii+235. Birkhäuser Boston Inc., Boston (1983). With the assistance of Musili, C., Nori, C., Previato, C., Stillman, M. ISBN 3-7643-3109-7

[Mum84] Mumford, D.: Tata Lectures on Theta II. Progress in Mathematics. Jacobian Theta Functions and Differential Equations, vol. 43, pp. xiv+272. Birkhäuser Boston Inc., Boston (1984). With the collaboration of Musili, C., Nori, M., Previato, E., Stillman, M., Umemura, H. ISBN 0-8176-3110-0

[Mum91] Mumford, D.: Tata Lectures on Theta III. Progress in Mathematics, vol. 97, pp. viii+202. Birkhäuser Boston Inc., Boston (1991). With the collaboration of Madhav Nori and Peter Norman. ISBN 0-8176-3440-1

[NO23] Nakagawa, K., Onuki, H.: QFESTA: efficient algorithms and parameters for FESTA using quaternion algebras. Cryptology ePrint Archive (2023)

[NO24] Nakagawa, K., Onuki, H.: SQIsign2D-east: a new signature scheme using 2-dimensional isogenies. In: Cryptology ePrint Archive (2024)

[Nic18] Nicholls, C.: Descent methods and torsion on Jacobians of higher genus curves. Ph.D. thesis. University of Oxford (2018)

[Oda69] Oda, T.: The first de Rham cohomology group and Dieudonné modules. Ann. Sci. l'École Normale Supérieure **2**(1), 63–135 (1969)

[Onu21] Onuki, H.: On oriented supersingular elliptic curves. Finite Fields Their Appl. **69**, 101777 (2021)

[OM22] Onuki, H., Moriya, T.: Radical isogenies on Montgomery curves. In: IACR International Conference on Public-Key Cryptography. Lecture Notes in Computer Science 13177, pp. 473–497. Springer (2022)

[ON24] Onuki, H., Nakagawa, K.: Ideal-to-isogeny algorithm using 2-dimensional isogenies and its application to SQIsign. Cryptology ePrint Archive (2024)

[PR23a] Page, A., Robert, D.: Clapotis: evaluating the isogeny class group action in polynomial time (2023)

[PR23b] Page, A., Robert, D.: Introducing Clapoti(s): evaluating the isogeny class group action in polynomial time (2023)

[PW24] Page, A., Wesolowski, B.: The supersingular endomorphism ring and one endomorphism problems are equivalent. In: Annual International Conference on the Theory and Applications of Cryptographic Techniques, pp. 388–417. Springer (2024)

[Pei20] Peikert, C.: He gives C-sieves on the CSIDH. In: Annual International Conference on the Theory and Applications of Cryptographic Techniques, pp. 463–492. Springer (2020)

[PT18] Pollack, P., Treviño, E.: Finding the four squares in Lagrange's theorem. Integers **18**, A15 (2018)

[Pri24] Pribanic, V.: Radical isogenies and modular curves. Adv. Math. Commun. **18**, 1748–1767 (2024)

[RS86] Rabin, M.O., Shallit, J.O.: Randomized algorithms in number theory. Commun. Pure Appl. Math. **39**(S1), S239–S256 (1986)

[RSSB16] Renes, J., Schwabe, P., Smith, B., Batina, L.: μKummer: efficient hyper-elliptic signatures and key exchange on microcontrollers. In: Gierlichs, B., Poschmann, A.Y. (eds.) CHES 2016. LNCS, vol. 9813, pp. 301–320. Springer, Heidelberg (2016). https://doi.org/10.1007/978-3-662-53140-2_15

[Ric36] Richelot, F.: Essai sur une méthode générale pour déterminer la valeur des intégrales ultra-elliptiques, fondée sur des transformations remarquables de ces transcendantes. C. R. Acad. Sci. Paris **2**, 622–627 (1836)

[Ric37] Richelot, F.: De transformatione Integralium Abelianorum primiordinis commentation. J. Reine Angew. Math. **16**, 221–341 (1837)

[Rob21] Robert, D.: Efficient algorithms for abelian varieties and their moduli spaces. HDR thesis. Université Bordeaux (2021). http://www.normalesup.org/~robert/pro/publications/academic/hdr.pdf. Slides: 2021-06-HDR-Bordeaux.pdf (1h, Bordeaux)

[Rob22a] Robert, D.: Evaluating isogenies in polylogarithmic time (2022)

[Rob22b] Robert, D.: Some applications of higher dimensional isogenies to elliptic curves (overview of results) (2022)

[Rob23a] Robert, D.: A note on optimising 2n-isogenies in higher dimension (2023). http://www.normalesup.org/~robert/pro/publications/notes/2023-06-optimising_isogenies.pdf

[Rob23b] Robert, D.: Breaking SIDH in polynomial time. In: Hazay, C., Stam, M. (eds.) EUROCRYPT 2023. LNCS, vol. 14008, pp. 472–503. Springer, Cham (2023). https://doi.org/10.1007/978-3-031-30589-4_17

[Rob23c] Robert, D.: The geometric interpretation of the Tate pairing and its applications (2023)

[RS06] Rostovtsev, A., Stolbunov, A.: Public-key cryptosystem based on isogenies. In: International Association for Cryptologic Research. Cryptology ePrint Archive (2006). http://eprint.iacr.org/2006/145

[Sch95] Schoof, R.: Counting points on elliptic curves over finite fields. J. Théor. Nombres Bordeaux **7**(1), 219–254 (1995)

[Sho99] Shoup, V.: Efficient computation of minimal polynomials in algebraic extensions of finite fields. In: Proceedings of the 1999 International Symposium on Symbolic and Algebraic Computation, pp. 53–58 (1999)

[Smi08] Smith, B.: Isogenies and the discrete logarithm problem in Jacobians of genus 3 hyperelliptic curves. In: Annual International Conference on the Theory and Applications of Cryptographic Techniques, pp. 163–180. Springer (2008). arXiv: 0806.2995 [math.NT]

[Stacks] T. Stacks Project Authors. Stacks Project (2018). https://stacks.math.columbia.edu

[Sut11] Sutherland, A.: Structure computation and discrete logarithms in finite abelian p-groups. Math. Comput. **80**(273), 477–500 (2011)

[Sut13a] Sutherland, A.: Isogeny volcanoes. Open Book Ser. **1**(1), 507–530 (2013)

[Sut13b] Sutherland, A.: On the evaluation of modular polynomials. Open Book Ser. **1**(1), 531–555 (2013)

[Tat67] Tate, J.T.: p-divisible groups. In: Proceedings of a Conference on Local Fields, pp. 158–183. Springer (1967)

[Tia20] Tian, S.: Translating the discrete logarithm problem on Jacobians of genus 3 hyperelliptic curves with (ℓ, ℓ, ℓ)-isogenies. arXiv: 2007.03172 [math.AG] (2020)

[Tia24] Tian, S.: Computing gluing and splitting (ℓ, ℓ)-isogenies. Designs Codes Cryptogr. 1–21 (2024)

[UJ18]   Urbanik, D., Jao, D.: SoK: the problem landscape of SIDH. In: Proceedings of the 5th ACM on ASIA Public-Key Cryptography Workshop, pp. 53–60 (2018)

[Vél71]   Vélu, J.: Isogénies entre courbes elliptiques. Compt. Rendu Acad. Sci. Paris Série A-B **273**, A238–A241 (1971)

[Voi21]   Voight, J.: Quaternion Algebras. Springer (2021)

[Wat69]   Waterhouse, W.: Abelian varieties over finite fields. Ann. Sci. Ecole Norm. Supp. **2**(4), 521–560 (1969)

[Wes22]   Wesolowski, B.: The supersingular isogeny path and endomorphism ring problems are equivalent. In: 2021 IEEE 62nd Annual Symposium on Foundations of Computer Science (FOCS), pp. 1100–1111. IEEE (2022)

[ZSPDB18]   Zanon, G.H., Simplicio, M.A., Pereira, G.C., Doliskani, J., Barreto, P.S.: Faster isogeny-based compressed key agreement. In: International Conference on Post-Quantum Cryptography. vol. 68, pp. 688–701. Springer, IEEE (2018)

[Zar74]   Zarhin, J.G.: A remark on endomorphisms of abelian varieties over function fields of finite characteristic. Math. USSR-Izv. **8**(3), 477 (1974)

Elliptic-Curve Factoring, Witnesses and Oracles

Jacek Pomykała[1,2]([✉]) [iD]

[1] Faculty of Cybernetics, Military University of Technology, Warsaw, Poland
jacek.pomykala@wat.edu.pl
[2] Faculty of Mathematics, Informatics, and Mechanics, University of Warsaw,
Warsaw, Poland
pomykala@mimuw.edu.pl

Abstract. The article concerns the selected approaches to integer factorization problem of positive integers n with the aid of elliptic curves E over $\mathbb{Z}_n$. It is focusing on the investigation of partially smooth numbers and their application to elliptic-curve factoring (ECF) approach by Lenstra and the ECF with oracle (ECFO), both in the probabilistic and deterministic approaches. We define the separating and non-separating decomposition witnesses for n, which play the significant role in both ECF and ECFO approaches and apply them for semiprime numbers $n = pq$ and admissible elliptic curves E over $\mathbb{Z}_n$.

Keywords: Integer factorization · Elliptic curves in residue rings · Fermat factoring · Frobenius traces modulo primes · Decomposition witnesses · Separating sets

2020 Mathematics Subject Classification: 11G05 · 11Y05 · 11M06 · 11N13 · 11N25

1 Introduction

It is widely believed that the integer factorization problem, that is a problem of finding a non-trivial factor of a positive integer n is a difficult computational question see [10, Section 5.5]. The best known rigorously proved deterministic algorithm is due to Harvey [24] with a slight improvement in [25], and runs in time $n^{1/5+o(1)}$, which builds on the $n^{2/9+o(1)}$-algorithm of Hittmeir [26]), which in turn improves the previous $n^{1/4+o(1)}$-algorithm of Pollard [49]. In fact, prior to [24], an $n^{1/5+o(1)}$-algorithm has also been known, however only conditional on the Generalised Riemann Hypothesis (GRH), (see [10, Section 5.5]).

There are also some distinguished methods of factoring special positive integers or implying the conditional, deterministic factorization algorithms and influencing the development of research in this field like Coppersmith method, smooth shifted prime $p - a$ methods, Fermat's factorization, quadratic sieve or the number field sieve method. The last gives heuristically the best (randomized) algorithm factoring arbitrary positive integer n. We refer to [10, Chapter 6] and [32]

© The Author(s), under exclusive license to Springer Nature Switzerland AG 2025
A. Dąbrowski et al. (Eds.): NuTMiC 2024, LNCS 14966, pp. 85–118, 2025.
https://doi.org/10.1007/978-3-031-82380-0_2

for an outline of these algorithms, see also [36] for some recent progress towards a rigorous version of the number field sieve. The particular cases $a = \pm 1$ were investigated in [49,61], and with the "arbitrary shift" $a = a_p$ in the interval $[-\sqrt{p}, \sqrt{p}]$ was the base of the elliptic-curve factorization algorithm [38,39].

The further progress towards the integer factorization problem comes from the investigation of polynomial time reductions of integer factoring, to computing the values of particular multiplicative functions $f(n)$ like Euler totient function $\varphi(n)$, sum power divisors function $\sigma_k(n)$ (see e.g. [5]) or $a_f(n)$ being the Fourier coefficients of eigen-form f of Hecke operators (see [2,7]).

It is known from the work of Miller [46] that under the assumption of GRH, computing the Euler function $\varphi(n)$ is deterministic polynomial time equivalent to computing a non-trivial factor of an integer n or proving that it is prime. Shoup [59, Section 10.4] has given an unconditional version of this reduction, however his algorithm is probabilistic.

For the progress towards the rigorous reduction in case when $a = \pm 1$ above we refer to [56,61,62] and [5] (where the additive combinatorics approach was used). It is worth noting that the almost prime numbers $n \in \mathbb{N}_s$ are in the particular interest in view of the application to cryptography (see [57]), where $s = 2$. The $s-$aspect is important in deterministic runtime complexity of factoring with Φ or $Dec\Phi$ oracles without the assumption of GRH, analyzed in [62] and [27]. The last work applies the algorithm (see Theorem 3.2 therein) that given n, $\Delta \leq n^{1/2}$ and $r < m < \Delta/\log^4 n$ such all prime divisors $p \mid n$ are in arithmetical progression $r \mod m$, it returns the nontrivial divisor of n in deterministic time $O\left(\sqrt{\frac{\Delta}{m}} \log^3 n\right)$.

Having at our disposal polynomially many values $f(m)$, $m < n$ of a (reasonable) multiplicative function f, the search for an algorithm that factors n has a natural interpretation of *oracle factoring*. Namely, given an oracle, which for every integer $m \geq 1$ outputs $f(m)$ we can factor n in polynomial polylog(n) time. We also recall, several other oracle factoring algorithms, such as

(i) a heuristic algorithm of Maurer [44] requiring certain $\varepsilon \log n$ oracle calls, which is based on the *elliptic curve factoring* algorithm of Lenstra [38,39],

(ii) probabilistic algorithm with the oracle being the order of elliptic curve over $\mathbb{Z}_n$ assuming GRH [30],

(iii) a rigorous algorithm of Coppersmith [8,9] requiring certain about $0.25 \log n$ oracle calls, which is based on an algorithm to find small solutions to polynomial congruences,

(iv) a deterministic subexponential algorithm of Źrałek with the Euler totient function oracle Φ [56,62].

(v) a rigorous algorithm with the oracle being the multiple M of an Euler totient function $\varphi(m)$ with $M \ll \exp\left((\log N)^{O(1)}\right)$ [54], which factors almost all positive integers $n \leq N$ in polynomial time polylog$N$,

(vi) a rigorous algorithm with the oracle being the multiple M of the order of elliptic curve $E : y^2 = x(x^2 - b)$ over $\mathbb{Z}_n$, which factors (in polylog(N) time) almost all positive integers $n \leq N$, with $M \ll N^{O(1)}$ or $M = N^{O(B)}$

depending on the oracle, where $B = polylog(N)$ is an upper bound for $b \geq 1$ above) [14],

see also [40, 41, 45, 48] and references therein.

Sutherland [60, Chapter 2.3] has designed a probabilistic factoring algorithm which uses an oracle that returns a multiple of a multiplicative order of integers modulo n, while the subexponential time deterministic reduction of integer factoring problem to discrete logarithm problem modulo composite (DLPC) was proved in [55]. Bach [1] established a similar result as in [60] for a discrete logarithm oracle in $\mathbb{Z}_n^*$.

The notions of *decomposition witness* in the classical $\mathbb{Z}_n^*$ approach to factoring with Φ oracle, called (see [27, 54, 62]) Miller-Rabin or Fermat-Euclid witnesses is very convenient in the investigation of reduction of factoring problem to computing the values of $\varphi(m)$, since it allows to enter into research the distribution of the Dirichlet character nonresidues $LN(\chi)$ modulo n (i.e. the least positive integer k such that $\chi(k) \notin \{0, 1\}$) and the application of large sieve machinery (see [3] and the references therein). Let $r \mid p - 1$ for prime r and p. The number $a \in \mathbb{Z}$ is called $r-$th power residue or nonresidue modulo p according to whether the congruence $x^r = a \mod p$ is solvable or not. Any $a \in \mathbb{Z}$, coprime to p is $r-$th power nonresidue modulo p if and only if $a^{\frac{p-1}{r}}$ is not congruent to 1 modulo p. This means that a is the Dirichlet character $\psi := (\chi^*)^{\frac{p-1}{r}}$ nonresidue, where χ^* is a generator of the dual group of $\mathbb{Z}_p^*$.

For a square-free, odd, composite number n such that $r \mid \varphi(n)$ we call the positive integer a the "separating witness for n" if a is $r-$th power nonresidue modulo some $p \mid n$ and $r-$th power residue modulo some prime $q \mid n/p$. Denote the least such a by $LS(r, n)$. Then n can be factored nontrivially in deterministic time $LS(r, n)(\log n)^{O(1)}$. This was applied in [17] for the deterministic reduction of factoring almost all square-free positive integers $n \leq N$ to computing the number of points of superelliptic curves C over $\mathbb{Z}_n$ given by the equations $C : y^r = x^3 + b_1 x + b_2$.

The probabilistic approach for the oracle which returns the exponents of points of elliptic curve E over residue class ring $\mathbb{Z}_n$ was investigated in [43] with the related error probability $184/225$ (improved to $53/80$ in [18]). In the last work also the conditional factorization of almost all $n \in \mathbb{N}_s$ was given.

The particular interest concerns the group structure of the elliptic curve E over $\mathbb{Z}_n$ ($r = 2$) and the oracles are either $OrdEll$ returning the values $E_n := \prod_{p\mid n}(p + 1 - a_E(p))$ or $MultEll(M)$ that returns the multiple of E_n not exceeding M, where $a_E(p)$ is the related Frobenius trace of E modulo p. Here the notion of Dirichlet character and elliptic-curve $\mathcal{P}-$signatures play the significant role, where $\mathcal{P}$ is the set of prime divisors of n. The first is defined by $S_\chi(\mathcal{P}) = S_\chi(n) := (\chi(p))_{p \in \mathcal{P}}$, while the second by $S_E(\mathcal{P}) = S_E(n) := (a_E(p))_{p \in \mathcal{P}}$ with the primes ordered in increasing order. Of special interest is the quadratic Dirichlet character defined by the Kronecker symbol $\chi_d = (\frac{d}{\cdot})$, where d is coprime to n, since then we can define the integer values $\rho(S_\chi) := \rho_\chi(n)$ equal to $\prod_{p\mid n} \chi(p)$ and $\rho(S_E) := \rho_E(n)$ equal to $\min_{p\mid n} \gcd(p + 1, a_E(p))$. Having the upper bounds, D for the least d with the suitable signature satisfying $\rho_{\chi_d}(n) = -1$ and Δ for

$\rho_E(n)$ we can factor n in deterministic/randomized time $D\Delta polylog(n)$, provided the related oracles are available for us.

The investigation of Dirichlet character signatures $S_\chi(\mathcal{P})$ and elliptic curve signature $S_E(\mathcal{P})$ was the aim of work done in [11–13] and [50,51] for the application of Dirichlet characters and large sieve inequalities in investigation of large modular subgroups generated by small integers sets. This research is not directly applied in this survey article, but since it is related to the asymptotic and rigorous approach of factorization (see Sects. 6,7,8), we state some results in this direction in the next section.

In the article we focus on probabilistic and deterministic (also the rigorous for almost all $n \in \mathbb{N}^*$) oracle factorization in terms of Δ and D described above. Let $E := E_{\bar{b}}$ be an elliptic curve over $\mathbb{Z}_n$ given by the Weierstrass equation $y^2 = x^3 + b_1 x + b_2$, where $\bar{b} = (b_1, b_2) \in \mathbb{Z}_n^2$ and $E_{\bar{b}}^d$ be the twisted curve given by the equation $y^2 = x^3 + b_1 d^2 x + b_2 d^3$. Referring to the above notation the condition $\rho_\chi(n) = -1$ (or in the asymptotic case for all $n \leq N$ with $o(N)$ exceptions, as N tends to infinity) is applied. It allows differentiation of values $\nu_2(\mathrm{ord}Q_p)$ and $\nu_2(\mathrm{ord}Q_q)$ of the related reductions of point $Q \in E(\mathbb{Z}_n)$ modulo p and q respectively, lying on the twisted curve $E^d := E_{(-b,0)}^d : y^2 = x(x^2 - bd^2)$ over $\mathbb{Z}_n$ in the rigorous approach [14].

On the other hand in the probabilistic approach we apply GRH to find the signature S_{χ_d} having exactly one -1 term as coordinate. Then applying the bound $\rho_E(n) \leq \Delta$ we deduce that the quotient $E_n/(E^d)_n$ in irreducible form differs from $(p+1-(d/p)a_E(p))/(p+1+(d/p)a_E(p))$ by the factor $\leq \Delta$ for some prime $p \mid n$, giving the randomized polynomial time complexity of the related factoring algorithm whenever $D\Delta = (\log N)^{O(1)}$ [30].

In contrast to the classical deterministic approach (with the group $G = \mathbb{Z}_n^*$) and the small witnesses $b \leq B$, the *elliptic decomposition witnesses* are defined as affine points $P = (x, y)$ lying on the elliptic curve E over $\mathbb{Z}_n$. Moreover focusing on the general factoring may consider the elliptic curves given by the Weierstrass equation $y^2 = (x - b_1)(x - b_2)(x - b_3)$ over $\mathbb{Z}_n$ and consider the partially smooth numbers orders of reduced elliptic curves $E(\mathbb{Z}_r)$ (see [18,52,53]). Concluding one can regard the elliptic *decomposition witnesses* as the useful tool joining the oracle factoring approach with the factoring without oralce, taking into account the partially smooth numbers. However unlike the classical approach here we will consider additionally the so called *nonseparating decomposition witnesses* $P \in E(\mathbb{Z}_n)$ [52,53].

The work is organized as follows. We begin with the basic facts on the elliptic curves over rationals and elliptic curves over residue ring $\mathbb{Z}_n$. Moreover we state the relevant results on Dirichlet characters and elliptic curves signatures. Furthermore we explain the extension of Fermat factoring for elliptic curves and state the relevant results.

Next we pass towards the factorization with oracles and introduce the suitable counting functions emerging in the rigorous approach. We present the ideas and some details on the probabilistic factoring with *OrdEll* oracle. Next we state the results on rigorous polynomial time asymptotic factoring with *MultEll* oracle supported by Sato-Tate [23] conjecture and then with the aid of the shifted sieve (see [29,58]) under GRH. At last we state the suitable asymptotic result, unconditionally.

Finally we focus on the factoring without an oracle. We define the *decomposition* (separating and nonseparating) *elliptic witnesses* and *admissible elliptic curves* over $\mathbb{Z}_n$. At the end we state the main results on deterministic time factorization both with *separating* and *nonseparating witnesses* and on the relevant *separating sets*.

Notation:

- Throughout the article we assume that $n \leq N$ if not stated otherwise
- $\omega(m)$, $\Omega(m)$ stand for the number of all distinct prime divisors and all prime divisors counting with multiplicity of positive integer m, respectively
- $\mathbb{N}^*$ stands for the set of square-free positive integers
- For positive integer n we denote by $\nu_2(m)$ the highest exponent in which 2 divides m
- $\mathbb{N}^*(N)$ and $\mathbb{N}^*(N,y)$ denote the set of numbers $n \in N^*$ not exceeding N and additionally free of prime divisors $\leq y$, respectively
- $\mathbb{N}_s \subseteq \mathbb{N}^*$ stands for the set of square-free positive integers having exactly s prime factors with the similar convention concerning $\mathbb{N}_s(N)$ and $\mathbb{N}_s(N,y)$, respectively
- $\mathcal{N} \subseteq \mathbb{N}^*$ is the subset of numbers coprime to 6 with the similar convention concerning $\mathcal{N}_s(N)$ and $\mathcal{N}_s(N,y)$, respectively
- For a given $B \geq 2$ by $s_B(m)$ we denote the largest divisor of m having all prime factors not exceeding B
- Throughout the article the symbol $o(1)$ refers to n or N tending to infinity
- $P^{\pm}(n)$ stands for the largest and smallest prime divisor of n
- $\pi(x)$ stand for the counting function of primes not exceeding x
- For elliptic curve E over rationals we denote by $a_E(n)$ the n-th coefficient of the related Hasse-Weil L-function
- For a square-free positive integer n and elliptic curve $E_{\bar{b}}$ over $\mathbb{Q}$ given by the short Weierstrass equation $y^2 = x^3 + b_1 x + b_2$, where $\bar{b} := (b_1, b_2) \in \mathbb{Z}^2$ we denote by $(E_{\bar{b}})_n$ the product $\Pi_{p|n} \# E_{\bar{b}}(\mathbb{Z}_p)$
- Throughout the paper we use the relevant notations polylog(n) (polylog$N$) and $(\log n)^{O(1)}$ $((\log N)^{O(1)})$ alternatively.

2 Background on Elliptic Curves over Residue Class Ring $\mathbb{Z}_n$

In this section we recall basic facts on elliptic curves over the ring $\mathbb{Z}_n$, where $n = \prod_{i=1}^s p_i \in \mathcal{N}_s$ (see [38,39]). The projective plane $\mathbb{P}^2(\mathbb{Z}_n)$ is defined to be the set of equivalence classes of primitive triples in $\mathbb{Z}_n^3$ (i.e., triples (x_1, x_2, x_3) with $\gcd(x_1, x_2, x_3, n) = 1$) with respect to the equivalence $(x_1, x_2, x_3) \sim (y_1, y_2, y_3)$ iff $(x_1, x_2, x_3) = u(y_1, y_2, y_3)$ for a unit $u \in \mathbb{Z}_n^*$. The equivalence class in $\mathbb{P}^2(\mathbb{Z}_n)$

represented by a point (x_1, x_1, x_3) is denoted by $(x_1 : x_2 : x_3)$. The subset $\{(x : y : 1), x, y \in \mathbb{Z}_n\}$ is identified with $\mathbb{Z}_n^2$ and is also called the set of finite points.

Let $E := E_{\bar{b}}$ over $\mathbb{Z}_n$ (denoted shortly by $E(\mathbb{Z}_n)$) be the set of all points in $\mathbb{P}^2(\mathbb{Z}_n)$ satisfying the homogeneous equation $y^2 z = x^3 + b_1 x z^2 + b_2 z^3$ of E, where $\Delta_{\bar{b}}$ is coprime to n. The point $O = (0 : 1 : 0)$, called the zero point belongs to $E(\mathbb{Z}_n)$. Let

$$V(E(\mathbb{Z}_n)) = \{(x, y) \in E(\mathbb{Z}_n)\} \cup \{O\}$$

be the set of finite points in $E(\mathbb{Z}_n)$ with the zero point O. Then for each point $(x : y : z) \in E(\mathbb{Z}_n) \setminus V(E(\mathbb{Z}_n))$ the $\gcd(z, n)$ is a nontrivial divisor of n.

More precisely for a prime divisor p_i, of n $(1 \le i \le s)$, the reduction of E modulo p_i induces a map $E(\mathbb{Z}_n) \to E(\mathbb{Z}_{p_i})$ and the induced mapping

$$\phi : E(\mathbb{Z}_n) \to E(\mathbb{Z}_{p_1}) \times \ldots \times E(\mathbb{Z}_{p_s}) \tag{1}$$

is a bijection by the Chinese remainder theorem. The set $E(\mathbb{Z}_n)$ is a group with the addition for which ϕ is a group isomorphism, which in general can defined using the so-called complete set of additions on E [39]. For our goal it is sufficient for addition of points in $E(\mathbb{Z}_n)$ to use addition formulas as over the field $\mathbb{Z}_p$.

Recall that for an elliptic curve $E : y^2 = x^3 + b_1 x + b_2$ over the field $\mathbb{Z}_p$, where $p > 3$, the set $E(\mathbb{Z}_p)$ of $\mathbb{Z}_p$-rational points on E with the addition of points given by the chord and tangent rule is an abelian group with the neutral element $O = (0 : 1 : 0)$ and the opposite point $-P = (x, -y)$ of $P = (x, y) \in E(\mathbb{Z}_p)$. Thus exactly P and $-P$ on E have the same x-coordinate. The sum of two non-zero points $P, Q \in E(\mathbb{Z}_p)$, where $Q \ne -P$, is given as follows. Let

$$\lambda = \frac{y_Q - y_P}{x_Q - x_P} \quad \text{if } Q \ne P, \tag{2}$$

and

$$\lambda = \frac{3x_P^2 + a}{2y_P} \quad \text{if } Q = P \tag{3}$$

be the coefficient of the line $y = \lambda(x - x_P) + y_P$ passing through P and Q and the tangent if $P = Q$, respectively. The sum $P + Q$ is given by

$$\begin{cases} x_{P+Q} = \lambda^2 - x_P - x_Q, \\ y_{P+Q} = \lambda(x_P - x_{P+Q}) - y_P. \end{cases} \tag{4}$$

Let E be an elliptic curve over $\mathbb{Z}_n$. For two finite points $P, Q \in E(\mathbb{Z}_n)$ let $(P_1, \ldots, P_s)$ and $(Q_1, \ldots, Q_s)$ be the images of P and Q in $E(\mathbb{Z}_{p_1}) \times \ldots \times E(\mathbb{Z}_{p_s})$ by the isomorphism (1). If we try to compute the sum $P + Q$ or doubling $2P$ using the above formula (4) over a field, we either compute the sum in $V(E(\mathbb{Z}_n))$ or obtain a nontrivial divisor of n. We have the following cases.

Doubling:

(i) $y_P \in \mathbb{Z}_n^* \iff 2P_i \ne O$ for all $1 \le i \le l$. Then $2P$ is a finite point and can be computed using (4) with λ satisfying (3).

(ii) $\gcd(y_P, n)$ is a nontrivial divisor of n $\iff$ $2P_i \neq O$ and $2P_j = O$ for some i, j.

(iii) $y_P = 0$ $\iff$ $2P = O$.

Addition:

(iv) $x_Q - x_P \in \mathbb{Z}_n^*$ $\iff$ $P_i \neq \pm Q_i$ for each i. Then $P + Q$ is finite and can be computed using (4) with λ satisfying (2).

(v) $\gcd(x_P - x_Q, n)$ is a nontrivial divisor of n $\iff$ $P_i \neq \pm Q_i$ and $P_j = \pm Q_j$ for some $1 \leq i, j \leq s$.

(vi) $x_P = x_Q$ $\iff$ $P_i = \pm Q_i$ for all i. Then $y_Q^2 = y_P^2$. If $P \neq \pm Q$, then $\gcd(y_Q - y_P, n)$ is a nontrivial divisor of n. (If $P = Q$, we have doubling.)

From the above it is easy to see that if for the finite points $P, Q \in E(\mathbb{Z}_n)$ we have $P_i + Q_i \neq O$ and $P_j + Q_j = O$ for some $i \neq j$, then computing $P + Q$ we obtain a nontrivial divisor of n. If $P = Q$, then $\gcd(y_P, n)$ is a nontrivial divisor of n in view of (ii). If $P \neq Q$, then $\gcd(x_P - x_Q, n)$ or $\gcd(y_P - y_Q, n)$ is a nontrivial divisor of n in view of (v) or (vi) above.

In what follows we assume that $B = B(N) < N$ tends to infinity together with N significantly slower than N but not faster than some power of $\log N$ and we try to compute the point multiplication $mP \in E(\mathbb{Z}_n)$ by a B−smooth scalar m. Applying the above formulas the computation of finite point $mP \in E(\mathbb{Z}_n)$ takes $O(\log m)$ additions in $E(\mathbb{Z}_n)$. By the Hasse bound, we can restrict ourselves to computing a multiple point $m_B P \in E(\mathbb{Z}_n)$ for B−smooth number $m_B = p_k^{e_k} \ldots 3^{e_3} 2^{e_2}$, where $p_i^{e_i} \leq (\sqrt{n}+1)^2$ for $1 \leq i \leq k$. Then during the computation of $m_B P$ we either find the nontrivial divisor of n or compute $m_B P$ in

$$\ll \log n \sum_{i \leq k} e_i \ll \left(\frac{B}{\log B} \right) \log N \ll B^{1+o(1)}$$

adding operations in $E(\mathbb{Z}_n)$.

Remark 1. Assume that $E = E_{\bar{b}}$, where $\Delta_{\bar{b}} \in \mathbb{Z}_n^*$, $P \in E(\mathbb{Z}_n)$ and for any prime $p \mid n$ the point $P_p \in E(\mathbb{Z}_p)$ is the point P reduced modulo p. Then during the computation of $m_B P \in E(\mathbb{Z}_n)$ we either factor n nontrivially or prove that all local orders $\operatorname{ord} P_p$ for $p \mid n$ are identical and can be computed them in deterministic time $B^{2+o(1)}$, as N tends to infinity. In section 12 we state the results on the heuristic subexponential time of factoring $n \leq N$ under the assumption of analogous conjecture as in [39], on partially smooth local orders of elliptic curves E over $\mathbb{Z}_n$ in the Hasse interval.

Let

$$L(s, E) = \prod_{p \mid \Delta_E} \left(\frac{1}{1 - a_E(p)p^{-s}} \right) \prod_{(p, \Delta_E)=1} \left(\frac{1}{1 - a_E(p)p^{-s} + p^{1-2s}} \right).$$

The series is convergent for $\mathrm{Re}(s) > 3/2$. It has the analytic continuation to the whole complex plane and the function $L(s, E) = \sum_{n=1}^{\infty} a_E(n) n^{-s}$ is called the

Hasse-Weil L-function of elliptic curve E over rationals. It satisfies the following functional equation

$$q^s \Gamma(s) L(s, E) = wq^{2-s} \Gamma(s-s) L(2-s, E),$$

where $w = \pm 1$ and q is called the conductor of elliptic curve E. We use the values of $a_E(n)$ as the querries for the oracle $\mathcal{O} = FrobEll$.

3   Dirichlet Characters and Elliptic Curves Signatures

For a given set of primes $\mathcal{P} = \{p_1, \ldots, p_m\}$, where $p_1 < p_2 < \ldots < p_m$, positive integer q and a Dirichlet character $\chi \mod q$ of order $r \mid \varphi(q)$, or an elliptic curve E over the rationals of conductor q we consider the signatures $S_\chi(\mathcal{P})$ and $S_E(\mathcal{P})$ defined by

$$S_\chi(\mathcal{P}) = (\chi(p_1), \ldots, \chi(p_m))$$

and

$$S_E(\mathcal{P}) = (a_{p_1}(E), \ldots, a_{p_m}(E)),$$

where a_{p_i} are the traces of $E \mod p_i$, and $p_i \in \mathcal{P}$ for $i = 1, 2 \ldots, m$.

The related $\mathcal{P}$-pattern $S = (s_{p_1}, s_{p_2}, \ldots, s_{p_m})$ consists of s_{p_i} being the complex roots of unity or zero and satisfies the periodicity condition $s_{p+kq} = s_p$ for Dirichlet character signature, or Hasse's condition $|s_p| < 2\sqrt{p}$ for elliptic curve E respectively, where $p \in \mathcal{P}$. In case when χ is the quadratic Dirichlet character we say that the related signature (pattern) is primitive if $S_\chi(\mathcal{P}) \in \{\pm 1\}^{|\mathcal{P}|}$. If $\mathcal{P}$ is the set of primes not exceeding B then the we write simply $S_\chi(\mathcal{P}) = S_\chi(B)$ and $S_E(\mathcal{P}) = S_E(B)$, respectively. Below we discuss the related distribution of Dirichlet characters and elliptic curves conductors with respect to the given pattern $S = S(B)$.

Definition 1. *Fix a positive integer r. A positive integer q is called*

(i) *(S, B, r)-character conductor if there exists a Dirichlet character $\chi \mod q$ of order r such that $S_\chi(B) = S$, where $q \equiv 1 \mod r$.*

(ii) *$(S, B, \mathcal{E})$-elliptic conductor of a given family of elliptic curves $\mathcal{E}$, if $S_E(B) = S$ for some elliptic curve $E \in \mathcal{E}$ of conductor q.*

In what follows we assume that $r \geq 2$ is fixed. We call the positive square-free integer n, (t, x)-balanced if it has t prime divisors and all they are in the intervall $(\frac{1}{2}x^{1/t}, x^{1/t}]$. Moreover if all prime divisors of q are congruent to 1 modulo r we call it (r, t, x)-balanced number. Let Q be a sufficiently large real number, r, s, t, $(s < t)$ be positive integers and $0 < \beta = \beta(Q) < 1$. In what follows we denote by $\mathbb{Q}(r, t, Q)$ the set of all (r, t, Q)-balanced positive integers. By $\mathbb{Q}(\beta; r, s, t, Q)$ we denote the set of distinct products $q = q'q''$ where $q' \in \mathbb{Q}(r, s, Q^\beta)$ and $q'' \in \mathbb{Q}(r, t-s, Q^{1-\beta})$. For arbitrary $\varepsilon > 0$ let us define the following constant:

$$L_0(\varepsilon) = \limsup_{P^+(q)=q^\varepsilon} \max_{a \mod q} \frac{\log P(q, a)}{\log q},$$

and $L_0 = \limsup_{\varepsilon \to 0} L_0(\varepsilon)$. It is known (see [6, Corollary 11]) that $L_0 < 12/5 + \delta$, for arbitrary $\delta > 0$.

3.1 Bounds for the Cardinality of the Set of Conductors Realizing the Fixed Pattern S

Below we state the result on the lower bound for (S, B, r)-character conductors $q \leq Q$, with the fixed pattern S of some primitive Dirichlet character χ of order r, where $B = B(Q) \leq \log Q$ is specified below. Namely we have

Theorem 1. *(see Theorem 2.1 [12])*

Let $Q > e^{e^c}$, $1 < \theta < 2$, $\frac{(\log \log Q)^{\theta}}{\log Q} < \beta = \beta(Q) < 1 - \frac{1}{\log \log \log Q}$ and $r \geq 2$ be a fixed positive integer. Moreover let $s < t$ be positive integers, where $t = t(Q)$, and $B = B_{\beta}(s, t, r, L_0) \leq \log Q$ be defined by: $B = \frac{\beta}{L_0} \log Q$ if $r = 2$ and if $r \geq 3$; $B = \log Q^{1-\beta}$ if $r \geq 3$ and r divides $t - s$; $B = \frac{2}{t-s} \log Q^{1-\beta}$ if r does not divide $t - s$ but 2 divides $t - s$. Then we have:

(i) Let $r = 2$ and $t = t(Q) = o((\log \log Q)^{\theta - 1})$. Given arbitrary $S = S(B) \in \{\pm 1\}^{\pi(B)}$ we have the inequality

$$\#\{q \in \mathbb{Q}(\beta; 2, 1, t, Q) : q \text{ is } (S, B, 2) - \text{character conductor}\} \geq Q^{1 - \beta(1 - \frac{1}{L_0}) + o(1)}$$

as $Q \to \infty$.

(ii) Let $r \geq 3$, $t = t(Q) < (1 + o(1))(\log \log \log Q)$. Assume that there exists a primitive Dirichlet character χ' mod q' such that $S_{\chi'}(B) = S$ for some (r, s, Q^{β})-balanced conductor q'. Then the number of (S, B, r)-primitive characters conductors q belonging to $\mathbb{Q}(\beta; r, s, t, Q)$ is $\geq Q^{(1-\beta)(1 - \frac{1+o(1)}{\log \log \log Q})}$, as $Q \to \infty$. Moreover if the condition

$$\left| \frac{\beta(t+1) - s(1+\beta)}{s(t-s)} \right| > \frac{\log 2}{\log Q}, \tag{5}$$

holds true, then we can relax the restriction on t as in case (i).

(iii) Let $B = \beta \log Q$, where $0 < \beta < 1/3$ is fixed. Then there exist a sequence of consecutive primes $\leq B$ not dividing r and a Dirichlet character χ mod q of order r, such that $S_{\chi}(B) = S$. Moreover

$$\#\{q \leq Q : q \text{ is } (S, B, r) - \text{character conductor}\} \geq Q^{1 - \beta + o(1)}$$

as $Q \to \infty$. Here and above the constant implied by by the symbol $o(1)$ depends on r.

Remark 2. The lower bound for the number of (S, B, r)-primitive characters conductors in Theorem 1 implies the same lower bound for the number of the related primitive characters ψ mod q.

The following result gives the lower bound for the set of conductors $N(S, Q)$ counting the positive numbers of $q \leq Q$ for which there exist an elliptic curve $E = E_{\bar{b}}$ of conductor q with the given $\mathcal{P}$-pattern S, in terms of the product $P := \prod_{p \in \mathcal{P}} p$. We have

Theorem 2. *(see Theorem 2.3 [13]) Let ε be arbitrary (sufficiently small) positive constant. Assume that $P \leq \varepsilon Q^{1/3}$ and $S = S(\mathcal{P})$ be realizable by some elliptic curve $E = E_{\bar{b}}$ of minimal discriminant $|\Delta_E| \leq \varepsilon Q$. Then we have the lower bound*

$$N(S,Q) \gg_\varepsilon \frac{Q^{5/6-\alpha}}{P^2},$$

where $\alpha = \frac{1}{3\sqrt{3}\log(3)}\left((2+\sqrt{3})\log\frac{2+\sqrt{3}}{2\sqrt{3}} - (2-\sqrt{3})\log\frac{2-\sqrt{3}}{2\sqrt{3}}\right) = 0.1688\ldots$. *Here $\gg_\varepsilon$ stands for the Vinogradov's symbol with the implied constant depending on ε.*

This means that there are many elliptic curves $E = E_{\bar{b}}$ of the bounded conductor $N_E \leq Q$ and the fixed signature $S_E(\mathcal{P}) = S$.

3.2   Bounds for the Cardinality of the Set of Conductors Realizing Almost Constant B−signatures

In what follows we consider the Dirichlet characters modulo $q \leq Q$ of order $r \mid \varphi(q)$ and elliptic curves $E = E_{\bar{b}}$ over $\mathbb{Q}$ of conductors $q \leq Q$. Let $A < B$ and $\mathcal{P}$ be the set of all primes $p \in [A, B]$, $LN(\chi)$ be the least character χ nonresidue and $LN(q,r) := \max_\chi LN(\chi)$, where χ runs over all Dirichlet characters modulo q of order $r \mid \varphi(q)$.

Definition 2. *Let a be a fixed positive integer. The signature $S_\chi(\mathcal{P})$ is called constant signature if $\chi(p) = 1$ for all prime $p \in \mathcal{P}$ coprime to q. Let $I = [A, B]$, $\mathcal{P} \subset I$, $0 < \delta < 1$ and a be a fixed positive integer. The signature $S_E(\mathcal{P})$ is called (a, I, δ)-constant if $|a_E(p)| \leq (1-\delta)\sqrt{A}$ for all $p \in \mathcal{P}$ and the set $\{a_E(p), p \in \mathcal{P}\}$ has cardinality a.*

Let $0 < \delta < 1$ be fixed, $\mathcal{P}$ be the set of all primes of the interval $I = [A, B]$ and $N_{ell}(d, I, \delta, Q)$ be the set of all positive integers $q \leq Q$ for which there exists an elliptic curve E of conductor q having a-constant signature $S_E = S_E(\mathcal{P})$ for some positive integer $a \leq d$. We have the following result.

Theorem 3. *([11, 13]) Let $d < (\log Q)^\Delta$, $I = [A, B]$, where $B = (\log Q)^\beta$, $B - A \geq B^\vartheta$, $\beta > 8$, $0 < \Delta < \frac{\beta}{8} - 1$, and $\frac{7}{8} + \frac{1+\Delta}{\beta} < \vartheta < 1$. Then there exists $\varepsilon = \varepsilon(\delta, \Delta, \vartheta) > 0$ such that*

$$N_{ell}(d, I, \delta, Q) \ll Q^{1-\varepsilon}.$$

Now we fix r and consider the positive integers $n \in \mathcal{N}_s^r(x, y)$ composed of primes $\equiv 1 \mod r$ and having s prime divisors (where $r \mid s$) that are $\leq x$ and free of prime divisors $< y$. Let $z = c\log x$ for some positive constant c.

Below we state the lower bound for the set $N_{char}(r, z; x, y)$ of numbers $n \in \mathcal{N}_s^r(x, y)$ for which there exist a Dirichlet character χ modulo n of order r (where $r \mid \varphi(n)$) with $LN(\chi) \geq z$ (that is $LN(n, r) \geq z$). In this connection for every prime $p \equiv 1 \mod r$ let $g_p(p)$ stand for the least primitive root   mod p and define (a unique) Dirichlet character $\chi_p = \chi \mod p$ of order r by the condition

$$\chi_p(g_p) = e^{2\pi i/r}. \tag{6}$$

For given n we define the Dirichlet character χ modulo n and order dividing r as follows

$$\chi := \chi_n := \prod_{p|n} \chi_p.$$

Let $\mathcal{P} = \mathcal{P}_z$ be the set of all primes not exceeding z. Then the characters χ_n have the least nonresidue $\geq z$ provided $r \mid s$. More precisely we have the following lower bound for $N_{char}(r, z; x, y)$.

Theorem 4. *(Theorem 7.6 [18]) Let $r \geq 2$, $y \leq x^{1/r}$, c be arbitrary positive constant and $z = c \log x$. Then there exists some positive constants $c_1 = c_1(c)$ and $c_2 = c_2(c)$ such that for any positive integer $s \equiv 0(\mod r)$, satisfying the condition:*

$$s \leq \min \left(\log x / \log(3y), c_1 \log \log x / \log \log \log x \right) \tag{7}$$

the following estimate holds true

$$N_{char}(r, z; x, y) \gg x^{1 - c_2 s \log s / \log \log x}, \tag{8}$$

provided $x > x_0(c_2)$.

4   Elliptic-Curve Analog of Fermat's Factorization

In this section we will focus on the particular case $n = pq \in \mathbb{N}_2$ of semiprime number. The classical Fermat's factoring method allows to efficiently factor the positive integer N provided $n = n_1 n_2$, where the absolute value of $n_1 - n_2$ is of order of magnitude $n^{1/4}$, since then we are able to find the related divisor close to $\sqrt{pq} = \sqrt{n}$. In this connection we consider the linear forms of type $F^{\pm} := F^{\pm}(a, b) = ap \pm bq$, where the coefficients a, b are related to the given elliptic curves $E(\mathbb{Z}_q)$ and $E(\mathbb{Z}_p)$ respectively (cf. [18]). It is worth adding here that the forms of this type $ap + bq$ with arbitrary integer coefficients a, b were considered in [35] and in [37] the factorization of n in deterministic time $O(n^{1/3 + o(1)})$ was proved.

   Let $E_p := p + 1 - a_E(p)$, $E_q := q + 1 - a_E(q)$ and $E_n := E_p E_q$. Letting $\{a, b\} = \{E_p, E_q\}$ we see that $\sqrt{nE_n} := \sqrt{(pE_q)(qE_p)}$ plays the role of $\sqrt{n}$ in the classical case. Namely by the geometric mean inequality we obtain that $ap + bq \leq \sqrt{abpq} = \sqrt{nE_n}$. Hence letting Δ be the difference between $ap + bq$ and $\sqrt{nE_n}$ the decomposition $n = pq$ can be found in deterministic time $polylog(n) \frac{\Delta^2}{(nE_n)^{1/4}}$. Separating the main part of $pE_q - qE_p$ and applying the triangle inequality one easily deduce the case (i) of the theorem below. The case (iii) follows from the Coppersmith factorization method [8,9] (cf. [16,53]). The case (ii) follows from the following

Lemma 1. *Let* $n = pq$, *where* $q \in [Mp, 2Mp]$, $a_E(q) > 1$ *and* $1 < d \mid \gcd(E_p, E_q)$. *If for some* $c > 0$ *we know the values of* n, E_n *and* $d \neq s \mid d^2$ *such that*

$$\left| \frac{p}{q} - \frac{E_p}{E_q} \left(\frac{d}{s} \right)^2 \right| < \frac{c}{D^2},$$

then p *and* q *can be computed in deterministic time* $polylog(n)(D^{-4}nM^2)$ *(which is* $< n^{1/2+o(1)}$ *provided* $D > n^{1/8}M^{1/2}$*).*

The following theorem reduces decomposition $n = pq$ to the knowledge about E_n and the approximation of $\frac{p}{q}$ by $\frac{E_p}{E_q}$ or $\frac{a_p(E)}{a_q(E)}$, respectively. More precisely we have

Theorem 5. *(cf. Theorem 3 [53]) Let* $n = pq$, *where* p, q *are sufficiently large,* $q \in [Mp, 2Mp]$, *where* $M < n^{1/4}$ *and* E_n *be given. Then we have*

(i) *Let* $A = \{a_p(E), a_q(E)\}$. *Assume that for some* $D = D(n) \geq 1/2$, *some* $a \in A$ *and* $b = A \setminus \{a\}$ *we have*

$$\left| \frac{p}{q} - \frac{a}{b} \right| \leq \frac{1}{D} \tag{9}$$

Then we can factor n *in deterministic time*

$$t = t(M, D, a_p(E), a_q(E)) = (\log n)^{O(1)} \left(M \left(1 + \left(\frac{|a_p(E)| + |a_q(E)|}{D} \right)^2 \right) \right). \tag{10}$$

(which is $< M \, polylog(n)$, *provided* $D > (nM)^{1/4}$*).*

(ii) *Let* $d \mid \gcd(E_p, E_q)$ *be given. Let* n, E_n *and* $d \mid \gcd(E_p, E_q) > 1$ *be given. Assume that for some* α *and* $|\theta_1| \leq 1$, $s \neq d$, $s \mid d^2$, *the following condition*

$$\left(\frac{s}{d} \right)^2 = 1 + \frac{1}{D} + \left(\frac{\alpha\theta_1}{D^2} \right), \tag{11}$$

holds true and

$$\frac{p}{q} - \frac{E_p}{E_q} = \frac{p}{q} \left(\frac{1}{D} + \frac{\beta\theta_2}{D^2} \right) \tag{12}$$

for some $\beta = \beta(\alpha)$ *and* $|\theta_2| \leq 1$. *Then we can detect* p, q *in deterministic time* $\frac{nM^2}{D^4}polylog(n)$. *If* d *is* $B = (\log n)^{O(1)} - smooth$ *number then one can efficiently detect its prime divisors and check the above condition in deterministic time* $2^{\Omega(d)}$.

(iii) *If* $M = O(1)$ *and* $\sqrt{n}/d = (\log n)^{O(1)}$ *then one can detect* p *and* q *in deterministic polylog(n) time.*

5   Towards the Nontrivial Factorization with Oracle

We begin with the related definition useful in the estimation of the size of set of numbers $n \leq N$ which can be factored efficiently by the given algorithm $\mathcal{A}$ and evaluate the cardinality of the set $\mathbb{N}_s(N, y)$. We start from the classical result that the number $\#\mathbb{N}(N, y)$ of positive integers $n \leq N$, free of prime factors not exceeding y is of the order

$$N \prod_{p \leq y} (1 - 1/p) \asymp N/\log y,$$

provided $y \leq N^{1/\log \log N}$. Moreover (see [28]) we have that for fixed $s \geq 2$ the number $\#\mathbb{N}_s(N)$ is asymptotically equal to

$$\frac{N}{(s-1)! \log N} (\log \log N)^{s-1}$$

(see Theorem 437 [28]). For the cardinality $\#\mathbb{N}_s(N, y)$ we have in turn

Theorem 6. *(see Lemma 6.2 [18]) For arbitrary fixed integer $s \geq 2$, any $\theta \in (0, 1)$, $c \in (0, 1)$ and y satisfying the following condition*

$$3 \leq y \leq \exp\left((\log N)^{\theta}\right) \tag{13}$$

we have

$$\frac{c}{2\,s!} \frac{N}{\log N} \left((1 - \theta) \log \log N\right)^{s-1} \leq \#\mathbb{N}_s(N, y)$$

$$\leq \frac{1}{c(s-1)!} \frac{N}{\log N} (\log \log N)^{s-1}, \tag{14}$$

where $N > N_0(c, \theta, s)$ is sufficiently large.

The $s-$aspect is particularly interesting for large $s = s(N)$ both in the cases of classical and elliptic-curve oracle factorization methods. In particular it is the case when the oracle $\mathcal{O}$ returns for arbitrary $n \in \mathbb{N}_s(N)$ and elliptic curve $E = E_{\bar{b}}$, the coefficient $a_E(n)$ of the $L-$function of elliptic curve E over $\mathbb{Q}$ (see [2,7] for more general approach). Here the number of oracle queries and the complexity of the related algorithm depend significantly on the value of s.

5.1   Lower Bound and Asymptotics for the Quantity of Polynomial Time Factorable Numbers

Let $\mathcal{A}$ be a deterministic algorithm and $\mathcal{O}$ an oracle. We let $F(N, \mathcal{A}, t_{\mathcal{A}})$ and $F_s(N, \mathcal{A}, t_{\mathcal{A}})$ be the counting functions of the numbers of the set $\mathbb{N}(N)$ and $\mathbb{N}_s(N)$ respectively, that can be factored by algorithm $\mathcal{A}$ in deterministic time $\leq t_{\mathcal{A}}$. Having at our disposal the oracle $\mathcal{O}$ the analogous counting functions are denoted by $F(N, \mathcal{A}, \mathcal{O}, t_{\mathcal{A}}, t_{\mathcal{O}})$ and $F_s(N, \mathcal{A}, \mathcal{O}, t_{\mathcal{A}}, t_{\mathcal{O}})$ respectively, where

$t_\mathcal{O}$ stands for the related number of queries to oracle $\mathcal{O}$. The key question is to prove the bounds for the above counting functions under the assumption that $t_\mathcal{A}, t_\mathcal{O} = polylogN$ (and may depend on s). The existence of the related algorithm $\mathcal{A}$ allowing to prove the nontrivial lower bound for $F(N, \mathcal{A}, t_\mathcal{A})$ was considered in works [22,31,51], the last giving the best lower bound estimate of order

$$\geq \frac{N}{\log N} (\log \log N)^{6/5+o(1)}.$$

On the other hand we have the following asymptotic inequality

$$F(N, \mathcal{A}, \mathcal{O}, t_\mathcal{A}, t_\mathcal{O}) \geq \mathbb{N}^*(N) - E(N, c, c')$$

for the oracle MultΦ that returns a multiple $\leq M$ of $\varphi(n)$ of size at most $M = O\left(\exp\left((\log N)^{c'}\right)\right)$, $t_\mathcal{A} = (\log N)^{c+5}$ and

$$E(N, c, c') \ll_{c,c'} N(\log N)^{-\frac{13}{2}c},$$

for any $c' > 0$ and $c \geq 4$ (see Theorem 6.1 [54]).

For the elliptic oracles $MultEll(N, M)$ that for any $n \in \mathbb{N}^*(N)$ returns the multiple M of E_n of size $N^{O(1)}$ the related results on determinisic polynomial time algorithm will be given in section 9. Below we outline shortly the more general $(\mathbb{N}, f)$−approach leading to the appropriate oracles $MultEll$, $FrobEll$, $OrdEll$ which are referred to below.

5.2 $(\mathbb{N}, f)$ and $(\mathbb{N}_s, f)$−Factoring

Let $f : \mathbb{N} \to \mathbb{Z}$ be a multiplicative function and $n \leq N$ be a number to be factored. The value $f(n)$, its divisor (e.g. the order of an element of the given group G) or the multiple $kf(n)$, not exceeding a given $M = M(N)$ can be regarded as the natural oracle querry related to f and n.

The problem of nontrivial decomposition of $n \in \mathbb{N}$ or $n \in \mathbb{N}_s$ is then called the $(\mathbb{N}, f)-$ or $(\mathbb{N}_s, f)-$factoring problem, respectively (see e.g. [27,47,54,62]). Also one can consider the more general framing in terms of the tuples $\bar{f}$ of functions and $\bar{n}$ of arguments, namely

- Let $\bar{f}$ be the tuple $\bar{f} = (f_1, f_2, ..., f_k)$ of multiplicative functions f_i and $\bar{n}$ the tuple of positive integers $\bar{n} := (n_1, n_2, ..., n_h)$ such that $(f_i(n_j))_{i \leq k, j \leq h}$ are known.
- The values $f_i(n_j)$ can be regarded as oracle querries (the classical example is $f_i := \sigma_i$) (see [5]).
- The typical approach considered here is to use the $(\mathbb{N}_s, \bar{f})$−factorization, where $f_i(n) = E_n^{t_i}$ is the order of t_i−twist of elliptic curve E over $\mathbb{Z}_n$, that is

$$E_n^t := \prod_{p|n}(p + 1 - a_{E^t}(p)) = \prod_{p|n}\left(p + 1 - \left(\frac{t}{p}\right)a_E(p)\right).$$

For such $\bar{f}$ the $(\mathbb{N}, \bar{f})$−factoring is more difficult, but can be handled in probabilistic polylog$N$ time (see [30]) and will be treated below in more detail.

– The oracle *FrobEll* relates to the results on $(\mathbb{N}, f)$–factorization for $f(n_j) = a_E(n_j)$, where $n_j := n^j$ were given in [2,7,19].

6   Probabilistic Oracle Factoring

Let $n \in \mathbb{N}$ be given and $\mathcal{P} = \mathcal{P}(n)$ being the set of all primes $p \mid n$.

Definition 3. *The signature $S_\chi(\mathcal{P})$ of Dirichlet character $\chi = \chi_d$ defined by the Legendre's symbols (d/p) sequence is given by*

$$S_\chi(\mathcal{P}) = \left(\frac{d}{p}\right)_{p \in \mathcal{P}}$$

with the primes of $\mathcal{P}$ in increasing order and we define

$$\rho_n(\chi) = \prod_{p \in \mathcal{P}} \chi(p) = \prod_{p \mid n} \chi(p).$$

Given $D \geq 2$ we say that n is $(char, D)$–admissible if $\rho_n(\chi) = -1$ for some $\chi = \chi_d$ with $d \leq D$ (in stronger version if additionally S_χ has exactly one coordinate equal to -1). It turns out that any $n \leq N$ is $(char, D)$–admissible for some $D = (\log N)^{O(1)}$ under GRH. Moreover selecting randomly an elliptic curve $E = E_{\bar{b}} \in E(\mathbb{Z}_n)$, any $n \in \mathbb{N}^*$ can be factored with oracle *OrdEll*, in time polylog$N$ with probability $1 - \varepsilon$, for every fixed $\varepsilon > 0$. The last conclusion is a consequence of the inequality $\rho_n(E) \leq D$ for D as above. We analyse this more carefully below.

6.1   Signatures $S_{\chi_d}(\mathcal{P})$ and Oracle Factoring with Probability $1 - \varepsilon$

Let $E(\mathbb{F}_p)$ be an elliptic curve over a finite field $\mathbb{F}_p$. Then for $d \in \mathbb{Z}_p^*$ its twist is defined as $E^d : dy^2 = x^3 + b_1 x + b_2$. There exists the isomorphism $E^d \ni (x, y) \to (x, \sqrt{d}y) \in E$, which is defined over $\mathbb{F}_p$ or $\mathbb{F}_{p^2}$ if d is a quadratic residue in $\mathbb{Z}_p$ or not, respectively; in the later case, E^d is called a quadratic twist of E. If E^d is a quadratic twist of E, then for $(x, y) \in \mathbb{Z}_p^2$ with $y \neq 0$ we have $(x, y) \in E(\mathbb{Z}_p) \iff (x, y) \notin E^d(\mathbb{Z}_p)$, the points shared by E and E^d are the 2-torsion points (with $y = 0$) and the point at infinity. Hence for the quadratic twist we have the following well-known formulas

$$|E(\mathbb{Z}_p)| + |E^d(\mathbb{Z}_p)| = 2p + 2$$

and

$$|E^d(\mathbb{Z}_p)| = p + 1 - \left(\frac{d}{p}\right) a_p,$$

where $\left(\frac{d}{p}\right)$ is the Legendre symbol ($d \in \mathbb{Z}_p^*$) and a_p is the trace of the Frobenius endomorphism $\varphi_p : E_p \to E_p$ satisfying Hasse's bound $|a_p| \leq 2\sqrt{p}$. For an ellitpic curve E over $\mathbb{Z}_n$ and its twists $E^d : dy^2 = x^3 + b_1 x + b_2$ for $d \in \mathbb{Z}_n^*$ we have therefore the following equalities

$$|E(\mathbb{Z}_n)| := E_n := \prod_{p|n}(p + 1 - a_p),$$

$$|E^d(\mathbb{Z}_n)| := (E^d)_n = \prod_{p|n}\left(p + 1 - \left(\frac{d}{p}\right)a_p\right).$$

We apply the following lemmas

Lemma 2. *Let p be a prime not dividing m. Assuming GRH, there exist $d \ll (\log mp)^2$ such that $\left(\frac{d}{p}\right) = -1$ and $\left(\frac{d}{m}\right) = 1$.*

This is just a consequence of Theorem 1.4 of [33], considering the coset aH of the subgroup $H \subset (\mathbb{Z}/pm)^*$ of the quadratic residues modulo pm, and a a quadratic non-residue modulo p and a quadratic residue modulo m.

Lemma 3. *(see Lemma 3 [30]) Let $D \in \mathbb{N}$, p a prime number and $S_{p,D}$ be the number of isomorphic classes of elliptic curves over $\mathbb{Z}_p$ with trace a such that $(a, p + 1) \leq D$. Then*

$$S_{p,D} \geq C \max\left\{\frac{p}{\log p} - \frac{p}{D \log p}\tau(p+1) - \tau(p+1)^2\frac{\sqrt{p}}{\log p}, \frac{p}{(\log p)^2}\right\},$$

for some explicit constant C.

Now let $n = p_1 \cdots p_r$ and $D \leq (\log n)^k$ for some fixed k. We select an elliptic curve $E := E_{\bar{b}} : y^2 = x^3 + b_1 x + b_2$ over $\mathbb{Z}_n$ at random. Let $a_{p_1} := a_{p_1}(E)$ and suppose that $\gcd(a_{p_1}, p_1 + 1) \leq D$. Take d as in Lemma 2 above. Then, $|E_{p_1}^d| = p_1 + 1 + a_{p_1}$ while $|E_{p_i}^d| = p_i + 1 - a_{p_i}$ for any $i = 2, \ldots, r$ and, hence

$$\frac{(E_{\bar{b}})_n}{(E_{\bar{b}}^d)_n} = \frac{p_1 + 1 - a_{p_1}}{p_1 + 1 + a_{p_1}}.$$

Letting

$$\frac{(E_{\bar{b}})_n}{(E_{\bar{b}}^d)_n} = \frac{a}{b}$$

where $\gcd(a, b) = 1$ we have that $a = (p_1 + 1 - a_{p_1})/d$, $b = (p_1 + 1 + a_{p_1})/d$. So we just have to mulptiply a by all the integers up to D in order to recover $p_1 + 1 - a_{p_1}$ and mulptiply b by all the integers up to D to recover $p_1 + 1 + a_{p_1}$. Adding both numbers we get $2(p_1 + 1)$ hence, we have found a factor of n.

If the algorithm does not return p_1, in $(\log n)^k$ steps, then stop, and select a new elliptic curve $E'_{\bar{B}} : y^2 = x^3 + B_1 x + B_2$ non isomorphic to $E_{\bar{b}}$. Note that we can assume $B_1 \not\equiv 0 \pmod p$ neither $B_2 \not\equiv 0 \pmod p$ for $p|n$ since otherwise we find a factor of n. Then, if we get an elliptic curve isomorphic to $E_{\bar{b}}$ over $\mathbb{Z}_p$ for some $p|n$ then $\lambda^4 B_1 \equiv b_1 \pmod p$ and $\lambda^6 B_2 \equiv b_2 \pmod p$, so $\lambda^4 \equiv \frac{b_1}{B_1}$ and

$\lambda^6 \equiv \frac{b_2}{B_2}$ (mod p), or $\lambda^2 \equiv \frac{b_2 B_1}{b_1 B_2}$ (mod p) and then $b_2^2 B_1^3 - b_1^3 B_2^2 \equiv 0$ (mod p) and we would find a factor. It is worth to remark that if $n|(b_2^2 B_1^3 - b_1^3 B_2^2)$ we can reject the curve, and the probability to find multiples of n is $\ll \frac{1}{n}$.

Now, repeating the process $K(\log n)^2$ times, for non isomorphic classes of elliptic curves we see, using Lemma 3, that the probability for the trace $a \in S_{p,D}$ can be bounded below independently of D by $\frac{p}{(\log p)^2}$, so the probability that none of them has trace $a \in S_{p,D}$ is bounded above by

$$\left(1 - \frac{c}{(\log p)^2}\right)^{K(\log n)^2} < \left(1 - \frac{c}{(\log p)^2}\right)^{K(\log p)^2} < e^{-\frac{K}{c}} < \frac{c}{K} < \varepsilon,$$

for $K \gg (1/\varepsilon)$. Finally note that checking that none of them is isomorphic to the previous ones would take $O\left(K^2 (\log n)^4\right)$ steps.

7   Deterministic, Conditional, Rigorous Approach

Here we investigate the average case running time of oracle factorization supported by the Sato-Tate conjecture with the oracle OrdEll that returns the order E_n of elliptic curve E over $\mathbb{Z}_n$ for $n \leq N$.

7.1   Elliptic Signatures and $(E, \Delta)-$Admissible Numbers

Similarly as in the definition of $(char, D)-$admissible number n, we use the value $\rho_n(E)$ to define

Definition 4. *Given $\Delta \geq 1$ we say that n is $(E, \Delta)-$admissible if $\rho_n(E) \leq \Delta$.*

Let $E = E_{\bar{b}}$ be a given elliptic curve over $\mathbb{Z}_n$. As a conclusion from the previous section we have that arbitrary square-free $n \leq N$ which is $(E, \Delta)-$admissible and $(char, D)-$admissible number can be factored in deterministic time $(D\Delta)(\log N)^{O(1)}$ with the aid of the values E_n and E_n^t, where t is at most $(\log N)^{O(1)}$. In what follows we assume that $E = E_{\bar{b}}$ over $\mathbb{Q}$ is an elliptic curve without complex multiplication (CM).

Definition 5. *Let $E/\mathbb{Q}$ be a given elliptic curve without CM and $\Delta \geq 1$. The prime number p is called (E, Δ) -admissible if*

$$\gcd(p + 1, a_E(p)) \leq \Delta.$$

Otherwise p is called (E, Δ)- exceptional. The positive integer n is called $(E, \Delta)-$exceptional if some of its prime factors $p \mid n$ is $(E, \Delta)-$exceptional.

The distribution of $(E, \Delta)-$admissible primes plays the significant role in the deterministic, asymmptotic oracle factorization. On average we expect to have the similar bound as for the Hasse interval $H_p = [p + 1 - 2\sqrt{p}, \ p + 1 + 2\sqrt{p}]$, namely that

$$\sum_{p \leq x} \gcd(p+1, a_E(p)) \leq c\pi(x)\log x(1 + o(1)), \tag{15}$$

for an elliptic curve $E/\mathbb{Q}$ without CM and $c = \frac{315}{4\pi^5}\zeta(3)$. Namely letting $q = \gcd(p+1, a_E(p))$ the LHS of (15) is

$$\leq \sum_{q \leq 2\sqrt{x}} q \sum_{p \leq x}^{\star} 1,$$

where $\star$ means that the summation is over $p \equiv -1 \mod q$, such that $a_E(p) \leq 2\sqrt{p}$, $a_E(p) \equiv 0 \mod q$. Since there are $\phi(q)$ admissible residue classes for the moduli q of primes p, the above sum should be equal in view of the Sato-Tate conjecture (see [23]) to

$$\frac{1}{q\phi(q)} \sum_{p \leq x,\, a_E(p) \in [-2\sqrt{p},\, 2\sqrt{p}]} 1 = \frac{2}{\pi} \int_{-1}^{1} \sqrt{1-\gamma^2}d\gamma \frac{\pi(x)}{q\phi(q)}(1 + o(1)),$$

as x tends to infinity. Now summing over q and using the asymptotic estimate

$$\sum_{q \leq t} \frac{1}{\phi(q)} = \frac{315}{2\pi^4}\zeta(3)\log t(1 + o(1)),$$

as t tends to infinity (see [15], vol 1, p. 134) we obtain the required bound.

7.2 Deterministic, Asymptotic Polylog$N$ Factoring

Let $\mathcal{P} = \mathcal{P}(n)$ and $S_d(\mathcal{P}) := S_{\chi_d}(\mathcal{P})$ be as above and assume that the primitive signature $S_d(\mathcal{P})$ contains exactly one coordinate equal to -1. Let N_n be the least positive integer d satisfying this property. Under GRH we have the bound $N_n \ll (\log n)^2$ (the more careful analysis concerning the least d such that $\rho_n(S_\chi) = -1$) with the related bound $N_n \leq \log N$ for all $n \in \mathcal{N}(N)$ besides at most $N^{1-c/\log\log N}$ exceptions will be considered in Sect. 9).

As we remarked above the polylog$N$ oracle factoring all $n \in \mathcal{N}(N)$ requires $(char, D)-$ and $(E, \Delta)-$admissibility condition for $D\Delta = (\log N)^{O(1)}$ under GRH. Below the $(E, \Delta)-$admissibility is relaxed by the average polylog(x) bound for $\gcd(p+1, a_E(p))$ for primes $p \leq x$, which is sufficient for our purpose.

Namely applying the inequality (15) with $\log x$ replaced by $(\log x)^\lambda$ we have the following upper bound for the number $Exc(N, E, \Delta)$ of $(E, \Delta)-$exceptional numbers $n \in \mathcal{N}$, $n \leq N$ which are free of prime divisors $\leq \log N$,

$$Exc(N, E, \Delta) \ll N/(\log\log N)^{\delta-\lambda},$$

where $\Delta = (\log N)^\delta$ $(\delta > \lambda)$. Letting $C(t) = \log^{2+o(1)} t$ in Theorem 4.5 of [18] and Lemma 2 of Sect. 6.2 we obtain the following

Theorem 7. *(cf. Theorem 4.5 [18]) Let $E/\mathbb{Q}$ be a given elliptic curve without CM and $1 \leq \lambda < \delta$. Assume GRH and that*

$$\pi^{-1}(x) \sum_{p \leq x} \gcd(p+1, a_E(p)) \ll \log^\lambda x, \qquad (16)$$

as x tends to infinity.

Then all $n \in \mathcal{N}$, $n \leq N$ with at most $O\big(N/(\log \log N)^{\delta - \lambda}\big) - exceptions$, can be factored nontrivially in $O\big(\log^{5+\delta+o(1)} N\big)$ deterministic time with at most $O\big(\log^{2+o(1)} N\big)$ queries to the oracle OrdEll, where the constant implied by the symbol O depends on $\delta - \lambda$.

8   Application of the Shifted Sieve for Special Family of Elliptic Curves

Here we further develop the *OrdEll* oracle factorization with the aid of the family of elliptic curves $\mathcal{E} = \mathcal{E}(B')$, where $B' = B'(N) > \log^3 N$. Instead of the assumption (16) above we consider the average one over reductions modulo p of the family $\mathcal{E}(B')$ of elliptic curves over $\mathbb{Q}$ without complex multiplication, defined by the Weierstrass equation $E_{b'} : y^2 = x^3 + x + b'$, $b \leq B'$, given by the condition (17) below. Let $R_t(m) = m/S_t(m)$, where $S_t(m)$ is the largest $t-$smooth divisor of m.

Theorem 8. *(see Theorem 5.1 [18]) Let $1 \leq \lambda < \delta - 1$, $B' = B'(x) \geq \log^{2+\delta} x$ and $K \geq 1$ be fixed. Let $w : [1, B'] \to \mathbb{R}_{\geq 0}$ be the given weighted function. Assume GRH and that*

$$\pi^{-1}(x) \sum_{p \leq x} \left| \sum_{b \leq B: \, a_p(E_b) \equiv 0 \mod d} w(b) - \frac{1}{d} \sum_{b \leq B: \, p \nmid b(4+27b^2)} w(b) \right| \ll \log^\lambda x, \quad (17)$$

holds true for arbitrary $d \mid R_K(p+1)$. Then, all except $O_{\delta-\lambda}\big(N/(\log \log N)^{\delta-\lambda}\big)$, $n \in \mathcal{N}(N)$ can be factored completely in $O\big(B'(N)(\log N)^{5+\pi(K)(1+\varepsilon)}\big)$ deterministic time with $O\big(B'(N) \log^2 N\big)$ queries to OrdEll oracle. Here ε can be arbitrarily small positive number, provided $K \geq K_0(\varepsilon)$ is sufficiently large (the term $\pi(K)(1+\varepsilon)$ above disappears if $K = 1$).

The proof of Theorem 8 is based on the shifted sieve (see [29]) in the following form

Lemma 4. *(see [58]) Let $\mathcal{B}$ be a finite set of positive integers, Γ a square-free integer, $w : \mathcal{B} \to \mathbb{R}_{\geq 0}$, $U : \mathcal{B} \to \mathbb{Z}$ and for $d \mid \Gamma$ we let*

$$| \mathcal{B}(d) | = \sum_{b \in \mathcal{B}, \, U(b) \equiv 0 \mod d} w(b)$$

and $|\mathcal{B}|:=|\mathcal{B}(1)|$. *Suppose that*

$$||\mathcal{B}(d)| - |\mathcal{B}|/d| \leq \Delta, \tag{18}$$

for all $d \mid \Gamma$. *Then*

$$\sum_{b \in \mathcal{B},\, \gcd(U(b),\Gamma)=1} w(b) \geq c_1 |\mathcal{B}| / \left(\log(\omega(\Gamma) + 1)\right)^2 - c_2 \omega(\Gamma)^2 \Delta, \tag{19}$$

for some absolute, positive constants c_1 *and* c_2.

We apply Lemma 4 for $\mathcal{B} = \mathcal{B}_p$ being the set of $b \leq B'(N)$ such that $6\Delta_b$ is coprime to n. Here $w(b)$ is a given weighted function, $\Gamma = R_K(p+1)$ and $U(b) = a_p(E_b^t)$ is the Frobenius trace modulo p of the twisted elliptic curve E_b^t ($t \leq D$). Similarly as for the proof of Theorem 7 we now define the related admissible and exceptional primes.

Definition 6. *Let* $E/\mathbb{Q}$ *be an elliptic curve without CM and* $\Delta \geq 1$. *The prime number* p *is called* (B', Δ, w, K) *-admissible if*

$$||\mathcal{B}_p(d)| - |\mathcal{B}_p|/d| \leq \Delta, \tag{20}$$

for all $d \mid R_K(p+1)$. *Otherwise it is called* (B', Δ, w, K)-*exceptional.*

Definition 7. *Let* $E = E_b$ *over* $\mathbb{Q}$ *be an elliptic curve without CM, and* p *be a fixed prime number greater than 3. Then* $n \in \mathcal{N}(N)$ *is called* (B', Δ, w, K, p)-*admissible if* $p \mid n$ *and* p *is* (B', Δ, w, K)-*admissible, where* $\Delta = \Delta(N)$. *If* n *is* (B', Δ, w, K, p)-*admissible for all primes* $p \mid n$ *then it is called* (B', Δ, w, K)-*admissible. Otherwise* n *it is called* (B', Δ, w, K)-*exceptional number.*

By the average condition (17) the number $Exc(x, B', \Delta, w, K)$ of exceptional primes $p \leq x$ is

$$Exc(x, B', \Delta, w, K) \ll \pi(x)/(\log x)^{\delta - \lambda}.$$

where $\Delta = \log^{\delta} x$, which implies that the cardinality $Exc(N, B', \Delta, w, K)$ of the related set of exceptional numbers $n \in \mathcal{N}(N)$, free of prime divisors $\leq N$ has order of magnitude $\ll N/(\log \log N)^{\delta - \lambda}$ which is smaller than the order of magnitude $|\mathcal{N}(N, y)|$ for $y = \log N$ provided $\delta > \lambda + 1$.

If $n \in \mathcal{N}(N, y)$ is admissible then it is divisible by some admissible prime $p \geq \log N$, hence applying Lemma 2 we find the twist E_b^t such that S_{χ_t} has exactly one term -1 corresponding to prime p in deterministic time $D \ll \log^2 N$ under GRH. Hence having the right t we have

$$(E_b)_n/(E_b^t)_n = \frac{(E_b)_p/k}{(E_b^t)_p/k} = \frac{\alpha}{\beta}$$

in the reduced form α/β. Therefore applying Lemma 4 above it remains to check all possible values for $k \mid 2\gcd(p+1, a_p(E_b^t)) \in \mathcal{L}_K$ to find both $p + 1 - a_p(E_b^t) =$

$p+1-(t/p)a_p(E_b)$ and $p+1+a_p(E_b^t) = p+1+(t/p)a_p(E_b)$ and hence their sum $2(p+1)$ in deterministic $D(N)B'(N)$polylog$N$=polylog$N$ time, whenever $B' = (\log N)^{O(1)}$. Since $\mathcal{L}_K$ is contained in the set of integers of absolute value $\leq 2\sqrt{p}$ that are $K-$smooth, we can bound its cardinality from above by $(\log N)^{\pi(K)(1+\varepsilon)}$ (see (1.19) of [21]). Hence for $K = O(1)$ we deduce that the prime divisor can be discovered in $|\mathcal{L}_K|$ polylog$N$ = polylog$N$ time. We remark that the condition $d \mid R_K(p+1)$ only slightly relax the assumption of uniform distribution of $a_p(E_b)$ in arithmetical progressions $a_p(E_b) \equiv 0 \mod d$, where $d \mid p+1$). The more careful analysis shows that in fact we obtain the complete factorization of $n \leq N$ in polynomial time polylog$N$. Finally we remark that the number of oracle querries is bouded by $D(N)B'(N) = B'(N)\log^2 N$.

9 Rigorous Asymptotic Approach with Supersingular Curves

Here we present the hybrid approach allowing to apply the Dirichlet character signatures in separating the values of $\nu_2(\text{ord}P_p)$ and $\nu_2(\text{ord}P_q)$ for some $p, q \mid n$. We do not assume that our oracle has an access to integers b with

$$\left(\frac{b}{n}\right) = -1$$

for the Jacobi symbol. Instead we estimate the proportion of integers $n \leq N$ for which this smallest value of b is large.

9.1 Definition of Oracles

First we describe a MultEll (N, B', M)-oracle which we assume is available to us. Let

$$E_b: \ y^2 = x(x^2 - b)$$

be an elliptic curve over $\mathbb{Q}$. By the Chinese remainder theorem we see that for a square-free n, the number of solutions $E(n, b)$ to the congruence

$$y^2 \equiv x(x^2 - b) \mod n, \qquad (x, y) \in \mathbb{Z}_n^2,$$

is given by

$$E(n, b) = \prod_{p|n} \#(E_b(\mathbb{Z}_p)), \tag{21}$$

provided that $\gcd(b, n) = 1$.

Definition 8 (MultEll(N, B', M)-oracle). *Given the parameters B', N and M, for each integer $n \leq N$ it returns*

- *a positive multiple*

$$k_b E(n, b) \leq M, \qquad k_b \in \mathbb{N},$$

of $E(n, b)$ given by (21) for every $b \leq B'$ with $\gcd(b, n) = 1$, if n square-free

– *an error message ⋆⋆⋆, if n is not square-free.*

Definition 9 (MultEll*(N, B', M)-oracle). *Given the parameters B', N and M, for each $n \leq N$, it returns*

– *a positive multiple*

$$kF(n, B') \leq M, \qquad k \in \mathbb{N},$$

of

$$F(n, B') = \prod_{b \leq B'} E(n, b),$$

if n square-free
– *an error message ⋆⋆⋆, if n is not square-free.*

9.2   Results
Theorem 9. *(see Theorem 1.2 [14]) Let*

$$2 < \gamma < 2 + \frac{\sqrt{257} - 15}{16}, \delta < 1/4$$

be fixed. Assume that for a sufficiently large integer N we are given a MultEll(N, B', M)-oracle where

$$B' = [(\log N)^\gamma], M = N^{O(1)}$$

Then there is a deterministic algorithm that finds a nontrivial factor of all integers $n \leq N$ with at most $N/(\log N)^\delta$ exceptions, in polynomial, deterministic time $O((\log N)^{\rho(\gamma)+o(1)})$, where

$$\rho(\gamma) = \max\left\{\gamma + 2, \frac{9\gamma - 17}{8(\gamma - 2)}\right\}$$

and $O((\log N)^\gamma)$ oracle queries.

Theorem 10. *(see Theorem 1.4 [14]) There are some positive constants c and C such that if for a sufficiently large integer N we are given a MultEll*(N, B', M)-oracle, where*

$$B' = [c \log N], M = N^{O(B')},$$

then there is a deterministic algorithm that finds a nontrivial factor of all integers $n \leq N$ with at most $N(\log N)^{-C/\log\log\log N}$ exceptions, in polynomial, deterministic time $\log^{3+o(1)} N$, with only one query of the MultEll*(N, B', M)-oracle.

We remark that our oracle outputs the results only for square-free integers. However in our scenario one can easily supplement our algorithms with a trial division for divisibility by integers d^2 with $d \leq N$ and notice that their number is $O(N/\log N)$.

9.3   Remarks on the Proofs

We recall that $\mathcal{N}(N, z)$ stands for the subset of $\mathcal{N}$ which consist of integers $\leq N$ and free of prime divisors $p \leq z$. Applying the familiar lower bound for the series of reciprocal primes one deduces that

Lemma 5. *Let $z \in [\log N, \exp((\log N)^{\delta})]$ and $\delta < 1/4$ be an arbitrary positive constant. Then we have*

$$\#\{n \in \mathcal{N}(N, z) : \ n = pqm, \ p \neq q \ primes, \ p, q \equiv 3 \mod 8, \ m \in \mathbb{Z}\}$$
$$= \#\mathcal{N}(N, z) + O(N(\log N)^{-\delta} \log \log N).$$

We recall that an integer a, relatively prime to a prime p, is called a *quadratic non-residue modulo n* if the congruence $x^2 \equiv a \mod n$ has no solution in integers.

Lemma 6. *Let $E_b : \ y^2 = x(x^2 - b)$ be an elliptic curve over $\mathbb{Z}_p$. If $p = 3 \mod 4$, then $E_b(\mathbb{Z}_p)$ has order $p+1$ and is cyclic or contain a noncyclic subgroup $\mathbb{Z}_2 \times \mathbb{Z}_2$ according to whether b is a quadratic non-residue or residue $\mod p$, respectively.*

For square-free n we denote by N_n the least b with $\gcd(b, n) = 1$ and such that for the Jacobi symbol we have

$$\left(\frac{b}{n}\right) = -1.$$

The following estimates are basically based on the application of large sieve machinery for the average upper bound for N_n. The following results is a special case of [4, Theorem 1].

Lemma 7. *For arbitrary $\gamma > 2$ and $\varepsilon > 0$ we have*

$$N_n \leq (\log x)^{\gamma}$$

for all but $O\left(x^{1/(\gamma-1-\varepsilon)}\right)$ odd, square-free, positive integers $n \leq x$.

We remark that the next result given by [34, Theorem 1.1] applies to more general settings of Kronecker symbol. We only need its part for square-free integers. Note that compared to Lemma 7 it gives a stronger bound on N_n but a weaker bound on the exceptional set.

Lemma 8. *There exists an absolute constant $C > 0$ such that*

$$N_n \ll \log n$$

for all but $O(x^{1-C/\log \log x})$ positive integers $n \in \mathcal{N}(x)$.

The ideas of proofs of Theorems 9 and 10 is based on separating the values $\nu_2(\mathrm{ord}P_p)$ and $\nu_2(\mathrm{ord}P_q)$, where $P = (dx, d^2y) \in E_b(\mathbb{Z}_n)$ of the curve $E_b^d : y^2 = x(x^2 - bd^2)$ modulo n and skillful evaluating of cardinality of sets of the related exceptional numbers.

10   Separating Elements and Partially Smooth Orders E_n of Elliptic Curves

Let $n = \prod_{i \leq s} p_i \in \mathcal{N}_s$ and

$$G = G_1 \times G_2 \times \ldots \times G_s \tag{22}$$

be the direct product of groups, where G_i are either the multiplicative groups $\mathbb{Z}^*_{p_i}$ or (additive) elliptic curve groups $E(\mathbb{Z}_{p_i})$ such that $6\Delta_E$ is coprime to n. Let $g \in G$ be represented by $(g_1, \ldots g_s)$, l be a prime number and $\mathrm{ord} g_i$ stand for the order of $g_i \in G_i$ $(1 \leq i \leq s)$. Let $\nu_l(m)$ stand for highest exponent of l dividing m.

10.1   Separating Primes and Witnesses

Definition 10. *We call l the separating prime for positive integer $n \in \mathbb{N}_s$ with respect to $g \in G$ if*

$$\nu_l(\mathrm{ord} g_i) \neq \nu_l(\mathrm{ord} g_j),$$

for some $1 \leq i \neq j \leq s$ and then g is called the separating l−witness for n.

The Fermat-Euclid (compositeness) witness $a \in \mathbb{Z}^*_n$ for n was introduced and investigated in relation to deterministic Pollard's $p - 1$ algorithm in [56, 62]. It satisfies the condition

$$\nu_l(\mathrm{ord}_p a) \neq \nu_l(\mathrm{ord}_q a),$$

for some prime $l \mid \mathrm{ord}_n a$. The more accurate definitions for the factorization of almost all $n \in \mathbb{N}^*$ in deterministic polylog$N$ time was applied and developed in [54] and [27] with the aid of Dirichlet's characters reinforced by application of the large and shifted sieve, respectively (see [29, 42, 58] and [51] for the approach when GRH is not assumed).

The efficient classical, deterministic factorization of $n \in \mathbb{N}^*$ refers to relatively small bounds $a \leq A = A(n)$ and $l \leq B = B(n)$ above (in practice of polylog$n$ order of magnitude). In general the small value of A is handled by estimate for the least Dirichlet characters nonresidue while B corresponds to the smoothness bound for the order $g \in G$ or order $|G|$ of the group (cf. [49, 61] in the classical case).

In the elliptic curve approach A and B−aspects are replaced by considering pairs (E, P), where $P \in E(\mathbb{Z}_n)$ and comparing the values

$$\nu_l(\mathrm{ord} P_p) \quad \text{and} \quad \nu_l(\mathrm{ord} P_q)$$

for primes $p, q \mid n$. For given $B \geq 2$ we consider two cases

 (i) the minimal prime l for which P is l−separating witness is $\leq B$
 (ii) the minimal prime l for which P is l−separating witness is $> B$.

In the first case we call the related (finite) point $P \in E(\mathbb{Z}_n)$ the $B-$separating witness for n. In the second case we call it $(B, \gamma)-$nonseparating witness, where $\gamma \in [0, 1/4)$ if

$$1 \leq \min_{p|n}(a_p(E), a_q(E)) \leq cn^{1/4 - \gamma/2}$$

and $\mathrm{ord}P_r > n^{1/2 - \gamma/2}$ for some $r \mid n$ (cf. [20] and [53]) for some (explicitly given) absolute positive constant c. Namely it is proved in [53] that if the maximal $B-$smooth factors of $\mathrm{ord}P_p$ and $\mathrm{ord}P_q$ are greater or equal to $n^{1/2 - \gamma/2}$, then we can factor n in deterministic polylog(n) time, provided $\gamma < 1/4$. We will clarify these concepts in the following sections. Also we will state the relevant results on the separating decomposition witnesses and sets respectively and analyse the particular case when $B = B(n)$ is a subexponential function.

10.2   Partially Smooth Numbers and Local Orders of Elliptic Curves

We recall that m is $B-$smooth if every prime divisor of m does not exceed B. From now on we assume everywhere that $2 \leq B \leq D$.

Definition 11. *Let n and $D \geq B > 1$ be fixed. Let $\beta \in [0, 1]$ and $s_B(m)$ stand for the largest $B-$ smooth divisor of m. The number $m \in \mathbb{N}$ is called $(B, \beta)-$smooth if $s_B(m) \geq m^\beta$. If additionally m is $D-$smooth then we call it $(B, \beta, D)-$smooth number.*

If we want to emphasize that m belongs to the interval I we say that m is $(I, B, \beta, D)-$smooth number. Let d^* denote the maximal square-free divisor of d. We let

Definition 12. *Let $\sigma \in [1, 2]$. We say that $(B, \beta, D)-$smooth number is $(B, \beta, D, \sigma)-$smooth if additionally we have*

$$\left(\frac{m}{s_B(m)}\right)^* \leq D^\sigma. \tag{23}$$

Let $Y \in \{(B), (B, \beta), (B, \beta, D), (B, \beta, D, \sigma)\}$. We say that the positive integer m is called $Y-partially$ smooth number (or shortly partially smooth if the related *smoothness* parameters B, β, D, σ are clear from the context). The restriction $\sigma \in [1, 2]$ is motivated by the possible improvement of $D^{2+o(1)}$ of the bound for the computational cost of decomposition $n = pq$, as mentioned in *Remark 1* of section 2.

The elliptic curve E over $\mathbb{Z}_n$ is called $Y-$admissible if its local order E_r is $Y-$partially smooth for some prime $r \mid n$.

11   Admissible Elliptic Curves and Decomposition Witnesses for Semiprime Numbers

From now on we consider the case of semiprime numbers $n = pq \in \mathcal{N}_2$. Let $\bar{P} \in E(\mathbb{Z}_n)$, $\bar{P} = (P, P')$, where P and P' are the related reduced points on the reduced elliptic curves $E(\mathbb{Z}_p)$ and $E(\mathbb{Z}_q)$ respectively. We consider two types of *decomposition witnesses* for semiprime $n = pq$ defined for $Y-$admissible elliptic curves E over $\mathbb{Z}_n$, where $Y \in \{(B), (B, \beta), (B, \beta, D), (B, \beta, D, \sigma)\}$. Both are significant for the rigorous evaluation of computational cost of decomposition $n = pq$.

11.1   Decomposition Witnesses and Admissible Curves

We start from the definition of separating witnesses $\bar{P} \in E(\mathbb{Z}_n)$, where E is $Y-$admissible, with $Y = (B)$. In the following statements w assume that $n \geq n_0$ is sufficiently large and $B = B(n)$. We let

Definition 13. *Let $n = pq \in \mathcal{N}_2$, E be an elliptic curve over $\mathbb{Z}_n$, $B \geq 2$ and E_r be a B-smooth number for some prime $r \in \{p, q\}$. The (finite) point $\bar{P} \in E(\mathbb{Z}_n)$ is called $B-$separating witness for n if*

$$s_B(ordP) \neq s_B(ordP'), \tag{24}$$

where $\varphi(\bar{P}) := (P, P')$.

The nonseparating witness $\bar{P}$ for n is defined by the negation of the above condition together with the additional one depending on the parameter $0 \leq \gamma < 1/4$ referring to elliptic curve E over $\mathbb{Z}_n$. We proceed it by the related definition of $(Y, \gamma)-$admissible elliptic curve E for $Y = (B)$.

Definition 14. *Let $0 \leq \gamma \leq 1/2$, $n = pq$, where $p < q < \vartheta p$ and $c = c(\vartheta)$ be fixed. The elliptic curve E is called $((B), \gamma)-$admissible if it is $(B)-$admissible and additionally we have*

$$\gcd(s_B(E_p), s_B(E_q)) \geq n^{1/2 - \gamma/2} \tag{25}$$

and

$$1 \leq \min(a_E(p), a_E(q)) \leq cn^{1/4 - \gamma/2}. \tag{26}$$

The last condition above relaxes the restriction for the elliptic curve E as γ decreases to 0. Moreover using the quadratic twist E^t of elliptic curve we can restrict ourselves to elliptic curves with positive product of the Frobenius traces $a_E(p)a_E(q)$. Now we define the nonseparating witness for n with respect to $Y = (B)$ and γ.

Definition 15. *Let $n = pq \in \mathcal{N}_2$, $0 \leq \gamma \leq 1/4$, $B \geq 2$ and E be $((B), \gamma)-$admissible elliptic curve over $\mathbb{Z}_n$. The point $\bar{P} = (P, P') \in E(\mathbb{Z}_n)$ is called $(B, \gamma)-$nonseparating witness for n if*

$$s_B(ordP) = s_B(ordP') \geq n^{\frac{1}{2} - \frac{\gamma}{2}}. \tag{27}$$

The following definition applies to Y being arbitrary element of the set defined above and $D-$separating witnesses which are not $B-$separating ones. Namely letting $Y = (B, \beta, D)$ we have

Definition 16. *Let $n = pq \in \mathcal{N}_2$, $B < D$, $0 \leq \beta \leq 1$ and E be (B, β, D)-admissible elliptic curve over $\mathbb{Z}_n$. The point $\bar{P} \in E(\mathbb{Z}_n)$ is called (B, D)-separating witness for n if it is $D-$separating witness but not $B-$separating witness for n.*

The condition of $(B, \beta)-$admissibility implies that $s_B(E_r) \geq cn^{\max(\beta/2, 1/2 - \gamma/2)}$ for some $c > 0$. Analogously one can define the separating set $\mathbf{X}$ for n as follows

Definition 17. *Let $B < D$. The set $\mathbf{X} \subseteq E(\mathbb{Z}_n)$ is called (E, B)-separating set for n if it contains some $B-$separating witness $\bar{P} \in E(\mathbb{Z}_n)$ for n. Moreover it is called $(E, B, D)-$separating set for n if it is (E, D)-separating but not (E, B)-separating set for n.*

By the definition of (B, β, D) or $(B, \beta, D, \sigma)-$partially smooth numbers we have that

$$\bar{q} := \left(\frac{E_r}{s_B(E_r)} \right)^* \leq \min\left(D^\sigma, E_r^{1-\beta} \right). \tag{28}$$

By the comment in section 2 we see that knowledge of the $(E, D)-$separating witness allows us to find the decomposition $n = pq$ in deterministic time $D^{2+o(1)}$. On the other hand the $(B, D)-$separating witness allows to factor n in deterministic time $D^{\sigma+o(1)}$ provided E is $(B, \beta, D, \sigma)-$admissible curve.

11.2   Polylog(n) Factoring with Nonseparating Witnesses

Unlike in the case of separating witnesses the case of nonseparating witnesses is valuable also for the oracle factoring approach. Namely considering the *DecOrdEll* oracle that returns the prime power factorization of E_n, the decomposition $n = pq$ can be found in deterministic polylog(n) with two oracle querries, provided the number of prime divisors of E_n is not too large (cf. [16]). In this connection we require the consistency of $(Y, \gamma)-$admissibility of E with the condition on $(B, \beta)-$admissibility, where $Y = (B, \beta)$. Henceforth we assume that

$$\gamma \in [0, \min(1 - \beta, 1/4)). \tag{29}$$

Since E_r is close to $n^{1/2}$ the border value $\gamma = 0$ corresponds to $\beta = 1$ above. On the other hand the constraint condition $\gamma < 1/4$ ensures that the related decomposition $n = pq$ can be computed in deterministic polylog(n) time.

Knowledge of the value of $E_r = r + 1 - a_E(r)$, for some $r \in \{p, q\}$ does not yet imply finding the solutions p, q of the related system of equations. However the decomposition $n = pq$ can be computed in deterministic polylog(n) time when the familiar *Coppersmith method* is applied (see [8,9]). Below we state the result on polylog(n) factoring $n \in \mathcal{N}_2$ depending on the partial information on the orders E_p and E_q with the aid of the notion of $(E, B, \gamma)-$nonseparating witness.

Theorem 11. *(see [53]) Let* $\gamma \in [0, \min(1 - \beta, 1/4))$, $n = pq$ *be sufficiently large, where* $p < q < \vartheta n^{1/2}$ *and* $P \in E(\mathbb{Z}_n)$. *Assume that we know the value* d *such that*

(i) $d := \gcd\left(s_B(E_p), s_B(E_q)\right) \geq n^{\frac{1}{2} - \frac{\gamma}{2}},$

    or

(ii) $d = s_B(\mathrm{ord}P) = s_B(\mathrm{ord}P') \geq n^{\frac{1}{2} - \frac{\gamma}{2}},$

where

$$1 \leq \min(a_E(p), a_E(q)) \leq cn^{\frac{1}{4} - \frac{\gamma}{2}}, \tag{30}$$

for some $c = c(\vartheta) > 0$. *Then the decomposition* $n = pq$ *can be discovered in deterministic polynomial time* $\mathrm{polylog}(n)$. *The same conclusion holds provided* $r/s_B(E_r) = (\log n)^{O(1)}$ *for some* $r \in \{p, q\}$.

The key point of the proof is the following. First we remark that the condition (ii) implies (i) since the local order $\mathrm{ord}P_r$ of point P divides the group order E_r for $r \in \{p, q\}$ so let us assume that the condition (i) holds true. Let $d = \gcd(E_p, E_q)$. Then $E_p = dr_p$, $E_q = dr_q$, where $\gcd(r_p, r_q) = 1$ and representing n in base d we have

$$n = c_0 + c_1 d + c_2 d^2. \tag{31}$$

Letting $t_r = a_E(r) - 1$ for $r \in \{p, q\}$, we obtain·

$$n = pq = (dr_p + a_E(p) - 1)(dr_q + a_E(q) - 1)$$
$$= (dr_p + t_p)(dr_q + t_q) = r_p r_q d^2 + (r_p t_q + r_q t_p)d + t_p t_q$$

where $t_p, t_q \geq 0$. Thus the coefficients r_p, r_q, t_p, t_q are uniquely defined by c_i ($i = 1, 2$) provided all they are in the interval $[0, d)$ and this follows from the bounds for $a_E(r)$, where $r \in \{p, q\}$. Now one can compute them in $\mathrm{polylog}(n)$ time solving the related system of equations, hence obtaining $p = dr_p + t_p$ and $q = dr_q + t_q$.

11.3 Large $G_{\mathbf{X}}$ Order Based Factoring

Let $G_{\mathbf{X}}$ be the subgroup of $E(\mathbb{Z}_n)$ generated by (the nonempty set) $\mathbf{X}$ and $G_{\mathbf{X}}^B$ be the homomorphic image of $G_{\mathbf{X}}$ under the homomorphism Φ_n^B sending the point $\bar{P} = (P, P') \in \mathbf{X}$ to the point to $\bar{Q} = (Q, Q') \in E(\mathbb{Z}_N)$, where $Q = \frac{E_p}{s_B(E_p)}P$ and $Q' = \frac{E_q}{s_B(E_q)}P'$.

Below we state the result on the sufficient condition for the set $\mathbf{X}$ to contain some (E, B)-separating witness provided both $E(\mathbb{Z}_p)$ and $E(\mathbb{Z}_q)$ are cyclic groups. As a consequence the obtain the decomposition $n = pq$ in deterministic time $|\mathbf{X}|B^{2+o(1)}$, as n tends to infinity.

Theorem 12. *([52]) Let* $n = pq$ *be sufficiently large, where* $p < q < \vartheta n^{1/2}$, $E(\mathbb{Z}_r)$ *is cyclic group for each* $r \in \{p, q\}$. *Assume that for a set* $\mathbf{X} \subseteq E(\mathbb{Z}_n)$ *we have that* $G_{\mathbf{X}}^B$ *is cyclic and for some* $0 \leq \eta < 1/2$ *and (explicitly computable) constant* $c = c(\vartheta)$, *we have*

$$|G_{\mathbf{X}}| \geq n^{1-\eta} \tag{32}$$

and

$$s_B(E_r) \geq cn^{2\eta}. \tag{33}$$

Then the set $\mathbf{X}$ *is* (E, B)*-separating set for* n *and in deterministic time* $|\mathbf{X}|B^{2+o(1)}$ *one can discover decomposition* $n = pq$.

We remark that unlike in the condition (i) of Theorem 11, here we have significantly weaker restriction on $s_B(E_r)$ if η is sufficiently small, but the condition (32) above should be referred to condition (ii) of Theorem 11 and inequality (30).

The proof of Theorem 12 is indirect. We assume on the contrary that the set $\mathbf{X}$ does not contain the separating witness $\bar{P} = (P, P')$. Then by the assumption that $G_{\mathbf{X}}^B$ is a cyclic group its order divides the least common multiple over $\bar{P} \in \mathbf{X}$ of the values

$$\mathrm{lcm}\left(\mathrm{ord}\Phi_n^B(P), \mathrm{ord}\Phi_n^B(P')\right) = \mathrm{lcm}\left(\mathrm{ord}\frac{E_p}{s_B(E_p)}P, \mathrm{ord}\frac{E_q}{s_B(E_q)}P'\right).$$

The first order in lcm above divides $\mathrm{ord}P$ and $\frac{E_p}{\gcd(E_p/s_B(E_p), E_p)}$ that is $\gcd(\mathrm{ord}P, s_B(E_p)) = s_B(\mathrm{ord}P)$. By the assumption we have that $s_B(\mathrm{ord}P) = s_B(\mathrm{ord}P')$, hence

$$\mathrm{lcm}\left(\mathrm{ord}\Phi_n^B(P), \mathrm{ord}\Phi_n^B(P')\right) = s_B(\mathrm{ord}P) = s_B(\mathrm{ord}P')$$

for all $\bar{P} \in \mathbf{X}$. Now it remains to apply the equality

$$|G_{\mathbf{X}}^B| = |G_{\mathbf{X}}|/|\ker \Phi_n^B|,$$

estimate the numerator from above and denominator from below using the conditions (32) and (33) to obtain the contradiction. The last assertion follows from *Remark 1* of section 2.

12 Collection of Results on (E, B, D)−factoring

Below we collect the results concerning the factoring with the aid of (E, B, D)−separating witnesses, (E, B, γ)− nonseparating witnesses and (E, B, D)−separating sets, in one theorem, where E is Y−admissible with $Y = (B)$ or $Y = (B, \beta, D)$.

Theorem 13. *(see [52]) Let* $n = pq$ *be sufficiently large, where* $p < q < \vartheta n^{1/2}$ *and* $B < D$. *Then we have*

(i) *Let* E *be* (B)−*admissible elliptic curve over* $\mathbb{Z}_n$ *and* $\bar{P}$ *be* (E, B)-*separating witness for* n. *Then one can find the prime power representation of its order and decompose* n *in deterministic time* $B^{2+o(1)}$. *If* $\bar{P}$ *is not* (E, B)-*separating witness then we have*

$$d := s_B(\mathrm{ord}P) = s_B(\mathrm{ord}P') \tag{34}$$

and then n can factored in deterministic computational time polylog(n) overhead provided the Eq. (30) holds true and $d > n^{1/2-\gamma/2}$ for some $0 \le \gamma < 1/4$ (which obviously implies that E is $((B),\gamma)-$admissible elliptic curve).

(ii) Let $\beta \in (0,1)$ and E be $(B,\beta,D)-$admissible such that $s_D(E_r)/s_B(E_r)$ is a prime power. If $\mathbf{X}$ is (E,B,D) separating set for n then one can decompose $n = pq$ in deterministic time $|\mathbf{X}|D^{1+o(1)}$.

(iii) Let E be $(B,\beta,D)-$admissible such that $s_D(E_r)/s_B(E_r)$ is a prime power. If $G_{\mathbf{X}} \subseteq E(\mathbb{Z}_n)$ is a cyclic group, $|G_{\mathbf{X}}| \ge n^{1-\gamma}$ for some $0 \le \gamma < \min(1/4, 1-\beta)$ and $\mathbf{X}$ is $(E,B,D)-$separating set for n, then one can compute the generator $\bar{P}_0 \in G_{\mathbf{X}}$ such that at least one of local orders ordP_0 or ordP_0' is greater or equal to $n^{1/2-\gamma/2}$ in deterministic time $|\mathbf{X}|D^{1+o(1)}$.

Moreover if

$$1 \le \min(a_E(p), a_E(q)) \le cn^{\frac{1}{4}-\frac{\gamma}{2}}, \tag{35}$$

for some explicit constant $c = c(\vartheta) > 0$, then one can decompose $n = pq$ in deterministic time $|\mathbf{X}|polylog(n)$.

The deterministic bound $D^{1+o(1)}$ above should be compared to the trivial bound $D^{2+o(1)}$ (with a natural extension for $(Y,\sigma)-$admissible curve with $Y = (B,\beta,D,\sigma)$). We remark that in the proof of the point (iii) above, the condition that $G_{\bar{X}}$ is cyclic allows to search for the related point $\bar{P}_0$ computing the sums of the points of $\bar{X}$. Moreover the generator $\bar{P}_0$ has some local order at least as large as $\sqrt{|G_{\mathbf{X}}|} \ge n^{1/2-\gamma/2}$ so the last conclusion follows from Theorem 11. The deterministic time $|\mathbf{X}|(B+D^{1+o(1)}) = |\mathbf{X}|D^{1+o(1)}$ above, follows from the search for the power of prime $\bar{q}$ defined by (28), since $\mathbf{X}$ contains some $D-$separating witness which is not $B-$separating witness for n.

12.1   Result on Subexponential B and D

Let $\mathcal{P} = \mathcal{P}(n)$ and for $\bar{b} = (b_1, b_2)$ consider the family $\mathcal{E}$ of elliptic curves given by the equation

$$E_{\bar{b}} : y^2 = (x - b_1)(x - b_2)(x + b_1 + b_2)$$

where b_1, b_2 and $-b_1-b_2$ are distinct elements of $\mathbb{Z}_n$ having the short Weierstrass form $E_{\bar{B}} : y^2 = x^3 + B_1 x + B_2$. Let y, b_1 be chosen randomly modulo n and $x \in \mathbb{Z}_n$ so that the Jacobi symbol $(x - b_1/n) = -1$. For randomly chosen $b_2 \notin \{b_1, -2b_1, -\frac{b_1}{2}\}$ mod n we check whether it satisfies the equation

$$\frac{y^2}{x - b_1} \equiv (x - b_2)(x + b_1 + b_2) \pmod n. \tag{36}$$

If not then we select another triple $(y, b_1, x) \in \mathbb{Z}_n^3$ until b_2 satisfies the required conditions. We have the following result (cf. [14, 52])

Theorem 14. *Let $n = pq$, where $p < q$. Given the pair $(E_{\bar{b}}, \bar{P})$, where $E_{\bar{b}}$ is defined above, $\bar{P} = (x, y) \in E(\mathbb{Z}_n)$, while E_p and E_q satisfy the condition*

$$\nu_2(E_p) \geq \nu_2(E_q).$$

Assume that

$$S_{\chi_1}(\mathcal{P}) = (-1, 1), \quad S_{\chi_2}(\mathcal{P}) = (1, 1)$$

where $\chi_1 := \left(\frac{x - b_1}{\cdot}\right)$ and $\chi_2 := \left(\frac{x - b_2}{\cdot}\right)$. Then $l = 2$ is a prime separating the local orders $\mathrm{ord}P$ and $\mathrm{ord}P'$.

This suggest the following steps of algorithm searching for the decomposition $n = pq$. Let $B = L(\alpha, \lfloor\sqrt{n}\rfloor))$ and $D = L(\alpha_0, \lfloor\sqrt{n}\rfloor))$, where $L(\alpha, r) := \exp\left(\alpha\sqrt{\log r \log\log r}\right)$ and $\alpha \leq \alpha_0 = 1/\sqrt{2}$.

1. Select randomly $T = (y, b_1, x) \in \mathbb{Z}_n^3$ as above (if the condition $(x - b_1/n) = -1$ is not satisfied then repeat a random selection of $T \in \mathbb{Z}_n^3$.
2. Select randomly $b_2 \notin \{b_1, -2b_1, -\frac{b_1}{2}\}$ mod n and check whether it satisfies the condition (36) above (if not then return to step 1). Repeat the selection of triples T defining $\lfloor L(\alpha_0, \lfloor\sqrt{n}\rfloor))\rfloor$ non-isomorphic elliptic curves $E_{\bar{b}}$ over $\mathbb{Z}_n$ each time performing steps 3–5.
3. Try to compute the multiple points $R := m_D\bar{P} \in E_{\bar{b}}(\mathbb{Z}_n)$, where m_D is the product of the suitable powers of primes $\leq D$ with $\bar{P} = (x, y)$, using the procedure mentioned in *Remark 1* of section 2. If all points R are finite then stop and return FAIL as output. Otherwise we pass to step 4.
4. Let B be the least positive integer such that $m_B\bar{P}$ is not finite. If $B \leq L(\alpha_0/2, \lfloor\sqrt{n}\rfloor)$ then computing the multiple points $(m_B/l^\nu)\bar{P}$ for prime $l \leq L(\alpha_0, \lfloor\sqrt{n}\rfloor)$ we find the order of $\bar{P} \in E(\mathbb{Z}_n)$ in deterministic time $L(2\alpha_0 + o(1), \lfloor\sqrt{n}\rfloor)$, otherwise pass to step 1.
5. If $\mathrm{ord}\bar{P} \geq n^{1/2 - \gamma/2}$ for $\gamma < 1/4$ then apply the procedure as pointed out in Theorem 11 and the comment there. If the nontrivial divisor d of n is discovered then stop and return d as output, otherwise return to step 1. If the nontrivial factor of n was not found for all triples T above then stop and return FAIL as output.

We conjecture that the expected number of trials generating the proper triples $T \in \mathbb{Z}_n^3$ needed to find a pair $(E, \bar{P})$, such that E_r is $L(\alpha, r)$–smooth for some $r \in \{p, q\}$ and $\nu_2(\mathrm{ord}P) \neq \nu_2(\mathrm{ord}P')$ is $L(1/2\alpha + o(1), r)$. Hence the expected time of factoring n with the aid of the above family $\mathcal{E}$ is

$$L\left(\frac{1}{2\alpha} + o(1), r\right) L(\alpha + o(1), r) = L\left(\frac{1}{2\alpha} + \alpha + o(1), r\right)$$

which minimizes the value of $1/(2\alpha) + \alpha$ for the choice $\alpha = \alpha_0$, giving the average-case time complexity of decomposition $n = pq$ (under GRH) equal to $L(2\alpha_0 + o(1), r)$ as in [39].

Acknowledgement. I would like to thank the anonymous reviewers for their careful reading and valuable comments concerning the article.

References

1. Bach, E.: Discrete logarithms and factoring. University of California at Berkeley (1984)
2. Bach, E., Charles, D.: The hardness of computing an eigenform. arXiv preprint arXiv:0708.1192 (2007)
3. Baier, S.: On the least n with χ (n)$\neq$ 1. Q. J. Math. **57**(3), 279–283 (2006)
4. Baier, S.: A remark on the least n with χ (n)$\neq$ 1. Arch. Math. **86**, 67–72 (2006)
5. Bystrzycki, R.: Detection of primes in the set of residues of divisors of a given number. In: Number-Theoretic Methods in Cryptology: First International Conference, NuTMiC 2017, Warsaw, Poland, 11–13 September 2017, Revised Selected Papers 1, pp. 178–194. Springer (2018)
6. Chang, M.C.: Short character sums for composite moduli. J. d'Analyse Mathématique **123**(1), 1–33 (2014)
7. Chow, A.: Applications of Fourier coefficients of modular forms. University of Toronto (Canada) (2015)
8. Coppersmith, D.: Small solutions to polynomial equations, and low exponent RSA vulnerabilities. J. Cryptol. **10**(4), 233–260 (1997)
9. Coppersmith, D.: Finding small solutions to small degree polynomials. In: International Cryptography and Lattices Conference, pp. 20–31. Springer (2001)
10. Crandall, R.E., Pomerance, C.: Prime numbers: a computational perspective, vol. 2. Springer (2005)
11. Dąbrowski, A., Pomykała, J.: On a Linnik problem for elliptic curves. Proc. Am. Math. Soc. **147**(9), 3759–3763 (2019)
12. Dąbrowski, A., Pomykała, J.: Signatures of Dirichlet characters and elliptic curves. J. Number Theory **220**, 94–106 (2021)
13. Dąbrowski, A., Pomykała, J., Pujahari, S.: On signatures of elliptic curves and modular forms. Ramanujan J. **60**(2), 505–516 (2023)
14. Dąbrowski, A., Pomykała, J., Shparlinski, I.E.: On oracle factoring of integers. J. Complex. **76**, 101741 (2023)
15. Dicson, L.: History of the theory of numbers. Co., New York (1934)
16. Dieulefait, L.V., Urroz, J.: Factorization and malleability of RSA moduli, and counting points on elliptic curves modulo N. Mathematics **8**(12), 2126 (2020)
17. Dryło, R., Pomykała, J.: Factoring n and the number of points of kummer hypersurfaces mod N, pp. 163–177 (2017)
18. Dryło, R., Pomykała, J.: Integer factoring problem and elliptic curves over the ring $\mathbb{Z}_n$. In: Colloquium Mathematicum, vol. 159, pp. 259–284. Instytut Matematyczny Polskiej Akademii Nauk (2020)
19. Drylo, R., Pomykala, J.: Smooth factors of integers and elliptic curve based factoring with an oracle. Banach Center Publications (2023). https://api.semanticscholar.org/CorpusID:265564322
20. Dryło, R., Pomykała, J.: Smooth factors of integers and elliptic curve based factoring with an oracle. Banach Center Publ. **126**, 73–88 (2023)
21. Granville, A.: Smooth numbers: computational number theory and beyond. Algorithmic number theory: lattices, number fields, curves and cryptography **44**, 267–323 (2008)
22. Hafner, J.L., McCurley, K.S.: On the distribution of running times of certain integer factoring algorithms. J. Algorithms **10**, 531–556 (1989). https://api.semanticscholar.org/CorpusID:37493151

23. Harris, M., Harris, M.: The sato-tate conjecture: introduction to the proof (2006). https://api.semanticscholar.org/CorpusID:201737404
24. Harvey, D.: An exponent one-fifth algorithm for deterministic integer factorisation. Math. Comput. **90**(332), 2937–2950 (2021)
25. Harvey, D., Hittmeir, M.: A log-log speedup for exponent one-fifth deterministic integer factorisation. Math. Comput. **91**(335), 1367–1379 (2022)
26. Hittmeir, M.: A babystep-giantstep method for faster deterministic integer factorization. Math. Comput. **87**(314), 2915–2935 (2018)
27. Hittmeir, M., Pomykała, J.: Deterministic integer factorization with oracles for Euler's totient function. Fund. Inform. **172**(1), 39–51 (2020)
28. Hunter, J.: G. h. hardy, and e. m. wright, an introduction to the theory of numbers (fourth edition) (clarendon press: Oxford university press, 1960), 421 pp., 42s. Proceedings of the Edinburgh Mathematical Society **12**, 161 (1961). https://api.semanticscholar.org/CorpusID:162247337
29. Iwaniec, H.: On the problem of Jacobsthal. Demonstratio Math. **11**(1), 225–232 (1978)
30. Jiménez Urroz, J., Pomykała, J.: Factoring numbers with elliptic curves. Ramanujan J. 1–9 (2024)
31. Knuth, D.E., Pardo, L.T.: Analysis of a simple factorization algorithm. Theor. Comput. Sci. **3**, 321–348 (1976). https://api.semanticscholar.org/CorpusID:38968900
32. Koblitz, N.: A Course in Number Theory and Cryptography, vol. 114. Springer, Cham (1994)
33. Lamzouri, Y., Li, X., Soundararajan, K.: Conditional bounds for the least quadratic non-residue and related problems. Math. Comput. **84**(295), 2391–2412 (2015)
34. Lau, Y.K., Wu, J.: On the least quadratic non-residue. Int. J. Number Theory **4**(03), 423–435 (2008)
35. Lawrence, F.W.: Factorisation of numbers. Q. J. Pure Appl. Math. **28**, 285–311 (1896)
36. Lee, J.D., Venkatesan, R.: Rigorous analysis of a randomised number field sieve. J. Number Theory **187**, 92–159 (2018)
37. Lehman, R.S.: Factoring large integers. Math. Comput. **28**(126), 637–646 (1974)
38. Lenstra, H.W., et al.: Elliptic curves and number-theoretic algorithms. Universiteit van Amsterdam Mathematisch Instituut (1986)
39. Lenstra, H.W., Jr.: Factoring integers with elliptic curves. Ann. Math. 649–673 (1987)
40. Lu, Y., Peng, L., Zhang, R., Hu, L., Lin, D.: Towards optimal bounds for implicit factorization problem. In: Selected Areas in Cryptography–SAC 2015: 22nd International Conference, Sackville, NB, Canada, 12–14 August 2015, Revised Selected Papers 22, pp. 462–476. Springer (2016)
41. Lu, Y., Zhang, R., Lin, D.: Improved bounds for the implicit factorization problem. Adv. Math. Commun. **7**(3), 243–251 (2013)
42. Martin, G.: The least prime primitive root and the shifted sieve. arXiv, Number Theory (1998). https://api.semanticscholar.org/CorpusID:7865146
43. Martin, S., Morillo, P., Villar, J.L.: Computing the order of points on an elliptic curve modulo n is as difficult as factoring N. Appl. Math. Lett. **14**(3), 341–346 (2001)
44. Maurer, U.M.: On the oracle complexity of factoring integers. Comput. Complex. **5**, 237–247 (1995)

45. May, A., Ritzenhofen, M.: Implicit factoring: on polynomial time factoring given only an implicit hint. In: International Workshop on Public Key Cryptography, pp. 1–14. Springer (2009)
46. Miller, G.L.: Riemann's hypothesis and tests for primality. In: Proceedings of the Seventh Annual ACM Symposium on Theory of Computing, pp. 234–239 (1975)
47. Morain, F., Renault, G., Smith, B.: Deterministic factoring with oracles. Appl. Algebra Eng. Commun. Comput. **34**(4), 663–690 (2023)
48. Nitaj, A., Ariffin, M.R.K.: Implicit factorization of unbalanced RSA moduli. J. Appl. Math. Comput. **48**, 349–363 (2015)
49. Pollard, J.M.: Theorems on factorization and primality testing. **76**(3), 521–528 (1974)
50. Pomykała, J.: On q-orders in primitive modular groups. Acta Arith. **166**, 397–404 (2014)
51. Pomykała, J.: On exponents of modular subgroups generated by small consecutive integers. Acta Arith. **176**, 321–342 (2016)
52. Pomykała, J., Jurkiewicz, M., Żołnierczyk, O., Prabucka, K.: Enhanced Performance of ECM for RSA Modulus via Generalized B-smoothness (2024, submitted)
53. Pomykała, J., Żołnierczyk, O.: Elliptic curve-integer factorizatiin and witnesses. In: Accepted for the conference ICCS 2024 (2024)
54. Pomykała, J., Radziejewski, M.: Integer factoring and compositeness witnesses. J. Math. Cryptol. **14**(1), 346–358 (2020)
55. Pomykała, J., Źrałek, B.: On reducing factorization to the discrete logarithm problem modulo a composite. Comput. Complex. **21**(3), 421–429 (2012)
56. Źrałek, B.: Using the smoothness of p-1 for computing roots modulo. arXiv preprint arXiv:0803.0471 (2008)
57. Rivest, R.L., Shamir, A., Adleman, L.: A method for obtaining digital signatures and public-key cryptosystems. Commun. ACM **21**(2), 120–126 (1978)
58. Shoup, V.: Searching for primitive roots in finite fields. In: Proceedings of the Twenty-Second Annual ACM Symposium on Theory of Computing, pp. 546–554 (1990)
59. Shoup, V.: A Computational Introduction to Number Theory and Algebra. Cambridge University Press, Cambridge (2009)
60. Sutherland, A.V.: Order computations in generic groups. Ph.D. thesis, Massachusetts Institute of Technology (2007)
61. Williams, H.C.: A p+1 method of factoring. Math. Comput. **39**(159), 225–234 (1982)
62. Źrałek, B.: A deterministic version of pollard's p-1 algorithm. Math. Comput. **79**(269), 513–533 (2010)

Elliptic Curves in Cryptography

On Chains of Pairing-Friendly Elliptic Curves

Maciej Grześkowiak[✉]

Faculty of Mathematics and Computer Science, Adam Mickiewicz University,
Uniwersytetu Poznańskiego 4, 61-614 Poznań, Poland
maciejg@amu.edu.pl

Abstract. Recently, many constructions of curves aim to implement different kinds of proof systems efficiently. For the protocol to be efficient, one must generate particular forms of prime numbers. This article presents an algorithm that finds desired prime numbers in polynomial time.

Keywords: Pairing-friendly elliptic curves · Chain of elliptic curves · Proof systems

1 Introduction

Let E be an elliptic curve defined over finite field $\mathbb{F}_p$, where p is a prime. Let $\#E(\mathbb{F}_p)$ be the order of group of $\mathbb{F}_p$-rational points of E.

Definition 1. *Fix $s \in \mathbb{N}$. An s-chain of elliptic curves is a list of distinct curves $E_1, \ldots, E_s$ defined over finite fields $\mathbb{F}_{p_1}, \ldots, \mathbb{F}_{p_s}$ respectively such that*

$$q \mid \#E_1(\mathbb{F}_{p_1}), \quad p_1 \mid \#E_2(\mathbb{F}_{p_2}), \quad \ldots, \quad p_{s-1} \mid \#E_s(\mathbb{F}_{p_s})$$

and $q, p_1, \ldots, p_s$ are primes.

If, in the above definition, we assume that p_s equals q, then the s-chain is called s-cycle of elliptic curves. As shown by Silverman and Stange, cycles of arbitrary lengths exist [14].

Let q be a divisor of $\#E(\mathbb{F}_p)$ such that q is prime to p. Hasse's theorem states $\#E(\mathbb{F}_p) = p + 1 - t$, where $|t| \le 2\sqrt{p}$, and $t \in \mathbb{Z}$ [13]. The embedding degree of E with respect to q is the smallest positive integer n such that $q \mid p^n - 1$, but q does not divide $p^d - 1$ for $d \mid n$ [5]. This condition is equivalent to $q > n$ divides $\Phi_n(p)$, where $\Phi_n(x)$ is the nth cyclotomic polynomial [3]. Elliptic curves over $\mathbb{F}_p$ that have a large subgroup of prime order q and a small embedding degree n are commonly referred to as *pairing-friendly* [5]. The construction method of a pairing-friendly curve relies on complex multiplication (CM) techniques. Cocks and Pinch [4] proposed an algorithm which, given a prime number q, a square-free integer $\Delta < 0$ and an integer n, constructs an elliptic curve such that q divides $\#E(\mathbb{F}_p)$ and whose embedding degree with respect to q is n, where

© The Author(s), under exclusive license to Springer Nature Switzerland AG 2025
A. Dąbrowski et al. (Eds.): NuTMiC 2024, LNCS 14966, pp. 121–135, 2025.
https://doi.org/10.1007/978-3-031-82380-0_3

$4p = t^2 - \Delta f^2$, $t, f \in \mathbb{Z}$. The algorithm's most consuming step is finding the equation of an elliptic curve E over $\mathbb{F}_p$. However, the authors didn't analyze the computational complexity of constructing a desired prime p. The second approach for generating desired curves with a fixed embedding degree n is by using a family of parametrized polynomials $(p_n(x), q_n(x), t_n(x))$ representing the field of definition, number of rational points, and trace, respectively [3]. Such the family of polynomials satisfies the following

$$q_n(x) = p_n(x) + 1 - t_n(x), \quad q_n(x) \mid \Phi_n(p_n(x))$$

and $4p_n(x) - t_n(x)^2 = \Delta f(x)^2$ for some small square-free integer $\Delta < 0$, $f(x) \in \mathbb{Z}[X]$. As an example, we briefly recall the idea of this algorithm for $n = 4$. We have

$$p_4(x) = x^2 + x + 1, \quad q_4(x) = x^2 + 1, \quad t_4(x) = x + 1.$$

The algorithm finds an integer x_0 such that $p = p_4(x_0)$ and $q = q_4(x_0)$ are simultaneously prime. Next, an elliptic curve E over $\mathbb{F}_p$ of order q is constructed using the CM method. However, while analyzing the computational complexity of the above algorithm, we can encounter some problems. To be more precise. We do not know if infinitely many integers x_0 exist such that $p = p_4(x_0)$ and $q = q_4(x_0)$ are simultaneously prime. However, there are some conjectures related to this problem [2, 15]. The discussion above motivated us to introduce the following definition.

Definition 2. *Given $n \in \mathbb{N}$ and a square-free integer $\Delta < 0$. Primes p and q are pairing-friendly with respect to n and Δ if there exist integers $f, t \in \mathbb{Z}$ such that*

$$q \mid p + 1 - t, \quad q \mid \Phi_n(p), \quad t^2 - 4p = \Delta f^2 < 0. \tag{1}$$

Given pairing-friendly primes p and q with respect to n, Δ and integers f, t as above, then there exists an ordinary elliptic curve E over $\mathbb{F}_p$ having cardinality $p + 1 - t$. From the theoretical point of view, it is interesting to know whether we can generate pairing-friendly primes with respect to n, Δ in polynomial time in $\log(\#E(\mathbb{F}_p))$. It is worth noting that the number n is small in cryptographic applications, say $n < 100$.

A recent area of interest in cryptography is zero-knowledge proof systems [7,8]. Roughly speaking, a zero-knowledge proof system is a protocol between two parties, called the prover and the verifier. The prover aims to convince the verifier that the prover knows a particular secret via a short proof that is cheap to verify and reveals no information about the secret. Lately, different kinds of zero-knowledge proof systems have been developed. One of the approaches to making such protocols efficient involves chains or cycles of pairing-friendly elliptic curves [6,7]. To be more precise, we introduce the following definition (compare with Definition 2 [7]).

Definition 3. *An s-chain of pairing-friendly elliptic curves is an s-chain of elliptic curves* $E_1/\mathbb{F}_{p_1}, \ldots, E_s/\mathbb{F}_{p_s}$ *with*

$$q \mid \#E_1(\mathbb{F}_{p_1}), \quad p_1 \mid \#E_2(\mathbb{F}_{p_2}), \quad \ldots, \quad p_{s-1} \mid \#E_s(\mathbb{F}_{p_s}),$$

where primes q, p_1 *are pairing-friendly with respect to* n_1 *and* Δ, *primes* p_i, p_{i+1} *are pairing-friendly with respect to* n_{i+1}, *and*

$$2^m \mid q - 1, \quad 2^m \mid p_i - 1, \quad i = 1, \ldots, s - 1.$$

for fixed positive integer m.

The last condition in the above definition regarding divisibility by 2^m is related to the effective implementation of the Fast Fourier Transforms in $\mathbb{F}_q, \mathbb{F}_{p_i}$ respectively. Such a possibility significantly speeds up the performance of the proposed scheme [7]. For this reason, we include this condition in the above definition. Throughout the paper we assume that $m \ll 1$.

We give an example of constructing the s-chain of pairing-friendly elliptic curves [6]. Set $s = 2$, and $f, g, h \in \mathbb{Z}[x]$,

$$f(x) = 36x^4 + 36x^3 + 18x^2 + 6x + 1, \quad g(x) = 36x^4 + 36x^3 + 24x^2 + 6x + 1,$$
$$h(x) = 5184x^8 + 10368x^7 + 12204x^6 + 8856x^5 + 4536x^4 + 1548x^3 +$$
$$363x^2 + 48x + 4.$$

If there is $x_0 \in \mathbb{Z}$ such that $q = f(x_0), p_1 = g(x_0)$, and $p_2 = h(x_0)$ are simultaneously prime numbers, then there exists two elliptic curves E_1 and E_2 defined over finite field $\mathbb{F}_{p_1}$ and $\mathbb{F}_{p_2}$ respectively, such that

$$q = \#E_1(\mathbb{F}_{p_1}), \quad p_1 \mid \#E_2(\mathbb{F}_{p_2}),$$

primes q, p_1 are pairing-friendly with respect to $n = 12$, and primes p_1, p_2 are pairing-friendly with respect to $n = 6$. For more construction of pairing-friendly elliptic curves see [7]. Note that estimating the complexity of the above algorithm's running time is possible under heuristic assumptions [2,15]. Now, we introduce the following definition.

Definition 4. *Given* $s, m, n_1, \ldots, n_s \in \mathbb{N}$ *and a square-free integer* $\Delta < 0$. *Primes* q *and* $p_1, \ldots, p_s$ *are s-chain-friendly with respect to* $n_1, \ldots, n_s$ *and* Δ, *if*

- *primes* q, p_1 *are pairing-friendly with respect to* n_1 *and* Δ,
- *for* $i = 1, \ldots, s - 1$, *primes* p_i, p_{i+1} *are pairing-friendly with respect to* n_{i+1}, *and* Δ.

Furthermore,

$$2^m \mid q - 1, \quad 2^m \mid p_i - 1, \quad i = 1, \ldots, s.$$

The above discussion shows that if we have given chain-friendly primes q and $p_1, \ldots, p_s$, we can construct the corresponding s-chain of pairing-friendly elliptic curves.

In the present paper, we introduce an algorithmic method for construct-ing chain-friendly primes. In particular, we propose an algorithm for generating pairing-friendly primes p and q with respect to n, Δ. We prove that our method for finding such primes is probabilistic and executes in polynomial time. To the best of the author's knowledge, such a result didn't appear in the literature.

The remaining part of the paper is organized as follows. In Sect. 2 we present the algorithm for generating pairing-primes with respect to n and Δ. A detailed analysis of our algorithms is presented in Sect. 3. Section 4 presents an algorith-mic method for constructing s-chain-friendly primes. An illustrative example is given in Sect. 5.

1.1   Notation

We denote by $\mathbb{Z} \subseteq \mathbb{Q} \subseteq \mathbb{R} \subseteq \mathbb{C}$ the ring of integers and the fields of rational, real, and complex numbers respectively. Throughout this paper, $\Delta < 0$ is a square-free rational integer, $K = \mathbb{Q}(\sqrt{\Delta})$ is the quadratic field with the corresponding ring of integers $\mathcal{O}_K = \{a + b\omega : a, b \in \mathbb{Z}\}$, and $\mathcal{O}_f = [1, f\omega]$, $f \in \mathbb{Z}$ is any order of K, where $\omega = \frac{1+\sqrt{\Delta}}{2}$ when $\Delta \equiv 1 \pmod 4$ and $\omega = \sqrt{\Delta}$ for $\Delta \equiv 2, 3 \pmod 4$. By $N(\alpha) = \alpha\overline{\alpha} = (a + b\omega)(a + b\overline{\omega})$ we denote the norm of an element $\alpha = a + b\omega \in \mathcal{O}_K$ with respect to $\mathbb{Q}$. That is

$$N(\alpha) = a^2 + ab + \tfrac{1-\Delta}{4}b^2 \quad \text{if} \quad \Delta \equiv 1 \pmod 4,$$
$$N(\alpha) = a^2 - \Delta b^2 \quad \text{if} \quad \Delta \equiv 2, 3 \pmod 4.$$

By $w(K)$ we denote the number roots of unity in K, and by $h(K)$ we denote the class number of the field K. Let $\mathfrak{f}$ be an ideal of $\mathcal{O}_K$. We say that two ideals $\mathfrak{A}, \mathfrak{B} \subseteq \mathcal{O}_K$, prime to $\mathfrak{f}$ are equivalent $\pmod{\mathfrak{f}}$, if there are $\alpha, \beta \in \mathcal{O}_K$ congruent to 1 $\pmod{\mathfrak{f}}$, so that $\alpha\mathfrak{A} = \beta\mathfrak{B}$. This equivalence splits the set of ideals of $\mathcal{O}_K$ into a finite number of classes. The set of all equivalence classes is called the group of narrow ray classes $\pmod{\mathfrak{f}}$. We denote it by $H_{\mathfrak{f}}^*(K)$, and by $h_{\mathfrak{f}}^*(K)$ we denote the number of elements in $H_{\mathfrak{f}}^*(K)$. Note that K is a totally complex field, so all non-zero numbers are totally positive and we obtain $H_{\mathfrak{f}}^*(K) = H_{\mathfrak{f}}(K)$, where $H_{\mathfrak{f}}(K)$ is called the group of ideal classes $\pmod{\mathfrak{f}}$. Let m be a positive integer. By $\mathcal{PT}$ we denote the number of bit operations necessary to carry out the deterministic primality test [1]. For simplicity, assume that $\mathcal{PT}$ takes no less than $O(\log^3 m)$ bit operations.

2   Algorithm for Pairing-Friendly Primes

The purpose of this section is to describe the algorithm which generates pairing-friendly primes p and q with respect to n and Δ. Fix $K = \mathbb{Q}(\sqrt{\Delta})$ with the corresponding ring of integers $\mathcal{O}_K = \{a + b\omega : a, b \in \mathbb{Z}\}$. The algorithm consists of the following three procedures. The first procedure is a slight modification of Procedure FINDPRIMEQ of [10]. The only difference is that we take $m = 1$ and $n = m_1$ in the presented procedure. We assume that for given $n_1, m \in \mathbb{N}$, we have computed $\gamma_1 \in \mathcal{O}_K$ such that $N(\gamma_1) \equiv 1 \pmod{\mathrm{lcm}(n_1, 2^m)}$. The fixed number m will be the input for all three procedures presented below.

Procedure. FINDPRIMEQ$(n_1, m, \gamma_1, \Delta, x)$. Given $n_1, m \in \mathbb{N}$, $\gamma_1 = f_1 + g_1\omega \in \mathcal{O}_K$ such that $|f_1|, |g_1| \leq \mathrm{lcm}(n_1, 2^m)$, $N(\gamma_1) \equiv 1 \pmod{\mathrm{lcm}(n_1, 2^m)}$, and $x \in \mathbb{R}$; this procedure finds $\alpha \in \mathcal{O}_K$ such that $q = N(\alpha) \equiv 1 \pmod{\mathrm{lcm}(n_1, 2^m)}$ is a prime, $x \leq q \leq 2x$, and $\mathrm{lcm}(n_1, 2^m) \mid q - 1$.

step 1. Compute $m_1 = \mathrm{lcm}(n_1, 2^m)$,
step 2. Choose $u \neq 0, v \neq 0$ at random in $\mathbb{Z}$,
  If $\Delta \equiv 1 \pmod 4$,

$$|u| \leq (\frac{\sqrt{1-\Delta}}{\sqrt{-\Delta}}(2x)^{1/2} - f_1)m_1^{-1}, \quad |v| \leq (\frac{2}{\sqrt{-\Delta}}(2x)^{1/2} - g_1)m_1^{-1}.$$

  If $\Delta \equiv 2, 3 \pmod 4$,

$$|u| \leq ((2x)^{1/2} - f)m_1^{-1}, \quad |v| \leq (\frac{1}{\sqrt{-\Delta}}(2x)^{1/2} - g)m_1^{-1}.$$

step 3. Compute $a = m_1 u + f_1$ and $b = m_1 v + g_1$
step 4. Compute

$$\begin{array}{ll} q = a^2 + ab + \frac{1-\Delta}{4}b^2 & \text{if} \quad \Delta \equiv 1 \pmod 4, \\ q = a^2 - \Delta b^2 & \text{if} \, \Delta \equiv 2, 3 \pmod 4. \end{array}$$

step 5. If $q < x$ or $q > 2x$, then go to step 2.
step 6. If q is a prime, then return $\alpha = a + b\omega$, q. Otherwise, go to step 2.

Remark 1. Note that if $m_1 = \mathrm{lcm}(n_1, 2^m)$, it is always possible to find $\gamma_1 = f_1 + g_1\omega \in \mathcal{O}_K$ satisfying $N(\gamma_1) \equiv 1 \pmod{m_1}$. For example, $f_1 = -1$ and $g_1 = 0$ is a possible solution. In general, the congruence $N(x) \equiv 1 \pmod{m_1}$ has more than one solution. For more details we refer the reader to [11] (see §47).

Lemma 1. *Fix $K = \mathbb{Q}(\sqrt{\Delta})$ with the corresponding ring of integers $\mathcal{O}_K$, and let $\mathfrak{f} = m_1\mathcal{O}_K$ with $m_1 = \mathrm{lcm}(n_1, 2^m)$. There exists $x_0 > 0$ such that for every $x \geq x_0$ and an arbitrary real $\lambda \geq 1$, procedure FINDPRIMEQ finds $\alpha = a + b\omega \in \mathcal{O}_K$ such that $q = N(\alpha) \equiv 1 \pmod{m_1}$ is a prime, $x \leq q \leq 2x$, where*

$$\begin{array}{ll} |a| \leq \frac{\sqrt{1-\Delta}}{\sqrt{-\Delta}}(2x)^{\frac{1}{2}}, |b| \leq \frac{2}{\sqrt{-\Delta}}(2x)^{\frac{1}{2}} & \text{if} \quad \Delta \equiv 1 \pmod 4, \\ |a| \leq (2x)^{\frac{1}{2}}, \qquad |b| \leq \frac{1}{\sqrt{-\Delta}}(2x)^{\frac{1}{2}} & \text{if} \, \Delta \equiv 2, 3 \pmod 4. \end{array}$$

with probability greater than or equal to $1 - e^{-\lambda}$ after repeating $[c_1\lambda(\log x)]$ steps of the procedure, where $c_1 = \frac{16\sqrt{1-\Delta}h_{\mathfrak{f}}^*(K)}{-\Delta m_1^2}$ when $\Delta \equiv 1 \pmod 4$, and $c_1 = \frac{16h_{\mathfrak{f}}^*(K)}{\sqrt{-\Delta}m_1^2}$ for $\Delta \equiv 2, 3 \pmod 4$. Every step of the procedure takes no more than $\mathcal{PT}$ bit operations.

Proof. See Theorem 2.1 [10].

We are now in a position to present our second procedure. We assume that α and q such that $q = N(\alpha) \equiv 1 \pmod{n_1}$ are the output of procedure FIND-PRIMEQ. In addition, w_{n_1}, the n_1-th root of unity modulo q, has been computed. Moreover, we suppose we that for a given $n_2, m \in \mathbb{N}$, we have calculated $\gamma_2 \in \mathcal{O}_K$ such that $N(\gamma_2) \equiv 1 \pmod{\operatorname{lcm}(n_2, 2^m)}$. The fixed number n_2 will be the input for the subsequent two procedures below.

Procedure. FINDINTEGERDELTA$(\alpha, q, n_2, m, \gamma_2, w_{n_1}, \Delta)$. Given $\alpha, \gamma_2 \in \mathcal{O}_K$, $\alpha = a + b\omega$, $\gamma_2 = f_2 + g_2\omega$ and w_{n_1} a primitive n_1-th root of unity $\pmod{q}$ such that $q = N(\alpha) \equiv 1 \pmod{n_1}$, $N(\gamma_2) \equiv 1 \pmod{\operatorname{lcm}(n_2, 2^m)}$; this procedure finds $\delta \in \mathcal{O}_K$ such that $N(\delta) \equiv w_{n_1} \pmod{q}$, $N(\delta) \equiv 1 \pmod{\operatorname{lcm}(n_2, 2^m)}$.

step 1. Compute $r \equiv a(-b)^{-1} \pmod{q}$.
step 2. Compute x and y modulo q.
  If $\Delta \equiv 1 \pmod 4$,

$$x \equiv (1 - (1 + w_{n_1})r)(1 - 2r)^{-1} \pmod{q},$$
$$y \equiv (w_{n_1} - 1)(1 - 2r)^{-1} \pmod{q}.$$

If $\Delta \equiv 2, 3 \pmod 4$

$$x \equiv (1 - w_{n_1})2^{-1} \pmod{q},$$
$$y \equiv (1 + w_{n_1})(2r)^{-1} \pmod{q}.$$

step 3. Compute $m_2 = \operatorname{lcm}(n_2, 2^m)$.
step 4. Compute $q' \equiv q^{-1} \pmod{m_2}$ and $m_2' \equiv m_2^{-1} \pmod{q}$.
step 5. Compute $k \equiv xq'q + f_2 m_2' m_2 \pmod{qm_2}$.
step 6. Compute $l \equiv yq'q + g_2 m_2' m_2 \pmod{qm_2}$.
step 7. Return $\delta = k + l\omega$.

Lemma 2. *Procedure* FINDINTEGERDELTA *takes no more than* $O(\log^3(2q))$ *bit operations.*

Proof. It is easily seen that the most consuming step of the procedure is computing a modular multiplicative inverse of an element modulo qm_2. This task can be done using the extended Euclidean algorithm which takes $O(\log^3(qm_2))$ bit operations. In this paper, we assume that $m, n_2 \ll 1$, so $qm_2 = q\operatorname{lcm}(n_2, 2^m) \ll 2q$. This finishes the proof.

Remark 2. Let $\Phi_n(x) \in \mathbb{Z}[x]$ be the n-th cyclotomic polynomial of degree $\varphi(n)$. We can use a probabilistic algorithm that factors $\Phi_n(x)$ over $\mathbb{F}_p[x]$ to calculate the roots of a polynomial $\Phi_n(x)$ over $\mathbb{F}_p$. The Cantor-Zassenhaus algorithm uses an expected number of $O(\varphi(n)^3 \log p)$ operations in $\mathbb{F}_p$ to factor $\Phi_n(x)$ [12][Th. 20.9]. In this paper, we consider the number $n < 100$ to be small; therefore, we assume that we can find a root of $\Phi_n(x)$ in our time-bound.

We are now in a position to present our third procedure. We assume that a prime q is the output of procedure FINDPRIMEQ and $\delta = k + l\omega \in \mathcal{O}_K$ is the output of procedure FINDINTEGERDELTA.

Procedure. FINDPRIMEP$(n_2, m, q, \delta, \Delta, \varepsilon, x)$. Given $n_2, m, q \in \mathbb{N}$, $0 < \varepsilon < 2/5$ and $x \in \mathbb{R}$; this procedure finds $\beta \in \mathcal{O}_K$ such that $p = N(\beta)$ is a prime, $N(\beta) \equiv N(\delta) \pmod{\text{lcm}(qn_2, 2^m)}$, $p \equiv 1 \pmod{n_2}$, and $2^m \mid p - 1$.

step 1. Compute $m_2 = \text{lcm}(n_2, 2^m)$.

step 2. Choose $s \neq 0, t \neq 0$ at random in $\mathbb{Z}$.
If $\Delta \equiv 1 \pmod 4$,

$$|s| \leq \frac{1}{2\sqrt{-\Delta}}(2m_2 x)^{(6+10\varepsilon)/(4-10\varepsilon)}, \quad |t| \leq \frac{1}{\sqrt{-\Delta}}(2m_2 x)^{(6+10\varepsilon)/(4-10\varepsilon)}.$$

If $\Delta \equiv 2, 3 \pmod 4$

$$|s| \leq \frac{1}{2}(2m_2 x)^{(6+10\varepsilon)/(4-10\varepsilon)}, \quad |t| \leq \frac{1}{2\sqrt{-\Delta}}(2m_2 x)^{(6+10\varepsilon)/(4-10\varepsilon)}$$

step 3. Compute $c = m_2 qs + k$ and $d = m_2 qt + l$.

step 4. Compute

$$\begin{aligned} p = c^2 + cd + \tfrac{1-\Delta}{4}d^2 & \quad \text{if} \quad \Delta \equiv 1 \pmod 4, \\ p = c^2 - \Delta d^2 & \quad \text{if } \Delta \equiv 2, 3 \pmod 4. \end{aligned}$$

step 5. If $p < (m_2 x)^2$ or $p > (2m_2 x)^{10/(2-5\varepsilon)}$, then go to step 2.

step 6. If p is a prime, then return $\beta = c + d\omega$, p. Otherwise, go to step 2.

Lemma 3. *Fix $K = \mathbb{Q}(\sqrt{\Delta})$ with the corresponding ring of integers $\mathcal{O}_K$, and fix $0 < \varepsilon < \frac{2}{5}$. Let $x \leq q \leq 2x$ be the output of procedure FINDPRIMEQ and $\delta \in \mathcal{O}_K$ be the output of procedure FINDINTEGERDELTA. Procedure FINDPRIMEP with the input consisting of $q, n_2, m \in \mathbb{N}$ has the following properties: there exists $x_0 > 0$ such that for every $x \geq x_0$, and for an arbitrary real $\lambda \geq 1$, and for any constant $A > 2$, the procedure finds $\beta \in \mathcal{O}_K$ such that,*

$$p = N(\beta) \text{ is a prime}, \quad (m_2 q)^2 \leq N(\beta) \leq (2m_2 x)^{10/(2-5\varepsilon)},$$

where $m_2 = \text{lcm}(n_2, 2^m)$, with probability greater than or equal to $1 - e^{-\lambda}$ after repeating $\lceil c_2 \lambda (\log 2m_2 x) \rceil$ steps of the procedure, where $c_2 = \frac{-80h(K)}{-(2-5\varepsilon)w(K)\Delta}$ when $\Delta \equiv 1 \pmod 4$, and $c_2 = \frac{-40h(K)}{(2-5\varepsilon)w(K)\Delta}$ if $\Delta \equiv 2, 3 \pmod 4$, for almost all α with the possible exception of at most $O(x(\log x)^{-A})$ values of α. Every step of the procedure takes no more than $\mathcal{PT}$ bit operations.

Proof. See Sect. 3.2.

We are now ready to proceed to the final stage of our algorithm construction. Given a square-free rational integer $\Delta < 0$, $n_1, n_2, m \in \mathbb{N}$ and $x \in \mathbb{R}$, $0 < \varepsilon < \frac{2}{5}$ perform the following algorithm to find pairing-friendly primes with respect to n_1 and Δ.

Algorithm 1(Pairing-friendly primes)

step 1. Find $\gamma_1 \in \mathcal{O}_K$ such that $N(\gamma_1) \equiv 1 \pmod{\operatorname{lcm}(n_1, 2^m)}$.
step 2. $\alpha = a + b\omega, q := \text{FINDPRIMEQ}(n_1, m, \gamma_1, \Delta, x)$.
step 3. Compute w_{n_1} a primitive n_1-th root of unity $\pmod q$.
step 4. Find $\gamma_2 \in \mathcal{O}_K$ such that $N(\gamma_2) \equiv 1 \pmod{\operatorname{lcm}(n_2, 2^m)}$.
step 5. $\delta = k + l\omega := \text{FINDINTEGERDELTA}(\alpha, q, n_2, m, \gamma_2, w_{n_1}, \Delta)$.
step 6. $\beta = c + d\omega, p := \text{FINDPRIMEP}(n_2, m, q, \delta, \Delta, \varepsilon, x)$.
step 7. Return p, q, α, β.

Theorem 1. *Fix* $K = \mathbb{Q}(\sqrt{\Delta})$ *with the corresponding ring of integers* $\mathcal{O}_K$. *Given positive integers* $m, n_1, n_2,$ *and* $\gamma_1, \gamma_2 \in \mathcal{O}_K$. *Algorithm 1 finds* $\alpha, \beta \in \mathcal{O}_K$ *such that* $q = N(\alpha)$, $p = N(\beta)$ *are pairing-friendly primes with respect to* n_1 *and* Δ *and* $2^m \mid q - 1$, $2^m \mid p - 1$, $q \equiv 1 \pmod{n_1}$, $p \equiv 1 \pmod{n_2}$.

Proof. See Sect. 3.1.

3   Analysis of Algorithm 1

To prove Theorem 1 and Theorem 2 we need two lemmas.

Lemma 4. *Let* $\Delta < 0$ *be a square-free integer. Fix* $K = \mathbb{Q}(\sqrt{\Delta})$ *with the corresponding ring of integers* $\mathcal{O}_K$. *Let* $\alpha = a + b\omega \in \mathcal{O}_K$ *be such that* $q = N(\alpha)$ *is a prime that does not divide* Δ. *Then there exists an integer* $r \equiv \omega \pmod{(\alpha)}$. *Further, the map*

$$\psi : \mathcal{O}_K / \alpha\mathcal{O}_K \longrightarrow \mathbb{Z}/q\mathbb{Z} \qquad (2)$$

defined by

$$(e + f\omega) + \alpha\mathcal{O}_K \longmapsto (e + fr) + q\mathbb{Z}$$

is the ring isomorphism. Moreover, $r \equiv a(-b)^{-1} \pmod q$.

Proof. See Lemma 1 [9].

Lemma 5. *Let* $\Delta < 0$ *be a square-free integer. Fix* $K = \mathbb{Q}(\sqrt{\Delta})$ *with the corresponding ring of integers* $\mathcal{O}_K$. *Let* $\alpha = a + b\omega \in \mathcal{O}_K$, *where* $q = N(\alpha) \equiv 1$ $\pmod n$ *is a prime that does not divide* Δ. *Let* w_n *be a primitive nth root of unity modulo* q, *and let* $r \equiv a(-b)^{-1} \pmod q$. *Let* k, l *be the solution of the system of linear equations over* $\mathbb{Z}/q\mathbb{Z}$

$$\begin{cases} \psi(1 + \alpha\mathcal{O}_K)k + \psi(\omega + \alpha\mathcal{O}_K)l = \ 1 + q\mathbb{Z} \\ \psi(1 + \alpha\mathcal{O}_K)k + \psi(\overline{\omega} + \alpha\mathcal{O}_K)l = w_n + q\mathbb{Z}, \end{cases} \qquad (3)$$

where ψ *is defined in (2). Let* $\delta = k + l\omega \in \mathcal{O}_K$, *then* $N(\delta) \equiv \omega_n \pmod q$. *Moreover,* $k \equiv (1 - (1 + \omega_n)r)(1 - 2r)^{-1} \pmod q$, $l \equiv (\omega_n - 1)(1 - 2r)^{-1} \pmod q$ *when* $\Delta \equiv 1 \pmod 4$, *and* $k \equiv (1 - \omega_n)2^{-1} \pmod q$ *and* $l \equiv (1 + \omega_n)(2r)^{-1}$ $\pmod q$ *if* $\Delta \equiv 2, 3 \pmod 4$.

Proof. See Lemma 2 [9].

3.1   Proof of Theorem 1

Proof. Assume that $n_1, m \in \mathcal{N}$ and

$$\gamma_1 = f_1 + g_1\omega \in \mathcal{O}_K, \quad N(\gamma_1) \equiv 1 \pmod{\operatorname{lcm}(n_1, 2^m)}$$

are procedure FINDPRIMEQ input. Let $m_1 = \operatorname{lcm}(n_1, 2^m)$ and let $q = N(\alpha)$, $\alpha = a + b\omega \in \mathcal{O}_K$ be a prime computed in step 4 of procedure FINDPRIMEQ. It is an elementary check that

$$q = N(\alpha) \equiv N(\gamma_1) \equiv 1 \pmod{m_1},$$

thus $2^m \mid q - 1$, $q \equiv 1 \pmod{n_1}$. If this is so, we may compute w_{n_1} a primitive n_1-th root of unity $\pmod{q}$ in step 3 of Algorithm 1.

Suppose that $w_{n_1} \pmod{q}$, $n_2, m \in \mathbb{N}$ and $\alpha, \gamma_2 \in \mathcal{O}_K$, where $\alpha = a + b\omega$, $\gamma_2 = f_2 + g_2\omega$, are procedure FINDINTEGERDELTA input. Assume that $r \pmod{q}$ and $x, y \pmod{q}$ are calculated in steps 1 and 2 of the procedure. Let $\eta = x + y\omega \in \mathcal{O}_K$. Lemmas 5 and 4 shows that $N(\eta) \equiv w_{n_1} \pmod{q}$. Let $m_2 = \operatorname{lcm}(n_2, 2^m)$ be computed in step 3 of the procedure. Since $(m_2, q) = 1$, we use the Chinese Remainder Theorem we find integers k and l such that

$$k \equiv f_2 \pmod{m_2}, \qquad\qquad l \equiv g_2 \pmod{m_2},$$
$$k \equiv x \pmod{q}, \qquad\qquad l \equiv y \pmod{q}.$$

in steps 4–6 of the procedure. Let $\delta = k + l\omega \in \mathcal{O}_K$ be the number returned by procedure FINDINTEGERDELTA. A short computation shows that

$$N(\delta) \equiv N(\eta) \equiv w_{n_1} \pmod{q}, \quad N(\delta) \equiv N(\gamma_2) \equiv 1 \pmod{m_2}.$$

Let $\delta \in \mathcal{O}_K$, $q, n_2, m \in \mathbb{N}$ be procedure FINDPRIMEP input. Let $m_2 = \operatorname{lcm}(n_2, 2^m)$ be calculated in step 1 of the procedure. If $p = N(\beta)$, where $\beta = c + d\omega \in \mathcal{O}_K$, is a prime computed in step 4 of the procedure, then

$$N(\beta) \equiv N(\delta) \equiv w_{n_1} \pmod{q}, \quad N(\beta) \equiv N(\delta) \equiv 1 \pmod{m_2},$$

so $2^m \mid p - 1$, $p \equiv 1 \pmod{n_2}$. Now, we show that primes p and q are pairing-friendly with respect to n_1 and Δ. The number β is the root of $x^2 - Tr(\beta)x + N(\beta)$, so

$$Tr(\beta)^2 - 4N(\beta) = d^2\Delta, \quad |Tr(\beta)| \leq 2\sqrt{p}.$$

By Lemma 2 [9]

$$Tr(\beta) \equiv Tr(\eta) \equiv 1 + w_{n_1} \pmod{q}, \quad N(\beta) \equiv N(\eta) \equiv w_{n_1} \pmod{q},$$

thus

$$N(\beta - 1) = N(\beta) + 1 - Tr(\beta) \equiv 0 \pmod{q}.$$

Consequently, p and q are pairing-friendly with respect to n_1 and Δ. This finishes the proof.

3.2   Proof of Lemma 3

Fix $0 < \varepsilon < \frac{2}{5}$. We define

$$\mathfrak{R} = \mathfrak{R}(y) = \{\beta \in \mathcal{O}_K, 0 < |\beta|^2 < y^{5/(2-5\varepsilon)}\}.$$

An element $\beta \in \mathcal{O}_K$ is said to be a prime number in K or prime, if an ideal $\beta\mathcal{O}_K$ is a prime ideal. Fix $\mathfrak{a}$ an integral ideal of $\mathcal{O}_K$ and $\delta \in \mathcal{O}_K$ such that $(\delta, \mathfrak{a})=1$. Denote by $\pi(\mathfrak{R}, \mathfrak{a}, \delta)$ the number of primes $\beta \in \mathfrak{R}$ such that $\beta \equiv \delta \pmod{\mathfrak{a}}$ and $\beta\mathcal{O}_K$ is a prime ideal of degree one of $\mathcal{O}_K$. T With notation as above, we have (see Lemma 3.3, [10]).

Lemma 6. *There exists x_0 such that for every $x \geq x_0$ and for every constant $B > 2$,*

$$\pi(\mathfrak{R}(x), \mathfrak{a}, \delta) = \frac{(2 - 5\varepsilon)w(K)}{5h(K)} \frac{x^{5/(2-5\varepsilon)}}{\varphi(\mathfrak{a})\log x} + O\left(\frac{x^{5/(2-5\varepsilon)}}{\varphi(\mathfrak{a})(\log x)^2}\right)$$

for all reduced residue classes $\delta \pmod{\mathfrak{a}}$, $(\delta, \mathfrak{a})=1$, and for all $\mathfrak{a}$ integral ideals of $\mathcal{O}_K$ with $N\mathfrak{a} \leq x$ with the possible exception of at most $O(x(\log x)^{-B})$ of $\mathfrak{a}$.

We are in a position to prove Theorem 3

Proof. Let $\beta = c + d\omega$. Denote by p_β the event that a randomly chosen $s, t \in \mathbb{Z}$ in step 2 are such that $N(\beta)$ is a prime in step 4 of procedure FINDPRIMEP. We compute the probability that in j trials p_β will occur. It is easily seen that β is chosen to satisfy $\beta \equiv \delta = k + l\omega \pmod{m_2 q \mathcal{O}_K}$, where $k, l \in \mathbb{Z}$, $|k| \leq 2xm_2$, $|l| \leq 2xm_2$ are computed in steps 5 and 6 of procedure FINDINTEGERDELTA. Moreover,

$$|\delta|^2 \leq \begin{cases} \left(k + \frac{l}{2}\right)^2 - \frac{\Delta l^2}{4} \leq (9 - \Delta)(xm_2)^2 & \text{if } \Delta \equiv 1 \pmod 4, \\ k^2 - \Delta l^2 \leq (4 - \Delta)(xm_2)^2 & \text{if } \Delta \equiv 2,3 \pmod 4 \end{cases} \tag{4}$$

Now, we define

$$A(x) = \{c + d\omega \in \mathcal{O}_K : c + d\omega \equiv \delta \pmod{m_2 q \mathcal{O}_K}\},$$

where $c, d \in \mathbb{Z}$, $c = m_2 qs + k$, $d = m_2 qt + l$ are computed in step 3 of procedure FINDPRIMEP, and $s, t \in \mathbb{Z}$,

$$|s| \leq \begin{cases} \frac{1}{2\sqrt{-\Delta}}(2m_2 x)^{(6+10\varepsilon)/(4-10\varepsilon)}, & \text{if } \Delta \equiv 1 \pmod 4, \\ \frac{1}{2}(2m_2 x)^{(6+10\varepsilon)/(4-10\varepsilon)} & \text{if } \Delta \equiv 2,3 \pmod 4 \end{cases} \tag{5}$$

and

$$|t| \leq \begin{cases} \frac{1}{\sqrt{-\Delta}}(2m_2 x)^{(6+10\varepsilon)/(4-10\varepsilon)}, & \text{if } \Delta \equiv 1 \pmod 4, \\ \frac{1}{2\sqrt{-\Delta}}(2m_2 x)^{(6+10\varepsilon)/(4-10\varepsilon)} & \text{if } \Delta \equiv 2,3 \pmod 4 \end{cases} \tag{6}$$

are selected in step 2 of the procedure. Thus, the number of elements of $A(x)$ is no greater than

$$\frac{4}{-\Delta}(2m_2x)^{(6+10\varepsilon)/(2-5\varepsilon)} \quad \text{if} \quad \Delta \equiv 1 \pmod 4,$$
$$\frac{1}{\sqrt{-\Delta}}(2m_2x)^{(6+10\varepsilon)/(2-5\varepsilon)} \quad \text{if} \quad \Delta \equiv 2,3 \pmod 4.$$

We define

$$P(x) = \{c + d\omega \in A(x) : N(c + d\omega)\text{-is a prime}\}.$$

From (4), (5) and (6), for $c + d\omega \in A(x)$, we obtain,

$$
\begin{aligned}
N(c + d\omega) &= |m_2q(s + t\omega) + \delta|^2 = |m_2q(s + t\omega)|^2 + |\delta|^2 + \Re(m_2q(s + t\omega)\bar{\delta}) \\
&\le (2m_2x)^2 \left(\left(s + \frac{t}{2}\right)^2 - \frac{\Delta t^2}{4} \right) + |\delta|^2 + \Re(m_2q(s + t\omega)\bar{\delta}) \\
&\le \frac{4 - \Delta}{-4\Delta}(2m_2x)^{10/(2-5\varepsilon)} + |\delta|^2 + \Re(m_2q(s + t\omega)\bar{\delta}) \\
&\le (2m_2x)^{10/(2-5\varepsilon)},
\end{aligned}
$$

for $x \ge x_0 = x_0(\varepsilon, \Delta)$ if $\Delta \equiv 1 \pmod 4$. From (4), (5) and (6), for $c+d\omega \in A(x)$, we have

$$
\begin{aligned}
N(c + d\omega) &= |m_2q(s + t\omega) + \delta|^2 \le (2m_2x)^2 \left(s^2 - \Delta t^2\right) + |\delta|^2 + \Re(m_2q(s + t\omega)\bar{\delta}) \\
&\le \frac{1}{2}(2m_2x)^{10/(2-5\varepsilon)} + |\delta|^2 + \Re(m_2q(s + t\omega)\bar{\delta}) \\
&\le (2m_2x)^{10/(2-5\varepsilon)},
\end{aligned}
$$

for $x \ge x_0 = x_0(\varepsilon, \Delta)$ if $\Delta \equiv 2,3 \pmod 4$. It is easily seen that $N(m_2q\mathcal{O}_K) \le (2m_2x)^2$. Lemma 6 shows that for sufficiently large x and any $A > 2$ the number of elements of $P(x)$ greater than or equal to

$$\frac{(2 - 5\varepsilon)w(K)}{10h(K)}\frac{(2m_2x)^{10/(2-5\varepsilon)}}{2\varphi(\mathfrak{a})\log(2m_2x)} \ge \frac{(2 - 5\varepsilon)w(K)}{10h(K)}\frac{(2m_2x)^{(6+10\varepsilon)/(2-5\varepsilon)}}{\log(2m_2x)}$$

with the possible exception of at most $O(x(\log x)^{-A})$ of ideals $m_2q\mathcal{O}_K$. Consequently, for sufficiently large x the probability that in j trials p_β does not occur is

$$
\begin{aligned}
\left(1 - \frac{c_1}{\log(2m_2x)}\right)^j &= \exp\left(j \log\left(1 - \frac{c_1}{\log(2m_2x)}\right)\right) \\
&\le \exp\left(\frac{-c_1 j}{\log(2m_2x)}\right) \le e^{-\lambda},
\end{aligned}
$$

for an arbitrary real $\lambda \ge 1$ and $j = c_2\lambda\log(2m_2x)$, where $c_2 = c_1^{-1}$ and c_1 is equal to

$$\frac{-(2-5\varepsilon)w(K)\Delta}{80h(K)} \quad \text{if} \quad \Delta \equiv 1 \pmod 4,$$
$$\frac{-(2-5\varepsilon)w(K)\Delta}{40h(K)} \quad \text{if} \quad \Delta \equiv 2,3 \pmod 4.$$

Consequently the probability that in j trials p_β does occur is greater than or equal to $1 - e^{-\lambda}$. So after repeating $[c_2\lambda \log(2m_2 x)]$ steps, the procedure finds $c + d\omega \in \mathcal{O}_K$ such that $N(c + d\omega)$ is a prime with probability greater than or equal to $1 - e^{-\lambda}$. The most time-consuming step of the algorithm is the deterministic primality test for the number q which takes no more than $\mathcal{PT}$ operations. This finishes the proof.

4   Algorithm for Chain-Friendly Primes

This chapter presents an algorithm that generates primes corresponding to a chain of pairing-friendly elliptic curves. Stated informally, our method generates pairing-friendly primes recursively, using procedures presented in the previous section to construct desired primes. Let us select a square-free integer and $\Delta < 0$ and $s \in \mathbb{N}$. Fix $K = \mathbb{Q}(\sqrt{\Delta})$ with the corresponding ring of integers $\mathcal{O}_K$. Given $m, n_1, \ldots, n_{s+1} \in \mathbb{N}$ we assume that, we have computed $\gamma_i \in \mathcal{O}_K$ such that $N(\gamma_i) \equiv 1 \pmod{\mathrm{lcm}(n_i, 2^m)}$, where $i = 1, \ldots, s + 1$.

Algorithm 2(the s-chain-friendly primes)
Given $0 < \varepsilon < 2/5$, and sufficiently large $x \in \mathbb{R}$. Given $s, m, n_i \in \mathbb{N}, \gamma_i \in \mathcal{O}_K$ such that $N(\gamma_i) \equiv 1 \pmod{\mathrm{lcm}(n_i, 2^m)}$, where $i = 1, \ldots, s+1$; the algorithm returns primes q and $p_1, \ldots, p_s$ that are s-chain-friendly with respect to $n_1, \ldots, n_s$ and Δ.

step 1. $\alpha = a + b\omega, q := \mathrm{FINDPRIMEQ}(n_1, m, \gamma_1, \Delta, x)$.
step 2. Compute w_{n_1} a primitive n_1-th root of unity $\pmod q$.
step 3. $\delta_1 = k_1 + l_1\omega := \mathrm{FINDINTEGERDELTA}(\alpha, q, n_2, m, \gamma_2, w_{n_1}, \Delta)$.
step 4. $\beta_1 = c_1 + d_1\omega, p_1 := \mathrm{FINDPRIMEP}(n_2, m, q, \delta_1, \Delta, \varepsilon, x)$.
step 5. For $i = 1$ to $s - 1$ do
    5.1 Compute $w_{n_{i+1}}$ a primitive n_{i+1}-th root of unity $\pmod{p_i}$.
    5.2 $\delta_{i+1} = k_{i+1} + l_{i+1}\omega := \mathrm{FINDINTEGERDELTA}(\beta_i, p_i, n_{i+2}, m, \gamma_{i+2}, w_{n_{i+1}}, \Delta)$.
    5.3 Take $x := p_i$.
    5.4 $\beta_{i+1} = c_{i+1} + d_{i+1}\omega, p_{i+1} := \mathrm{FINDPRIMEP}(n_{i+2}, m, p_i, \delta_{i+1}, \Delta, \varepsilon, x)$.
step 6. Return $q, p_1, \ldots, p_s, \alpha, \beta_1, \ldots, \beta_s$.

Theorem 2. *Fix $K = \mathbb{Q}(\sqrt{\Delta})$ with the corresponding ring of integers $\mathcal{O}_K$. Given $s, m, n_i \in \mathbb{N}$ and $\gamma_i \in \mathcal{O}_K$ such that $N(\gamma_i) \equiv 1 \pmod{\mathrm{lcm}(n_i, 2^m)}$, where $i = 1, \ldots, s + 1$. Algorithm 2 finds $\alpha, \beta_j \in \mathcal{O}_K$ such that $q = N(\alpha)$, $p_1 = N(\beta_1)$ are pairing-friendly primes with respect to n_1 and Δ, $p_j = N(\beta_j)$, $p_{j+1} = N(\beta_{j+1})$ are pairing-friendly primes with n_{j+1} and Δ. Furthermore,*

$$2^m \mid q - 1, \quad 2^m \mid p_j - 1, \quad q \equiv 1 \pmod{n_1} \quad p_j \equiv 1 \pmod{n_{j+1}}$$

for $j = 1, \ldots, s$. In particular, Algorithm 2 generates s-chain-friendly primes $q, p_1, \ldots, p_s$ with respect to $n_1, \ldots, n_s$ and Δ such that

$$(m_2 q)^2 \le p_1 \le (m_2 q)^{10/(2-5\varepsilon)}, \quad (m_{k+2} p_k)^2 \le p_{k+1} \le (m_{k+2} p_k)^{10/(2-5\varepsilon)},$$

where $m_{k+2} := \mathrm{lcm}(n_{k+2}, 2^m)$ and $k = 1, \ldots, s - 1$.
Proof. See Sect. 4.1.

4.1 Proof of Theorem 2

Proof. Let's assume that steps 1 through 4 of Algorithm t have been executed. Let primes q be procedure FINDPRIMEQ output in step 1, and let p_1 be procedure FINDPRIMEP output. Theorem 1 shows that q, p_1 are pairing-friendly primes with respect to n_1, Δ, and

$$2^m \mid q - 1, \quad 2^m \mid p_1 - 1, \quad q \equiv 1 \pmod{n_1} \quad p_1 \equiv 1 \pmod{n_2}. \tag{7}$$

The proof proceeds along the same lines as the proof of Theorem 1 from this moment. We start with $i = 1$ in step 5 of Algorithm 2. Lemma 3 shows

$$(m_2 q)^2 \le p_1 \le (m_2 q)^{10/(2-5\varepsilon)}.$$

From (7), $p_1 \equiv 1 \pmod{n_2}$, so we may compute w_{n_2}, a primitive n_2-th root of unity $\pmod{p_1}$, in step 5.1 of Algorithm 2. Let p_1, $w_{n_2} \pmod{p_i}$, n_3, $m \in \mathbb{N}$ and $\beta_i, \gamma_3 \in \mathcal{O}_K$ are the input to procedure FINDINTEGERDELTA in step 5.2. It is an elementary check that the procedure returns

$$\delta_2 = k_2 + l_2 \omega, \quad |k_2| \le m_3 p_1, \quad |l_2| \le m_3 p_1, \quad m_3 = \operatorname{lcm}(n_3, 2^m).$$

If $n_3, m, p_1 \in \mathbb{N}$, $\delta_3, \gamma_3 \in \mathcal{O}_K$ and $\Delta, x, \varepsilon \in \mathbb{R}$ are to procedure FINDPRIMEP, then the output is a prime $p_2 = N(\beta_2)$, where $\beta_2 \in \mathcal{O}_K$. Moreover, Theorem 1 and Lemma 3 shows that p_1, p_2 are pairing-friendly primes with respect to n_2, Δ, and

$$p_2 \equiv 1 \pmod{n_3} \quad (m_3 p_1)^2 \le p_2 \le (m_3 p_1)^{10/(2-5\varepsilon)}.$$

For $i = 2, \ldots, s-1$, we now apply the above arguments again, with p_i replaced by p_{i+1}, to obtain pairing-friendly primes p_i, p_{i+1} with respect to n_{i+1}, Δ such that

$$p_{i+1} \equiv 1 \pmod{n_{i+2}} \quad (m_{i+2} p_i)^2 \le p_{i+1} \le (m_{i+2} p_i)^{10/(2-5\varepsilon)},$$

where $m_{i+2} := \operatorname{lcm}(n_{i+2}, 2^m)$. We leave the details to the reader. This finishes the proof.

5 An Illustrative Example

Let us fix $\Delta = -3$. We define

$$K = \mathbb{Q}(\sqrt{-3}), \quad \mathcal{O}_K = \left\{ a + b\omega : a, b \in \mathbb{Z}, \ \omega = \frac{1 + \sqrt{-3}}{2} \right\}$$

The Given $s = 3$, $m = 10$, $n_1 = 12$, $n_2 = 7$, $n_3 = 5$, we apply Algorithm 2 to find 2-chain-friendly primes q, p_1, p_2 with respect to n_1, n_2 and $\Delta = -3$, (see Definition 4). We start by selecting $\gamma_i \in \mathcal{O}_K$ such that

$$N(\gamma_i) \equiv 1 \pmod{m_i}, \quad m_i = \operatorname{lcm}(n_i, 2^m), \quad i = 1, 2, 3.$$

For instance we take $\gamma_i = 1 + (m_i - 1)\omega$. Thus,

$$\gamma_1 = 1 + 3071\omega, \quad \gamma_2 = 1 + 7167\omega, \quad \gamma_2 = 1 + 5119\omega.$$

If $x = 2^{31}$, then procedure FINDPRIMEQ$(n_1, m, \gamma_1, \Delta, x)$ returns $\alpha\mathcal{O}_K$ and a prime q in step 1, where

$$\alpha = a + b\omega = 82837 + 58799\omega, \quad q = 15190023733$$

Next, the algorithm computes $w_{n_1} = w_{12}$ a primitive 12-th root of unity $\pmod{q}$ in step 2,

$$w_{12} \equiv 2180948271 \pmod{q}.$$

In step 3 of the algorithm, procedure FINDINTEGERDELTA$(\alpha, q, n_2, m, \gamma_2, w_{n_1}, \Delta)$ calculates

$$\delta_1 = k_1 + l_1\omega = 94259143987 + 71893063759\omega \in \mathcal{O}_K.$$

If we take $\varepsilon = 0.001$, then procedure FINDPRIMEP$(n_2, m, q, \delta_1, \Delta, \varepsilon, x)$ returns $\beta = c_1 + d_1\omega \in \mathcal{O}_K$ and a prime p_1 in step 4, such that

$$\beta_1 = 7427494894803196611120413235 + 6608975653374697961048915013\omega$$

and

$$p_1 = 147934372524546254329385362532374725427392842206696894 9.$$

We are at the point where we can perform steps 5.1 through 5.4 of Algorithm 2. The algorithm computes $w_{n_2} = w_7$ a primitive 7-th root of unity $\pmod{p_1}$ in step 5.1,

$$w_7 \equiv 221053556086393153931598821979288788098190894511091785 \pmod{p_1}.$$

Next, procedure FINDINTEGERDELTA$(\beta_1, p_1, n_3, m, \gamma_3, w_{n_2}, \Delta)$ calculates $\delta_2 = k_2 + l_2\omega \in \mathcal{O}_K$

$$k_2 = 663958212899998887268246028303486598727490211491777353 2161,$$
$$l_2 = 754281972838599394465000080918398991814383650165155198 4639.$$

Taking $x = p_1$ and $\varepsilon = 0.001$, procedure FINDPRIMEP$(n_3, m, p_1, \delta_2, \Delta, \varepsilon, x)$ returns $\beta_2 = c_2 + d_2\omega \in \mathcal{O}_K$ and a prime p_2 such that,

$$c_2 = 565165862159793298479281958223746473159251566025823080 62$$
$$1303489412690625778868441955098971131036892401856488941,$$
$$d_2 = 652746074992879892535326063009785278912615585345407629 69$$
$$8713749368808999031460895826298983641869258648636600919,$$

and

$p_2 = 360564205959485788057216071193186996911578746250376472888120017668144869392676620202244739516116472165085211680517007580292673956667384892662904497259715214958899243553811284908883107024623738046284809704374938771960622821$

in step 5.4 of the algorithm.

References

1. Agrawal, M., Kayal, N., Saxena, N.: Primes is in P. Ann. Math. **160**(2), 781–793 (2004)
2. Bateman, P., Horn, R.: A heuristic asymptotic formula concerning the distribution of prime numbers. Math. Comput. **16**, 363–367 (1962)
3. Bellés-Muñoz, M., Urroz, J., Silva, J.: Revisiting cycles of pairing-friendly elliptic curves. In: CRYPTO 2023. LNCS, vol. 14082, pp. 3–37 (2023)
4. Blake, I.F., Seroussi, G., Smart, N.P.: Advances in Elliptic Curve Cryptography. Cambridge University Press, Cambridge (2005)
5. Freeman, D., Scott, M., Teske, E.: A taxonomy of pairing-friendly elliptic curves. J. Cryptology **23**(2), 224–280 (2010)
6. Costello, C., et al.: Geppetto: versatile verifiable computation. In: 2015 IEEE Symposium on Security and Privacy, San Jose, CA, USA, pp. 253–270 (2015)
7. El Housni, A., Guillevic, A.: A survey of elliptic curves for proof systems. Des. Codes Cryptogr. **91**, 3333–3378 (2023)
8. Groth, J.: On the size of pairing-based non-interactive arguments. In: EURO-CRYPT 2016. LNCS, vol. 9666, pp. 305–326 (2016)
9. Grześkowiak, M.: Algorithms for pairing-friendly primes. In: Pairing-Based Cryptography Pairing 2013. LNSC, vol. 8365, pp. 215–228 (2014)
10. Grześkowiak, M.: An algorithmic construction of finite elliptic curves of order divisible by a large prime. Fundamenta Informaticae **136**, 331–343 (2015)
11. Hecke, E.: Lectures on the Theory of Algebraic Numbers. Springer, Cham (1981)
12. Shoup, V.: A Computational Introduction to Number Theory and Algebra. Cambridge University Press, Cambridge (2012)
13. Silverman, J.: The Arithmetic of Elliptic Curves. Springer, Cham (1985)
14. Silverman, J., Stange, K.: Amicable pairs and aliquot cycles for elliptic curves. Experiment Math. **20**(3), 329–357 (2011)
15. Schinzel, A., Sierpiński, W.: Sur Certaines Hypothèses Concernment Les Nombres Premiers. Acta Arith. **4**, 185–208 (1958)

Faster Algorithms for Isogeny Computations over Extensions of Finite Fields

Shiping Cai[1], Mingjie Chen[2]([✉]) [iD], and Christophe Petit[3,4] [iD]

[1] School of Mathematics, Sun Yat-sen University, Guangzhou, China
shiping.cai@ulb.be
[2] COSIC, KU Leuven, Leuven, Belgium
mjchennn555@gmail.com
[3] University of Birmingham, Birmingham, UK
[4] Université libre de Bruxelles, Brussels, Belgium

Abstract. Any isogeny between two supersingular elliptic curves can be defined over $\mathbb{F}_{p^2}$, however, this does not imply that computing such isogenies can be done with field operations in $\mathbb{F}_{p^2}$. In fact, the kernel generators of such isogenies are defined over extension fields of $\mathbb{F}_{p^2}$, generically with extension degree linear to the isogeny degree. Most algorithms related to isogeny computations are only efficient when the extension degree is small. This leads to efficient algorithms used in isogeny-based cryptographic constructions, but also limits their parameter choices at the same time. In this paper, we consider three computational subroutines regarding isogenies, focusing on cases with large extension degrees: computing a basis of ℓ-torsion points, computing the kernel polynomial of an isogeny given a kernel generator, and computing the kernel generator of an isogeny given the corresponding quaternion ideal under the Deuring correspondence. We then apply our algorithms to the constructive Deuring correspondence algorithm from [EPSV23] in the case of a generic prime characteristic, achieving around 30% speedup over [EPSV23].

1 Introduction

Isogeny-based cryptography is a key candidate for post-quantum cryptography, developed in response to potential quantum threats to current cryptographic systems. Protocols based on isogenies distinguish themselves from other post-quantum cryptography candidates by their compactness. As suggested by its name, the central objects in isogeny-based cryptography are isogenies between supersingular elliptic curves.

As a well-studied topic in both algorithmic number theory and elliptic curve cryptography, plenty of algorithms exist regarding elliptic curves and isogenies computations such as the classic Vélu's formula [V71] and square-root Vélu [BDFLS20], which are both algorithms for computing an isogeny (meaning computing its codomain and evaluation on points) given the domain curve and the kernel. However, the swift advancement in isogeny-based cryptography has led to emerging scenarios that demand more efficient algorithms. For instance, the ideal to isogeny translation is the computational bottleneck for [DFKL+20, DFLLW23, GPS17] but received attention only in recent few years.

The current lack of efficient algorithms is particularly evident in computations over extension fields of $\mathbb{F}_{p^2}$. While it is possible to work exclusively within $\mathbb{F}_{p^2}$ by judiciously selecting parameters for protocol design, this approach can be restrictive, as seen in the case of [DFKL+20], potentially necessitating the use of a larger field characteristic p. Recently, there are a few papers that discuss algorithms in extension fields of larger degrees such as [EPSV23, BGDS23] and [SEMR23]. In particular, the motivation of [SEMR23] is exactly to relax the constrains on parameter choices for [DFKL+20], and as a result, they achieve better verifying speed without degrading the signing speed as shown by their reference SageMath implementation. In this paper, we build upon these advancements and further explore computational subroutines critical for isogeny-based cryptographic schemes in extension fields.

1.1 Contributions

Let p be a prime > 3, E be a supersingular elliptic curve defined over $\mathbb{F}_{p^2}$, and ℓ be an odd prime. Let k be the extension degree w.r.t. (p, ℓ) (Definition 2).

In this paper, we look at three computational subroutines:

1. **Torsion basis generation** In Sect. 3, we provide a new algorithm (Algorithm 1) that finds a basis of $E_0[\ell]$ where E_0 is the curve $y^2 = x^3 + x$ over $\mathbb{F}_{p^2}$ for $p \equiv 3 \bmod 4$ with the extra conditions that $\ell \equiv 1 \bmod 4$ and $\mathrm{ord}(p) = \ell - 1$ in $(\mathbb{Z}/\ell\mathbb{Z})^\times$. This algorithm takes $\widetilde{O}(\ell)$ operations in $\mathbb{F}_{p^2}$ and surpasses other known methods when $k > 5$ according to our experiments.

2. **Kernel polynomial computation** In Sect. 4, we provide a new algorithm (Algorithm 3) that computes the kernel polynomial of a kernel subgroup H given a minimal polynomial w.r.t. H (see Definition 4). This algorithm mainly follows [EPSV23, Algorithm 4], and it takes $O(k^2) + \widetilde{O}(\ell^3/k^2) + \widetilde{O}(\ell)$ operations in $\mathbb{F}_{p^2}$. Theoretical analysis indicates that the cost of our algorithm outperforms [EPSV23, Algorithm 4] when $k^3 > \ell^2$. The experimental results from running our SageMath implementation suggest that the improvement of our algorithm over [EPSV23, Algorithm 4] is only clear when k is large enough, e.g., $k > 300$.

3. **Ideal kernel points generation** In Sect. 5, we provide an algorithm (Algorithm 4) that computes the kernel of the corresponding isogeny given a left $\mathcal{O}_0$-ideal where $\mathcal{O}_0$ is a quaternion maximal order isomorphic to $\mathrm{End}(E_0)$. It combines Algorithm 1 with a simple yet effective trick (Lemma 17) that we observe. We incorporate Algorithm 4 to [EPSV23, Algorithm 2]. This speeds up the kernel points generation step by around 40% in the constructive Deuring correspondence algorithm from [EPSV23] and speeds up the full algorithm by around 30%.

We also provide a SageMath [Dev23] implementation to measure the performance of our algorithms in this work. The code is available at https://github.com/wennycai/isoext/tree/master.

2   Preliminaries

In this section, we provide a brief overview of the background necessary for this work.

2.1   Quaternion Algebras, Supersingular Elliptic Curves, and Isogenies

Quaternion Algebra. Let p be a prime and let $\mathcal{B}_{p,\infty}$ denote the unique (up to isomorphism) quaternion algebra ramified precisely at p and ∞. We fix a $\mathbb{Q}$-basis $\langle 1, \mathbf{i}, \mathbf{j}, \mathbf{k} \rangle$ of $\mathcal{B}_{p,\infty}$ that satisfies $\mathbf{i}^2 = -q$, $\mathbf{j}^2 = -p$ and $\mathbf{k} = \mathbf{ij} = -\mathbf{ji}$ for some integer q. A *fractional ideal* I in $B_{p,\infty}$ is a $\mathbb{Z}$-lattice of rank 4. We denote by $n(I)$ the *norm* of I as the largest rational number such that $n(\alpha) \in n(I)\mathbb{Z}$ for any $\alpha \in I$. An *order* $\mathcal{O}$ is a subring of $B_{p,\infty}$ that is also a fractional ideal. An order is called *maximal* when it is not contained in any other larger order. A fractional ideal is *integral* if it is contained in its *left order* $\mathcal{O}_L(I) = \{\alpha \in \mathcal{B}_{p,\infty} \mid \alpha I \subset I\}$, or equivalently in its *right order* $\mathcal{O}_R(I) = \{\alpha \in \mathcal{B}_{p,\infty} \mid I\alpha \subset I\}$.

Supersingular Elliptic Curves and Their Isogenies. Let E, E_1, E_2 be elliptic curves defined over a finite field of characteristic p. An isogeny from E_1 to E_2 is a non-constant rational map that is simultaneously a group homomorphism. An isogeny is called as *cyclic* if it is not $[k]\varphi'$ for any integer k and φ' another isogeny from E_1 to E_2. An isogeny from a curve E to itself is an *endomorphism*. The set $\mathrm{End}(E)$ of all endomorphisms of E forms a ring under addition and composition. $\mathrm{End}(E)$ is either an order in an imaginary quadratic field and E is called *ordinary*, or a maximal order in $\mathcal{B}_{p,\infty}$, in which case E is called *supersingular*.

Let $\phi : E_1 \to E_2$ be an isogeny and let its degree to be odd. According to [Gal12, Lemma 25.1.16], ϕ can be written as

$$\phi(x, y) = \left(\frac{A(x)}{\psi^2(x)}, \frac{B(x, y)}{\psi^3(x)} \right),$$

where $\psi(x)$ is a polynomial of degree $(\deg \phi - 1)/2$. We refer to this form as *standard form* of an isogeny ϕ in this paper.

For an supersingular elliptic curve over $\mathbb{F}_q$ where q is a power of p, we denote its *quadratic twist* over $\mathbb{F}_q$ by E^t which is a curve isomorphic to E over a quadratic extension of $\mathbb{F}_q$.

As an important integer that will be used throughout the paper, we give precise definition of *extension degree*.

Lemma 1. *Let E be a supersingular elliptic curve defined over $\mathbb{F}_{p^2}$, let ℓ be an odd prime and e be a positive integer, the following three descriptions determine the same integer.*

1. *The smallest integer k such that $E[\ell^e] \subseteq E(\mathbb{F}_{p^{2k}})$ or $E^t[\ell^e] \subseteq E^t(\mathbb{F}_{p^{2k}})$ where E^t is the quadratic twist of E over $\mathbb{F}_{p^{2k}}$.*
2. *The degree of the minimal polynomial of $x(P)$ over $\mathbb{F}_{p^2}$ for any order ℓ^e point in $E[\ell^e]$.*
3. *Order of p^2 in $(\mathbb{Z}/\ell^e\mathbb{Z})^{\times}$, i.e., the smallest integer such that $p^{2k} \equiv 1 \bmod \ell^e$.*

Proof. The equivalences follow from the definition of a quadratic twist, the structure of $E(\mathbb{F}_{p^{2k}})$ and $E^t(\mathbb{F}_{p^{2k}})$, and the fact that $\#E(\mathbb{F}_{p^{2k}}) \times \#E^t(\mathbb{F}_{p^{2k}}) = (p^{2k} - 1)^2$ ([Sch87, Lemma 4.8]).   □

Definition 2. *Given p and ℓ^e, we call the integer defined from Lemma 1 the extension degree w.r.t. (p, ℓ^e) and we denote it by k.*

The Deuring Correspondence. Fix a supersingular elliptic curve E_0, and an order $\mathcal{O}_0 \simeq \mathrm{End}(E_0)$. The curve/order correspondence allows one to associate to each outgoing isogeny $\varphi : E_0 \to E_1$ an integral left $\mathcal{O}_0$-ideal, and every such ideal arises in this way (see [Koh96] for instance). Through this correspondence, the ring $\mathrm{End}(E_1)$ is isomorphic to the right order of this ideal. This isogeny/ideal correspondence is defined in [Wat69], and in the separable case, it is explicitly given as follows.

Definition 3. *Given I an integral left $\mathcal{O}_0$-ideal coprime to p, we define the I-torsion $E_0[I] = \{P \in E_0(\overline{\mathbb{F}}_{p^2})|\alpha(P) = 0 \text{for all } \alpha \in I\}$. To I, we associate the separable isogeny φ_I of kernel $E_0[I]$. Conversely given a separable isogeny φ, the corresponding ideal is defined as $I_\varphi = \{\alpha \in \mathcal{O}_0 \mid \alpha(P) = 0 \text{ for all } P \in \ker(\varphi)\}$.*

We summarize properties of the Deuring correspondence in Table 1, borrowed from [DFKL+20].

Table 1. The Deuring correspondence, a summary [DFKL+20].

Supersingular j-invariants over $\mathbb{F}_{p^2}$	Maximal orders in $\mathcal{B}_{p,\infty}$
$j(E)$ (up to Galois conjugacy)	$\mathcal{O} \cong \mathrm{End}(E)$ (up to isomorphism)
(E_1, φ) with $\varphi : E \to E_1$	I_φ integral left $\mathcal{O}$-ideal and right $\mathcal{O}_1$-ideal
$\theta \in \mathrm{End}(E_0)$	Principal ideal $\mathcal{O}\theta$
$\deg(\varphi)$	$n(I_\varphi)$

2.2 Kernel Polynomial and Minimal Polynomial of a Kernel Subgroup

Let $E/\mathbb{F}_{p^2}$ be a supersingular elliptic curve, the *kernel polynomial* defining a finite subgroup $H \leq E$, or equivalently an isogeny with kernel H, is the unique monic squarefree polynomial h_H whose set of roots is precisely the set of x-coordinates of nonzero points in H. A related concept is the *minimal polynomial of H*, whose definition is introduced in [EPSV23] and we recall here.

Definition 4 [EPSV23, *Definition 15*]. *Let E be an elliptic curve over a field K and $f \in K[X]$ a monic squarefree polynomial. The subgroup defined by f is the subgroup H of E generated by the set of points $\{P \in E \backslash \{0\}|f(x(P)) = 0\}$. In this situation, we say that f is a defining polynomial for H, and if f is furthermore irreducible, we refer to f as a minimal polynomial of H.*

Let ℓ be an odd prime and $H \leq E$ be a cyclic subgroup of order ℓ defined over $\mathbb{F}_{p^2}$. H gives rise to a unique cyclic isogeny $\varphi : E \to E'$ such that $\ker(\varphi) = H$ up to post-composition, and φ is defined over $\mathbb{F}_{p^2}$. Note that as discussed in [EPSV23, Section 2.2], in fact any isogeny between two supersingular elliptic curves can be defined over $\mathbb{F}_{p^2}$, by working with isomorphism representatives on which the p^2-power Frobenius is a scalar.

We denote the kernel polynomial of H by h_H. $h_H(x) \in \mathbb{F}_{p^2}[x]$ as H is defined over $\mathbb{F}_{p^2}$, and h_H is of degree $\frac{\ell-1}{2}$ by its definition. Let $H = \langle P \rangle$, and let $f(x)$ denote the minimal polynomial of $x(P)$ over $\mathbb{F}_{p^2}$, denote its degree by k, clearly $f(x) \mid h_H(x)$. Let π denote the p^2-power Frobenius map, $x(\pi(P)) = x(P)^{p^2}$ is another root of $f(x)$, and therefore $\pi(P)$ is another generator of H and there exists $\lambda \in (\mathbb{Z}/\ell\mathbb{Z})^\times$ such that $\pi(P) = [\lambda]P$.

Clearly, there is a free and transitive action of the group $(\mathbb{Z}/\ell\mathbb{Z})^\times/\{\pm 1\}$ on the set $X := H - 0_E/\{\pm 1\}$ by scalar multiplication. The subgroup $\langle \lambda \rangle/\{\pm 1\}$ partitions X into $\frac{\ell-1}{2k}$ orbits, each orbit is of length k. Let a be a primitive root in $(\mathbb{Z}/\ell\mathbb{Z})^\times$, then

$$h_H(x) = \prod_{i=0}^{\frac{\ell-1}{2k}-1} \left(\prod_{j=0}^{k-1} (x - x([a^i \lambda^j]P)) \right). \tag{1}$$

This follows from the fact that $\{a^i\}_{i=0}^{\frac{\ell-1}{2k}-1}$ is a set of coset representatives of $\langle \lambda \rangle/\{\pm 1\}$ in $(\mathbb{Z}/\ell\mathbb{Z})^\times/\{\pm 1\}$, and $\{\lambda^j\}_{j=0}^{k-1}$ is a set of representatives of $\langle \lambda \rangle/\{\pm 1\}$. For each fixed i, the inner layer product $\prod_{j=0}^{k-1}(x - x([a^i \lambda^j]P))$ is the minimal polynomial of $x([a^i]P)$ over $\mathbb{F}_{p^2}$ of degree k.

2.3 Torsion Basis Generation Algorithms over Extension Fields

In this section, we revisit the following problem of computing a torsion basis of supersingular elliptic curves.

Problem 5. Let E be a supersingular elliptic curve over $\mathbb{F}_{p^2}$, ℓ be a prime and e be a positive integer, compute a $\mathbb{Z}/\ell^e\mathbb{Z}$-basis of $E[\ell^e]$.

Torsion basis generation in isogeny-based cryptography is a fundamental step that is involved in the efficiency of lots of constructions. In most isogeny-based schemes, one prefers choosing a good characteristic p to make sure the extension degree k w.r.t. (p, ℓ^e) equals to 1. This choice leaves solving Problem 5 for general p, ℓ^e less studied. In what follows, we summarize two recent algorithms that considers solving Problem 5 in extension fields. We note that in both works, the torsion basis generation algorithms are not their main focuses but rather a tool developed for further computational tasks.

Method 1. The authors of [EPSV23] use a straightforward method to compute the torsion basis, which is by sampling random elements in $E(\mathbb{F}_{p^{2k}})$ and multiplying the points by a cofactor $\frac{(p^k \pm (-1)^k)^2}{\ell^e}$ until a $\mathbb{Z}/\ell^e\mathbb{Z}$-linearly independent pair P, Q is found. This method costs $O(k \log p)$ operations in $\mathbb{F}_{p^{2k}}$ [EPSV23, Section 3.3].

Method 2. This method is introduced in [BGDS23, Appendix A] and we briefly recall the main ideas here. Let π be the p^2-power Frobenius endomorphism on E and $\Phi_k(x)$ denote the k-th cyclotomic polynomial. Since $\Phi_k(x)|x^k-1$, there exists an endomorphism θ_k on E such that $(\pi^k - \mathrm{id}_E) = \Phi_k(\pi) \circ \theta_k$. For any point $P \in E(\mathbb{F}_{p^{2k}})$, $\theta_k(P) \in \ker(\Phi_k(x))$, then sampling random points P and computing $[\frac{\# \ker(\Phi_k(\pi))}{\ell^e}]\theta_k(P)$ would give rise to a basis of $E[\ell^e]$ similar to the idea of the previous method. This method costs $O(\varphi(k) \log p)$ operations in $\mathbb{F}_{p^{2k}}$ [BGDS23, Appendix A], where $\varphi(k)$ represents the value of the Euler totient function.

2.4 Constructive Deuring Correspondence

The constructive Deuring correspondence problem is the following:

Problem 6. Given a maximal order $\mathcal{O}$ in $\mathcal{B}_{p,\infty}$, compute a supersingular elliptic curve $E/\mathbb{F}_{p^2}$ such that $\mathrm{End}(E) \cong \mathcal{O}$.

A natural strategy to tackle this, which is also used in [EPSV23], goes as follows:

Step 0: Fix a base curve $E_0/\mathbb{F}_{p^2}$ with a known effective endomorphism ring $\mathcal{O}_0$.

Step 1 (KLPT): Construct an ideal I connecting $\mathcal{O}_0$ and $\mathcal{O}$ of suitable norm.

Step 2 (IdealToIsogeny): Compute the isogeny corresponding to I as $\varphi_I : E_0 \to E$.

[EPSV23] provides the most efficient algorithm for computing this correspondence in the literature.

3 Torsion Basis Generation

Let p be a prime such that $p \equiv 3 \bmod 4$. Let E_0 be a supersingular elliptic curve over $\mathbb{F}_p$ defined by the equation $y^2 = x^3 + x$. There exists an automorphism τ on E_0 given by the map $\tau : (x,y) \mapsto (-x, \sqrt{-1}y)$ where $\sqrt{-1}$ is the square root of -1 in $\mathbb{F}_{p^2}$. In this section, we propose a new method that finds a basis of $E_0[\ell]$ for an odd prime ℓ when ℓ satisfies the additional conditions that:

- $\ell \equiv 1 \bmod 4$;
- $\mathrm{ord}(p) = \ell - 1$ in $(\mathbb{Z}/\ell\mathbb{Z})^\times$, this is equivalent to $\mathrm{ord}(p^2) = \frac{\ell-1}{2}$ in $(\mathbb{Z}/\ell\mathbb{Z})^\times$ given the previous condition, i.e., the extension degree is $\frac{\ell-1}{2}$.

The main idea of our method is to extract the minimal polynomials of $x(P), x(Q)$ where P,Q form a basis of $E[\ell]$ from the denominator of the x-coordinate of a degree ℓ endomorphism θ on E and its dual $\hat{\theta}$. Since the extension degree $k = \frac{\ell-1}{2}$, the x-coordinates of P,Q are defined over $\mathbb{F}_{p^{\ell-1}}$. However, our method allows us to only work in the base field $\mathbb{F}_{p^2}$ to obtain the minimal polynomials. This significantly reduces the cost when the extension degree is large.

3.1   New Algorithm for Computing $E_0[\ell]$

Lemma 7. *Let ℓ be an odd prime such that $\operatorname{ord}(p) = \ell - 1$ in $(\mathbb{Z}/\ell\mathbb{Z})^\times$. Let $\theta = \left(\frac{\psi_x(x)}{\phi_x(x)}, \frac{\psi_y(x)}{\phi_y(x)} y \right)$ be an endomorphism of degree ℓ defined over $\mathbb{F}_{p^2}$ in its standard form (see definition in Sect. 2.1). Let R be any generator of $\ker\theta$. Then $\phi_x(x) = f^2(x)$ where $f(x)$ is the minimal polynomial of $x(R)$.*

Proof. Since $\deg\theta = \ell$ and ℓ is odd, we have $\phi_x(x) \in \mathbb{F}_{p^2}[x]$ is of degree $\ell - 1$ and is a square. Since $x(P)$ is a root of $\phi_x(x)$, then $f(x) \mid \phi_x(x)$. Note that $\operatorname{ord}(p) = \ell - 1$ implies that $f(x)$ is of degree $\frac{\ell-1}{2}$ (see Lemma 1), then it follows immediately that $\phi_x(x) = f^2(x)$. □

Algorithm 1. $\mathsf{TorsionBasisE}_0(\ell)$

Input: A prime $\ell \equiv 1 \bmod 4$ such that $\operatorname{ord}(p) = \ell - 1$ in $(\mathbb{Z}/\ell\mathbb{Z})^\times$.
Output: Minimal polynomials of x-coordinates of a basis P, Q of $E_0[\ell]$.
1: Compute a, b such that $a^2 + b^2 = \ell$ by Cornacchia's algorithm [Coh93, Section 1.5.2].
2: Compute the x-coordinates of the standard form of $\theta = [a] + \tau[b]$ and $\hat\theta = [a] - \tau[b]$, which we denote by $\frac{A(x)}{F(x)}$ and $\frac{B(x)}{\tilde F(x)}$ respectively.
3: Compute the square roots of $F(x)$ and $\tilde F(x)$ as $f_P(x)$ and $f_Q(x)$.
4: **Return** $f_P(x), f_Q(x)$.

Lemma 8. *Let p, E_0 and τ be as introduced in the beginning of Sect. 3. Let ℓ be an odd prime with $\ell \equiv 1 \bmod 4$ such that $\operatorname{ord}(p) = \ell - 1$ in $(\mathbb{Z}/\ell\mathbb{Z})^\times$. Algorithm 1 returns the minimal polynomials the x-coordinates of a basis of $E_0[\ell]$, and its takes $\widetilde{\mathcal{O}}(\ell)$ operations in $\mathbb{F}_{p^2}$.*

Proof. The correctness follows from Lemma 7 and the fact that a generator of $\ker(\theta)$ and a generator of $\ker(\hat\theta)$ are $\mathbb{Z}/\ell\mathbb{Z}$-linearly independent. Indeed, let $\ker(\theta) = \langle R \rangle$, since $\hat\theta = 2a - \theta$, $\ker(\hat\theta) = \langle R \rangle$ if and only if $\ell \mid 2a$ which is impossible by our choice of a.

As for complexity, the bit-complexity for Step 1 is $O(\log^2 \ell)$ [FS16, Section 3.1]. Step 2 takes $\widetilde{O}(\ell)$ operations in $\mathbb{F}_p$ [ACL+23, Lemma 2.7]. At Step 3, since $F(x) = f_P^2(x)$ from Lemma 7 and $f_P(x)$ is irreducible over $\mathbb{F}_{p^2}[x]$, we have $\gcd(F(x), F'(x)) = f_P(x)$, where $F'(x)$ is the derivative of $F(x)$. Hence, Step 3 requires $\widetilde{O}(\ell)$ operations in $\mathbb{F}_{p^2}$ using FFT-based polynomial arithmetic. □

Generalization of the torsion basis computation algorithm (Algorithm 1). At first glance, it may seem like our method is very restrictive as it is for only one curve E_0 and the prime ℓ needs to satisfy extra conditions. In fact, the curve E_0 plays a major role in isogeny-based cryptography. Assuming that $p \bmod \ell$ is random in $(\mathbb{Z}/\ell\mathbb{Z})^\times$, then the condition $\operatorname{ord}(p) = \ell - 1$ holds with probability $\frac{\varphi(\ell-1)}{\ell-1}$ where φ is the Euler's totient function. As ℓ considered in isogeny-based cryptography is significantly smaller than p, we compute the ratio $\frac{\varphi(\ell-1)}{\ell-1}$ for all primes $\ell < 5000$ such that $\ell \equiv 1 \bmod 4$, and they have $\frac{1}{5}$ as a lower bound. This means that for a random pair (p, ℓ) with $p \equiv 3 \bmod 4$ and $\ell \equiv 1 \bmod 4$ and $\ell < 5000$, there is at least a chance of $\frac{1}{5}$ that our algorithm can be applied.

Our method can be generalized to other curves E as long as E has a known endomorphism τ such that ℓ splits in $\mathbb{Z}[\tau]$. We can also remove the condition that $\operatorname{ord}(p^2) = \frac{\ell-1}{2}$ in $(\mathbb{Z}/\ell\mathbb{Z})^{\times}$ and even extend to computing a basis for the ℓ^e-torsion. The main idea still applies but the algorithm now requires factoring $F(x)$ and $\tilde{F}(x)$ as the minimal polynomials of the x-coordinates of a torsion basis is only a factor of their square roots.

3.2 Experiments and Performance

We implement Algorithm 1 in SageMath, and compare with the results running the SageMath implementations of the two methods introduced in Sect. 2.3. The comparison between the three methods are given in Table 2.

We choose $\ell \in \{5, 13, 17, 41, 97, 193\}$, and for each ℓ, we randomly generate a 256-bit p that satisfies the conditions that $p \equiv 3 \bmod 4$ and $\operatorname{ord}(p) = \ell - 1$ in $(\mathbb{Z}/\ell\mathbb{Z})^{\times}$. Note that in our case when $\ell \equiv 1 \bmod 4$, we have that $E_0[\ell] \subseteq E_0(\mathbb{F}_{p^{2(\ell-1)}})$ while $E_0^t[\ell] \subseteq E_0^t(\mathbb{F}_{p^{\ell-1}})$. Therefore, it is more efficient for Method 1 to work with the twist curve E_0^t and then sends back the torsion basis to E_0 via their isomorphism. However, this trick does not apply to Method 2 as Method 2 works with a curve on which the p^2-power Frobenius is defined.

Note that Algorithm 1 outputs minimal polynomials of $x(P)$ and $x(Q)$ while the other two methods give the x, y coordinates for both points. For a fair comparison, we also recover P and Q from $f_P(x)$ and $f_Q(x)$. Note that $\mathbb{F}_{p^{\ell-1}}$ can be constructed as an extension of $\mathbb{F}_{p^2}$ with modulus equals to $f_P(x)$ or $f_Q(x)$, which means the x-coordinate of torsion basis is the generator β of $\mathbb{F}_{p^{\ell-1}}$. The y-coordinate of torsion basis can be recovered by computing a square root on the twisted curve.

Table 2. Timings of torsion basis generation for different method. Since method 1 is a probabilistic algorithm, we run 10 times for each method and take the average as the final runtime. All of these benchmarks are running in SageMath 10.0 on a laptop with an Intel Core i7-12700H processor.

Method	$\ell = 5$	$\ell = 13$	$\ell = 17$	$\ell = 41$	$\ell = 97$	$\ell = 193$
Method 1 [EPSV23]	0.14 s	1.26 s	3.17 s	16.50 s	196.88 s	737.17 s
Method 2 [BGDS23]	0.46 s	5.30 s	13.46 s	109.73 s	1080.62 s	7369.61 s
This work	0.19 s	0.86 s	1.82 s	6.86 s	44.11 s	190.90 s

According to Table 2, we give a logarithmic runtime picture, where the runtime of torsion basis generation is represented on a logarithmic scale, as shown in Fig. 1. As ℓ increases, the growth rate of runtime for the three algorithms slightly decelerates. Overall, Algorithm 1 significantly outperforms both Method 1 and

Method 2. Therefore, for large primes $\ell \equiv 1 \pmod 4$ with $\mathrm{ord}(p) = \ell - 1$ in $(\mathbb{Z}/\ell\mathbb{Z})^{\times}$, using Algorithm 1 for generating a torsion basis of $E_0[\ell]$ is much more efficient.

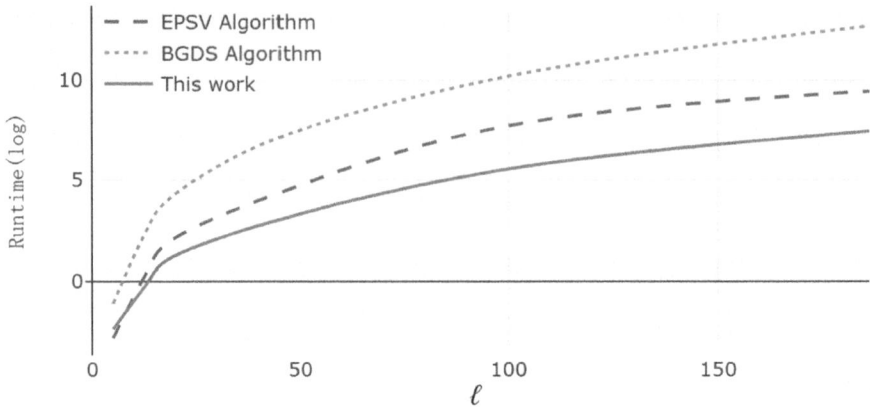

Fig. 1. We plot the following graphs with data points provided in Table 2. The x-axis represents the values of ℓ. The y-axis represents the logarithm of the runtime (in seconds) from Table 2. This figure is mainly to illustrate the growth trend of the runtime of the three algorithms as ℓ increases.

4   Computing Kernel Polynomials

Given a kernel generator $P \in E$ of prime order ℓ that lives in an extension field $\mathbb{F}_{p^{2k}}$ over $\mathbb{F}_{p^2}$, the time complexity of computing ϕ defined by the kernel subgroup $\langle P \rangle$ using Velu's formula is $\widetilde{O}(\ell)$ operations over $\mathbb{F}_{p^{2k}}$. An alternative method is to use Kohel's formulas. Given the kernel polynomial of the isogeny ϕ, it takes $\widetilde{O}(\ell)$ operations over $\mathbb{F}_{p^2}$ to compute ϕ. [EPSV23, Algorithm 4] proposes a new method to compute the kernel polynomial of ϕ given P, which takes $O(k^2) + O(\ell k) + \widetilde{O}(\ell)$ operations in the field $\mathbb{F}_{p^2}$. This leads to an algorithm that computes the isogeny ϕ given P which can be faster than Velu's formulas in some cases. Moreover, when it comes to evaluating a point $P' \in \mathbb{F}_{p^{2k'}}$, Velu's formulas perform this evaluation in the composition of the two fields $\mathbb{F}_{p^{2k}}$ and $\mathbb{F}_{p^{2k'}}$ whereas Kohel's formulas perform this evaluation in $\mathbb{F}_{p^{2k'}}$. Therefore, in the application scenarios when evaluations of points defined over different field extensions are needed, it can be preferable to use Kohel's formula as well.

In this section, we further improve on [EPSV23, Algorithm 4] that computes the kernel polynomial of φ given P, precisely, the x-coordinate of P. Their main idea is to compute all the irreducible factors of the kernel polynomial and then multiply them together, and each irreducible factor is computed via Shoup's algorithm given a root of the irreducible factor. The main idea of our new algorithm is a new method that computes the irreducible factors that outperforms Shoup's algorithm for some parameter choices.

4.1   Kernel Polynomial Computation Algorithm

Let us recall that from Sect. 2.2, $\pi(P) = [\lambda]P$ with $\lambda \in (\mathbb{Z}/\ell\mathbb{Z})^\times$. The subgroup $\langle\lambda\rangle/\{\pm 1\} \subseteq (\mathbb{Z}/\ell\mathbb{Z})^\times/\{\pm 1\}$ is the unique subgroup of order k in $(\mathbb{Z}/\ell\mathbb{Z})^\times/\{\pm 1\}$ and we denote it by $\mathcal{S}$. Our new algorithm starts with finding a pair of integers (a, b) that is *good* in the sense of the following definition.

Definition 9. *We say a pair of integers (a, b) is (ℓ, k)-good if it satisfies the following:*

(1) modulo ℓ, a, b are primitive roots in $(\mathbb{Z}/\ell\mathbb{Z})^\times$,
(2) viewed as elements in $(\mathbb{Z}/\ell\mathbb{Z})^\times/\{\pm 1\}$, b and a are in the same coset with respect to the subgroup $\mathcal{S}$,
(3) $\gcd(a, b) = 1$.

Any integer a gives rise to an endomorphism on a supersingular elliptic curve E which is simply the multiplication-by-a map often denoted by $[a]$. We introduce a new notation $[a]_x(x)$ which is the x-coordinate of $[a]$ when written as rational maps.

Let us write $[a]_x(x) = \frac{a_1(x)}{a_2(x)}$ with $a_1(x), a_2(x) \in \mathbb{F}_{p^2}[x]$ be coprime polynomials with degrees bounded above by $O(a^2)$, and $[b]_x(x) = \frac{b_1(x)}{b_2(x)}$ with $b_1(x), b_2(x) \in \mathbb{F}_p[x]$ be coprime polynomials with degrees bounded above by $O(b^2)$. Let $f(x) = \sum_{i=0}^k c_i x^i$ with $c_i \in \mathbb{F}_{p^2}$ be the minimal polynomial of $x(P)$ over $\mathbb{F}_{p^2}$. Then

$$f([a]_x(x)) = \frac{\sum_{i=0}^k c_i a_2^{k-i}(x) a_1^i(x)}{a_2^k(x)} =: \frac{d_{fa}(x)}{n_{fa}(x)},$$

$$f([b]_x(x)) = \frac{\sum_{i=0}^k c_i b_2^{k-i}(x) b_1^i(x)}{b_2^k(x)} =: \frac{d_{fb}(x)}{n_{fb}(x)}.$$

Lemma 10. *Let a, b be two integers satisfying the conditions (2) and (3) in Definition 9, then $\gcd(d_{fa}(x), d_{fb}(x))$ is the minimal polynomial of $x([a^{-1}]P)$ where here a^{-1} is the inverse of a modulo ℓ.*

Proof. Since $[a]([a^{-1}]P) = P$, we have that $[a]_x(x([a^{-1}]P)) = x(P)$ and it is a root of $f(x)$. Therefore, $x([a^{-1}]P)$ is a root of $f([a]_x(x))$, and it means that $d_{fa}(x([a^{-1}]P)) = 0$. Similarly, we can show that $d_{fb}(x([b^{-1}]P)) = 0$. Since a, b are in the same coset of $(\mathbb{Z}/\ell\mathbb{Z})^\times/\{\pm 1\}$ modulo $\langle\lambda\rangle/\{\pm 1\}$, $x([a^{-1}]P)$ and $x([b^{-1}]P)$ have the same minimal polynomial, let us denote it by $f_1(x)$. Then this implies that $f_1(x) \mid \gcd(d_{fa}(x), d_{fb}(x))$.

We now show that $d_{fa}(x)$ and $d_{fb}(x)$ have no other common roots besides those shared with $f_1(x)$. Consider the two sets $[a^{-1}](\mathcal{S} \cdot P)$ (the orbit of P of the $\mathcal{S}$ action multiplied with $[a^{-1}]$) and $[b^{-1}](\mathcal{S} \cdot P)$, they are equal by the conditions on a, b. The roots of $d_{fa}(x)$ are x-coordinates of $[a^{-1}](\mathcal{S} \cdot P)\tilde{+}E[a]$ (where the addition $\tilde{+}$ is performed over all possible pairwise combinations by summing each element from the first set with each element from the second set), and the roots of $d_{fb}(x)$ are x-coordinates of $[b^{-1}](\mathcal{S} \cdot P) \tilde{\mp} E[b]$. Since $\gcd(a, b) = 1$, one can see that there are no common elements in these two sets except for $[a^{-1}](\mathcal{S} \cdot P)$. This shows that $f_1(x) = \gcd(d_{fa}(x), d_{fb}(x))$. □

Heuristic 11. *We can find desired* $a, b \in [-B, B] - \{0\}$ *with* B *a small multiple of* $\frac{\ell-1}{4k}$.

We provide a justification of the Heuristic above.

- There must be two integers in $[-B, B] - \{0\}$ belong to the same coset whenever $B > \frac{\ell-1}{4k}$ since then $2B > \frac{\ell-1}{2k}$ and $\frac{\ell-1}{2k}$ is the number of cosets.
- The chance that the other two conditions are satisfied for a random pair in $[-B, B] - \{0\}$ is reasonably high, we expect to obtain one such pair by adding a small constant multiple to $\frac{\ell-1}{4k}$.

Remark 12. Asymptotically, we expect to find the desired $a, b \in [-B, B] - \{0\}$ when B is a small multiple of $((\ell - 1)/k)^{1/2}$ by the birthday paradox. This can be used to improve the complexity of Algorithm 2 and then Algorithm 3. But we do not use this bound since we are dealing with ℓ and k where ℓ/k is not very large.

Algorithm 2. KernelPolynomialFromMinipoly(E,f,ℓ)

Input: An elliptic curve $E/\mathbb{F}_{p^2}$, a prime ℓ and a minimal polynomial f of $x(P)$ of degree k where P is a point of order ℓ.
Output: The kernel polynomial h w.r.t. the kernel subgroup $\langle P \rangle$.
1: Set $m \leftarrow (\ell - 1)/2k$ and $f_0 \leftarrow f$.
2: Find smallest pair (a, b) (smallest in terms of $\max\{|a|, |b|\}$) that is (ℓ, k)-good. Let $B = \max\{|a|, |b|\}$.
3: **For** $i = 1$ to $m - 1$ **do**
4:     Compute $[a]_x(x) = \frac{a_1(x)}{a_2(x)}$ and $[b]_x(x) = \frac{b_1(x)}{b_2(x)}$.
5:     Construct polynomials $d_{fa}(x) \leftarrow a_2^k(x) f_{i-1}([a]_x(x))$, $d_{fb}(x) \leftarrow b_2^k(x) f_{i-1}([b]_x(x))$.
6:     Set $f_i(x) \leftarrow \gcd(d_{fa}(x), d_{fb}(x))$.
7: **end For**
8: Compute $h \leftarrow \prod_{i=0}^{m-1} f_i$ using a product tree.
9: **Return** h.

Lemma 13. *Assuming Heuristic 11, Algorithm 2 is correct and it takes* $\widetilde{O}(\frac{\ell^3}{k^2}) + \widetilde{O}(\ell)$ *operations in* $\mathbb{F}_{p^2}$.

Proof. The correctness of Algorithm 2 follows from Lemma 10 and Eq. (1). The smallest (a, b) pair that is (ℓ, k) good can be found by an exhaustive search, i.e., by enumerating all (a, b) pairs in each coset such that $|a|, |b| \leq B$ and check whether they are (ℓ, k)-good. The cost of finding (a, b) is $\widetilde{O}(\log^3(\ell) B^2)$ bit operations. The cost of Step 4 - 6 is $\widetilde{O}(k B^2)$ using FFT as explained in Remark 14,

---

Algorithm 3. KernelPolynomialFromIrrationalX(E, ζ, ℓ)

Input: Elliptic curve $E/\mathbb{F}_{p^2}$, extension $\mathbb{F}_{p^{2k}}/\mathbb{F}_{p^2}$, x-coordinate $\zeta \in \mathbb{F}_{p^{2k}}$ of an order-ℓ point $P \in E$ lying in an eigenspace of the p^2-power Frobenius on E.
Output: The kernel polynomial $h_{\langle P \rangle} \in \mathbb{F}_{p^2}[x]$ defining the subgroup of E generated by P.
  1: Find the minimal polynomial $f \in \mathbb{F}_{p^2}[x]$ of ζ over $\mathbb{F}_{p^2}$ using Shoup's algorithm.
  2: Return KernelPolynomialFromMinipoly(E, f, ℓ).

---

which is $\widetilde{O}(\frac{\ell^2}{k})$ according to Heuristic 11. Therefore the loop takes $\widetilde{O}(\frac{\ell^3}{k^2})$ operations in $\mathbb{F}_{p^2}$. Finally, the product tree in the last step can be computed in $\widetilde{O}(\ell)$ operations in $\mathbb{F}_{p^2}$, using FFT based arithmetic, same as [EPSV23, Lemma 16].

$\square$

Remark 14. We explain how to compute $d_{fa}(x)$, $d_{fb}(x)$ and $f_i(x)$ here in more detail. If $2^{\lfloor \log_2^{k \cdot a^2 + 1} \rfloor} | p^2 - 1$, then there always exist $2^{\lfloor \log_2^{k \cdot a^2 + 1} \rfloor}$-th roots of unity over $\mathbb{F}_{p^2}$, we denote the set of these units to be μ_a. For $i = 1, \ldots, m - 1$, we compute each $d_{fa}(x)$ in Algorithm 2 by first computing $Y_a := \{[a]_x(\eta), \eta \in \mu_a\}$, we then compute the evaluation of $f_{i-1}(x)$ at each element in Y_a in $\widetilde{O}(k)$ operations over $\mathbb{F}_{p^2}$. Taking these evaluation values, the inverse FFT algorithm can be used to recover the coefficients of $d_{fa}(x)$ in $\widetilde{O}(kB^2)$ operations in $\mathbb{F}_{p^2}$. The construction of $d_{fb}(x)$ is similar. Then the final gcd computation can be done in $\widetilde{O}(kB^2)$ operations in $\mathbb{F}_{p^2}$ with FFT.

Lemma 15. *Algorithm 3 is correct and takes $O(k^2) + \widetilde{O}(\frac{\ell^3}{k^2}) + \widetilde{O}(\ell)$ operations in $\mathbb{F}_{p^2}$.*

Proof. The complexity of Algorithm 3 is $O(k^2)$ for running Shoup's algorithm plus the complexity of Algorithm 2.

Remark 16. Let us recall here that the corresponding algorithm [EPSV23, Algorithm 4] takes $O(k^2) + O(\ell k) + \widetilde{O}(\ell)$ operations in $\mathbb{F}_{p^2}$. Therefore, the difference is only in the middle term and we expect to outperform [EPSV23, Algorithm 4] when $\frac{\ell^3}{k^2} < \ell k$, i.e., when $k^3 > \ell^2$. We test this observation with our experiments in next section.

4.2   Experiments and Performance

We explain our strategy for generating the parameters k, ℓ, p. We first fix the extension degree k, then we find a prime ℓ such that $\frac{\ell-1}{2k} > 1$. We then find a pair (a, b) that is (ℓ, k)-good (Definition 9). Finally we look for primes of the form $2^r f \pm 1$ (for $p \equiv 1 \bmod 4$ and $p \equiv 3 \bmod 4$ respectively) with $r = \lfloor \log_2^{k \cdot \max(a,b)^2 + 1} \rfloor$ by randomly generating f and testing whether $2^r f \pm 1$ is a prime.

In the case of $p \equiv 1 \bmod 4$, the 2^r-th root of unity is an element in $\mathbb{F}_p$. However, since we consider the elliptic curve over $\mathbb{F}_{p^2}$, computing $[a]_x(x)$ and

$[b]_x(x)$ still involves operations on $\mathbb{F}_{p^2}$ and there is not much difference in the case of $p \equiv 1 \bmod 4$ and $p \equiv 3 \bmod 4$. For convenience, we only discuss the case of $p \equiv 1 \bmod 4$ in this work. We use the above strategy to find 13 parameters for our experiments, with k ranging from 24 to 940. Besides, for these ℓ and k we can always find the smallest $a, b \in \{\pm 2, \pm 3\}$. Table 3 provides comparisons of computing the kernel polynomial for a given x-coordinate of an order-ℓ point.

From Remark 16, we expect that our algorithm can achieve better performance if $k^3 > \ell^2$. However, from Table 3, experiments suggests that our algorithm is faster than the previous method in [EPSV23] if $k > 300$ regardless of $k^3 > \ell^2$ or not. This is potentially caused by the fact that we have an extra $\log$ factor in $\widetilde{O}(\frac{\ell^3}{k^2})$ that was not taken into consideration in the comparison in Remark 16, therefore we perform worse when k is relatively small. On the other hand, the (a, b) pair we encounter in practice are much smaller than the upper bound we gave in the complexity analysis of Algorithm 2, this explains why we outperform [EPSV23] when k is large enough in practice (Fig. 2).

Table 3. Timings of computing the kernel polynomial from the x-coordinate of generator for ℓ-isogeny, running in SageMath 10.1 on a computer with an Intel(R) Core(TM) i9-12900K processor. For comparison, we also record the runtime of this procedure in [EPSV23]. Both algorithms run in a same environment.

(ℓ, k)	[EPSV23]	Algorithm 3	Speed Up
$(97, 24)$	0.04 s	0.09 s	-125.00%
$(431, 43)$	0.29 s	0.56 s	-93.10%
$(1499, 107)$	1.67 s	2.51 s	-50.30%
$(1201, 200)$	2.96 s	3.04 s	-2.7%
$(4423, 201)$	10.03 s	11.90 s	-18.64%
$(1009, 252)$	3.11 s	4.31 s	-38.59%
$(3461, 346)$	12.05 s	11.41 s	5.31%
$(10781, 385)$	47.51 s	26.50 s	44.22%
$(5641, 470)$	31.96 s	23.90 s	25.22%
$(2017, 504)$	11.38 s	10.29 s	9.58%
$(24001, 600)$	146.88 s	71.42 s	51.38%
$(28387, 747)$	196.60 s	78.25 s	60.20%
$(28201, 940)$	306.89 s	136.20 s	55.62%

5 Application to the Constructive Deuring Correspondence

In this section, we consider applications of our new algorithms to the constructive Deuring correspondence algorithm in [EPSV23]. The relevant part is Step 2

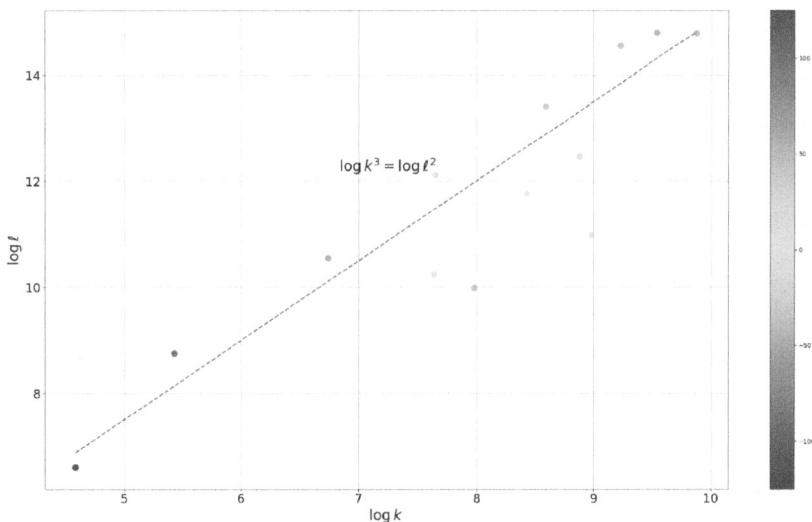

Fig. 2. Heatmap of timing comparison about computing kernel polynomials in Table 3. On the right side of this log-log picture is a colorbar from -100 (resp. blue) to 100 (resp. red), which indicates the percentage of speed up. The dotted red line represents the case when $k^3 = \ell^2$, i.e. $2\log \ell = 3\log k$. Hence, dots above the line correspond to the case when $k^3 > \ell^2$, and those below correspond to the case when $k^3 < \ell^2$. The color of the dots in the figure represents the speedup percentage. (Color figure online)

in the general strategy as defined in Sect. 2.4, i.e., the **IdealToIsogeny** step. Even though both of our two algorithms TorsionBasis$E_0(\ell)$ (Algorithm 1) and KernelPolynomialFromIrrationalX(E, ζ, ℓ)(Algorithm 3) can be used in theory when the parameter requirements are met, we only incorporate our Algorithm 1 to the algorithm in [EPSV23] as the effect of the SageMath implementation of Algorithm 3 becomes clear only when k is bigger than 300, and this never happens for the extension degrees considered in [EPSV23].

Specifically, we use our Algorithm 1 in IdealToKernelGens ([EPSV23, Algorithm 2]). The cost of this algorithm consists of two main parts, one is computing a basis of $E_0[\ell^e]$ which costs $(k \log p)$ operations in $\mathbb{F}_{p^{2k}}$, the other one is evaluating a basis of $\mathcal{O}_0$ on the torsion basis points which costs $O(\log p)$ operations in $\mathbb{F}_{p^{2k}}$. We replace their method for generating a torsion basis algorithm for E_0 with ours whenever parameters p, ℓ satisfy our conditions. Besides this, we also observe one trick that could speed up their computation which is supported by the following lemma.

Lemma 17. *Let E be a supersingular elliptic curve, ℓ be a prime, and e be a positive integer. Let $P \in E[\ell^e]$ be of order ℓ^e, let $\gamma \in \mathrm{End}(E)$ be an endomorphism whose degree is ℓ^e, then $\gamma(P)$ is of order ℓ^e with probability $\frac{\ell}{\ell+1}$.*

Proof. There exists an explicit isomorphism of $E[\ell^e]$ and $\mathbb{Z}/\ell^e\mathbb{Z} \times \mathbb{Z}/\ell^e\mathbb{Z}$ such that under this isomorphism $\gamma = L$ where L is the map

$$L : \mathbb{Z}/\ell^e\mathbb{Z} \times \mathbb{Z}/\ell^e\mathbb{Z} \to \mathbb{Z}/\ell^e\mathbb{Z}$$
$$(m, n) \quad\quad \mapsto m.$$

A point $(m, n) \in \mathbb{Z}/\ell^e\mathbb{Z} \times \mathbb{Z}/\ell^e\mathbb{Z}$ is of order ℓ^e if and only if at least one of m and n is of order ℓ^e, and therefore that are $2 \times (\ell^e - \ell^{e-1})\ell^e - (\ell^e - \ell^{e-1})^2$ such elements. $L(m, n)$ is of order ℓ^e if and only m is of order ℓ^e and there are $(\ell^e - \ell^{e-1})\ell^e$ such choices. Therefore, the desired probability is

$$\frac{(\ell^e - \ell^{e-1})\ell^e}{2 \times (\ell^e - \ell^{e-1})\ell^e - (\ell^e - \ell^{e-1})^2} = \frac{\ell}{\ell + 1}.$$

$\square$

In view of this lemma, it is of high probability that knowing one element of order ℓ^e in $E_0[\ell^e]$ suffices to find the kernel generator (see step 9 of [EPSV23, Algorithm 2]). Therefore, it is unnecessary to generate the full torsion basis as in step 8 of [EPSV23, Algorithm 2]. We introduce Algorithm 4 as a variant of [EPSV23, Algorithm 2] to include our Algorithm 1 and this new observation, Algorithm 4 deviates from [EPSV23, Algorithm 2] from step 8.

The asymptotic runtime of Algorithm 4 is the same as that of [EPSV23, Algorithm 2], but it performs better in practice. The performance depends heavily on the prime factors of $N(J) = \prod_{i=1}^{r} \ell_i^{e_i}$, so it is hard to give a simple estimate of how faster the new algorithm is than the old one, in particularly so for the basis generation step. If we only consider the trick of computing one less torsion basis, the speedup over [EPSV23, Algorithm 2] is $\sum_{i=1}^{r} \frac{2\ell_i}{\ell_i+1}$ times.

Implementations. To measure the performance of our improvements, we compare the runtime of the constructive Deuring correspondence algorithm in [EPSV23] with our version replacing [EPSV23, Algorithm 2] by Algorithm 4.

We first run the comparison on the 3 parameters mentioned in [EPSV23]. For each parameter, we record the primes p and the norms of the ideals J connecting E_0[1] to E in Appendix A. The results are summarized in Table 4, and the table indicates that our work can significantly reduce the cost of the constructive Deuring correspondence computation, achieving approximately a 1.3x acceleration. However, we note that the requirements on ℓ are met only around 3 times in the whole iterations for parameter p_2 and 1 time for p_{3923}. Both of them need around 70 times iterations for kernel points generation in total. As a consequence, for both primes, Algorithm 1 only brings around 1% speed up.

In order to better measure the performance of the new torsion basis generation algorithm, we randomly generate three 256-bit primes and adjust cost

---

[1] Note that when $p \equiv 1 \bmod 4$, we take E_0 to be the curve defined as in [EPSV23, Section 3.1] and our torsion basis generation algorithm can be adapted to it as well.

Algorithm 4. IdealToKerGens(J, E_0)

Input: left $\mathcal{O}_0$-ideal J of norm $N = \prod_{i=1}^{r} \ell_i^{e_i}$, curve E_0 with effective endomorphism ring $\text{End}(E_0) \cong \mathcal{O}_0$.

Output: $\{G_1, \ldots, G_r\}$, a generating set of $\ker \varphi_J$ with $\text{ord}(G_i) = \ell^{e_i}$.

1: Compute $\alpha \in \text{End}(E_0)$ such that $J = \mathcal{O}_0\alpha + \mathcal{O}_0 N$ under the isomorphism $\text{End}(E_0) \cong \mathcal{O}_0$.
2: let $(\phi_1, \ldots, \phi_4)$ be a basis of $\text{End}(E_0) \cong \mathcal{O}_0$ consisting of efficiently evaluatable endomorphisms.
3: Write $\bar{\alpha}$ as a fraction of the form $(c_1\phi_1 + \ldots + c_4\phi_4)/t$, where $c_1, c_2, c_3, c_4 \in \mathbb{Z}$ and $t \in \mathbb{Z}_{\geq 1}$.
4: **For** $i \in \{1, \ldots, r\}$ **do**
5: Set $v_i = \nu_{\ell_i}(t)$ to be the ℓ_i-adic valuation of t.
6: Let $c_j^{(i)} = c_j(t/\ell^{v_i})^{-1} \bmod \ell_i^{e_i+v_i}$ for $j \in \{1, \ldots, 4\}$.
7: Define $\gamma_i = c_1^{(i)}\phi_1 + \cdots + c_4^i\phi_4$.
8: **If** $p \equiv 3 \bmod 4$ **and** $\ell_i \equiv 1 \bmod 4$ **and** $e_i + v_i = 1$ **then**
9:   $f(x), f'(x) \leftarrow$ **TorsionBasis**(ℓ_i).
10:   Compute ℓ_i-torsion point P from $f(x)$.
11:   Compute $G_i \leftarrow \gamma_i(P)$.
12:   **If** $\ell^{e_i-1}G_i = \infty_{E_0}$ **then**
13:    Compute ℓ_i-torsion point Q from $f'(x)$.
14:    Compute $G_i \leftarrow \gamma_i(Q)$.
15:   **end If**
16: **else**
17:   Find ℓ_i-torsion point $P \in E_0$.
18:   Compute $G_i \leftarrow \gamma_i(P)$.
19:   **If** $\ell^{e_i-1}G_i = \infty_{E_0}$ **then**
20:    Find ℓ_i-torsion point $Q \in E_0$.
21:    Compute $G_i \leftarrow \gamma_i(Q)$.
22:   **end If**
23: **end If**
24: **end For**
25: **Return** $\{G_1, \ldots, G_r\}$

Table 4. Timings of the IdealToKerGens algorithms (referred to as *kernel* in the table) and the full constructive Deuring correspondence computations (referred to as *Deuring* in the table) for different primes, using SageMath 10.0 on a laptop with an Intel Core i7-12700H processor. Since the IdealToKerGens algorithms are probabilistic, we execute 10 times benchmarks with random maximal orders and take the average as the final result for each characteristic.

Primes	Kernel			Deuring		
	[EPSV23, Alg 2]	Algorithm 4	Speed Up	[EPSV23]	This work	Speed Up
p_{3923}	274.97 s	167.05 s	39.25%	520.07 s	404.60 s	22.20%
p_1	174.07 s	107.83 s	38.05%	352.42 s	266.92 s	24.26%
p_2	477.32 s	299.57 s	37.24%	689.76 s	512.63 s	25.68%

model proposed in [EPSV23, Section 4.2] to allow more ℓ in the norm of the ideal (which is an input of the **KLPT** step, see Sect. 2.4) that satisfies the conditions of Algorithm 1. Specially, we consider all primes $\ell < 150$ and add it to the norm if it satisfies that $\ell \equiv 1 \bmod 4$ and that the order of p in $(\mathbb{Z}/\ell\mathbb{Z})^{\times}$ is $\ell - 1$. The runtime of constructive Deuring correspondence computation for p_3, p_4, p_5 is provided in Table 5. According to the results, after adjusting cost model to use new algorithm for torsion basis generation more frequently, one can get a significant speed up during the computation of the constructive Deuring correspondence.

Table 5. Timings of the IdealToKerGens algorithms (referred to as *kernel* in the table) and the full constructive Deuring correspondence computations (referred to as *Deuring* in the table) for p_3, p_4 and p_5, using SageMath 10.0 on a laptop with an Intel Core i7-12700H processor. Since we want to measure the speedup brought by Algorithm 1 and the trick implied by Lemma 17, we add two invariants to our work. One is IdealToKerGens without using Algorithm 1 (referred to as *trick only*), the other one is IdealToKerGens with the trick and Algorithm 1, which is Algorithm 4. Since the kernel point generation algorithm is probabilistic, we execute 10 times benchmarks with random maximal orders and take the average as the final result.

		p_3		p_4		p_5	
		Kernel	Deuring	Kernel	Deuring	Kernel	Deuring
[EPSV23]		2217.06 s	3205.94 s	1424.43 s	2007.57 s	1371.94 s	2318.64 s
This work	trick only	1801.41 s	2614.69 s	1086.72 s	1608.23 s	928.96 s	1823.63 s
	Algorithm 4	1144.33 s	1957.88 s	832.24 s	1353.75 s	787.53 s	1678.88 s
Speed Up	trick only	18.75%	18.44%	23.71%	19.89%	32.29%	21.35%
	Algorithm 4	48.39%	38.93%	41.57%	32.57%	42.60%	27.59%

Acknowledgements. We thank Lorenz Panny for helpful discussions which sparked the first algorithm. We thank Kaizhan Lin and the anonymous reviewers for helpful suggestions. Mingjie Chen and Christophe Petit are partly supported by EPSRC through grant number EP/V011324/1.

A   Experimental Parameters for Constructive Deuring Correspondence

The primes used in Sect. 5 are given as follows.

$p_{3923} = 23759399264157352358673788613307970528646815114090876784643387662192449945599$

$p_1 = 11956566944641502957704189594909498993478297403838643406058180334130656750161$

$p_2 = 37670568336551536389503919665937491111216122470333837677213877442445311999999$

$p_3 = 88881583528687251695085351202716893361162950661419911309645115899579883741851$

$p_4 = 82460884298062985073154572827668830392815810118105762484859341606469999173287$

$p_5 = 95008112981120315885997172640930988762706307039235815007466698894954688225259$

For each prime, we also give the norm of the equivalent left ideal J. Our optimization focus on kernel point generations, so we provide the top three costly torsion groups during this procedure which was also mentioned in [EPSV23, Table 2].

- For prime p_{3923}, the top three costly torsion groups to work are $E_0[3^{68}] \in E_0(\mathbb{F}_{q^{27}})$, $E_0[109] \subseteq E_0(\mathbb{F}_{q^{27}})$ and $E_0[4733] \subseteq E_0(\mathbb{F}_{q^{26}})$.

$$N(J) = 2^{65} \cdot 3^{66} \cdot 5^4 \cdot 7^2 \cdot 11^2 \cdot 13^2 \cdot 17^2 \cdot 19^2 \cdot 23 \cdot 29^2 \cdot 31 \cdot 37^2 \cdot 41 \cdot 43 \cdot 47 \cdot 53$$
$$\cdot 61 \cdot 67 \cdot 73 \cdot 79 \cdot 97 \cdot 101 \cdot 109 \cdot 127 \cdot 131 \cdot 139 \cdot 151 \cdot 157 \cdot 197 \cdot 239 \cdot 241$$
$$\cdot 263 \cdot 271 \cdot 281 \cdot 283 \cdot 307 \cdot 331 \cdot 397 \cdot 461 \cdot 521 \cdot 563 \cdot 599 \cdot 607 \cdot 619 \cdot 743$$
$$\cdot 827 \cdot 941 \cdot 1153 \cdot 1301 \cdot 2179 \cdot 2357 \cdot 2393 \cdot 3061 \cdot 3361 \cdot 3907 \cdot 3923 \cdot 4733$$
$$\cdot 8273 \cdot 9199 \cdot 9661 \cdot 10069 \cdot 10753 \cdot 11719 \cdot 12517 \cdot 17033 \cdot 26489 \cdot 58897$$
$$\cdot 62731 \cdot 107641.$$

- For prime p_1, the top three costly torsion groups are $E_0[461] \subseteq E_0(\mathbb{F}_{q^{23}})$, $E_0[691] \subseteq E_0(\mathbb{F}_{q^{23}})$ and $E_0[13789] \subseteq E_0(\mathbb{F}_{q^{18}})$.

$$N(J) = 2^5 \cdot 3^5 \cdot 5^2 \cdot 7^2 \cdot 11 \cdot 13^{19} \cdot 17 \cdot 19^2 \cdot 29^{18} \cdot 31 \cdot 37 \cdot 41 \cdot 43^{18} \cdot 47 \cdot 53 \cdot 61$$
$$\cdot 79 \cdot 89 \cdot 97 \cdot 101 \cdot 103 \cdot 113 \cdot 131 \cdot 137 \cdot 151 \cdot 157 \cdot 181 \cdot 193 \cdot 199 \cdot 239$$
$$\cdot 277 \cdot 281 \cdot 331 \cdot 401 \cdot 419 \cdot 421 \cdot 443 \cdot 457 \cdot 461 \cdot 541 \cdot 601 \cdot 617 \cdot 691 \cdot 739$$
$$\cdot 811 \cdot 919 \cdot 1621 \cdot 2473 \cdot 2741 \cdot 3299 \cdot 3373 \cdot 4049 \cdot 4933 \cdot 6823 \cdot 8609$$
$$\cdot 11953 \cdot 13789 \cdot 13921 \cdot 15467 \cdot 15679 \cdot 25969 \cdot 33161 \cdot 41681 \cdot 91837.$$

- For prime p_2, the top three costly torsion groups are $E_0[859] \subseteq E_0(\mathbb{F}_{q^{33}})$, $E_0[1321] \subseteq E_0(\mathbb{F}_{q^{33}})$ and $E_0[409] \subseteq E_0(\mathbb{F}_{q^{34}})$.

$$N(J) = 2^{42} \cdot 3^5 \cdot 5^8 \cdot 7^2 \cdot 11^4 \cdot 13 \cdot 17 \cdot 19^4 \cdot 23 \cdot 29 \cdot 31 \cdot 37 \cdot 41 \cdot 43 \cdot 47^3 \cdot 53$$
$$\cdot 59 \cdot 61 \cdot 67^6 \cdot 73 \cdot 79 \cdot 97 \cdot 101^3 \cdot 103 \cdot 113^3 \cdot 137^3 \cdot 139 \cdot 157 \cdot 181^2 \cdot 197$$
$$\cdot 223 \cdot 239 \cdot 241 \cdot 271 \cdot 277^3 \cdot 281 \cdot 307^3 \cdot 311 \cdot 349 \cdot 397 \cdot 409 \cdot 421^3 \cdot 449$$
$$\cdot 547 \cdot 691 \cdot 829 \cdot 859 \cdot 907 \cdot 919 \cdot 1013 \cdot 1103 \cdot 1171 \cdot 1321 \cdot 1597 \cdot 2341$$
$$\cdot 2647 \cdot 2777 \cdot 3271 \cdot 3739 \cdot 4513 \cdot 5419 \cdot 6091 \cdot 9007 \cdot 10267 \cdot 11981$$
$$\cdot 20641 \cdot 26083 \cdot 32957 \cdot 52627.$$

- For prime p_3, before adjusting cost model, the top three costly torsion groups are $E_0[11161] \subseteq E_0(\mathbb{F}_{q^{45}})$, $E_0[2351] \subseteq E_0(\mathbb{F}_{q^{47}})$ and $E_0[1223] \subseteq E_0(\mathbb{F}_{q^{47}})$.

$$N(J) = 2^4 \cdot 3^5 \cdot 5^4 \cdot 11 \cdot 13^2 \cdot 17^2 \cdot 19 \cdot 23 \cdot 29 \cdot 31 \cdot 37 \cdot 41 \cdot 43 \cdot 47 \cdot 53 \cdot 59 \cdot 61$$
$$\cdot 67 \cdot 71 \cdot 73 \cdot 79 \cdot 83 \cdot 89 \cdot 97 \cdot 101 \cdot 103 \cdot 109 \cdot 113 \cdot 137 \cdot 149 \cdot 157 \cdot 173 \cdot$$
$$181 \cdot 191 \cdot 193 \cdot 197 \cdot 199 \cdot 239 \cdot 241 \cdot 251 \cdot 257 \cdot 271 \cdot 311 \cdot 353 \cdot 397 \cdot 431$$
$$\cdot 463 \cdot 521 \cdot 601 \cdot 631 \cdot 677 \cdot 769 \cdot 953 \cdot 1063 \cdot 1117 \cdot 1223 \cdot 1301 \cdot 1453 \cdot$$
$$1621 \cdot 1783 \cdot 1801 \cdot 2351 \cdot 2521 \cdot 2857 \cdot 4673 \cdot 5581 \cdot 6043 \cdot 7937 \cdot 8087 \cdot$$
$$9103 \cdot 9829 \cdot 10729 \cdot 11161 \cdot 12161 \cdot 13441 \cdot 17209 \cdot 17807 \cdot 29017 \cdot 46901$$
$$\cdot 47269 \cdot 47441 \cdot 77969 \cdot 85021 \cdot 89839 \cdot 180503.$$

– For prime p_4, before adjusting cost model, the top three costly torsion groups are $E_0[1723] \subseteq E_0(\mathbb{F}_{q^{41}})$, $E_0[1559] \subseteq E_0(\mathbb{F}_{q^{41}})$ and $E_0[3361] \subseteq E_0(\mathbb{F}_{q^{40}})$.

$$N(J) = 2^5 \cdot 3^4 \cdot 5^2 \cdot 7^2 \cdot 13^2 \cdot 17 \cdot 19 \cdot 23 \cdot 29 \cdot 31 \cdot 37^2 \cdot 41 \cdot 43 \cdot 47 \cdot 53 \cdot 59 \cdot 61$$
$$\cdot 67 \cdot 71 \cdot 73 \cdot 79 \cdot 83 \cdot 89 \cdot 97 \cdot 101 \cdot 109 \cdot 113 \cdot 127 \cdot 139 \cdot 157 \cdot 163 \cdot 181 \cdot$$
$$193 \cdot 229 \cdot 233 \cdot 239 \cdot 241 \cdot 277 \cdot 311 \cdot 379 \cdot 397 \cdot 401 \cdot 409 \cdot 463 \cdot 677 \cdot 769$$
$$\cdot 859 \cdot 877 \cdot 881 \cdot 937 \cdot 1217 \cdot 1301 \cdot 1321 \cdot 1423 \cdot 1489 \cdot 1559 \cdot 1597 \cdot 1723$$
$$\cdot 1873 \cdot 1973 \cdot 2377 \cdot 2887 \cdot 3361 \cdot 3461 \cdot 3499 \cdot 3571 \cdot 3697 \cdot 3769 \cdot 3877 \cdot$$
$$4231 \cdot 4993 \cdot 5023 \cdot 5437 \cdot 5669 \cdot 5881 \cdot 6211 \cdot 6449 \cdot 6781 \cdot 8317 \cdot 8677 \cdot$$
$$10501 \cdot 12757 \cdot 15227 \cdot 19441 \cdot 19793 \cdot 29389 \cdot 64577.$$

– For prime p_5, the top three costly torsion groups are $E_0[83] \subseteq E_0(\mathbb{F}_{q^{41}})$, $E_0[2029] \subseteq E_0(\mathbb{F}_{q^{39}})$ and $E_0[859] \subseteq E_0(\mathbb{F}_{q^{39}})$.

$$N(J) = 2^4 \cdot 3^4 \cdot 5^3 \cdot 7^3 \cdot 11^2 \cdot 13 \cdot 17 \cdot 19 \cdot 23 \cdot 29 \cdot 31 \cdot 37 \cdot 41 \cdot 43 \cdot 47 \cdot 53 \cdot 59$$
$$\cdot 61 \cdot 67 \cdot 71 \cdot 73 \cdot 79 \cdot 83 \cdot 89 \cdot 97 \cdot 101 \cdot 103 \cdot 109 \cdot 113 \cdot 127 \cdot 131 \cdot 137 \cdot$$
$$151 \cdot 157 \cdot 193 \cdot 199 \cdot 223 \cdot 229 \cdot 233 \cdot 239 \cdot 257 \cdot 349 \cdot 409 \cdot 541 \cdot 601 \cdot 631$$
$$\cdot 661 \cdot 701 \cdot 727 \cdot 751 \cdot 859 \cdot 1009 \cdot 1213 \cdot 1471 \cdot 1489 \cdot 1601 \cdot 1657 \cdot 2029$$
$$\cdot 2341 \cdot 2441 \cdot 2609 \cdot 2801 \cdot 3361 \cdot 3559 \cdot 4159 \cdot 4831 \cdot 5077 \cdot 7193 \cdot 9649 \cdot$$
$$9929 \cdot 13441 \cdot 13907 \cdot 13913 \cdot 14293 \cdot 21937 \cdot 22573 \cdot 27961 \cdot 38851 \cdot 52667$$
$$\cdot 59621 \cdot 69193 \cdot 162499 \cdot 170047 \cdot 208141 \cdot 288493.$$

References

[ACL+23]   Arpin, S., Chen, M., Lauter, K., Scheidler, R., Stange, K., Tran, H.: Orienteering with one endomorphism. La Mat. **2**, 06 (2023)

[BDFLS20] Bernstein, D.J., De Feo, L., Leroux, A., Smith, B.: Faster computation of isogenies of large prime degree. In: Galbraith, S. (ed.) Progress in ANTS XIV: Proceedings of the Fourteenth Algorithmic Number Theory Symposium, pp. 39–55. Mathematical Sciences Publishers, Auckland (2020)

[BGDS23]   Banegas, G., Gilchrist, V., Le Dévéhat, A., Smith, B.: Fast and frobenius: rational isogeny evaluation over finite fields. In: Aly, A., Tibouchi, M. (eds.) Progress in Cryptology – LATINCRYPT 2023, pp. 129–148. Springer, Cham (2023)

[Coh93]   Cohen, H.: A Course in Computational Algebraic Number Theory. Graduate Texts in Mathematics, vol. 138. Springer, Berlin (1993)

[Dev23]   The Sage Developers. Sagemath, the sage mathematics software system (version 10) (2023). https://www.sagemath.org/

[DFKL+20] De Feo, L., Kohel, D., Leroux, A., Petit, C., Wesolowski, B.: SQISign: compact post-quantum signatures from quaternions and isogenies. In: Moriai, S., Wang, H. (eds.) Advances in Cryptology – ASIACRYPT 2020, pp. 64–93. Springer, Cham (2020)

[DFLLW23] De Feo, L., A., Longa, P., Wesolowski, B.: New algorithms for the deuring correspondence: towards practical and secure SQISign signatures. In: Hazay, C., Stam, M. (eds.) Advances in Cryptology – EUROCRYPT 2023, pp. 659–690. Springer, Cham (2023)

[EPSV23] Eriksen, J.K., Panny, L., Sotáková, J., Veroni, M.: Deuring for the people: supersingular elliptic curves with prescribed endomorphism ring in general characteristic. Cryptology ePrint Archive, Paper 2023/106 (2023). Accepted in LuCaNT 2023

[FS16] Fité, F., Sutherland, A.V.: Sato-Tate groups of $y^2 = x^8 + c$ and $y^2 = x^7 - cx$. In: Frobenius distributions: Lang-Trotter and Sato-Tate conjectures. Contemporary Mathematics, vol. 663, pp. 103–126. American Mathematical Society, Providence (2016)

[Gal12] Galbraith, S.D.: Mathematics of Public Key Cryptography. Cambridge University Press (2012)

[GPS17] Galbraith, S.D., Petit, C., Silva, J.: Identification protocols and signature schemes based on supersingular isogeny problems. In: Advances in Cryptology–ASIACRYPT 2017: 23rd International Conference on the Theory and Applications of Cryptology and Information Security, Hong Kong, China, 3–7 December 2017, Part I, pp. 3–33. Springer (2017)

[Koh96] Kohel, D.R.: Endomorphism Rings of Elliptic Curves over Finite Fields. Ph.D. thesis, University of California, Berkeley (1996)

[Sch87] Schoof, R.: Nonsingular plane cubic curves over finite fields. J. Combin. Theory Ser. A **46**(2), 183–211 (1987)

[SEMR23] Santos, M.C.-R., Eriksen, J.K., Meyer, M., Reijnders, K.: Aprèssqi: extra fast verification for SQIsign using extension-field signing. Cryptology ePrint Archive, Paper 2023/1559 (2023). https://eprint.iacr.org/2023/1559

[V71] Vélu, J.: Isogénies entre courbes elliptiques. Comptes Rendus l'Acad. Sci. Paris A **273**, 238–241 (1971)

[Wat69] Waterhouse, W.C.: Abelian varieties over finite fields. In: Annales scientifiques de l'École normale supérieure, vol. 2, pp. 521–560 (1969)

Constructing Families of Pairing-Friendly Elliptic Curves with Small ρ-Value and Rational Solutions of Some Systems of Polynomial Equations

Robert Dryło[✉]

Institute of Mathematical Economics, Warsaw School of Economics,
al. Niepodległości 162, 02-554 Warsaw, Poland
rdrylo@sgh.waw.pl

Abstract. One uses parametric families $(r(x), t(x), q(x))$ of pairing-friendly elliptic curves to obtain such curves with ρ-value closer to 1, and one of the goals is to obtain families with ρ close to 1. The Brezing-Weng method and its extension allow to construct complete, complete with variable discriminant, and sparse families. To construct families using these methods one chooses a number field K, which contains a kth primitive root of unity ζ_k, and a polynomial $r(x) \in \mathbb{Q}[x]$ such that $K \cong \mathbb{Q}[x]/(r(x))$, and then obtains $t(x)$ and $q(x)$ in $\mathbb{Q}[x]$ according to these methods. Obtained triple (r, t, q) satisfies part of conditions from the definition of a family, and if this triple will be useful to obtain parameters of elliptic curves depends among others if one can expect to obtain infinitely many primes or prime powers as values of $q(x)$ for $x \in \mathbb{N}$. In this paper for a number field K as above of the extension degree $n = [K : \mathbb{Q}] > 2$ we will consider triples (z, ζ_k, α) of elements in K such that z is a primitive element of $K/\mathbb{Q}$, and α has a suitable property such that (z, ζ_k, α) determines a triple of polynomials (r, t, q) as above, where r is the minimal polynomial of z over $\mathbb{Q}$, and t and q are obtained in a suitable way. Note that if $\sqrt{-d} \in K$ for a square-free $d \in \mathbb{Z}_{>0}$, then one can use triples $(z, \zeta_k, \sqrt{-d})$ to search for complete families. For a fixed primitive element γ of $K/\mathbb{Q}$ we will use coordinates of elements in K in the basis $1, \ldots, \gamma^{n-1}$. Let $1 \le \rho_0 < 2$ be a given upper bound on ρ-value. We will give conditions on coordinates of z and α in K such that if (z, ζ_k, α) satisfy these conditions, then one can obtain from it a triple (r, t, q) with $\rho \le \rho_0$. These conditions are given by ranks of some matrices, and equivalently can be given by minors of these matrices, hence one can obtain a system of polynomial equations on coordinates of elements in K. However for larger n because of degrees of polynomials in this system it can be difficult to determine solutions of this system.

Keywords: Bilinear pairings · pairing-friendly curves · complex multiplication method · Brezing-Weng method · systems of polynomial equations

A. Dąbrowski et al. (Eds.): NuTMiC 2024, LNCS 14966, pp. 156–171, 2025.
https://doi.org/10.1007/978-3-031-82380-0_5

1   Introduction

Elliptic curves with a subgroup of large prime order and a small embedding degree are used for implementing cryptographic schemes based on bilinear pairings (see, e.g., [3,4,7,19,22]). An elliptic curve is commonly called pairing-friendly if it can be used for secure and efficient implementation of such schemes. Let E be an elliptic curve over $\mathbb{F}_q$, and $r \neq \operatorname{char} \mathbb{F}_q$ be a prime number which divides the order $|E(\mathbb{F}_q)|$. The embedding degree with respect to r is defined as $k = \min\{i \in \mathbb{Z}_{>0} : r \mid q^i - 1\}$. The field $\mathbb{F}_{q^k}$ is the smallest field extension of $\mathbb{F}_q$ which contains all rth roots of unity. Bilinear pairings used in cryptography map a pair of two r-torsion points on E to the group of rth roots of unity in $\mathbb{F}_{q^k}$. Supersingular elliptic curves have embedding degrees ≤ 6 (see, e.g., [15]). Pairing-friendly ordinary elliptic curves are rare and require special construction. Choosing embedding degree k and parameters of an elliptic curve one should consider attacks on the discrete logarithm problem in a field of extension degree k which are based on the index calculus method or the tower number field sieve (TNFS) (see [13] and [18]).

Constructions of pairing-friendly ordinary elliptic curves are based on the complex multiplication (CM) method. To construct such a curve first for a given k one obtains parameters (r, t, q) of an ordinary elliptic curve E over $\mathbb{F}_q$ with the trace t such that r is prime, divides the order $|E(\mathbb{F}_q)|$, and k is the embedding degree with respect to r. Additionally these parameters are determined to have suitably small CM discriminant d, where $d \in \mathbb{Z}_{>0}$ is square-free such that $4q - t^2 = dy^2$ for $y \in \mathbb{Z}$. This condition on d is needed to construct a curve using the CM method, which is efficient for suitably small CM discriminants. The parameter $\rho = \frac{\log_2 q}{\log_2 r}$ is approximately equal to the ratio of bit lenghts of $|E(\mathbb{F}_q)|$ and r. For security given by r operations on a curve may be more efficient if bit lenghts of r and $|E(\mathbb{F}_q)|$ are close, thus ρ-value close to 1 may be more suitable for some applications. The Cocks-Pinch (see [15, Theorem 4.1]) and the Dupont-Enge-Morain [10] methods can be used to obtain parameters (r, t, q) of ordinary elliptic curves with a given embedding degree k and CM discriminant d for which usually $\rho \approx 2$ (see also [5,17]). To obtain pairing-friendly ordinary elliptic curves with ρ-value closer to 1 one can use parametric families. A triple of polynomials $(r(x), t(x), q(x))$ in $\mathbb{Q}[x]$ parameterizes a family of ordinary elliptic curves with embedding degree k and CM discriminant d if these polynomials satisfy suitable conditions such that one expects that for infinitely many $x \in \mathbb{N}$ we will obtain a prime power $q(x)$, a prime $r(x)$, and $(r(x), t(x), q(x))$ will be as above parameters of an elliptic curve E over $\mathbb{F}_{q(x)}$ with trace $t(x)$, embedding degree k with respect to $r(x)$, and CM discriminant d. Such curves for large x have ρ-values close to the ρ-value of the family $\rho = \deg q(x)/\deg r(x)$. A triple of polynomials with the above property we will also called a family.

Miyaji, Nakabayashi and Takano [21] used families to characterize ordinary elliptic curves of prime order with embedding degrees $k = 3, 4, 6$. Families of elliptic curves with order of the form small cofactor times a prime with embedding degree $k = 3, 4, 6$ were also considered in [16,23]. These families have $\rho = 1$. Families with $\rho = 1$ were also given by Freeman [14] for $k = 10$, and by Barreto-Naehrig [2] for $k = 12$. Families with $1 \leq \rho < 2$ which are called complete can be construted using the Brezing-Weng method [6] (see also [1]). In this method one chooses a number field K, which contains kth primitive roots of unity and $\sqrt{-d}$ for a CM discriminant d. Then polynomials $r(x), t(x)$ and $y(x) \in \mathbb{Q}[x]$ are chosen such that $K \cong \mathbb{Q}[x]/(r(x))$, $\bar{t} = \zeta_k + 1$, and $\bar{y} = (\zeta_k - 1)/\sqrt{-d} \in K$, where $\zeta_k \in \mathbb{Q}[x]/(r(x))$ is a kth primitive root of unity, and bar denotes the class modulo $r(x)$. One takes $q(x) = \frac{1}{4}(t(x)^2 + dy(x)^2)$, and checks among others using some conjectures if one may expect to obtain for $x \in \mathbb{N}$ infinitely many values $q(x)$ which are primes or prime powers. If K is a suitable cyclotomic field and $r(x)$ is a cyclotomic polynomial, then one obtains so-called cyclotomic families. In general, for a number field K and monic $r(x) \in \mathbb{Q}[x]$ we have $\mathbb{Q}[x]/(r(x)) \cong K$ if and only if $r(x)$ is the minimal polynomial of a primitive element of $K/\mathbb{Q}$. Kachisa, Schaefer, and Scott [20] to search for families with small ρ-value used minimal polynomials of primitive elements of $K/\mathbb{Q}$ obtained by choosing various coordinates of these element in a fixed basis of $K/\mathbb{Q}$. Freeman, Scott and Teske in [15] among others gave classification of constructions of families of pairing-friendly elliptic curve, and gathered known and gave new contructions of families. According to definitions from [15] we have complete, complete with variable discriminant, and sparse families, which depends on the CM equation $4(x) - t(x)^2 = dy^2$ which should have infinitely many integer solutions (x, y). From Siegel's theorem [25, Theorem IX.4.3] we obtain the following necessary condition for existence of these solution. The left side of the CM equation has to be of the form $4q(x) - t(x)^2 = g(x)y(x)^2$ for some polynomials $g(x), y(x) \in \mathbb{Q}[x]$, where g is of degree ≤ 2. If $g = d$, then a family is complete. If $\deg g = 1$, then a family is called complete with variable discriminant. If $\deg g = 2$, we can assume that g has no double root and has positive leading coefficient, and we say that a family is sparse. The above families of Miyaji et al. [21] and Freeman [14] are sparse. The equality $4q(x) - t(x)^2 = g(x)y(x)^2$ was used in [8] to extend the Brezing-Weng method as follows to construct complete families with variable discriminants and sparse families. Let K be number field which contains kth primitive roots of unity, and let $K \cong \mathbb{Q}[x]/(r(x))$ for a polynomial $r(x) \in \mathbb{Q}[x]$. Search for nonzero polynomial $g(x) \in \mathbb{Q}[x]$, and choose polynomials $t(x)$ and $y(x)$ in $\mathbb{Q}[x]$ such that

$$\deg g \leq 2, \sqrt{-\bar{g}} \in \mathbb{Q}[x]/(r(x)), \bar{t} = \zeta_k + 1, \text{ and } \bar{y} = (\zeta_k - 1)\sqrt{-\bar{g}}, \qquad (1)$$

where $\zeta_k \in \mathbb{Q}[x]/(r(x))$ is a kth primitive root of unity. Take $q(x) = \frac{1}{4}(t(x)^2 + g(x)y(x)^2)$. If the obtained triple of polynomials $(r(x), t(x), q(x))$ will be useful to obtain parameters of elliptic curves depends among others if $q(x)$ will have suitable properties as above. If $\sqrt{-d} \in K$, then for $g = d$ we have the Brezing-Weng method for complete families. For more constructions of families see [11–13, 18, 24, 26].

In Sects. 2 and 3 we recall some results related to pairing-friendly elliptic curves and families. In Sect. 4 we give the following contribution of this paper, which is based on the preprint [9]. We assume that K is a number field of the extension degree $n = [K : \mathbb{Q}] > 2$, which contains a kth primitive root of unity ζ_k. We will consider triples of polynomials (r, t, q) in $\mathbb{Q}[x]$ such that $\mathbb{Q}[x]/(r) \cong K$, r is monic, $q = \frac{1}{4}(t^2 + gy^2)$, where g, t and y satisfy (1), and degrees of t and y are $< n$. For such a triple (r, t, q) let $\rho = \deg q / \deg r$. Let z be a primitive element of $K/\mathbb{Q}$, let $r \in \mathbb{Q}[x]$ be the minimal polynomial of z over $\mathbb{Q}$, and let $f : K \to \mathbb{Q}[x]/(r)$ be the field isomorphism such that $f(z) = \bar{x}$. If for $\alpha \in K \setminus \{0\}$ there exists polynomial $g \in \mathbb{Q}[x]$ of degree ≤ 2 such that $\bar{g} = -f(\alpha^2)$, then we will say that (z, ζ_k, α) determines the triple of polynomials (r, t, q) in $\mathbb{Q}[x]$, where $q = \frac{1}{4}(t^2 + gy^2)$ for polynomials $t, y \in \mathbb{Q}[x]$ of degree $< n$ such that $\bar{t} = f(\zeta_k) + 1$, and $\bar{y} = f((\zeta_k - 1)/\alpha)$.

Let γ be a fixed primitive element $K/\mathbb{Q}$. We will use the basis $1, \gamma, \ldots, \gamma^{n-1}$ of $K/\mathbb{Q}$. Let $1 \leq \rho_0 < 2$ be a given upper bound on ρ-value.

First we will consider the case of complete families. If $\sqrt{-d} \in K$ for square-free $d \in \mathbb{Z}_{>0}$, then we will give conditions on coordinates of primitive elements z of $K/\mathbb{Q}$ such that $(z, \zeta_k, \sqrt{-d})$ determines a triple (r, t, q) with $\rho \leq \rho_0$. These conditions are given by ranks of some matrices, or equivalently by minors of these matrices (see Theorem 7). These minors give a system of polynomial equations in the projective space $\mathbb{P}^{n-2}(\mathbb{Q})$ such that if $z = \sum_{i=1}^{n-1} x_i \gamma_i$ is a primitive element of $K/\mathbb{Q}$ for $X = (x_1 : \ldots : x_{n-1}) \in \mathbb{P}^{n-2}(\mathbb{Q})$, then X is a solution of this system if and only if $(z, \zeta_k, \sqrt{-d})$ determines a triple (r, t, q) with $\rho \leq \rho_0$. However, determining solutions of this system in practice may be difficult for larger n, because of degrees of polynomials in this system. We will show how the Barreto-Naehrig family [2] with embedding degree $k = 12$ and $\rho = 1$ can be obtained using this method in Example 8.

Next we consider complete families with variable discriminant and sparse families. First we assume that polynomials r and t in $\mathbb{Q}[x]$ are given such that $K \cong \mathbb{Q}[x]/(r)$, t is as in (1), and $2 \deg t / n \leq \rho_0$. We give a system of polynomial equations in $\mathbb{P}^{n-1}(\mathbb{Q})$ such that if $X = (x_0 : \ldots : x_{n-1}) \in \mathbb{P}^{n-1}(\mathbb{Q})$ satisfies this system, then for $\alpha = \sum_{i=0}^{n-1} x_i \bar{x}^i$ we have $\bar{g} = -\alpha^2$ for $g \in \mathbb{Q}[x]$ of degree ≤ 2, and g allows to obtain a polynomial q as above such that the triple (r, t, q) has $\rho \leq \rho_0$ (see Theorem 9).

We also give conditions on coordinates of primitive elements z of $K/\mathbb{Q}$ and $\alpha \in K \setminus \{0\}$ such that for z and α which satisfy these conditions one can obtain from (z, ζ_k, α) a triple (r, t, q) with $\rho \leq \rho_0$. These conditions are given by ranks of some matrices, or equivalently by minors of these matrices. These minors give a system of polynomial equations in $\mathbb{P}^{n-2}(\mathbb{Q}) \times \mathbb{P}^{n-1}(\mathbb{Q})$ such that if $z = \sum_{i=1}^{n-1} x_i \gamma^i$ is a primitive element of $K/\mathbb{Q}$ for $X = (x_1 : \ldots : x_{n-1}) \in \mathbb{P}^{n-2}(\mathbb{Q})$, and (X, X') is a solution of this system, then for $\alpha = \sum_{i=0}^{n-1} x_i' \gamma^i$, where $X' = (x_0' : \ldots : x_{n-1}') \in \mathbb{P}^{n-1}(\mathbb{Q})$, the triple (z, ζ_k, α) determines the triple of polynomials (r, t, q) with $\rho \leq \rho_0$ (see Theorem 11). We show how Freeman's family [14] with embedding degree $k = 10$ and $\rho = 1$ can be obtained using this method (see Example 12).

2   Pairing-Friendly Elliptic Curves and the Cocks-Pinch Algorithm

In this section we recall basic properties of pairing-friendly elliptic curves, and the Cocks-Pinch algorithm to construct such curves.

Let $E/\mathbb{F}_q$ be an elliptic curve, and $r \neq \text{char}\,\mathbb{F}_q$ be a prime number. Bilinear pairings used in cryptography map pairs of r-torsion points on E to the group of rth roots of unity in an extension field of $\mathbb{F}_q$. These pairings have values in the smallest field extension $\mathbb{F}_{q^k}$ which contains all rth roots of unity. The extension degree $k = \min\{i \in \mathbb{Z}_{>0} : r \mid q^i - 1\}$ is called the embedding degree with respect to r.

Let $t = q + 1 - |E(\mathbb{F}_q)|$ be the trace of E over $\mathbb{F}_q$. The trace satisfies the Hasse bound $|t| \leq 2\sqrt{q}$, or equivalently $(\sqrt{q} - 1)^2 \leq |E(\mathbb{F}_q)| \leq (\sqrt{q} + 1)^2$.

Hence the bit lenght of the order $|E(\mathbb{F}_q)|$ is close to $\log_2 q$. The parameter $\rho = \frac{\log_2 q}{\log_2 r}$ is approximately equal to the ratio of bit lenghts of $|E(\mathbb{F}_q)|$ and r. For some applications ρ-value close to 1 may be more suitable because then for security level given by r operations on a curve may be more efficient.

The qth power Frobeniuse endomorphism of E satisfies the characteristic equation $x^2 - tx + q = 0$ with discriminant $t^2 - 4q \leq 0$. An elliptic curve E is called ordinary if and only if $\gcd(t, q) = 1$. If E is ordinary, then $t^2 - 4q = -dy^2$, where $d, y \in \mathbb{Z}$ and $d > 0$ is square-free. This integer d is called the CM discriminant of an ordinary elliptic curve $E/\mathbb{F}_q$. If E is ordinary, then the endomorphism ring $\text{End}(E)$ is isomorphic to an order in the quadratic imaginary field $\mathbb{Q}(\sqrt{-d})$. For each integer $t \in [-2\sqrt{q}, 2\sqrt{q}]$ such that $\gcd(t, q) = 1$ there exists an ordinary elliptic curve over $\mathbb{F}_q$ with the trace t, and can be constructed using the complex multiplication (CM) method. Such an ordinary elliptic curve over $\mathbb{F}_q$ can be obtained from reduction of an elliptic curve in characteristic 0 with the endomorphism ring isomorphic to the maximal order $\mathcal{O}_K$ in the imaginary quadratic field $K = \mathbb{Q}(\sqrt{-d})$, where d is the square-free part of $4q - t^2$. The j-invariants of elliptic curves with the endomorphism ring isomorphic to $\mathcal{O}_K$ are exactly roots of the Hilbert class polynomial $H_K(x) \in \mathbb{Z}[x]$ of the field K. In practice computation of the Hilbert class polynomial is efficient for suitably small CM discriminants d (for example, $d < 10^{12}$). Therefore the CM method is efficient for suitably small d.

By parameters of an ordinary elliptic curve with an embedding degree k and discriminant d we will mean a triple (r, t, q) of integers such that q is a prime power, r is prime, $r \nmid q$, and there exists an elliptic curve $E/\mathbb{F}_q$ with the trace t and CM discriminant d such that r divides $|E(\mathbb{F}_q)|$, and k is embedding degree with respect to r.

We will denote by bar equivalence classes modulo r. The embedding degree with respect to r is equal to k if and only if $\bar{q}$ is a kth primitive root of unity in $\mathbb{F}_r$. For parameters (r, t, q), d, and k as above we have

(1) $\zeta_k = \bar{q} = \bar{t} - 1$ is a kth primitive root of unity in $\mathbb{F}_r$, in particular $k \mid r - 1$.
(2) $-d$ is a quadratic residue modulo r, and $\bar{y} = \pm(\zeta_k - 1)/\sqrt{-\bar{d}}$, where $4q - t^2 = dy^2$.

The following Cocks-Pinch algorithm (see [15, Theorem 4.1]) allows to obtain parameters (r, t, q) of an elliptic curve with embedding degree k and discriminant d.

Algorithm 1. Input: An embedding degree k and CM discriminant d.
Output: Parameters (r, t, q) of an elliptic curve with embedding degree k and discriminant d or no output.

1. Choose a prime r such that $k \mid r - 1$, and $-d$ is a quadratic residue modulo r.
2. Determine a kth primitive root of unity $\zeta_k \in \mathbb{F}_r$, and $\sqrt{-d} \in \mathbb{F}_r$.
3. Choose integers $t_0, y_0 \in [0, r)$ such that $\bar{t}_0 = \zeta_k + 1$ and $\bar{y}_0 = (\zeta_k - 1)/\sqrt{-d}$.
4. For a given bound $b \geq 0$ and integers $i, j \in [-b, b]$ let $t = t_0 + ir$, $y = y_0 + jr$, and $q = \frac{1}{4}(t^2 + dy^2)$. If q is prime, then return (r, t, q).

Elliptic curves obtained by the Cocks-Pinch method usually have $\rho \approx 2$. To construct pairing-friendly elliptic curves with $\rho < 2$ one can use parametric families.

3 Families of Pairing-Friendly Elliptic Curves and the Brezing-Weng Algorithm

In this section we recall basic properties of families of pairing-friendly elliptic curves from [15], and the extension of the Brezing-Weng algorithm from [8], which allows also construct complete families with variable discriminant and sparse families.

One may say in not precise way that a triple $(r(x), t(x), q(x))$ of polynomials in $\mathbb{Q}[x]$ parameterizes a family of elliptic curves with embedding degree k and discriminant d if one expects that for infinitely many $x \in \mathbb{N}$ the values $(r(x), t(x), q(x))$ are parameters of an elliptic curve with embedding degree k and CM discriminant d. Below we recall the definition of a family.

The following definition is related to the Bouniakowsky-Schinzel conjecture for a polynomial in $\mathbb{Z}[x]$ to take infinitely many prime values for $x \in \mathbb{N}$.

Definition 2 ([15, Def. 2.5]). We say that a polynomial $f(x) \in \mathbb{Q}[x]$ represents primes if the following conditions are satsfied:

(1) f is nonconstant,
(2) f has positive leading coefficient,
(3) f is irreducible,
(4) $f(x) \in \mathbb{Z}$ for some $x \in \mathbb{Z}$,
(5) $\gcd(\{f(x) : x, f(x) \in \mathbb{Z}\}) = 1$.

We say that a polynomial $f(x) \in \mathbb{Q}[x]$ is integer-valued if $f(x) \in \mathbb{Z}$ for each $x \in \mathbb{Z}$.

Recall that for $k \in \mathbb{Z}_{>0}$ the kth cyclotomic polynomial $\Phi_k(x) = \prod_{\zeta_k \in \mathbb{C}}(x - \zeta_k) \in \mathbb{Z}[x]$, where the product is for all kth primitive roots of unity $\zeta_k \in \mathbb{C}$. We have $x^k - 1 = \prod_{\ell \mid k} \Phi_\ell(x)$. For an algebraically closed field F with $\operatorname{char} F = 0$ or $\gcd(\operatorname{char} F, k) = 1$ kth primitive roots of unity in F are exactly roots of Φ_k in F.

Definition 3. Let $k, d \in \mathbb{Z}_{>0}$, and d be square-free. A triple of polynomials $(r(x), t(x), q(x))$ in $\mathbb{Q}[x]$ parametrizes a family of elliptic curves with embedding degree k and discriminant d if the following conditions are satisfied:

(1) $q(x) = p(x)^\ell$ for some $\ell > 0$ and $p(x)$ represents primes,
(2) $r(x)$ is nonconstant, irreducible, integer-valued, and has positive leading coefficient,
(3) $r(x)$ divides $q(x) + 1 - t(x)$,
(4) $r(x)$ divides $\Phi_k(t(x) - 1)$,
(5) the CM equation $4q(x) - t(x)^2 = dy^2$ has infinitely many integer solutions (x, y).

A triple $(r(x), t(x), q(x))$ we will also call a family if it parametrizes a family of elliptic curves with embedding degree k and discriminant d. For a family $(r(x), t(x), q(x))$ the ρ-value is defined by $\rho = \frac{\deg q(x)}{\deg r(x)}$. If for infinitely many $x \in \mathbb{N}$ the values $(r(x), t(x), q(x))$ are parameters of elliptic curves, then ρ-values of these curves tend to the ρ-value of the family when x tends to ∞.

A family $(r(x), t(x), q(x))$ with discriminant d is called complete if $4q(x) - t(x)^2 = dy(x)^2$ for some polynomial $y(x) \in \mathbb{Q}[x]$. Complete families can be constructed by the Brezing-Weng algorithm [6]. According to properties of the CM equation, one also defines complete families with variable discriminant and sparse families.

Recall that if $f(x) \in \mathbb{Q}[x]$ is of degree ≥ 3 and has no multiple roots, then from Siegel's theorem [25, Theorem IX.4.3] the curve $y^2 = f(x)$ has only finitely many points in $\mathbb{Z}^2$. Hence if $f, h \in \mathbb{Q}[x]$, $\deg f \geq 3$, and f has no multiple roots, then the curve $y^2 = f(x)h(x)^2$ has only finitely many points in $\mathbb{Z}^2$. We obtain this writing $af(x)h(x)^2 = f_1(x)h_1(x)^2$ for some $a \in \mathbb{N}_{>0}$ and $f_1, h_1 \in \mathbb{Z}[x]$. If there were infinitely many integer points on the curve $ay^2 = f_1(x)h_1(x)^2$, then for some $b \mid a$ the curve $by^2 = f_1(x)$ would have infinitely many integer points, which is impossible from Siegel's theorem. Thus if the CM equation $4q(x) - t(x)^2 = dy^2$ has infinitely many integer solutions, then its left side is of the form $4q(x) - t(x)^2 = g(x)y(x)^2$ for polynomials $g(x), y(x) \in \mathbb{Q}[x]$ such that $\deg g(x) \leq 2$.

Complete families with variable discriminant and sparse families were defined in [8] as follows to extend the Brezing-Weng algorithm to construct such families. A family $(r(x), t(x), q(x))$ is complete with variable discriminant if $4q(x) - t(x)^2 = (ax + b)y(x)^2$ for some $y(x) \in \mathbb{Q}[x]$, $a, b \in \mathbb{Q}$ and $a \neq 0$. For such family and each square-free $d > 0$ substituting $x = (dx^2 - b)/a$ we obtain the triple $(r((dx^2 - b)/a), t((dx^2 - b)/a), q((dx^2 - b)/a))$ such that $4q((dx^2 - b)/a) - t((dx^2 - b)/a)^2 = d(xy(x))^2$.

A family $(r(x), t(x), q(x))$ is sparse if $4q(x) - t(x)^2 = g(x)y(x)^2$ for $g(x), y(x) \in \mathbb{Q}[x]$, where g has degree 2, no double root, and a positive leading coefficient.

For a polynomial $r(x) \in \mathbb{Q}[x]$ we will denote by a bar classes in the residue ring $\mathbb{Q}[x]/(r(x))$. If $(r(x), t(x), q(x))$ is a family with embedding degree k, and $4q(x) - t(x)^2 = g(x)y(x)^2$, where $g, y \in \mathbb{Q}[x]$, and $\deg g \leq 2$, then

(1) $\zeta_k = \bar{t} - 1 = \bar{q}$ is a kth primitive root of unity in the residue field $\mathbb{Q}[x]/(r(x))$,
(2) $\sqrt{-\bar{g}} \in \mathbb{Q}[x]/(r(x))$,
(3) $\bar{y} = (\zeta_k - 1)/\sqrt{-\bar{g}}$.

The following extension of the Brezing-Weng algorithm allows to construct complete families with variable discriminant and sparse families.

Algorithm 4. ([8]) Input: An embedding degree k.
Output: A family (r, t, q) of elliptic curves with embedding degree k or no output.

(1) Choose a number field K which contains kth primitive roots of unity.
(2) Choose a polynomial $r \in \mathbb{Q}[x]$ such that $\mathbb{Q}[x]/(r) \cong K$ and r satisfies (2) in Definition 3.
(3) Determine a kth primitive root of unity $\zeta_k \in \mathbb{Q}[x]/(r)$.
(4) Find a polynomial $g \in \mathbb{Q}[x]$ of degree ≤ 2 with positive leading coefficient such that $-\bar{g}$ is a square in $\mathbb{Q}[x]/(r)$.
(5) Let t and y be the polynomials in $\mathbb{Q}[x]$ of degrees $< \deg r$ such that $\bar{t} = \zeta_k + 1$ and $\bar{y} = (\zeta_k - 1)/\sqrt{-\bar{g}}$.
(6) Let $q = \frac{1}{4}(t^2 + gy^2)$.
(7) If the triple (r, t, q) satisfies conditions such that one can expects to obtain parameters $(r(x), t(x), q(x))$ of elliptic curves for infinitely many $x \in \mathbb{N}$, then return (r, t, q).

If $\sqrt{-d} \in K$, and in the above algorithm we take $g = d$, then we obtain the Brezing-Weng algorithm [6] for complete families with discriminant d. Families obtained by Algorithm 4 have $\rho = \max\{2 \deg t, \deg g + 2 \deg y\}/\deg r \leq 2$, and if family is complete, then $\rho < 2$.

4 Families with Small ρ-Value and Rational Solutions of Some Systems of Polynomial Equations

Let K be a number field such that kth primitive roots of unity are in K, and let $n = [K : \mathbb{Q}] > 2$ be the extension degree of $K/\mathbb{Q}$. Let $1 \leq \rho_0 < 2$ be a given upper bound on ρ-value, and let $\ell = \max\{i \in \mathbb{N} : 2i/n \leq \rho_0\}$. Let $\zeta_k \in K$ be a fixed kth primitive root of unity.

For a primitive element z of $K/\mathbb{Q}$ and the pair (z, ζ_k) one can use the Brezing-Weng method as follows. Let $r \in \mathbb{Q}[x]$ be the minimal polynomial of z over $\mathbb{Q}$, and let $f : K \to \mathbb{Q}[x]/(r)$ be the field isomorphism such that $f(z) = \bar{x}$. Let $t \in \mathbb{Q}[x]$ be of degree $< n$ such that $\bar{t} = f(\zeta_k) + 1$. Then we will say that (z, ζ_k) determines the pair of polynomials (r, t).

If we want to construct complete families and $\sqrt{-d} \in K$, where $d \in \mathbb{Z}_{>0}$ is square-free, let $y \in \mathbb{Q}[x]$ be of degree $< n$ such that $\bar{y} = f((\zeta_k - 1)/\sqrt{-d})$, and let $q = \frac{1}{4}(t^2 + dy^2)$. Then we will say that $(z, \zeta_k, \sqrt{-d})$ determines the triple of polynomials (r, t, q).

If we want to construct complete families with variable discriminant or sparse families, we can search for nonzero $\alpha \in K$ such that $-f(\alpha^2) = \bar{g}$ for a polynomial

$g \in \mathbb{Q}[x]$ of degree ≤ 2, and take $y \in \mathbb{Q}[x]$ of degree $< n$ such that $\bar{y} = f((\zeta_k - 1)/\alpha)$, and $q = \frac{1}{4}(t^2 + gy^2)$. We will say that (z, ζ_k, α) determines the triple of polynomials (r, t, q). Obviously, if $\sqrt{-d} \in K$ and $\alpha = \sqrt{-d}$, then $g = d$ and we obtain $q = \frac{1}{4}(t^2 + dy^2)$ as in the above case of complete families.

For obtained triple (r, t, q) let $\rho = \deg q / \deg r$. In case of complete families we have $\rho = 2\max\{\deg t, \deg y\}/n$, and $\rho \leq \rho_0$ iff $\deg t$ and $\deg y$ are $\leq \ell$. In case of complete families with variable discriminant and sparse families we have $\rho = \max\{2 \deg t, \deg g + 2 \deg y\}/n$. If $\deg g = 1, 2$, we will obtain $\rho \leq \rho_0$ iff $\deg t \leq \ell$ and $\deg y \leq \ell - 1$.

Let $\lambda, \mu \in \mathbb{Q}$, and $\lambda \neq 0$. Note that if (z, ζ_k, α) determines a triple (r, t, q), then $(\lambda z + \mu, \zeta_k, \alpha)$ determines the triple $(\lambda^n r(\lambda^{-1}(x - \mu)), t(\lambda^{-1}(x - \mu)), q(\lambda^{-1}(x - \mu)))$. One can check this as follows. Let $z_1 = \lambda z + \mu$, and for any polynomial $h \in \mathbb{Q}[x]$ let $h_1 = h(\lambda^{-1}(x - \mu))$. We have $h(z) = h_1(z_1)$. Clearly $\lambda^n r_1(x)$ is the minimal polynomial of z_1 over $\mathbb{Q}$. We have $\zeta_k + 1 = t(z) = t_1(z_1)$, $g(z) = g_1(z_1)$ and $(\zeta_k - 1)/\sqrt{-g(z)} = y(z) = y_1(z_1)$. Hence from (z_1, ζ_k, α) we obtain the triple $(\lambda^n r_1, t_1, q_1)$.

Let γ be a fixed primitive element of $K/\mathbb{Q}$. We will give a condition on coordinates $x_0, \ldots, x_{n-1} \in \mathbb{Q}$ in the basis $1, \gamma, \ldots, \gamma^{n-1}$ of primitive elements $z = \sum_{i=0}^{n-1} x_i \gamma^i$ of $K/\mathbb{Q}$ if for some $\alpha \in K$ one can use (z, ζ_k, α) to obtain a triple of polynomials (r, t, q) with $\rho \leq \rho_0$. If (z, ζ_k, α) gives a triple of polynomials, then $(\lambda z + \mu, \zeta_k, \alpha)$ for $\lambda, \mu \in \mathbb{Q}$ and $\lambda \neq 0$ gives a triple of polynomials with the same ρ-value. Therefore we will assume that the coordinate $x_0 = 0$, and will consider classes λz for primitive elements $z = x_1 \gamma + \ldots + x_{n-1} \gamma^{n-1}$, $\lambda \in \mathbb{Q} \setminus \{0\}$, and $X = (x_1, \ldots, x_{n-1}) \in \mathbb{Q}^{n-1}$.

Let $\mathrm{span}_{\mathbb{Q}}(1, \ldots, z^j)$ be the subspace spanned by $1, z, \ldots, z^j$ over $\mathbb{Q}$ for $z \in K \setminus \{0\}$ and $j \in \mathbb{N}$. If (z, ζ_k) determines as above the pair of polynomials (r, t), then we have $\deg t \leq \ell$ iff $\zeta_k \in \mathrm{span}_{\mathbb{Q}}(1, \ldots, z^\ell)$.

There exist homogeneous polynomials $f_{ij}(x_1, \ldots, x_{n-1}) \in \mathbb{Q}[x_1, \ldots, x_{n-1}]$ such that

$$(\sum_{i=1}^{n-1} x_i \gamma^i)^j = \sum_{i=0}^{n-1} f_{ij}(x_1, \ldots, x_{n-1})\gamma^i$$

for $(x_1, \ldots, x_{n-1}) \in \mathbb{Q}^{n-1}$ and $j \in \mathbb{N}$. We have $\deg f_{ij} = j$. Note that for $z = \sum_{i=1}^{n-1} x_i \gamma^i$ and $X = (x_1, \ldots, x_{n-1}) \in \mathbb{Q}^n$ we have z is a primitive element of $K/\mathbb{Q}$ if and only if $\det[f_{ij}(X)]_{i,j=0,\ldots,n-1} \neq 0$.

For a given element $a \in K$ with coordinates $a = a_0 + a_1 \gamma + \ldots + a_{n-1}\gamma^{n-1}$, $a_i \in \mathbb{Q}$, we will give a condition on coordinates $X = (x_1, \ldots, x_{n-1}) \in \mathbb{Q}^{n-1}$ of primitive elements $z = \sum_{i=1}^{n-1} x_i \gamma^i$ such that $a \in \mathrm{span}_{\mathbb{Q}}(1, z, \ldots z^j)$ for $j < n$. Let

$$A_{j,a}(X) = \begin{bmatrix} 1 & f_{01}(X) & \cdots & f_{0,j}(X) & a_0 \\ 0 & f_{11}(X) & \cdots & f_{1,j}(X) & a_1 \\ \vdots & & & & \\ 0 & f_{n-11}(X) & \cdots & f_{n-1,j}(X) & a_{n-1} \end{bmatrix}$$

for $X = (x_1, \ldots, x_{n-1}) \in \mathbb{Q}^{n-1}$.

Proposition 5. *For a primitive element $z = \sum_{i=1}^{n-1} x_i \gamma^i$ of $K/\mathbb{Q}$, $X = (x_1, \ldots, x_{n-1}) \in \mathbb{Q}^{n-1}$, and $0 \leq j < n$ we have $a \in \mathrm{span}_{\mathbb{Q}}(1, \ldots, z^j)$ if and only if $\mathrm{rank}\, A_{j,a}(X) < j+2$, or equivalently all $(j+2) \times (j+2)$ minors of the matrix $A_{j,a}(X)$ are equal to 0.*

Proof. Columns $1, \ldots, j+1$ are coordinates of $1, \ldots, z^j$. Thus these columns are linearly independent for a primitive element z of $K/\mathbb{Q}$. The last column is the linear combinations of these columns if and only if $\mathrm{rank}\, A_{j,a}(X) < j+2$. The second equivalence is obvious.

Minors in the above proposition are homogeneous polynomials of degrees $1 + \ldots + j = j(j+1)/2$. Therefore determining zeros of a system with polynomials of such degrees for larger n and j can be a hard problem.

Let $U = \{X \in \mathbb{P}^{n-2}(\mathbb{Q}) : \det[f_{ij}(X)]_{i,j=0,\ldots,n-1} \neq 0\}$. For $X = (x_1 : \ldots : x_{n-1}) \in \mathbb{P}^{n-2}(\mathbb{Q})$ we have $X \in U$ if and only if $z = \sum_{i=1}^{n-1} x_i \gamma^i$ is a primitive element of $K/\mathbb{Q}$.

Corollary 6. *For $X = (x_1 : \ldots : x_{n-1}) \in U$ let $z = \sum_{i=1}^{n-1} x_i \gamma^i$, and (r, t) be the pair of polynomials determined by (z, ζ_k) as above. We have $\deg t \leq \ell$ if and only if all $(\ell+2) \times (\ell+2)$ minors of the matrix $A_{\ell, \zeta_k}(X)$ are equal to 0.*

Proof. The polynomial t is the unique polynomial in $\mathbb{Q}[x]$ of degree $< n$ such that $t(z) = \zeta_k + 1$. Thus $\deg t \leq \ell$ iff $\zeta_k \in \mathrm{span}_{\mathbb{Q}}(1, \ldots, z^\ell)$. Hence the corollary follows from the above proposition.

4.1 Complete Families with $\rho \leq \rho_0$

Theorem 7. *Assume that $\sqrt{-d} \in K$ for square-free $d \in \mathbb{Z}_{>0}$. Let $X = (x_1 : \ldots : x_{n-1}) \in U$, and $z = \sum_{i=1}^{n-1} x_i \gamma^i$. Let (r, t, q) be the triple of polynomials with parameter ρ determined by $(z, \zeta_k, \sqrt{-d})$. We have $\rho \leq \rho_0$ if and only if ranks of matrices $A_{\ell, \zeta_k}(X)$ and $A_{\ell, (\zeta_k - 1)/\sqrt{-d}}(X)$ are equal to 0, or equivalently all $(\ell+2) \times (\ell+2)$ minors of these matrices are equal to 0.*

Proof. We have $q = \frac{1}{4}(t^2 + dy^2)$ for unique polynomials t and y in $\mathbb{Q}[x]$ of degree $< n$ such that $t(z) = \zeta_k + 1$ and $y(z) = (\zeta_k - 1)/\sqrt{-d}$. We have $\rho \leq \rho_0$ if and only if t and y are of degree $\leq \ell$, or equivalently ζ_k and $(\zeta_k - 1)/\sqrt{-d}$ are in $\mathrm{span}_{\mathbb{Q}}(1, \ldots, z^\ell)$. Thus the theorem follows from the above proposition.

Example 8. As an application of the above method we will show how the Barreto-Naehrig family (see [2] or [15, Example 6.8]) with embedding degree $k = 12$, discriminant $d = 3$, and $\rho = 1$ can be obtained using the above method. Let us determine complete families (r, t, q) with $k = 12$, $d = 3$, and $\rho = 1$ such that $\mathbb{Q}[x]/(r) \cong K = \mathbb{Q}(\zeta_{12})$ for a 12th primitive root of unity ζ_{12}. The 12th cyclotomic field has the extension degree $[\mathbb{Q}(\zeta_{12}) : \mathbb{Q}] = 4$, and $\sqrt{-3} \in \mathbb{Q}(\zeta_{12})$. The coordinates of ζ_{12} and $(\zeta_{12} - 1)/\sqrt{-3}$ in the basis $1, \zeta_{12}, \zeta_{12}^2, \zeta_{12}^3$ are $(0, 1, 0, 0)$ and $(-1/3, 1/3, 2/3, -2/3)$. For $\rho_0 = 1$ we have $\ell = 2$. The matrices $A_{2, \zeta_{12}}(X)$ and $A_{2, (\zeta_{12}-1)/\sqrt{-3}}(X)$ from Theorem 7 are 4×4 for $X = (x_1, x_2, x_3) \in \mathbb{Q}^3$. Thus we want to determine all $X = (x_1 : x_2 : x_3) \in \mathbb{P}^2(\mathbb{Q})$ such that $\det A_{2,\zeta_{12}}(X) = 0$ and $\det A_{2,(\zeta_{12}-1)/\sqrt{-3}}(X) = 0$. These determinants give equations

$$- x_1^2 x_3 + 2x_1 x_2^2 - 2x_1 x_3^2 + x_2^2 x_3 = 0,$$

$$2x_1^3 + 4x_1^2 x_2 + 5x_1^2 x_3 + 4x_1 x_2 x_3 + 2x_1 x_3^2 + 3x_2^2 x_3 + 4x_2 x_3^2 = 0.$$

These curves have the following common points in $\mathbb{P}^2(\mathbb{Q})$: $(-2 : 0 : 1), (-1 : -1 : 1), (0 : 0 : 1), (1 : -1 : 1), (0 : 1 : 0)$. Only the points $(1 : -1 : 1)$ and $(-1 : -1 : 1)$ give primitive elements of K. The first point determines the class of the family

$$r = x^4 + 2x^3 + 6x^2 - 4x + 4,$$

$$t = \frac{1}{6}x^2 + \frac{2}{3}x + \frac{5}{3},$$

$$q = \frac{1}{36}x^4 + \frac{1}{18}x^3 + \frac{1}{3}x^2 + \frac{5}{9}x + \frac{7}{9}.$$

Substituting $-6x - 2$ to the above polynomials we obtain $r(-6x - 2) = 36 r_{BN}, t(-6x - 2) = t_{BN}$, and $q(-6x - 2) = q_{BN}$ for the Barreto-Naehrig family

$$r_{BN} = 36x^4 + 36x^3 + 18x^2 + 6x + 1,$$

$$t_{BN} = 6x^2 + 1,$$

$$q_{BN} = 36x^4 + 36x^3 + 24x^2 + 6x + 1.$$

The second point determines the triple $r = x^4 + 2x^3 + 2x^2 + 4x + 4$, $t = \frac{1}{2}x^2 + 1$, $q = \frac{1}{12}x^4 + \frac{1}{6}x^3 + \frac{2}{3}x^2 + \frac{1}{3}x + \frac{1}{3}$, but $q(x) \notin \mathbb{Z}$ for any $x \in \mathbb{Q}$. This can be checked as follows.

Let $f(x) \in \mathbb{Q}[x]$, and let $Nf \in \mathbb{Z}[x]$ for $N \in \mathbb{Z}_{>0}$. Recall from [15, Sec. 2.1] that if $f(a) \in \mathbb{Z}$ for some $a \in \mathbb{Z}$, then $f(b) \in \mathbb{Z}$ for some $b \in \{0, \ldots, N - 1\}$, because writing $a = cN + b$, where b is the remainder from division by N and $c \in \mathbb{Z}$, we have from the binomial formula $f(a) = f(cN + b) = f(b) + \text{integer}$, thus $f(b) \in \mathbb{Z}$. If for nonconstant polynomial $g \in \mathbb{Z}[x]$ we have $g(\frac{x_0}{y_0}) \in \mathbb{Z}$ for relatively prime $x_0, y_0 \in \mathbb{Z}$, then it is easy to see that y_0 divides the leading coefficient of g.

We have $12q \in \mathbb{Z}[x]$. If $q(x_0/y_0) \in \mathbb{Z}$, where $x_0, y_0 \in \mathbb{Z}$ are relatively prime, then y_0 divides the leading coefficient of $12q$, which is equal to 1, hence $y_0 = \pm 1$. It is impossible that $q(x) \in \mathbb{Z}$ for some $x \in \mathbb{Z}$, because $q(x) \notin \mathbb{Z}$ for each $x \in \{0, \ldots, 11\}$.

4.2   Complete Families with Variable Discriminant or Sparse Families with $\rho \leq \rho_0$

First assume that we are given a pair of polynomials (r, t) in $\mathbb{Q}[x]$ such that r is irreducible of degree n and monic, $\bar{t} = \zeta_k + 1$ for a kth primitive root of unity ζ_k in $\mathbb{Q}[x]/(r)$, and $\deg t \leq \ell$. In Corollary 6 it is shown how such pairs (r, t) can be obtained from solutions of some system of polynomial equations. First we will give a method to search for non-zero elements $\alpha \in \mathbb{Q}[x]/(r)$ for which there exist polynomials $g, y \in \mathbb{Q}[x]$ such that $\deg g \leq 2$, $\deg y \leq \ell - 1$, $\bar{g} = -\alpha^2$, and $\bar{y} = (\zeta_k - 1)/\alpha$. Let $q = \frac{1}{4}(t^2 + gy^2)$. We can also say that $(\bar{x}, \zeta_k, \alpha)$ determines (r, t, q). This triple (r, t, q) has $\rho \leq \rho_0$.

If non-zero $\alpha \in \mathbb{Q}[x]/(r)$ satisfies the condition

$$\alpha^2 \in \mathrm{span}_\mathbb{Q}(1, \bar{x}, \bar{x}^2) \text{ and } (\zeta_k - 1)/\alpha \in \mathrm{span}_\mathbb{Q}(1, \ldots, \bar{x}^{\ell-1}), \tag{2}$$

then for the polynomial q as above the triple (r, t, q) has $\rho \leq \rho_0$. Let $X = (x_0, \ldots, x_{n-1}) \in \mathbb{Q}^n \setminus \{0\}$ and $\alpha = \sum_{i=0}^{n-1} x_i \bar{x}^i$. Let $w_0, \ldots, w_{n-1} \in \mathbb{Q}[x_0, \ldots, x_{n-1}]$ be quadratic forms such that

$$\alpha^2 = \sum_{i=0}^{n-1} w_i(X) \bar{x}^i.$$

Let $L_\alpha : \mathbb{Q}[x]/(r) \to \mathbb{Q}[x]/(r)$ be the multiplication by $\alpha \neq 0$, and let M_α be the matrix of L_α in the basis $1, \bar{x}, \ldots, \bar{x}^{n-1}$. The coordinates $X' = (x'_0, \ldots, x'_{n-1}) \in \mathbb{Q}^n$ of $\alpha^{-1} = x'_0 + x'_1 \bar{x} + \ldots + x'_{n-1} \bar{x}^{n-1}$ satisfy the system of linear equations $M_\alpha [x'_0, \ldots, x'_{n-1}]^T = [1, 0, \ldots, 0]^T$, where T denotes transposition. From Cramer's rule we have $x'_i = \det M_{\alpha,i} / \det M_\alpha$, where $M_{\alpha,i}$ is the matrix obtained by replacing the ith column in M_α by $[1, 0, \ldots, 0]^T$. One can write $\det M_{\alpha,i}$ as a homogeneous polynomial of variables $X = (x_0, \ldots, x_{n-1})$ of degree $n - 1$ for $\alpha = \sum_{i=0}^{n-1} x_i \bar{x}^i$. Hence there exist homogeneous polynomials $v_0, \ldots, v_{n-1} \in \mathbb{Q}[x_0, \ldots, x_{n-1}]$ of degrees $n - 1$ such that

$$(\zeta_k - 1)/\alpha = \frac{1}{\det M_\alpha} \sum_{i=0}^{n-1} v_i(X) \bar{x}^i.$$

Theorem 9. *Let (r, t) be a pair of polynomials as above. Let $X = (x_0 : \ldots : x_{n-1}) \in \mathbb{P}^{n-1}(\mathbb{Q})$ and $\alpha = \sum_{i=0}^{n-1} x_i \bar{x}^i$. If X satisfies the system of polynomial equations $w_3 = \ldots = w_{n-1} = v_\ell = \ldots = v_{n-1} = 0$, then α determines a polynomial q such that the triple (r, t, q) has $\rho \leq \rho_0$.*

Proof. This system is satisfied if and only if condition (2) is satisfied. Thus we have $\rho \leq \rho_0$.

Remark 10. Let (r, t) be a pair of polynomials as above, where $\deg t \leq \ell$. One can also search as follows for a polynomial q as above such that (r, t, q) has $\rho \leq \rho_0$ (see also [23] or [15, Sec. 6]). From condition $r \mid q + 1 - t$ we have $q + 1 - t = hr$ for some $h \in \mathbb{Q}[x]$, thus $q = hr + t - 1$. We want to find h such that $4q - t^2 = 4(hr + t - 1) - t^2 = 4hr - (t - 2)^2 = gy^2$ for $g, y \in \mathbb{Q}[x]$, where $\deg g \leq 2$. Assume we are looking for h of degree e. We want to obtain $\rho \leq \rho_0$, thus we are looking for h such that $\deg(gy^2)/n \leq \rho_0$. Let $h = h_e x^e + \ldots + h_0$, where $h_i \in \mathbb{Q}$ are unknown coefficients. The coefficients at the powers of x of the polynomial $4hr - (t - 2)^2$ are polynomials of degree

≤ 1 in variables $h_0, \ldots, h_e$ over $\mathbb{Q}$. If $\deg y > 0$, then gy^2 has a multiple root, thus discriminant of $4hr - (t-2)^2$ is equal to 0, this gives an equation satisfied by coefficients $h_0, \ldots, h_e$. If $e_1 = \deg(gy^2) < e$, then the coefficients of the polynomial $4hr - (t-2)^2$ at $x^{e_1+1}, \ldots, x^e$ must be equal to 0.

Let K be a number field of the extension degree $n = [K : \mathbb{Q}] > 2$ such that a kth primitive root $\zeta_k \in K$. We will assume that z is a primitive element of $K/\mathbb{Q}$, and $\alpha \in K \setminus \{0\}$. If the following condition is satisfied

$$\zeta_k \in \mathrm{span}_{\mathbb{Q}}(1, \ldots, z^\ell), \alpha^2 \in \mathrm{span}_{\mathbb{Q}}(1, z, z^2), \text{ and } (\zeta_k - 1)/\alpha \in \mathrm{span}_{\mathbb{Q}}(1, \ldots, z^{\ell-1}), \tag{3}$$

then (z, ζ_k, α) determnies as in the beginning of this section the triple (r, t, q) in $\mathbb{Q}[x]$ with $\rho \leq \rho_0$. Let γ be a fixed primitive element of $K/\mathbb{Q}$. Let $X = (x_1, \ldots, x_{n-1}) \in \mathbb{Q}^{n-1}$, $X' = (x'_0, \ldots, x'_{n-1}) \in \mathbb{Q}^n$, $z = \sum_{i=1}^{n-1} x_i \gamma^i$, and $\alpha = \sum_{i=0}^{n-1} x'_i \gamma^i$. We will give conditions on coordinates X of z and X' of α equivalent to condition (3). Similarly as above there exist homogeneous polynomials $f_{ij} \in \mathbb{Q}[x_1, \ldots, x_{n-1}]$, and $w_i, v_i \in \mathbb{Q}[x'_0, \ldots, x'_{n-1}]$ such that

$$z^j = \sum_{i=0}^{n-1} f_{ij}(X)\gamma^i \text{ for } j \in \mathbb{N},$$

$$\alpha^2 = \sum_{i=0}^{n-1} w_i(X')\gamma^i,$$

$$(\zeta_k - 1)/\alpha = \frac{1}{\det(L_\alpha)} \sum_{i=0}^{n-1} v_i(X')\gamma^i,$$

where $L_\alpha : K \to K$ is the map of multiplication by α. Let $A_{\ell, \zeta_k}(X)$ be the matrix as in Corollary 6. Let

$$A_{2,w}(X, X') = \begin{bmatrix} 1 & f_{01}(X) & f_{0,2}(X) & w_0(X') \\ 0 & f_{11}(X) & f_{1,2}(X) & w_1(X') \\ \vdots & & & \\ 0 & f_{n-11}(X) & f_{n-1,2}(X) & w_{n-1}(X') \end{bmatrix},$$

and

$$A_{\ell-1,v}(X, X') = \begin{bmatrix} 1 & f_{01}(X) & \cdots & f_{0,\ell-1}(X) & v_0(X') \\ 0 & f_{11}(X) & \cdots & f_{1,\ell-1}(X) & v_1(X') \\ \vdots & & & & \\ 0 & f_{n-11}(X) & \cdots & f_{n-1,\ell-1}(X) & v_{n-1}(X') \end{bmatrix}.$$

Theorem 11. *Let* $(X, X') \in U \times \mathbb{P}^{n-1}(\mathbb{Q})$, $z = \sum_{i=1}^{n-1} x_i \gamma^i$, *and* $\alpha = \sum_{i=0}^{n-1} x'_i \gamma^i$. *Assume that* $\mathrm{rank}\, A_{\ell, \zeta_k}(X) < \ell + 2, \mathrm{rank}\, A_{2,w}(X, X') < 4$, *and* $\mathrm{rank}\, A_{\ell-1,v}(X, X') < \ell+1$, *or equivalently the following minors are equal to 0: all* $(\ell + 2) \times (\ell + 2)$ *minors of* $A_{\ell, \zeta_k}(X)$, *all* 4×4 *minors of* $A_{2,w}(X, X')$, *and all* $(\ell + 1) \times (\ell + 1)$ *minors of* $A_{\ell-1,v}(X, X')$, *Then* (z, ζ_k, α) *determines a triple of polynomials* (r, t, q) *with* $\rho \leq \rho_0$.

Proof. Condition (3) is satisfied if and only the above conditions on ranks are satisfied. Thus we will obtain $\rho \leq \rho_0$.

Example 12. As an application of the above method let us determine families (r, t, q) with embedding degree $k = 10$ and $\rho = 1$ such that $\deg r = 4$, $4q - t^2$ is of degree 2 and has no double root. The family with these properties was given by Freeman (see [14] or [15, Sec. 5.3]). Let $\zeta_{10} \in \mathbb{C}$ be a 10th primitive root of unity. The 10th cyclotomic field has the extension degree $[\mathbb{Q}(\zeta_{10}) : \mathbb{Q}] = 4$. For $\rho_0 = 1$ we have $\ell = 2$. We will use the basis $1, \zeta_{10}, \zeta_{10}^2, \zeta_{10}^3$. Let $X = (x_1, x_2, x_3) \in \mathbb{Q}^3$ and $z = x_1 \zeta_{10} + x_2 \zeta_{10}^2 + x_3 \zeta_{10}^3$. For the above families we do not need to use variables $x_0', \ldots, x_3'$, because if (z, ζ_{10}, α) determines such family, then the assumption on $4q - t^2$ implies that $(\zeta_k - 1)/\alpha \in \mathbb{Q}$, hence $\alpha = a(\zeta_k - 1)$ for some $a \in \mathbb{Q}$. Thus instead of the matrix $A_{2,w}$, we can use the matrix $A_{2,(\zeta_k-1)^2}(X)$, because rank $A_{2,(\zeta_k-1)^2}(X) < 4$ if and only if $(\zeta_k - 1)^2 \in \text{span}_{\mathbb{Q}}(1, z, z^2)$ for $z = x_1 \zeta_{10} + x_2 \zeta_{10}^2 + x_3 \zeta_{10}^3$ and $X = (x_1, x_2, x_3) \in \mathbb{Q}^3$. The matrices $A_{2,\zeta_k}(X)$ and $A_{2,(\zeta_k-1)^2}(X)$ are 4×4, thus have ranks less than 4 if and only if their determinants are equal to 0. Hence we have the following equations

$$\det A_{2,\zeta_{10}}(X) = -x_1^2 x_3 + 2x_1 x_2^2 + 2x_1 x_2 x_3 + 2x_1 x_3^2 + x_2^3 + x_2^2 x_3 = 0,$$

$$\det A_{2,(\zeta_{10}-1)^2}(X) = -2x_1^2 x_2 - 5x_1 x_2^2 - 4x_1 x_2 x_3 - 2x_1 x_3^2 - 2x_2^3 - x_2^2 x_3 - x_3^3 = 0.$$

These two curves have in $\mathbb{P}^2(\mathbb{Q})$ the following common points: $(0 : -1 : 1), (-1/2 : 1 : 0)$, and $(1 : 0 : 0)$, but the first point does not determine a primitive element of $K/\mathbb{Q}$. The point $(-1/2 : 1 : 0)$ determines the triple of polynomials

$$r = x^4 + \frac{3}{2}x^3 + \frac{9}{4}x^2 + \frac{7}{8}x + \frac{11}{16},$$

$$t = \frac{8}{5}x^2 + \frac{6}{5}x + \frac{13}{5},$$

$$q = \frac{16}{25}x^4 + \frac{24}{25}x^3 + \frac{76}{25}x^2 + \frac{44}{25}x + \frac{51}{25}.$$

Substituting to these polynomials $-\frac{5}{2}x - 1$ we obtain $r(-\frac{5}{2}x - 1) = \frac{25}{16}r_F$, $t(-\frac{5}{2}x - 1) = t_F$, and $q(-\frac{5}{2}x - 1) = q_F$ for Freeman's family

$$r_F = 25x^4 + 25x^3 + 15x^2 + 5x + 1,$$

$$t_F = 10x^2 + 5x + 3,$$

$$q_F = 25x^4 + 25x^3 + 25x^2 + 10x + 3.$$

The point $(1 : 0 : 0)$ does not give a family, because it determines the triple $(\zeta_{10}, \zeta_{10}, \zeta_{10} - 1)$, hence $g = -(x - 1)^2$ has negative leading coefficient.

References

1. Barreto, P.S.L.M., Lynn, B., Scott, M.: Constructing elliptic curves with prescribed embedding degrees. In: Security in Communication Networks-SCN 2002. LNCS, vol. 2576, pp. 263–273. Springer, Berlin (2002)
2. Barreto, P.S.L.M., Naehrig, M.: Pairing-friendly elliptic curves of prime order. In: Selected Areas in Cryptography–SAC 2005. LNCS, vol. 3897, pp. 319–331. Springer, Berlin (2006)
3. Boneh, D., Franklin, M.: Identity-based encryption from the Weil pairing. In: Advances in Cryptology–Crytpo 2001. LNCS, vol. 2139, pp. 213–229. Springer, Berlin (2001). Full version: SIAM J. Comput. **32**, 586–615 (2003)
4. Boneh, D., Lynn, B., Shacham, H.: Short signatures from the Weil pairing. In: Advances in Cryptology–Asiacrypt 2001. LNCS, vol. 2248, pp. 514–532. Springer, Berlin (2002). Full version: J. Cryptol. **17**, 297–319 (2004)
5. Boneh, D., Rubin, K., Silverberg, A.: Finding composite order ordinary elliptic curves using the Cocks-Pinch method. J. Number Theory **131**(5), 832–841 (2011)
6. Brezing, F., Weng, A.: Elliptic curves suitable for pairing based cryptography. Des. Codes Cryptogr. **37**, 133–141 (2005)
7. Cha, J.C., Cheon, J.H.: An identity-based signature from gap Diffie-Hellman groups. In: Public-Key Cryptography-PKC 2003. LNCS, vol. 2567, pp. 18–30. Springer, Berlin (2003)
8. Dryło, R.: On constructing families of pairing-friendly elliptic curves with variable discriminant. In: Indocrypt-2011. LNCS, vol. 7107, pp. 310–319. Springer, Berlin (2011)
9. Dryło, R.: On constructing optimal families of pairing-friendly elliptic curves. Preprint IMPAN (no. 715) (2010). https://www.impan.pl/pl/wydawnictwa/preprinty-impan
10. Dupont, R., Enge, A., Morain, F.: Building elliptic curves with arbitrary small MOV degree over finite prime fields. J. Cryptol. **18**, 79–89 (2005)
11. Fotiadis, G., Konstantinou, E.: Generating pairing-friendly elliptic curve parameters using sparse families. J. Math. Cryptol. **12**(2), 83–99 (2018)
12. Fotiadis, G., Konstantinou, E.: More sparse families of pairing-friendly elliptic curves. In: International Conference on Cryptology and Network Security, pp. 384–399. Springer, Cham (2014)
13. Fotiadis, G., Konstantinou, E.: TNFS resistant families of pairing-friendly elliptic curves. Theoret. Comput. Sci. **800**, 73–89 (2019)
14. Freeman, D.: Constructing pairing-friendly elliptic curves with embedding degree 10. In: Algorithmic Number Theory Symposium–ANTS-VII. LNCS, vol. 4076, pp. 452–465. Springer, Berlin (2006)
15. Freeman, D., Scott, M., Teske, E.: A taxonomy of pairing-friendly elliptic curves. J. Cryptol. **23**, 224–280 (2010)
16. Galbraith, S., McKee, J., Valença, P.: Ordinary abelian varieties having small embedding degree. Finite Fields Appl. **13**, 800–814 (2007)
17. Grześkowiak, M.: Algorithms for pairing-friendly primes. In: International Conference on Pairing-Based Cryptography, pp. 215–228. Springer, Cham (2013)
18. Guillevic, A.: A short-list of pairing-friendly curves resistant to special TNFS at the 128-bit security level. In: IACR International Conference on Public-Key Cryptography, pp. 535–564. Springer, Cham (2020)
19. Joux, A.: A one round protocol for tripartite Diffie-Hellman. In: Algorithmic Number Theory Symposium–ANTS-IV. LNCS, vol. 1838, pp. 385–393. Springer, Berlin (2000). Full version: J. Cryptol. 17, 263–276 (2004)

20. Kachisa, E., Schaefer, E., Scott, M.: Constructing Brezing-Weng pairing friendly elliptic curves using elements in the cyclotomic field. In: Pairing-Based Cryptography–Pairing 2008. LNCS, vol. 5209, pp. 126–135. Springer, Berlin (2008)
21. Miyaji, A., Nakabayashi, M., Takano, S.: New explicit conditions of elliptic curves traces for FR-reduction. IEICE Trans. Fundam. **E84-A**, 1234–1243 (2001)
22. Paterson, K.: ID-based signatures from pairings on elliptic curves. Electron. Lett. **38**, 1025–1026 (2002)
23. Scott, M., Barreto, P.S.L.M.: Generating more MNT elliptic curves. Des. Codes Cryptogr. **38**, 209–217 (2006)
24. Scott, M., Guillevic, A.: A new family of pairing-friendly elliptic curves. In: Arithmetic of Finite Fields: 7th International Workshop, WAIFI 2018, Bergen, Norway, 14–16 June 2018, Revised Selected Papers 7, pp. 43–57. Springer (2018)
25. Silverman, J.: The Arithmetic of Elliptic Curves. Springer, Berlin (1986)
26. Yoon, K.: A new method of choosing primitive elements for Brezing-Weng families of pairing-friendly elliptic curves. J. Math. Cryptol. **9**(1), 1–9 (2015)

Number Theory

Algebraic Equipage for Learning with Errors in Cyclic Division Algebras

Cong Ling and Andrew Mendelsohn[(✉)]

Department of EEE, Imperial College London, London, UK
{c.ling,andrew.mendelsohn18}@imperial.ac.uk

Abstract. In *Noncommutative Ring Learning With Errors From Cyclic Algebras*, a variant of Learning with Errors from cyclic division algebras, dubbed 'Cyclic LWE', was developed, and security reductions similar to those known for the ring and module case were given, as well as a Regev-style encryption scheme. In this work, we make a number of improvements to that work: namely, we describe methods to increase the number of cryptographically useful division algebras, demonstrate the hardness of CLWE from ideal lattices obtained from non-maximal orders, and study Learning with Rounding in cyclic division algebras.

Keywords: CLWE · Structured LWE · Cyclic Division Algebras

1 Introduction

With the advent of quantum computation[1], new avenues of cryptographic attack have arisen, such as cryptanalysis using the famed 'Shor's Algorithm' to factor integers in polynomial time [36]. The cryptographic community has responded to these developments by searching for quantum-resistant protocols: one of the most prominent of these efforts relies on the hardness of solving lattice-based problems, which appear no easier to solve using quantum than classical algorithms.

One of the most popular of these problems is the Learning with Errors (LWE) problem, which in its simplest form (informally) runs as follows: take a secret vector of integers modulo a prime, $s \in \mathbb{Z}_q^n$, a uniformly random vector $a \leftarrow \mathbb{Z}_q^n$, and an error $e \leftarrow \mathbb{Z}_q$ chosen according to some distribution χ, and output the pair $(a, \langle a, s \rangle + e \bmod q)$. There are two problems to solve: firstly, the *search* LWE problem is to recover s from a number of LWE samples; secondly, the *decision* LWE problem is to decide whether m independent samples are chosen either according the LWE distribution, or uniformly at random, from $\mathbb{Z}_q^n \times \mathbb{Z}_q$. In [32], Regev gave a reduction from approximate GapSVP to decision LWE, guaranteeing it a certain level of hardness, for certain parameters.

---

[1] For background, see [23].

This work was supported in part by the Engineering and Physical Sciences Research Council (EPSRC) under Grant Nos. EP/X037010/1 and EP/Y037243/1.

A. Dąbrowski et al. (Eds.): NuTMiC 2024, LNCS 14966, pp. 175–205, 2025.
https://doi.org/10.1007/978-3-031-82380-0_6

The LWE problem was later defined for rings (RLWE) [22] and modules (MLWE) [20], amongst a wide range of other algebraic structures such as orders [8] and group rings [10]. In the ring case, the simplest form of an RLWE sample is defined as follows: given a number field K with ring of integers $\mathcal{O}_K$, a prime q, and a secret $s \in \mathcal{O}_{K_q}$, where $\mathcal{O}_{K_q}$ denotes the set of equivalence classes of $\mathcal{O}_K$ modulo q, sample $a \leftarrow \mathcal{O}_{K_q}$ uniformly at random and $e \leftarrow \mathcal{O}_{K_q}$ according to some error distribution, and output $(a, a \cdot s + e) \in \mathcal{O}_{K_q} \times \mathcal{O}_{K_q}$. This, and the other forms of LWE mentioned above, are all forms of *structured* LWE.

To see this, consider the example of $K = \mathbb{Q}[x]/(x^n + 1)$ for power-of-two n, and fixing the usual basis, writing $a = \sum_i a_i x^i$, $s = \sum_i s_i x^i$ and taking a prime q, one can write $a \cdot s + e \mod q \in \mathcal{O}_{K_q}$ over the integers as

$$\begin{pmatrix} a_0 & -a_{n-1} & \cdots & -a_1 \\ a_1 & a_0 & \cdots & -a_2 \\ \vdots & \vdots & \ddots & \vdots \\ a_{n-1} & a_{n-2} & \cdots & a_0 \end{pmatrix} \begin{pmatrix} s_0 \\ s_1 \\ \vdots \\ s_{n-1} \end{pmatrix} + \begin{pmatrix} e_0 \\ e_1 \\ \vdots \\ e_{n-1} \end{pmatrix} \mod q$$

The motivation to design structured forms of LWE arises due to the relative inefficiency of general matrix multiplication, and the large public keys of LWE-based schemes. By choosing particular algebraic structures from which to sample elements for LWE-style instances, one can obtain greater efficiency, albeit often as a tradeoff with the security level of the underlying 'hard problem'. For example, the structure of RLWE transposes the matrix-vector multiplications into polynomial multiplications, which can be computed with greater efficiency. In addition, the structure of such schemes allows one to store fewer bits of information to reconstruct public keys, as will be seen below. Differently structured LWE variants offer differing balances of security and efficiency; particularly desirable would be a scheme with the efficiency of RLWE and the security of MLWE.

In [15], a novel noncommutative LWE construction was created, relying on the structure of cyclic division algebras (CDAs), called 'Cyclic LWE', or CLWE. This work uses the concept of an *order* in an algebra, that is, subrings which are also lattices. Analogously to the commutative ring case, ideals in orders form lattices, and these similarities between orders in CDAs and rings of integers in number fields were used to develop computationally-intractable lattice problems from maximal order ideals, and give a security reduction establishing hardness results for search and decision CLWE from such problems. In addition, a Regev-style encryption scheme was given. Since the domain over which RLWE is implemented is the maximal order of a number field, RLWE is a special case of CLWE, obtained when the algebra is (trivially) a number field; and a single sample of CLWE is loosely equivalent to multiple correlated MLWE samples.

Our Results. In this paper we continue the development of CLWE, that is, LWE from *cyclic division algebras*. Our contributions are four-fold:

1. We discuss the construction of CDAs, focusing on the creation of *non-norm elements*. These elements are fundamental to constructing cryptographically-

secure CDAs for CLWE; see [15, Section 3.2] for an attack on CLWE in non-division cyclic algebras. We generalize a theorem from [15] to increase the number of valid CDAs for CLWE, and give concrete examples of such algebras. In particular, we prove that one can construct CDAs from cyclotomic fields of arbitrary conductor, rather than just prime-power conductor.

2. We discuss obtaining appropriate prime moduli for CLWE. Recall that for RLWE, one chooses a prime completely split in a given number field; for CLWE, one has further constraints on the choice of modulus - we often want the prime to split in an extension also. We provide methods for finding such primes, and give examples of algebras together with appropriate moduli that can be used for CLWE.

3. We analyse the security of CLWE instantiated in suborders of the (so-called) natural order, proving a reduction from problems on ideal lattices, for invertible ideals obtained from such suborders. We then adapt methods from [31] to relax the invertibility condition on the ideals used.

4. We generalize Learning with Rounding (LWR) to cyclic algebras, named Cyclic LWR (CLWR). We adapt the security proof of [6] to hold in this setting, establishing a link between CLWE and CLWR. Our proof holds in the case of power-of-two degree with super-polynomial modulus; the chief difficulty is the analysis of the statistical distance of the relevant distributions. Moving to a Learning with Rounding-based scheme can increase efficiency, since one no longer has to sample an error term.

We present this suite of algebraic results as a step toward taking CLWE from cryptographic theory to reality, allowing greater flexibility of parameter choices for the algebras and orders used, facilitating the further development of CLWE.

Previous Work. Cyclic divison algebras have been used extensively in coding theory: they were introduced by Sethuraman et al. [35] and developed by Lahtonen [17] and Oggier [30] (amongst others). The appeal of such an algebraic object was the ability of coding theorists to use them to construct so-called 'perfect codes', exploiting properties of the discriminant and determinant of the algebra, which proved useful when applied to multiple-antenna communication.

Regarding the literature on structured LWE, samples of M/RLWE encode multiple samples of LWE by fixing a $\mathbb{Z}$-basis and viewing ring multiplication as matrix-vector multiplication of the matrices obtained with respect to the fixed basis. In addition to these, polynomial LWE [37], order LWE [8], group ring LWE [10] and others are alternative forms of structuring LWE.

Paper Organisation. After preliminaries, in Sect. 3 we discuss non-norm elements, in Sect. 4 prime moduli for CLWE, in Sect. 5 give a reduction from ideals in non-maximal orders, and in Sect. 6 study CLWR.

2 Preliminaries

Lattices. An n-dimensional lattice is a discrete additive subgroup of $\mathbb{R}^n$. One can consider a lattice $\mathcal{L}$ to be the set of integer linear combinations of a set of vectors $B = \{\mathbf{b}_1, \ldots, \mathbf{b}_k\}$ that are linearly independent, for some $k \leq n$, written $\mathcal{L}(B) = \left\{ \sum_{i=1}^k z_i \mathbf{b}_i : z_i \in \mathbb{Z} \right\}$. All lattices in this work will have $k = n$.

The above notion can also be generalised to vector spaces over fields. Let V be a finite-dimensional vector space over a number field K with ring of integers $\mathcal{O}_K$. An $\mathcal{O}_K$-*lattice* in V is a subspace $\mathcal{L} \subset V$ such that $\mathcal{L}$ is a finitely-generated $\mathcal{O}_K$-module. Equivalently, $\mathcal{L}$ is a finitely-generated torsion-free $\mathcal{O}_K$-module. $\mathcal{L}$ is *full* if it contains a K-basis of V, so $V = K \cdot \mathcal{L}$. Taking a basis B of V, the $\mathcal{O}_K$-linear span of B is a full lattice.

Definition 1. Let $\mathcal{L}$ be a lattice, and $\mathbb{R}^n$ be endowed with inner product $\langle \cdot, \cdot \rangle$. Then the set $\mathcal{L}^* = \{ v \in \mathbb{R}^n : \langle \mathcal{L}, v \rangle \subset \mathbb{Z} \}$ is called the *dual lattice* of $\mathcal{L}$.

Informally, examples of lattice problems are finding a shortest vector in the lattice (SVP), finding the closest vector to a given point (CVP), deciding whether the shortest vector has size less than one or greater than a parameter β (GapSVP), or outputting n independent sufficiently short vectors (SIVP), when one knows only a certain basis of the lattice. These are believed to be hard and such problems have been used to ground the security of problems that are capable of being used to construct quantum-resistant cryptographic schemes. Such schemes are often based off approximate variants of the above problems; for example, approx-SVP is the problem of finding a lattice vector of norm at most a factor of ξ larger than the shortest vector, for some specified value $\xi > 0$.

Discrete Gaussians. For $\mathbb{R}^n$ equipped with (Euclidean) norm $\| \cdot \|$, and $r > 0$, we define the *Gaussian function* $\rho_r : \mathbb{R}^n \to (0,1]$ by $\rho_r(\boldsymbol{x}) = \exp\left(-\pi \|\boldsymbol{x}\| / r^2\right)$.

The spherical Gaussian distribution D_r over $\mathbb{R}^n$ outputs a vector $\mathbf{v}$ with probability proportional to $\rho_r(\mathbf{v})$, and an elliptical Gaussian $D_{\mathbf{r}}$ can be sampled as follows: fix a basis $\mathbf{b}_1, \ldots, \mathbf{b}_n$ of $\mathbb{R}^n$, and a vector $\mathbf{r} = (r_1, \ldots, r_n)$. Sample $x_i \leftarrow D_{r_i}$ (independently for $i \neq j$) and output $\sum_{i=1}^n x_i \mathbf{b}_i$.

The discrete Gaussian distribution $D_{\mathcal{L},r}$, defined over a lattice $\mathcal{L}$, outputs $\mathbf{x}$ with probability $\frac{\rho_r(\mathbf{x})}{\rho_r(\mathcal{L})}$ for each $\mathbf{x} \in \mathcal{L}$.

The *smoothing parameter*, defined below, will be used throughout this work:

Definition 2. Let $\mathcal{L}$ be a lattice and $\varepsilon > 0$. Then the *smoothing parameter* $\eta_\varepsilon(\mathcal{L})$ of $\mathcal{L}$ is the smallest $r > 0$ such that $\rho_{1/r}\left(\mathcal{L}^* / \{\mathbf{0}\}\right) \leq \varepsilon$.

The *statistical distance* between two distributions D, D' over a discrete set S is defined $\Delta(D, D') = \frac{1}{2} \sum_{x \in S} |D(x) - D'(x)|$. We may denote the uniform distribution over S by $U(S)$. The following is a useful lemma:

Lemma 1 [26, Lemma 4.1]. *For a lattice $\mathcal{L}$ over $\mathbb{R}^n, \varepsilon > 0, r \geq \eta_\varepsilon(\mathcal{L})$, and $\boldsymbol{x} \in \mathbb{R}^n$, the statistical distance between $(D_r + \boldsymbol{x}) \bmod \mathcal{L}$ and the uniform distribution modulo $\mathcal{L}$ is bounded above by $\varepsilon/2$. Equivalently, $\rho_r(\mathcal{L} + \boldsymbol{x}) \in \left[\frac{1-\varepsilon}{1+\varepsilon}, 1\right] \cdot \rho_r(\mathcal{L})$.*

Algebraic Number Fields. Let K be a number field and $\mathcal{O}_K$ its ring of integers. A Dedekind domain is an integrally closed, Noetherian domain in which every prime ideal is maximal. In a Dedekind domain every fractional ideal has a unique factorization into prime ideals. Recall $\mathcal{O}_K$ is a Dedekind domain.

Let L be a finite extension of K, $\mathfrak{q}$ a prime ideal of $\mathcal{O}_L$, and $\mathfrak{p}$ a prime ideal of $\mathcal{O}_K$. We say $\mathfrak{q}$ lies above $\mathfrak{p}$ if $\mathfrak{q} \cap \mathcal{O}_K = \mathfrak{p}$. Moreover, $\mathfrak{p}\mathcal{O}_L$ is an ideal of $\mathcal{O}_L$, which is a Dedekind domain, so $\mathfrak{p}\mathcal{O}_L$ has a unique factorization into prime ideals:

$$\mathfrak{p}\mathcal{O}_L = \prod_{i=1}^{g} \mathfrak{q}_i^{e_i} = \mathfrak{q}_1^{e_1}...\mathfrak{q}_g^{e_g}, \tag{1}$$

for $\mathfrak{q}_i \subset \mathcal{O}_L$ prime all lying above $\mathfrak{p}$. We call $e_{\mathfrak{q}_i|\mathfrak{p}} = e_i$ the *ramification index* of $\mathfrak{q}_i$ over $\mathfrak{p}$, and $f_{\mathfrak{q}_i|\mathfrak{p}} = f_i = [\mathcal{O}_L/\mathfrak{q}_i : \mathcal{O}_K/\mathfrak{p}]$ the *inertial degree* of $\mathfrak{p}$ in $\mathfrak{q}_i$. Now suppose L/K is Galois. Then all the e_i and f_i are equal and $\mathrm{Gal}(L/K)$ acts transitively on the set of primes lying above any fixed prime of $\mathcal{O}_K$, and $\mathcal{O}_L/\mathfrak{q}_i$ and $\mathcal{O}_K/\mathfrak{p}$ are finite fields.

Proposition 1. *The ramification index and inertial degree are multiplicative over towers of number fields; i.e. for ideals $\mathbf{Q}/\mathfrak{q}/q$ in $\mathcal{O}_M/\mathcal{O}_L/\mathcal{O}_K$, $e_{\mathbf{Q}|q} = e_{\mathbf{Q}|\mathfrak{q}}e_{\mathfrak{q}|q}$ and $f_{\mathbf{Q}|q} = f_{\mathbf{Q}|\mathfrak{q}}f_{\mathfrak{q}|q}$.*

Definition 3. An ideal $\mathfrak{p} \subset \mathcal{O}_K$ *ramifies* in L if $e > 1$, and is *unramified* if $e = 1$. Alternatively, we say that L/K is unramified at $\mathfrak{p}$. If $e = 1$ and $f = 1$, $\mathfrak{p}$ *splits completely*. If $f = g = 1$, then $e = [L : K]$ and $\mathfrak{p}$ is *totally ramified*.

Primes in Cyclotomic Fields. Here we consider the ramification of primes p in cyclotomic fields of the form $\mathbb{Q}(\zeta_n)$, for some primitive nth root of unity ζ_n.

Case 1: p does not divide n. Then

Proposition 2 [25, Proposition 7.7]. *Let $K = \mathbb{Q}(\zeta_n)$. Let p be a rational prime, $\gcd(p, n) = 1$ and f be the lowest integer such that $p^f \equiv 1 \mod n$. Then we have $f = f_{\mathfrak{p}|p}$ for any prime ideal $\mathfrak{p}$ of K lying above p.*

Let p be a prime not dividing n. Then p is unramified in $\mathbb{Q}(\zeta_n)$, since the only ramified primes are those dividing the discriminant, whose prime factors are precisely those of n. This is equivalent to saying that $e = 1$. So $fg = [\mathbb{Q}(\zeta_n) : \mathbb{Q}] = \phi(n)$, where ϕ is the Euler totient function, and one can use the above proposition to find f, and hence g.

Case 2: p divides n. Write $n = p^a \cdot n'$, where $\gcd(p, n') = 1$. Then f is the lowest positive integer such that $p^f \equiv 1 \mod n'$. Also, $e = \phi(p^a)$, and $g = \phi(n')/f$ [38].

We will need Dirichlet's theorem on arithmetic progressions:

Theorem 1. *Let $a, n \in \mathbb{Z}_{\geq 1}$ and $\gcd(a, n) = 1$. Then the natural density of primes p such that $p \equiv a \mod n$ in the set of all primes of $\mathbb{Z}$ is $1/\phi(n)$.*

Example: Take q such that p^a is the highest power of p that divides $q-1$. Then $q \equiv 1 \bmod p^a$, so by the theorem there are infinitely many such q.

The following map between fields will be useful:

Definition 4. Let L be a number field of degree n over K. Fix a basis of L/K. Multiplication by an element α of L is a linear map, so has an associated matrix m_α. The *norm* is defined as the map $N_{L/K} : L \to K$, given by $\alpha \mapsto \det(m_\alpha)$. Equivalently, for Galois extensions, $N_{L/K}(\alpha) = \prod_i^n \sigma_i(\alpha)$, where $\sigma_i \in \mathrm{Gal}(L/K)$.

Local Fields. We will also need some p-adic theory. For more background information, the interested reader is directed to [34].

Definition 5. A field equipped with a metric is complete if every Cauchy sequence of elements converges. A valuation on a field K is a homomorphism $\nu : K \to \mathbb{R}$ such that $\nu(xy) = \nu(x) + \nu(y)$, $\nu(x) = \infty$ if and only if $x = 0$, and $\nu(x+y) \geq \min(\nu(a), \nu(b))$. If the image of ν is $\mathbb{Z}$ then ν is said to be a *discrete valuation*. If ring R has field of fractions K, and K is equipped with a discrete valuation such that $R = \{x : x \in K \text{ and } \nu(x) \geq 0\}$, then we call R a *discrete valuation ring* (DVR).

DVRs are PIDs and have precisely one proper maximal ideal.

The p-Adic Valuation: Let $p \in \mathbb{Z}$. Define $\nu_p(x) = r$, for $x = p^r a$ with $a = \frac{b}{c}$ with $\gcd(p,b) = \gcd(p,c) = 1$. Likewise for a Dedekind domain A with field of fractions K and a prime ideal $\mathfrak{p}$, $\nu_\mathfrak{p}(x) := r$, for $(x) = \mathfrak{p}^r \mathbf{bc}^{-1}$ where $\mathbf{b}, \mathbf{c}$ are fractional ideals coprime to $\mathfrak{p}$. This valuation is discrete. The completion of $\mathbb{Q}$ with respect to ν_p is denoted $\mathbb{Q}_p$ and called the rational field of p-adic numbers.

Definition 6. A *local* field is a field K that is complete with respect to a discrete valuation, and has finite residue field. Denote the ramification index of such a field's unique prime in a finite extension L by $e_{L/K}$.

Thus $\mathbb{Q}_p$ is an example of a local field. We can generate further examples as follows: let K be a number field, and $\mathfrak{q}$ a prime ideal of $\mathcal{O}_K$ lying above q. Denote the completion of K by $\mathfrak{q}$ by $K_\mathfrak{q}$. This is also a local field, and is a finite extension of $\mathbb{Q}_q$. The following gives important properties of such fields:

Proposition 3. *Let K be a local field.*

(i) $\mathcal{O}_K$ *is a discrete valuation ring with a unique proper prime (maximal) ideal.*
(ii) *A generator, π, of the maximal ideal, $\mathfrak{m}$, is called a uniformizer. This element is irreducible.*
(iii) *The group of units of $\mathcal{O}_K$ is denoted $\mathcal{O}_K^\times = U$. The nth unit group is defined*
$$U^n = \{u \in \mathcal{O}_K^\times : u \equiv 1 \bmod \mathfrak{m}^n\}. \text{ We have } U/U^n = (\mathcal{O}_K/\mathfrak{m}^n)^\times.$$

Definition 7. We say an extension of local fields L/K is *tamely* ramified if $char(\bar{K}) \nmid e_{L/K}$, where $\bar{K}$ is the residue field of K.

Proposition 4 [28, Chapter 2, Proposition 5.3]. *Let $K_{\mathfrak{q}}$ be as above. Then we have $K_{\mathfrak{q}}^{\times} \cong \mu_{q-1} \times \langle \pi_{\mathfrak{q}} \rangle \times U_{\mathfrak{q}}^1$, where μ_{q-1} is a cyclic group of order $q-1$, $\pi_{\mathfrak{q}}$ is a uniformizer of $K_{\mathfrak{q}}$, and $U_{\mathfrak{q}}^1$ is the first unit group of $K_{\mathfrak{q}}$.*

Theorem 2 (Hasse Norm Theorem [16]**).** *Let L/K be a cyclic extension of number fields. A nonzero element of K is a local norm at all primes $\mathfrak{p} \in L$ lying over $p \in K$, for all $p \in K$, if and only if it is a norm of an element in L.*

Theorem 3 (Local Reciprocity Map [13]**).** *Let $\mathfrak{Q}$ be a prime ideal of $\mathcal{O}_L$ and $\mathfrak{q}$ be a prime ideal of $\mathcal{O}_K$, both lying above q, so $L_{\mathfrak{Q}}/K_{\mathfrak{q}}$ is a finite abelian extension of local fields. Then there is an isomorphism*

$$\Theta : K_{\mathfrak{q}}^{\times} / N_{L_{\mathfrak{Q}}/K_{\mathfrak{q}}}(L_{\mathfrak{Q}}^{\times}) \rightarrow \mathrm{Gal}(L_{\mathfrak{Q}}/K_{\mathfrak{q}}).$$

Orders. An order $\mathcal{O}$ in a number field K is a subring which is also a lattice. The maximal order (with respect to inclusion) is the ring of integers, $\mathcal{O}_K$. The behaviour of ideals in orders is significantly determined by the *conductor ideal*:

Definition 8. The conductor of an order $\mathcal{O} \subset \mathcal{O}_K$ is defined

$$\mathfrak{c} = \mathfrak{c}_{\mathcal{O}} = \{x \in K : x\mathcal{O}_K \subset \mathcal{O}\}.$$

We need the following lemma on the behaviour of ideals in orders:

Lemma 2 [11]. *Let $\mathcal{O}$ be an order in K with conductor $\mathfrak{c}$.*

1. *For each $\mathcal{O}_K$-ideal $\mathfrak{a}$ coprime to $\mathfrak{c}$, $\mathfrak{a} \cap \mathcal{O}$ is an $\mathcal{O}$-ideal coprime to $\mathfrak{c}$ and the natural ring homomorphism $\mathcal{O}/(\mathfrak{a} \cap \mathcal{O}) \rightarrow \mathcal{O}_K/\mathfrak{a}$ is an isomorphism.*
2. *For each $\mathcal{O}$-ideal $\mathfrak{b}$ that is coprime to $\mathfrak{c}$, $\mathfrak{b}\mathcal{O}_K$ is an $\mathcal{O}_K$-ideal coprime to $\mathfrak{c}$ and the natural ring homomorphism $\mathcal{O}/\mathfrak{b} \rightarrow \mathcal{O}_K/\mathfrak{b}\mathcal{O}_K$ is an isomorphism.*
3. *The nonzero ideals coprime to $\mathfrak{c}$ in $\mathcal{O}_K$ and in $\mathcal{O}$ are in bijection by $\mathfrak{a} \mapsto \mathfrak{a} \cap \mathcal{O}$ and $\mathfrak{b} \mapsto \mathfrak{b}\mathcal{O}_K$ and these bijections are multiplicative.*

2.1 Cyclic Division Algebras

Let L/K be a degree d extension of number fields with cyclic Galois group. This means that there is automorphism of L which generates $\mathrm{Gal}(L/K)$ - denote this element of the Galois group by θ. Consider the following direct sum:

$$\mathcal{A} = L \oplus uL \oplus ... \oplus u^{d-1}L,$$

where u is an auxiliary element satisfying 1) $u^d = \gamma$, where $\gamma \in K^{\times}$, and 2) for all $x \in L$, we have $xu = u\theta(x)$. $\mathcal{A}$ is a *cyclic algebra*, and property 2) gives it a non-commutative multiplication operation. To ensure every element of $\mathcal{A}$ has a multiplicative inverse, we require that γ is a *non-norm element*:

Definition 9. An element α of K is *non-norm* (or satisfies the *non-norm condition*) if there does not exist an element $x \in L$ such that $\alpha^i = N_{L/K}(x)$, for $0 < i < [L : K]$. Equivalently, α is non-norm if $\alpha \notin N_{L/K}(N^{\times})$ for all proper intermediate subfields $K \subset N \subset L$ [29].

Proposition 5 [1, Theorem 11.12, p. 184]. *The cyclic algebra $\mathcal{A}$ is a division algebra if and only if γ is a non-norm element.*

Then $\mathcal{A}$ is called a *cyclic division algebra* (CDA) with non-norm element γ. We denote this algebra by $\mathcal{A} = (L/K, \theta, \gamma)$. In this scenario, $\mathcal{A}$ is a K-algebra. We note that if γ is a norm, $\mathcal{A}$ is isomorphic to a matrix algebra over K.

(Maximal) Orders. Here we discuss integral structures lying within CDAs. The chief references are [33] and [19]. We denote a general cyclic K-algebra by A, and the CDA defined above by $\mathcal{A}$.

Definition 10. An $\mathcal{O}_K$-*order* in A is a full $\mathcal{O}_K$-lattice $\mathcal{O}$ in A which is a subring of A. Thus a $\mathbb{Z}$-*order* $\mathcal{O}$ in an algebra A is a finitely generated $\mathbb{Z}$−module that contains a $\mathbb{Q}$−basis of A, and is also a unital subring of A. If a set is called an order, it is understood to be an $\mathcal{O}_K$-order. Every order in A contains $\mathcal{O}_K$.

In any cyclic K-algebra a maximal order always exists (with respect to inclusion), and every order is contained within a maximal order. We primarily use the following order of $(L/K, \theta, \gamma)$, where $\gamma \in \mathcal{O}_K$, which we call the *natural order*:

$$\Lambda = \bigoplus_{i=0}^{d-1} u^i \mathcal{O}_L = \mathcal{O}_L \oplus u\mathcal{O}_L \oplus u^2\mathcal{O}_L \oplus ... \oplus u^{d-1}\mathcal{O}_L$$

When γ is a unit, it is possible that the natural order is also maximal, i.e. that Λ is not contained in any other non-trivial order. When $K = \mathbb{Q}(\zeta_n)$ and $\gamma = \zeta_n$, for n a prime power, Λ was shown to be maximal in [15].

Definition 11. Let $\mathcal{L}$ be a full $\mathcal{O}_K$-lattice in A. The *left order* of $\mathcal{L}$ is defined

$$\mathcal{O}_l(\mathcal{L}) = \{a \in A : a\mathcal{L} \subset \mathcal{L}\}.$$

The right order is defined analogously. They are both $\mathcal{O}_K$-orders in A. We note the following properties:

1. If $\mathcal{O}$ is an order, then $\mathcal{O}_l(\mathcal{O}) = \mathcal{O}_r(\mathcal{O}) = \mathcal{O}$.
2. If $\mathcal{L} \subset \mathcal{O}$ for some lattice $\mathcal{L}$ and order $\mathcal{O}$, then $\mathcal{O} \subset \mathcal{O}_l(\mathcal{L})$.

If $\mathcal{I}$ and $\mathcal{J}$ are full $\mathcal{O}_K$-lattices in A, $\mathcal{I}\mathcal{J} = \left\{\sum_{i=1}^{j} x_i y_i : x_i \in \mathcal{I}, y_i \in \mathcal{J}, j \in \mathbb{N}\right\}$ is a full $\mathcal{O}_K$-lattice, and we say $\mathcal{I}\mathcal{J}$ is a *proper* product if $\mathcal{O}_r(\mathcal{I}) = \mathcal{O}_l(\mathcal{J})$.

Ideals in Maximal Orders. We will see how the ideal theory of orders changes depending on whether a given order is maximal or not. The theory in the case of maximal orders is much better behaved than for non-maximal orders.

Definition 12. Let $\mathcal{O}$ be an order in A. A *left integral ideal* in $\mathcal{O}$ is an additive subgroup $\mathcal{I}$ such that $\mathcal{O}\mathcal{I} \subset \mathcal{I}$. A *left fractional ideal* is a subset of the form $\lambda\mathcal{I}$, for some left integral ideal $\mathcal{I}$ and $\lambda \in \mathcal{O}_K \setminus \{0\}$. Right ideals may be defined analogously, and an ideal which is both left and right is called *two-sided*. Every integral ideal is a fractional ideal. An ideal in $\mathcal{O}$ may be called an $\mathcal{O}$-ideal.

For any lattice $\mathcal{L}$, $\mathcal{L}$ is a left $\mathcal{O}_l(\mathcal{L})$-ideal and a right $\mathcal{O}_r(\mathcal{L})$-ideal, and the left and right orders of $\mathcal{L}$ are the largest orders satisfying this property. Let $\mathcal{I}$ be a left $\mathcal{O}$-ideal. Then it can be seen that $\mathcal{I}$ is two-sided if and only if $\mathcal{O} \subset \mathcal{O}_r(\mathcal{I})$. When $\mathcal{O}$ is maximal, this is equivalent to $\mathcal{O}_l(\mathcal{I}) = \mathcal{O}_r(\mathcal{I})$. We say an ideal is *full* if it is full as a lattice, and it turns out any ideal in an order of a division algebra is full. Following [19], we denote by $\mathrm{Frac}_2(\mathcal{O})$ the set of full two-sided $\mathcal{O}$-ideals.

The Case of Two-Sided Ideals. In the following, $\mathcal{I}$ and $\mathcal{J}$ will be full $\mathcal{O}_K$-lattices. Since the product of left $\mathcal{O}$-ideals is a left $\mathcal{O}$-ideal, and since $\mathcal{O}_l(\mathcal{I}\mathcal{J}) \supset \mathcal{O}_l(\mathcal{I})$, $\mathcal{O}_r(\mathcal{I}\mathcal{J}) \supset \mathcal{O}_r(\mathcal{J})$, two-sided ideals are closed under multiplication.

Definition 13. Define the *inverse of* $\mathcal{I}$ to be $\mathcal{I}^{-1} = \{\alpha \in A : \mathcal{I}\alpha\mathcal{I} \subset \mathcal{I}\}$.

Proposition 6 [33, (22.6), (22.7), (23.3)]. $\mathcal{I}^{-1}$ *is a full $\mathcal{O}_K$-lattice in A, and*

$$\mathcal{O}_l\left(\mathcal{I}^{-1}\right) \supset \mathcal{O}_r(\mathcal{I}), \quad \mathcal{O}_r\left(\mathcal{I}^{-1}\right) \supset \mathcal{O}_l(\mathcal{I}), \quad \mathcal{I}\mathcal{I}^{-1} \subset \mathcal{O}_l(\mathcal{I}), \text{ and } \mathcal{I}^{-1}\mathcal{I} \subset \mathcal{O}_r(\mathcal{I}).$$

If $\mathcal{I}$ is a two-sided ideal of a maximal order $\mathcal{O}$, $\mathcal{I}^{-1}$ is also an $\mathcal{O}$-ideal such that $\mathcal{I}\mathcal{I}^{-1} = \mathcal{I}^{-1}\mathcal{I} = \mathcal{O}$.

Thus in a maximal order the full two-sided ideals have unique inverses, with identity element being the order itself.

As usual, if $\mathcal{I}, \mathcal{J}$ are $\mathcal{O}$-ideals such that $\mathcal{I} \supset \mathcal{J}$, we say $\mathcal{I}$ *divides* $\mathcal{J}$ and write $\mathcal{I} \mid \mathcal{J}$. If $\mathfrak{P}$ is a two-sided ideal in $\mathcal{O}$, we say $\mathfrak{P}$ is *prime* if $\mathfrak{P} \mid \mathcal{I}\mathcal{J}$ implies $\mathfrak{P} \mid \mathcal{I}$ or $\mathfrak{P} \mid \mathcal{J}$ for any two-sided ideals $\mathcal{I}, \mathcal{J}$. In CDAs the prime ideals perform a similar role as in number fields; we have the following theorem:

Theorem 4 [33, Theorem 22.10]. *Suppose $\mathcal{O}$ is a maximal $\mathcal{O}_K$-order in A. Then $\mathrm{Frac}_2(\mathcal{O})$ is an abelian group with respect to multiplication, and any proper ideal factors into a product of powers of prime ideals in a unique way.*

This theorem does not hold in the non-maximal order case, although ideal multiplication remains commutative. The theorem also does not hold for one-sided ideals: while the product of two one-sided ideals is again a one-sided ideal, the inverse of a one-sided $\mathcal{O}$-ideal is not in general an $\mathcal{O}$-ideal, even if $\mathcal{O}$ is maximal.

Finally, we record an important property of prime ideals of maximal orders of central simple K-algebras:

Proposition 7 [33, Theorem 32.1]. *The prime ideals of a maximal order coincide with its maximal two-sided ideals. There is a bijective correspondence between the nonzero prime ideals $\mathfrak{p} \subset \mathcal{O}_K$ and the prime ideals $\mathfrak{P} \subset \mathcal{O}$ such that $\mathfrak{p} = \mathfrak{P} \cap \mathcal{O}_K$. Moreover, we have $\mathfrak{p}\mathcal{O} = \mathfrak{P}^{e_\mathfrak{p}}$ for some integer $e_\mathfrak{p} \geq 1$.*

The Case of One-Sided Ideals. We now develop the one-sided ideal theory. In the following, $\mathcal{I}$ will be a full $\mathcal{O}_K$-lattice. The product of two left $\mathcal{O}$-ideals is again a left $\mathcal{O}$-ideal, and $\mathcal{O}_l(\mathcal{I}\mathcal{J}) \supset \mathcal{O}_l(\mathcal{I})$ and $\mathcal{O}_r(\mathcal{I}\mathcal{J}) \supset \mathcal{O}_r(\mathcal{J})$. So we have closure under multiplication of left (or right) ideals in a fixed order.

The following proposition is a one-sided analogue of Proposition 6.

Proposition 8. $\mathcal{I}^{-1}$ is a full $\mathcal{O}_K$ lattice in A. We have $\mathcal{O}_l\left(\mathcal{I}^{-1}\right) \supset \mathcal{O}_r(\mathcal{I}), \mathcal{O}_r\left(\mathcal{I}^{-1}\right) \supset \mathcal{O}_l(\mathcal{I}), \mathcal{II}^{-1} \subset \mathcal{O}_l(\mathcal{I})$, and $\mathcal{I}^{-1}\mathcal{I} \subset \mathcal{O}_r(\mathcal{I})$.

Definition 14. Say a lattice $\mathcal{I}$ is *left invertible* if $\mathcal{II}^{-1} \subset \mathcal{O}_l(\mathcal{I})$. Right invertibility is defined similarly, and for one-sided ideals one can replace lattice with ideal in the definition. In general, left and right invertibility are not equivalent.

Definition 15. An ideal $\mathcal{I}$ of a maximal order is called *normal*. If $\mathcal{I}$ is normal, $\mathcal{I}$ is integral if it is integral for its left order.

Proposition 9 [33, §21.2, §22.9]. Let $\mathcal{I}$ be a full $\mathcal{O}_K$-lattice in A. Then $\mathcal{O}_l(\mathcal{I})$ is a maximal order if and only if $\mathcal{O}_r(\mathcal{I})$ is. If $\mathcal{I}$ is normal with left order $\mathcal{O}$ and right order $\mathcal{O}'$, then $\mathcal{I}$ is integral as a left $\mathcal{O}$-ideal if and only if it is integral as a right $\mathcal{O}'$-ideal, i.e., $\mathcal{I} \subset \mathcal{O}$ if and only if $\mathcal{O} \subset \mathcal{O}'$.

Lemma 3 [33, Theorem 22.15]. Let $\mathfrak{m}$ be a maximal left ideal of a maximal order $\mathcal{O}$. Then $\mathfrak{P} = \{\alpha \in \mathcal{O} : \alpha\mathcal{O} \subset \mathfrak{m}\}$ is a prime ideal of $\mathcal{O}$.

The lemma means that each maximal one-sided ideal $\mathfrak{m}$ of a maximal order has an unique associated prime ideal $\mathfrak{P}$. While the above results are insufficient to form a group on one-sided ideals, one can describe the *Brandt groupoid* instead, which is the set of normal ideals of A with multiplication restricted to proper products (we do not expand on this here). We make the following observations for $\mathcal{I}$ with $\mathcal{O} = \mathcal{O}_l(\mathcal{I})$ and $\mathcal{O}' = \mathcal{O}_r(\mathcal{I})$: if $\mathcal{J} \subset \mathcal{O}$ is two-sided, then $\mathcal{J}\mathcal{I}$ is a right $\mathcal{O}'$-ideal, $\mathcal{I}^{-1}\mathcal{J}$ is a left $\mathcal{O}'$-ideal, and $\mathcal{I}^{-1}\mathcal{J}\mathcal{I}$ is a two-sided $\mathcal{O}'$-ideal.

Quotients of Lattices. We now develop useful ideal-theoretic notions for non-maximal orders, generalising lemmas from [31, Section 2] for use in Sect. 5.

Definition 16. Let $\mathcal{L}, \mathcal{L}' \subset A$ be lattices. Define the 'lattice quotient'

$$(\mathcal{L} : \mathcal{L}')_l = \{x \in A : x\mathcal{L}' \subset \mathcal{L}\} \text{ and } (\mathcal{L} : \mathcal{L}')_r = \{x \in A : \mathcal{L}'x \subset \mathcal{L}\}.$$

Note that $\mathcal{O}_l(\mathcal{L}) = (\mathcal{L} : \mathcal{L})_l$ and $\mathcal{O}_r(\mathcal{L}) = (\mathcal{L} : \mathcal{L})_r$. Observe that the lattice quotient satisfies additive closure. Moreover, $\mathcal{I}\mathcal{L}' \subset \mathcal{L}$ if and only if $\mathcal{I} \subset (\mathcal{L} : \mathcal{L}')_l$, for sets $\mathcal{I} \subset A$. Finally, we have $(\mathcal{L} : \mathcal{L}')_l(\mathcal{L}' : \mathcal{L}'')_l \subset (\mathcal{L} : \mathcal{L}'')_l$, since $(\mathcal{L} : \mathcal{L}')_l(\mathcal{L}' : \mathcal{L}'')_l\mathcal{L}'' \subset (\mathcal{L} : \mathcal{L}')_l\mathcal{L}' \subset \mathcal{L}$.

Lemma 4. Let $\mathcal{O}$ be an order, and $\mathcal{I}, \mathcal{I}'$ be fractional two-sided $\mathcal{O}$-ideals such that $\mathcal{I}'$ is invertible. Then $(\mathcal{I} : \mathcal{I}')_l = \mathcal{I}\mathcal{I}'^{-1}$.

Proof. $\mathcal{I}\mathcal{I}'^{-1}\mathcal{I}' = \mathcal{I} \Rightarrow \mathcal{I}\mathcal{I}'^{-1} \subset (\mathcal{I} : \mathcal{I}')_l$. Then $\mathcal{I} = \mathcal{I}\mathcal{I}'^{-1}\mathcal{I}' \subset (\mathcal{I} : \mathcal{I}')_l\mathcal{I}' \subset \mathcal{I}$. So $(\mathcal{I} : \mathcal{I}')_l \subset \mathcal{I}\mathcal{I}'^{-1}$. Thus $\mathcal{I}\mathcal{I}'^{-1} \subset (\mathcal{I} : \mathcal{I}')_l \subset \mathcal{I}\mathcal{I}'^{-1}$ gives the result. $\square$

Lemma 5. Let $\mathcal{L}, \mathcal{L}' \subset A$ be lattices. Then $(\mathcal{L} : \mathcal{L}')_l = (\mathcal{L}'\mathcal{L}^\vee)^\vee$.

Proof. Let $x \in A$. We have $x \in (\mathcal{L}'\mathcal{L}^\vee)^\vee \iff \mathrm{Tr}(x\mathcal{L}'\mathcal{L}^\vee) \subset \mathbb{Z} \iff x\mathcal{L}' \subset \mathcal{L}^{\vee\vee} = \mathcal{L} \iff x \in (\mathcal{L} : \mathcal{L}')_l$. $\square$

Lemma 6. *We have $\mathcal{O}_l(\mathcal{L}) \subset \mathcal{O}_l((\mathcal{L} : \mathcal{L}')_l)$ and $\mathcal{O}_l(\mathcal{L}') \subset \mathcal{O}_r((\mathcal{L} : \mathcal{L}')_l)$. Moreover, $(\mathcal{L} : \mathcal{L}')_l$ is a left ideal in $\mathcal{O}_l(\mathcal{L})$ and a right ideal in $\mathcal{O}_l(\mathcal{L}')$.*

Proof. For the latter statements: $\mathcal{O}_l(\mathcal{L})(\mathcal{L} : \mathcal{L}')_l = (\mathcal{L} : \mathcal{L})_l (\mathcal{L} : \mathcal{L}')_l \subset (\mathcal{L} : \mathcal{L}')_l$, and $(\mathcal{L} : \mathcal{L}')_l \mathcal{O}_l(\mathcal{L}') = (\mathcal{L} : \mathcal{L}')_l(\mathcal{L}' : \mathcal{L}')_l \subset (\mathcal{L} : \mathcal{L}')_l$. For the former, note that the previous line implies that $\mathcal{O}_l(\mathcal{L}) \subset \mathcal{O}_l((\mathcal{L} : \mathcal{L}')_l)$ and $\mathcal{O}_l(\mathcal{L}') \subset \mathcal{O}_r((\mathcal{L} : \mathcal{L}')_l)$. $\square$

Lemma 7. *Suppose $\mathcal{L} \subset \mathcal{L}'$ are lattices in $\mathcal{A}$. Then $(\mathcal{L} : \mathcal{L}')_l$ is integral in $\mathcal{O}_l(\mathcal{L})$ and in $\mathcal{O}_l(\mathcal{L}')$.*

Proof. $\mathcal{L} \subset \mathcal{L}' \Rightarrow 1 \in (\mathcal{L}' : \mathcal{L})_l$. Then $(\mathcal{L} : \mathcal{L}')_l \subset (\mathcal{L} : \mathcal{L}')_l(\mathcal{L}' : \mathcal{L})_l \subset (\mathcal{L} : \mathcal{L})_l = \mathcal{O}_l(\mathcal{L})$, and $(\mathcal{L} : \mathcal{L}')_l \subset (\mathcal{L}' : \mathcal{L})_l(\mathcal{L} : \mathcal{L}')_l \subset (\mathcal{L}' : \mathcal{L}')_l = \mathcal{O}_l(\mathcal{L}')$. $\square$

If $\mathcal{O}$ is a non-maximal order contained in $\mathcal{O}'$, then $(\mathcal{O} : \mathcal{O}')_l$ is not an invertible left $\mathcal{O}$-ideal. For suppose there exists a (right) $\mathcal{O}$-ideal $\mathcal{I}$ such that $\mathcal{I}(\mathcal{O} : \mathcal{O}')_l = \mathcal{O}$. Then $\mathcal{O}' = \mathcal{O}\mathcal{O}' = \mathcal{I}(\mathcal{O} : \mathcal{O}')_l\mathcal{O}' = \mathcal{I}(\mathcal{O} : \mathcal{O}')_l = \mathcal{O}$, a contradiction.

Definition 17. The left 'pseudoinverse' of a two-sided $\mathcal{O}$-ideal $\mathcal{I}$ is $(\mathcal{O} : \mathcal{I})_l$.

We record two properties of pseudoinverses: first, if $\mathcal{I}$ is a two-sided ideal of $\mathcal{O}$, $(\mathcal{O} : \mathcal{I})_l$ is a two-sided ideal of $\mathcal{O}$. To see this, note is a left ideal by Lemma 6. Consider $x \in (\mathcal{O} : \mathcal{I})_l$ and $a \in \mathcal{O}$. Then $xa\mathcal{I} \subset x\mathcal{I} \subset \mathcal{O}$.

Second, if $\mathcal{I}$ is a two-sided invertible $\mathcal{O}$-ideal, by Lemma 4 $(\mathcal{O} : \mathcal{I})_l = \mathcal{I}^{-1}$.

Embedding CDAs into $\mathbb{R}^m$. Consider a CDA $\mathcal{A} = (L/K, \theta, \gamma)$ with $[L : K] = d$. Fixing the L-basis of $\mathcal{A}$, $\{u^i\}_{i \geq 0}$, we can express an element as the linear map $\phi(x)$ given by left multiplication on the u^i. For example, if $x = \oplus_{i=0}^{d-1} u^i x_i \in \mathcal{A}$,

$$\phi(x) = \begin{pmatrix} x_0 & \gamma\theta(x_{d-1}) & \ldots & \gamma\theta^{d-1}(x_1) \\ x_1 & \theta(x_0) & \ldots & \gamma\theta^{d-1}(x_2) \\ \ldots & \ldots & \ldots & \ldots \\ x_{d-1} & \theta(x_{d-2}) & \ldots & \theta^{d-1}(x_0) \end{pmatrix}.$$

If we denote the n embeddings $K \hookrightarrow \mathbb{C}$ by α, we can extend these to embeddings of L (which, in an abuse of notation, we also denote by α). Since all the nd embeddings of L are obtained by extending the set of L-automorphisms $\{\alpha \circ \theta^i\}_{\alpha,i}$ to embeddings of L, we may form a vector in $\mathbb{R}^{nd^2}$ from x by concatenating the vectorized images of the $\alpha(\phi(x))$ for all $\alpha \in \text{Emb}(K)$. Then the image of any discrete additive subgroup of $\mathcal{A}$ is a lattice in $\mathbb{R}^{nd^2}$. When γ is a unit, this embedding is equivalent to extending the canonical (Minkowski) embedding of L coefficientwise to algebra elements. We then define three norms on $\mathcal{A}$: we set $\|x\|_p^p = \sum_{\alpha \in \text{Emb}(K)} \sum_{i,j} |\alpha(\phi(x)_{i,j})|^p$, and $\|x\|_\infty = \max_{\alpha,i,j} |\alpha(\phi(x)_{i,j})|$, where $\phi(x)_{i,j}$ denotes the i, jth entry of $\phi(x)$, and finally we set $\|x\|_{2,\infty} = \max_{\alpha,j} \sqrt{\left(\sum_{i=0}^{d-1} |\alpha \circ \theta^j(x_i)|^2\right)}$. We may denote $\|\cdot\|_2$ by $\|\cdot\|$.

Let the trace $\text{Tr}(\cdot)$ of $x \in \mathcal{A}$ be defined $\text{Tr}(x) = T_{K/\mathbb{Q}} \circ \text{trace}(\phi(x))$, where $T_{K/\mathbb{Q}}$ is the field trace. This map is symmetric. The dual of an ideal $\mathcal{I}$ is the set

$$\mathcal{I}^\vee = \{x \in \mathcal{A} : \text{Tr}(x\mathcal{I}) \subset \mathbb{Z}\}.$$

The codifferent ideal of Λ is $\Lambda^\vee$. We may denote $\Lambda/q\Lambda$ by Λ_q.

The CLWE Problem. In [15], LWE was instantiated using the natural order inside a CDA. We first define the following distribution:

Definition 18 (The CLWE Distribution). Let L/K be a Galois extension of number fields of dimension $[L : K] = d$, $[K : \mathbb{Q}] = n$ with cyclic Galois group generated by θ. Let $\mathcal{A} := (L/K, \theta, \gamma)$ be a cyclic division algebra with element u such that $u^d = \gamma \in \mathcal{O}_K$. Let Λ be the natural order of $\mathcal{A}$. For an error distribution ψ over $\oplus_{i=0}^{d-1} u^i L_{\mathbb{R}}$, integer modulus $q \geq 2$, and a secret $s \in \Lambda_q^\vee$, a sample from the CLWE distribution $\Pi_{q,s,\psi}$ is obtained by sampling $a \leftarrow \Lambda_q$ uniformly at random, $e \leftarrow \psi$, and outputting $(a, b) = (a, (a \cdot s)/q + e \bmod \Lambda^\vee) \in \left(\Lambda_q, \oplus_{i=0}^{d-1} u^i L_{\mathbb{R}}/\Lambda^\vee \right)$.

The public value a has matrix representation $\phi(a)$, and to construct this matrix only the first column need be stored by a user. As for RLWE, via the matrix representation one can see that CLWE is a form of structured LWE. We now define search and decision problems, where Ψ is a family of error distributions.:

Definition 19. Let $\Pi_{q,s,\psi}$ be a CLWE distribution for parameters $q \geq 2, s \in \Lambda_q^\vee$, and error distribution $\psi \in \Psi$. The search CLWE problem, denoted $\text{CLWE}_{q,s,\psi}$, is to recover s from a collection of independent samples from $\Pi_{q,s,\psi}$.

Definition 20. Let Υ be a distribution on a family of error distributions over $\oplus_{i=0}^{d-1} u^i L_{\mathbb{R}}$ and U_Λ the uniform distribution on $\left(\Lambda_q, \left(\oplus_{i=0}^{d-1} u^i L_{\mathbb{R}} \right)/\Lambda^\vee \right)$. The decision CLWE problem, $\text{DCLWE}_{q,\Upsilon}$, is given a collection of independent samples from either $\Pi_{q,s,\psi}$ for a random choice of $(s, \psi) \leftarrow U\left(\Lambda_q^\vee\right) \times \Upsilon$, or from U_Λ, to decide which is the case with non-negligible advantage.

3   Non-norm Elements for General Cyclotomic Fields

In [15], the authors used the work of [27] to construct CDAs containing nth roots of unity for prime-power n. Recall the method runs as follows: let $n = p^r$ for some prime p and $r \in \mathbb{N}$, and set $K = \mathbb{Q}(\zeta_n)$. Let $\ell = 1 \bmod n$, $\ell \neq 1 \bmod pn$ be a prime, and set $M = \mathbb{Q}(\zeta_n, \zeta_\ell) = \mathbb{Q}(\zeta_{n\ell})$. One then fixes a subfield of degree d over K inside M by taking the fixed field of σ^d, where d divides n and $\text{Gal}(M/K) = \langle \sigma \rangle$, and the resulting algebra $(L/K, \theta, \zeta_n)$ is division, where $\text{Gal}(L/K) = \langle \theta \rangle$ - that is to say, ζ_n is non-norm. In this section, we extend this method to composite n. The main result of this section is the following theorem:

Theorem 5. *Let $n \in \mathbb{N}_{\geq 2}$. Set $K = \mathbb{Q}(\zeta_n)$, where ζ_n is a primitive nth root of unity. Then there exist infinitely many cyclic Galois extensions L/K of degree n such that ζ_n satisfies the non-norm condition.*

Proof. Write $n = p_1^{e_1}...p_k^{e_k}$, for $p_1, ..., p_k$ pairwise coprime. Pick $\ell \in \mathbb{Z}$ prime such that $p_1^{e_1}...p_k^{e_k} \mid \ell - 1$, and such that e_i is the highest power of p_i such that this is true, for all i, so we have $\ell - 1 = p_1^{e_1}...p_k^{e_k} p_{k+1}^{e_{k+1}}...p_r^{e_r}$ for some primes $p_{k+1}, ..., p_r$ distinct from the p_j for $1 \leq j \leq k$, and integers $e_i \geq 1$. By Theorem 1, there are infinitely many such primes q. Note that $\gcd(n, \ell) = 1$.

Now consider $M = K(\zeta_\ell) = \mathbb{Q}(\zeta_n, \zeta_\ell)$. We have $\mathrm{Gal}(M/K) \cong (\mathbb{Z}/\ell\mathbb{Z})^*$. Let σ be a generator of $\mathrm{Gal}(M/K)$. Then σ^n fixes an extension, denoted L, of degree n over K. We have $\mathrm{Gal}(L/K) \cong \mathrm{Gal}(M/K)/\mathrm{Gal}(M/L)$, which implies that $\mathrm{Gal}(L/K)$ is also cyclic. We will use Theorem 2 to prove the result by localizing at a certain prime ideal. Let the prime ideal $\mathbf{Q}$ lie above $\mathfrak{Q}$ lie above $\mathfrak{q}$ lie above ℓ in the tower of fields $M/L/K/\mathbb{Q}$. Proposition 2 implies that the inertial degree $f_{\mathfrak{q}|\ell} = 1$. This is equivalent to $[\mathcal{O}_K/\mathfrak{q} : \mathbb{F}_\ell] = 1$.

Now, $\ell \mid n\ell$ exactly once, since $\gcd(\ell, n) = 1$. Then $f_{\mathbf{Q}|\ell}$ is the smallest f such that $\ell^f \equiv 1 \bmod n$. But $\ell \equiv 1 \bmod n$, so $f_{\mathbf{Q}|\ell} = 1$. The inertial degree is multiplicative, that is $f_{\mathbf{Q}|\ell} = f_{\mathbf{Q}|\mathfrak{q}} f_{\mathfrak{q}|\ell}$, so $f_{\mathbf{Q}|\mathfrak{q}} = 1$ also. We have found $f_{\mathbf{Q}|\mathfrak{q}}$ by considering the tower $M/K/\mathbb{Q}$; we now consider the tower $M/L/K$, which yields that $f_{\mathbf{Q}|\mathfrak{q}} = f_{\mathbf{Q}|\mathfrak{Q}} f_{\mathfrak{Q}|\mathfrak{q}}$. Hence $f_{\mathbf{Q}|\mathfrak{q}} = 1$ implies $f_{\mathfrak{Q}|\mathfrak{q}} = 1$ also.

Denote the completions of the fields L, K, and $\mathbb{Q}$ by the corresponding valuations by $L_{\mathfrak{Q}}$, $K_{\mathfrak{q}}$, and $\mathbb{Q}_\ell$ respectively. Then, since $g = 1$ in extensions of local fields, we combine this with the above discussion on the inertial degrees to conclude that $L_{\mathfrak{Q}}/K_{\mathfrak{q}}$ is totally ramified. Further, since the characteristic of the residue field of $K_{\mathfrak{q}}$ is ℓ, but total ramification of $L_{\mathfrak{Q}}/K_{\mathfrak{q}}$ means that the ramification index $e_{L_{\mathfrak{Q}}/K_{\mathfrak{q}}} = [L_{\mathfrak{Q}} : K_{\mathfrak{q}}] = n$, we obtain that $L_{\mathfrak{Q}}/K_{\mathfrak{q}}$ is tamely ramified. Using the local reciprocity map, we obtain $[K_{\mathfrak{q}}^\times : N_{L_{\mathfrak{Q}}/K_{\mathfrak{q}}}(L_{\mathfrak{Q}}^\times)] = |\mathrm{Gal}(L_{\mathfrak{Q}}/K_{\mathfrak{q}})| = [L_{\mathfrak{Q}} : K_{\mathfrak{q}}] = n$. Moreover, a standard result of local fields is that for a totally and tamely ramified finite extension of local number fields, $N_{L_{\mathfrak{Q}}/K_{\mathfrak{q}}}(L_{\mathfrak{Q}}^\times)$ contains a uniformizer $\pi_{\mathfrak{q}}$ of $K_{\mathfrak{q}}$, as well as $U_{\mathfrak{q}}^1$ (see [13, page 115]). Pulling together the above in conjunction with the local reciprocity map, we obtain the following commutative diagram:

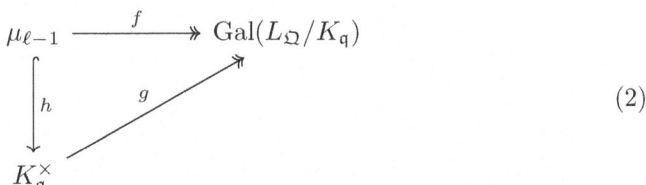

$$(2)$$

Here g is the homomorphism defined by the local reciprocity map, with kernel $N_{L_{\mathfrak{Q}}/K_{\mathfrak{q}}}(L_{\mathfrak{Q}}^\times)$ and image $\mathrm{Gal}(L_{\mathfrak{Q}}/K_{\mathfrak{q}}) \cong \mathbb{Z}/p_1^{e_1}\mathbb{Z} \times \ldots \times \mathbb{Z}/p_k^{e_k}\mathbb{Z}$, h is the inclusion of $\mu_{\ell-1}$ into $K_{\mathfrak{q}}^\times \cong \mu_{\ell-1} \times \langle \pi_{\mathfrak{q}} \rangle \times U_{\mathfrak{q}}^1$, and f is the natural surjection from the CRT decomposition of $\mu_{\ell-1} \cong \prod_{i=1}^r \mathbb{Z}/p_i^{e_i}\mathbb{Z}$ onto $\mathbb{Z}/p_1^{e_1}\mathbb{Z} \times \ldots \times \mathbb{Z}/p_k^{e_k}\mathbb{Z}$.

We have $f = g \circ h$, where f, g and h are homomorphisms. Since f maps non-identity elements of order dividing n non-trivially, by commutativity of the diagram g also maps such elements non-trivially, so they are not contained in $\ker(g)$. But by the local reciprocity map, $\ker(g) = N_{L_{\mathfrak{Q}}/K_{\mathfrak{q}}}(L_{\mathfrak{Q}}^\times)$; so $\zeta_n^i \notin N_{L_{\mathfrak{Q}}/K_{\mathfrak{q}}}(L_{\mathfrak{Q}}^\times)$, for $0 < i < p_1^{e_1}\ldots p_k^{e_k}$, as required. The Hasse norm theorem gives the result. $\quad\square$

Corollary 1. *Let $n = p_1^{e_1}\ldots p_k^{e_k}$ and $K = \mathbb{Q}(\zeta_{p_1^{e_1}\ldots p_k^{e_k}})$, where the p_i are pairwise coprime. There exists a cyclic Galois extension E/K of any index d dividing n, such that ζ_n is non-norm in E/K, i.e. ζ_n is not a norm for $0 < i < d$.*

Proof. Mutatis mutandis, identical to [15].   □

These results allow us to create many new CDAs with roots of unity as non-norm elements. We give two examples.

Examples 1. Set $n = 3^2 \cdot 2^6 = 576$. Note $\ell = 577$ is prime, and $\ell - 1 = 576$ is divisible by 576, with neither 27 nor 128 dividing $\ell - 1$. So 577 satisfies the conditions of Theorem 5. Observe $M := \mathbb{Q}(\zeta_n, \zeta_\ell) = \mathbb{Q}(\zeta_{576}, \zeta_{577})$ is an extension of $K := \mathbb{Q}(\zeta_{576})$ of degree 576. Let $\mathrm{Gal}(M/K) = \langle \sigma \rangle$. Then, as in the proof above, $\sigma^n = e$ fixes an extension of degree $n = 576$ over K - in this case the extension is M itself. By the theorem ζ_{576} is a non-norm element in $\mathbb{Q}(\zeta_{576}, \zeta_{577})/\mathbb{Q}(\zeta_{576})$. Apply Corollary 1 to fix a degree 2 cyclic Galois extension of K, denoted E, which is possible since 2 divides 576. This extension also has the property that ζ_{576} is a non-norm element over K. Let $\mathrm{Gal}(E/K)$ be generated by θ. Then the cyclic algebra $(E/K, \theta, \zeta_{576})$ has dimension $2^2 \cdot 192 = 768$ over $\mathbb{Q}$, and is division.

2. Set $n = 3^3 \cdot 2^5 = 864$. Note that $\ell = 16,417$ is prime, $16,416 \equiv 0 \bmod 864$, and neither 81 nor 64 divide $\ell - 1 = 16,416$. So ℓ satisfies the required properties to apply Theorem 5. Observe $M := \mathbb{Q}(\zeta_n, \zeta_\ell) = \mathbb{Q}(\zeta_{864}, \zeta_{16,417})$ is a cyclic extension of $K := \mathbb{Q}(\zeta_{864})$, of degree $16,416$. Let $\mathrm{Gal}(M/K) = \langle \sigma \rangle$. Then, as in the proof above, $\sigma^n = \sigma^{864}$ fixes an extension of degree $n = 864$ over K, denoted L. The theorem indicates that ζ_{864} is a non-norm element in $L/\mathbb{Q}(\zeta_{864})$. We can then apply Corollary 1 to fix a cyclic Galois extension of K of degree 2 over K, since 2 divides $16,416$. This extension also has the property that ζ_{864} is a non-norm element over K. Call this extension E and let $\mathrm{Gal}(E/K) = \langle \theta \rangle$. Then the cyclic algebra $(E/K, \theta, \zeta_{864})$ has dimension $2^2 \cdot 288 = 1152$ over $\mathbb{Q}$, and is division.

Maximality of the Natural Order. We note that [15, Theorem 3] states that the natural order in the above prime-power case ($n = p^r$) is maximal. It can be shown with an identical proof that when one takes $n = p_1^{a_1}...p_l^{a_l}$, as long as ℓ satisfies $\ell \equiv 1 \bmod n$ and $p_i n \nmid \ell - 1$ for $i = 1,...,l$, the result still holds: note that the ramification of ideals above ℓ in $\mathcal{O}_K$ in $\mathcal{O}_M$ is the same as before (since the inertial degree depends on $\ell \equiv 1 \bmod n$ to equal 1), the ramification index is $\phi(\ell)$, and $g_{M/K} = [M : K]/\phi(\ell) = 1$; this can be summarised by noting that the relevant properties depend on the prime ℓ, and not the degree of K. However, if γ is not a unit, the natural order is not (in general) maximal.

4   Cryptographic Moduli for CDAs from Any Cyclotomics

Here we provide methods to find the primes needed to construct CDAs over cyclotomic fields of composite conductor for CLWE, and give concrete examples. To build algebras meeting the requirements of the previous theorem, we need to find a small degree extension L of $K = \mathbb{Q}(\zeta_n)$ and a non-norm element for L/K. As explained above, this is done by taking a large prime ℓ satisfying certain conditions, taking the compositum of K and $\mathbb{Q}(\zeta_\ell)$, and taking an intermediate extension L of K of desired degree. If chosen well, ζ_n will be a non-norm element.

Recall that ℓ must satisfy two conditions: firstly, that $\ell \equiv 1 \bmod n$, and secondly, that $\ell \not\equiv 1 \bmod p_i \cdot n$, for $i = 1, ..., k$, where p_i is a prime in the prime factorization of $n = p_1^{e_1}...p_k^{e_k}$. We can find such an ℓ by considering the arithmetic progression $1 + p_1^{e_1}...p_k^{e_k} + \sum_{i=0}^{m} p_1^{e_1+1}...p_k^{e_k+1}$, for $m = 0, 1, 2, ...$ Theorem 1 implies there are infinitely many primes in this progression, and for our parameters searching elements in the progression can be done efficiently.

Similarly to RLWE [22][2], the security proof of CLWE holds for primes q completely split in $\mathcal{O}_K$. Moreover, to enable efficient multiplication, it will be convenient to have q also completely split in L. We can use the fact that if q splits completely in M, it splits completely in all subfields. Thus the most naive approach to find a prime that splits completely in L and K is to find a prime that splits completely in $M = \mathbb{Q}(\zeta_{n\ell})$. This is equivalent to $q \equiv 1 \bmod \ell n$.

This approach is limited, however. Since $\ell \equiv 1 \bmod n$, we have $\ell > n$; since $q \equiv 1 \bmod \ell n$, we have $q > \ell n > n^2$. In general, we want $[K : \mathbb{Q}] = \phi(n)$ to be large, and $[L : K]$ small. As we want q not to be too large, this forces n to be small. Moreover, the larger n gets, the smaller ℓ is required to be. For example, take $n = 100$. Then $[K : \mathbb{Q}] = 40$, and, if we can find an appropriate non-norm element, we could take a degree 5 extension of K for an algebra of degree 1000 over $\mathbb{Q}$. 101 is prime, so we can take $\ell = 101$; then we have $q \geq 100 \cdot 101 + 1 = 10101$. Since 10101 is not prime, the next smallest possible option for q is $2\ell n + 1 = 20203$, which is also not prime. The search continues in this manner, with the size of q rapidly growing.

The Quadratic Case. We first consider extensions L/K of degree 2, and write L as the compositum of K and a quadratic subfield E of $E' := \mathbb{Q}(\zeta_\ell)$. Since E' is cyclotomic with prime conductor, E is the *unique* quadratic subfield of E'. This can be written explicitly as follows [18, Theorem 9.3]:

$$E = \begin{cases} \mathbb{Q}(\sqrt{\ell}), & \text{if } \ell \equiv 1 \bmod 4 \\ \mathbb{Q}(\sqrt{-\ell}), & \text{if } \ell \equiv 3 \bmod 4. \end{cases}$$

The discriminant of this field, d_E, is also well known, being $d_E = \ell$ if $\ell \equiv 1 \bmod 4$, else $d_E = 4\ell$ if $\ell \equiv 3 \bmod 4$. A final fact is that a prime q splits completely in E if and only if d_E is a quadratic residue modulo q, i.e. there exists $x \in \mathbb{Z}/q\mathbb{Z}$ such that $d_E \equiv x^2 \bmod q$ [28, Proposition 8.5].

A prime q splits in L if and only if q splits in both K and E. Thus to find primes that split in both L and K we search for primes q such that $q \equiv 1 \bmod n$, with d_E becoming a quadratic residue modulo q. Finally, to ensure we can find non-norm elements, we require that $\gcd(n, 2) \neq 1$ (cf. [15, Theorem 10]).

Here we run through an example. Set $n = 320$. With $d = 2$, we obtain a degree 512 algebra. Since $1601 \equiv 1 \bmod 320$ and is prime, we can set $\ell = 1601$. Then $\ell \equiv 1 \bmod 4$, so $d_E = \ell = 1601$. There are many small primes that are

---

[2] Subsequent works have allayed this restriction on the modulus.

congruent to 1 mod 320 and have d_E as a quadratic residue; for example, 4481 suffices. It is prime, and $4481 = 320 \cdot 14 + 1$. Furthermore, we have $1213^2 \equiv 1601$ mod 4481. Below is a table of results for quadratic extensions of $\mathbb{Q}(\zeta_n)$. Listed are n, $[\mathcal{A} : \mathbb{Q}]$, ℓ, example value(s) of q, and 'min log rop'. This last quantity is obtained by running the lattice estimator[3] [2] with similar parameters for the secret, error, and default number of samples as Kyber512, but replacing the lattice dimension and modulus with the corresponding entries of the table. We list the minimum of the base-2 logarithms over the given values of q in the row. The 'meaning' of the rop results is to estimate the number of ring operations required to solve the corresponding LWE instances (Table 1).

Table 1. The 'Naive' Quadratic Case

n	$[\mathcal{A} : \mathbb{Q}]$	ℓ	q	min log rop
320	512	1601	4481, 7681, 9601, 13121	120.6
324	432	1621	3889, 6481, 8101, 10369	104.9
352	640	353	3169, 6337, 11617, 13729	151.2
400	640	401	4001, 4801, 14401	150.5
432	576	433	3889, 8209, 12097, 15121	133.8
448	768	449	4481, 8513, 10753	190.6
484	880	1453	3389, 11617, 13553, 15973	215.9
576	768	577	7489, 10369, 13249, 14401	185.0
640	1024	641	7681, 9601, 12161, 13441	265.3
648	864	7129	3889, 6481, 9721, 10369	220.2
864	1152	2593	3457, 10369	316.4

The Naive Method for Higher Values of d. In this section we address extensions L such that $[L : K] > 2$. Unlike in the quadratic case we cannot, in general, write down explicitly what a low-degree subfield of $\mathbb{Q}(\zeta_\ell)$ is. We return to the naive method explained above, finding $q \equiv 1$ mod ℓn. Though cumbersome, one can still obtain some limited results by this method, although the resulting algebras are slightly small since the largest n that can be taken is 141, if one imposes, say, $q < 20000$. A simple method to find appropriate values of ℓ is to fix prime ℓ and set $n = \ell - 1$. Some results are below (Table 2):

---

[3] Commit 564470e.

Table 2. The 'Naive' Case for $d > 2$

n	d	$[\mathcal{A} : \mathbb{Q}]$	ℓ	q	min log rop
25	5	500	101	5051	130.5
52	6	864	53	8269	225.2
58	6	1008	59	10267	267.0
60	6	720	61	7321	184.0
66	6	720	67	4423	194.4
78	6	936	79	6163	256.0
82	4	640	83	13613	151.3
82	6	1440	83	13613	414.7
138	4	704	139	19183	162.7

Refined Naive Method. There is, however, a refinement on the above method by which we relax the condition that $q \equiv 1 \bmod \ell n$. Again write L as the compositum of K and a subfield E of $E' = \mathbb{Q}(\zeta_\ell)$. Since $\ell > 2$ is prime, the maximal real subfield $E'^+ = \mathbb{Q}(\zeta_\ell + \zeta_\ell^{-1})$ of E' has degree $\frac{\ell-1}{2}$ over $\mathbb{Q}$, and Galois group isomorphic to $\mathrm{Gal}(E'/\mathbb{Q})/\{\pm 1\}$. The primes that ramify in E'^+ are the primes dividing the discriminant of E', which is only ℓ. This makes it easy to select unramified primes. The inertial degree is also well known in the maximal real subfield: it is the smallest integer f such that $q^f \equiv \pm 1 \bmod \ell$. So if E is contained in E'^+, we need to find prime q such that $q \equiv \pm 1 \bmod \ell$ and $q \equiv 1 \bmod n$.

So which subfields are contained in E'^+? Since $\mathrm{Gal}(E'/\mathbb{Q})$ is cyclic and $\mathrm{Gal}(E'/E'^+)$ is one of its subgroups, $\mathrm{Gal}(E'/E'^+)$ is cyclic, so the quotient of the two groups is cyclic and isomorphic to $\mathrm{Gal}(E'^+/\mathbb{Q})$, of order $\frac{\ell-1}{2}$. The subgroups of a cyclic group $\mathbb{Z}/m\mathbb{Z}$ correspond to divisors d of m, and there is a single subgroup of order d for each divisor. Hence in $\mathrm{Gal}(E'^+/\mathbb{Q})$ there is a subgroup for each divisor of $\frac{\ell-1}{2}$, and by the Galois correspondence there is a unique Galois subfield of E'^+ with degree over $\mathbb{Q}$ equal to that divisor. Since this is also a subfield of E', this subfield is also the unique subfield of the given degree in E'. Note also that any Galois extension of $\mathbb{Q}$ is either totally real or totally imaginary, so odd-degree subfields of $\mathbb{Q}(\zeta_\ell)$ are totally real.

In light of this, our approach is as follows: we look for small values of n, so that we can find small values of ℓ, and hence find a value of q in the appropriate range. We will also look for values of ℓ such that our desired values of $d = [L : K]$ divide $\frac{\ell-1}{2}$, so that E will be totally real and we can apply this 'refined naive method' in our hunt for q. A final constraint to note is that we must have $\gcd(n, d) \geq 2$, else we will not have ζ_n be a non-norm element.

Here we run through the previous example again. We have $n = 100$, $[K : \mathbb{Q}] = 40$, and want a degree 5 extension of K to obtain an algebra of degree 1000 over $\mathbb{Q}$. Since 101 is prime, we can take $\ell = 101$. This time, instead of requiring $q \equiv 1 \mod \ell n$, we need $q \equiv \pm 1 \mod \ell$ and $q \equiv 1 \mod n$. Note that $5 \mid \frac{101-1}{2} = 50$, so a subfield of degree 5 of the maximal totally real subfield exists. One finds that 10301 satisfies both of these conditions: $10301 = 100 \cdot 103 + 1$, and $10301 = 101 \cdot 102 - 1$. Furthermore, 10301 is of the appropriate size; where the naive method failed, the tweaked version yielded a small prime.

Below is a table, Table 3, giving examples of valid primes using this method. Listed are n, the degree $d = [L : K]$, $[\mathcal{A} : \mathbb{Q}]$, and possible values for ℓ and q.

Table 3. The Refined 'Naive' Case for $d > 2$

n	d	$[\mathcal{A} : \mathbb{Q}]$	ℓ	q	min log rop
32	5	400	97	18913	91.8
45	5	600	181	16651	138.4
70	5	600	71	9941	145.9
70	7	1176	71	9941	326.3
90	5	600	2791	5581	155.2
100	5	1000	101	10301	264.1
102	3	288	103	10711	72.1
130	5	1200	131	17291	318.7

Completely Splitting Primes in Subfields of Cyclotomics. We now consider primes in subextensions of higher degree. The results in Tables 4 and 5 are based off the following theorem:

Theorem 6 [24, Theorem 30]. *Let ℓ be a prime, $E' = \mathbb{Q}(\zeta_\ell)$, and denote its unique subfield of degree d over $\mathbb{Q}$ by E'_d. Let q be a prime coprime to ℓ. Then q splits completely in E'_d if and only if q is a dth power modulo ℓ.*

This allows us to generalize the method of the quadratic case to larger degree subfields of E'. As examples, we list results for cubic and quartic extensions of K: we list, in addition to the previous values, an x for which $q \equiv x^d \mod \ell$. We illustrate the quartic case with extensions of $\mathbb{Q}(\zeta_{128})$, which yields $[\mathcal{A} : \mathbb{Q}] = 1024$.

5   Hardness of CLWE from Ideal Lattices in Suborders

In this section we define LWE in suborders of Λ, and obtain a security reduction analogous to the maximal order case. In order for this proof to hold, we initally restrict our ideals to those coprime to the ideal generated in a suborder of Λ by the conductor ideal of an $\mathcal{O}_K$-suborder of $\mathcal{O}_L$, as described below. Similarly to CLWE, one step of the reduction requires a restricted secret space.

Table 4. The Cubic Case

n	$[\mathcal{A}:\mathbb{Q}]$	ℓ	x	q	min log rop
111	648	4441	2152	12211	155.1
171	972	6841	2112	3079	291.8
183	1080	8053	1422	7321	301.1
201	1188	4423	2246	4021	363.9
360	864	2521	461	6121	232.3

Table 5. The Quartic Case

n	ℓ	x	q	min log rop
128	2689	275	3329	304.8
128	641	147	3457	303.8
128	3457	780	4481	295.7
128	12161	10957	4993	292.3
128	4481	571	6529	287.8

Constructing Non-maximal Orders. Here we construct families of non-maximal orders in $\mathcal{A} = (L/K, \theta, \gamma)$ where $K = \mathbb{Q}(\zeta_n)$, L is constructed as in [15], and $\gamma = \zeta_n$. We do this as follows: let $\mathcal{O}' \subsetneq \mathcal{O}_L$ be an order in L. Define

$$\mathcal{O} := \bigoplus_{i=0}^{d-1} u^i \mathcal{O}' = \mathcal{O}' \oplus u\mathcal{O}' \oplus \ldots \oplus u^{d-1}\mathcal{O}'.$$

This is an additively closed subset of Λ. Since $\mathcal{O}'$ is multiplicatively closed and $\gamma \in \mathcal{O}'$, we conclude that $\mathcal{O}$ is a subring of Λ, and so discrete. Since $\mathcal{O}' \cdot \mathbb{Q} = L$, we have $\mathcal{O} \cdot \mathbb{Q} = \mathcal{A}$. Thus we have

Proposition 10. *Let $\mathcal{A}$ and $\mathcal{O}$ be as above. Then $\mathcal{O}$ is a suborder of $\mathcal{A}$.*

Ideal Lattice Problems in Non-maximal Orders. As shown above, when an order is maximal it has a 'nice' ideal theory closely related to the ideals of $\mathcal{O}_K$. When an order is not maximal, many of these properties (e.g. two-sided ideals forming an abelian group) may be lost. However, we can still define the standard lattice problems on lattices obtained from embeddings of these ideals.

Definition 21. Let $\mathcal{A}$ be a cyclic algebra, let $\mathcal{I}$ be ideal of an order $\mathcal{O}$, and let $0 < \delta < \lambda_1(\mathcal{I})/2$. Then the $\mathcal{A}$-BDD$_{\mathcal{O},\mathcal{I},\delta}$ problem, on input $y = x + e$ for $x \in \mathcal{I}$ and $e \in \bigoplus_{i=0}^{d-1} u^i L_{\mathbb{R}}$ satisfying $\|e\|_{2,\infty} \leq \delta$, is to compute x.

Definition 22. For any $q \geq 2$, the $q\mathcal{A}$-BDD$_{\mathcal{O},\mathcal{I},d}$ problem is as follows: given an instance of the $\mathcal{A}$-BDD$_{\mathcal{O},\mathcal{I},\delta}$ problem $y = x + e$ with solution $x \in \mathcal{I}$ and error $e \in \bigoplus_{i=0}^{d-1} u^i L_{\mathbb{R}}$ satisfying $\|e\|_{2,\infty} \leq \delta$, output $x \bmod q\mathcal{I}$.

Lemma 8. *For any $q \geq 2$ and order ideal $\mathcal{I}$, there is a deterministic polynomial time reduction from $\mathcal{A}$-BDD$_{\mathcal{O},\mathcal{I},\delta}$ to $q\mathcal{A}$-BDD$_{\mathcal{O},\mathcal{I},\delta}$*

Proof. Adapted from [32, Lemma 3.5], which is lattice preserving.   □

Cyclic Order LWE. We call CLWE over a possibly non-maximal order Cyclic Order LWE, COLWE. We define the COLWE distribution analogously to CLWE:

Definition 23. Let L/K be a Galois extension of number fields of dimension $[L : K] = d$ with cyclic Galois group generated by θ. Let $\mathcal{A} := (L/K, \theta, \gamma)$ be the resulting cyclic algebra with center K and invariant u with $u^d = \gamma \in \mathcal{O}_K$. Let $\mathcal{O} \subset \Lambda$ be a non-maximal order of $\mathcal{A}$. For an error distribution ψ over $\oplus_{i=0}^{d-1} u^i L_\mathbb{R}$, an integer modulus $q \geq 2$, and a secret $s \in \mathcal{O}_q^\vee$, a sample from the COLWE distribution $\Pi_{\mathcal{O},q,s,\psi}$ is obtained by sampling $a \leftarrow \mathcal{O}_q$ uniformly at random, $e \leftarrow \psi$, and outputting $(a, b) = (a, (a \cdot s)/q + e \mod \mathcal{O}^\vee) \in \mathcal{O}_q \times (\oplus_{i=0}^{d-1} u^i L_\mathbb{R})/\mathcal{O}^\vee$.

Definition 24. Let Ψ be a family of error distributions over $\oplus_{i=0}^{d-1} u^i L_\mathbb{R}$. The search COLWE problem, denoted by $\text{COLWE}_{\mathcal{O},q,s,\psi}$, is to recover s from a collection of independent samples from $\Pi_{\mathcal{O},q,s,\psi}$ for any $s \in \mathcal{O}_q^\vee$ and $\psi \in \Psi$.

Definition 25. Let Υ be a distribution on a family of error distributions over $\oplus_{i=0}^{d-1} u^i L_\mathbb{R}$. Let $U_\mathcal{O}$ denote the uniform distribution on $\left(\mathcal{O}_q, \left(\oplus_{i=0}^{d-1} u^i L_\mathbb{R} \right) /\mathcal{O}^\vee \right)$. Then, the decision COLWE problem, $\text{DCOLWE}_{\mathcal{O},q,\Upsilon}$, is given a collection of independent samples from either $\Pi_{q,s,\psi}$ for a random choice of $(s, \psi) \leftarrow U \left(\mathcal{O}_q^\vee \right) \times \Upsilon$ or from $U_\mathcal{O}$, decide which is the case with non-negligible advantage.

Security Reductions. A proof of the hardness of search COLWE from BDD over ideals in non-maximal orders requires one to restrict to invertible ideals (as is done for group ring LWE and for OLWE). We note in passing that a proof of the hardness of search COLWE from BDD over one-sided ideals in non-maximal orders is also plausible, possibly requiring further restrictions to the valid ideals.

The Technical Lemmas. We adapt the method of [8]. Let $\text{ass}_\mathcal{O}(\mathcal{I}) = \{\mathfrak{p}_i : \mathcal{I} \subset \mathfrak{p}_i\}$ be the *associated primes* of the ideal $\mathcal{I}$, where the $\mathfrak{p}_i$ are prime ideals of $\mathcal{O}$.

Lemma 9. *Let $\mathcal{I}$ be an invertible ideal of non-maximal order $\mathcal{O}$ and $\mathcal{J}$ be an integral ideal of $\mathcal{O}$. Then there exists a $t \in \mathcal{I} \cap \mathcal{O}_K$ such that the ideal $t \cdot \mathcal{I}^{-1} \subset \mathcal{O}$ is coprime to $\mathcal{J}$, and we can compute such a t efficiently given $\mathcal{I}$ and $\text{ass}_\mathcal{O}(\mathcal{J})$.*

Proof. Let $\{\mathfrak{p}_1, ..., \mathfrak{p}_r\} = \text{ass}_\mathcal{O}(\mathcal{J})$ and $t \in (\mathcal{I} \setminus \bigcup_i \mathfrak{p}_i \mathcal{I}) \cap \mathcal{O}_K$. Suppose $t \cdot \mathcal{I}^{-1} + \mathcal{J} \neq \mathcal{O}$. So $t \cdot \mathcal{I}^{-1} + \mathcal{J} \subset \mathcal{M}$ for maximal ideal $\mathcal{M} \subset \mathcal{O}$. Maximal ideals are prime, so $t \cdot \mathcal{I}^{-1}$ lies in an associated prime of $\mathcal{J}$. This implies $t \in \mathfrak{p}_i \mathcal{I}$ for some $\mathfrak{p}_i \in \text{ass}_\mathcal{O}(\mathcal{J})$, a contradiction. To construct such a t, take an $\mathcal{O}_K$ element in $\mathcal{I} \setminus \mathfrak{p}_i \mathcal{I}$ for all i, and compute the preimage under the CRT (see [33, §2.3]). □

Lemma 10. *Let $\mathcal{O}$ be as above. Let $\mathcal{I}, \mathcal{J}$ be ideals of $\mathcal{O}$, with $\mathcal{I}$ invertible, and $t \in \mathcal{I} \cap \mathcal{O}_K$ chosen such that $t \cdot \mathcal{I}^{-1}$ and $\mathcal{J}$ are coprime as ideals, and let $\mathcal{P}$ be an arbitrary fractional ideal of $\mathcal{O}$. Then, the function $\chi_t : \mathcal{A} \to \mathcal{A}$ defined as $\chi_t(x) = t \cdot x$ induces a module isomorphism from $\mathcal{P}/\mathcal{J} \cdot \mathcal{P} \to \mathcal{I} \cdot \mathcal{P}/\mathcal{I} \cdot \mathcal{J} \cdot \mathcal{P}$. Furthermore, if $\mathcal{J} = \langle q \rangle$ for an unramified prime $q \in \mathbb{Z}$ we can efficiently compute the inverse.*

Proof. The standard argument which relies on coprimality of the ideals. □

Hardness of the Search Problem from Invertible Ideals. We first state a lemma from [15], which enables the quantum step of the proof to hold:

Lemma 11. *There is an efficient quantum algorithm that given any nd^2 dimensional lattice $\mathcal{L} := \sigma_{\mathcal{A}}(\mathcal{I})$ for some ideal $\mathcal{I} \subset \mathcal{O}$, $0 < \delta < \lambda_1(\mathcal{L}^*)/(2\sqrt{2nd})$, and an oracle that solves $\mathcal{A}\text{-}BDD_{\mathcal{O},\mathcal{L}^*,\delta}$ with all but negligible probability, outputs an independent sample from $D_{\mathcal{L},\sqrt{d}\omega(\sqrt{\log(nd)})/\sqrt{2}\delta}$.*

We now prove an important lemma:

Lemma 12. *Let $\mathcal{A} = (L/K, \theta, \gamma)$ be a CDA constructed as above, and $\mathcal{O} \subset \Lambda$ be a non-maximal order. There is a ppt. algorithm that given an unramified prime $q \geq 2$, an invertible fractional $\mathcal{O}$-ideal $\mathcal{I}^\vee$, a $q\mathcal{A}\text{-}BDD_{\mathcal{O},\mathcal{I}^\vee,\alpha q \cdot \omega(\sqrt{\log(nd)})/\sqrt{2nd} \cdot r}$ instance $y = x + e$, a parameter $r \geq \sqrt{2}q \cdot \eta(\mathcal{I})$, and $D_{\mathcal{I},r'}$ samples with $r' \geq r$, outputs samples within negligible statistical distance of the COLWE distribution $\Pi_{\mathcal{O},q,s,\Sigma}$ for a secret $s = \chi_t(x \bmod q\mathcal{I}^\vee) \in \mathcal{O}_q^\vee$, where χ_t is as in Lemma 9 and Σ is an error distribution such that if $|\gamma| = 1$ the resulting error e'' has Gaussian marginal distribution in its i, j^{th} coordinate with parameter $r_{i,j} \leq \alpha$.*

Proof. First compute $t \in \mathcal{I}$ such that $\mathcal{I}^{-1} \cdot t$ and $q\mathcal{O}$ are coprime using the Lemma 9. We now create a sample from the COLWE distribution as follows: sample $z \leftarrow D_{\mathcal{I},r'}$, $e' \leftarrow D_{\alpha/\sqrt{2}}$, and compute a pair

$$(a,b) = \left(\chi_t^{-1}(z \bmod q\mathcal{I}), (z \cdot y)/q + e' \bmod \mathcal{O}^\vee\right) \in \mathcal{O}_q \times \bigoplus_{i=0}^{d-1} u^i L_\mathbb{R}/\mathcal{O}^\vee$$

We show that (a, b) is within negligible statistical distance of the COLWE distribution and s is uniformly random. First, note that $r \geq q \cdot \eta(\mathcal{I})$ so Lemma 1 implies the statistical distance between $z \bmod q\mathcal{I}$ and the uniform distribution is at most 2ε. As χ_t is bijective, $a = \chi_t^{-1}(z \bmod q\mathcal{I})$ is statistical distance 2ε of the uniform distribution over $\mathcal{O}_q$, as required. We now show that $b = a \cdot s/q + e''$, for an error e'' and uniform s, conditioned on some fixed a. We have

$$b = (z \cdot y)/q + e' \bmod \mathcal{O}^\vee$$
$$= (z \cdot x)/q + (z \cdot e)/q + e' \bmod \mathcal{O}^\vee,$$

so since $z = t \cdot a \bmod \mathcal{O}_q^\vee$ and t lies in the center of $\mathcal{A}$ it follows that $(z \cdot x)/q = (a \cdot t \cdot x)/q = (a \cdot s)/q \bmod \mathcal{O}^\vee$ for $s := \chi_t(x \bmod q\mathcal{I}^\vee)$ (this only holds for invertible ideals). Hence s is uniformly random over $\mathcal{O}_q^\vee$, if x is uniform over $\mathcal{I}^\vee$, since χ_t is a bijection. The analysis of the error is as in [15, Lemma 11]. $\square$

Combining Lemma 12 and Lemma 11, we arrive at the following:

Theorem 7. *Given a $COLWE_{\mathcal{O},q,\Sigma_\alpha}$ oracle for input $\alpha \in (0,1)$, $q \in \mathbb{Z}_{\geq 2}$, an ideal $\mathcal{I} \subset \mathcal{O}$, $r \geq \sqrt{2}q \cdot \eta(\mathcal{I})$ satisfying $r' = r \cdot \omega(\sqrt{\log N})/(\alpha q) > \sqrt{2N}/\lambda_1(\mathcal{I}^\vee)$, and polynomially many samples from the discrete Gaussian $D_{\mathcal{I},r}$ there exists an efficient quantum algorithm that outputs an independent sample from $D_{\mathcal{I},r'}$.*

Corollary 2. *Let $\mathcal{A}, \mathcal{O}, \alpha$ and q be as above. Then there is a polynomial-time quantum reduction from $\mathcal{A}\text{-}SIVP_\xi$ to $COLWE_{\mathcal{O},q,\Sigma_\alpha}$ when $\sqrt{8N}d\xi = \omega(\sqrt{dn})/\alpha$.*

Search to Decision Reduction. Here we adapt the standard search-to-decision reduction for structured LWE. Consider the following CRT-style decomposition:

Lemma 13 [30]. *Let Λ be the natural order of a cyclic division algebra $\mathcal{A} = (L/K, \theta, \gamma)$ with $\gamma \in \mathcal{O}_K$ and let $\mathcal{I}$ be an ideal of $\mathcal{O}_K$ which splits completely as $\mathcal{I} = \mathfrak{q}_1 \ldots \mathfrak{q}_n$ as an ideal of $\mathcal{O}_K$. Then, we have the isomorphism*

$$\Lambda/\mathcal{I}\Lambda \cong \mathcal{R}_1 \times \ldots \times \mathcal{R}_n$$

where $\mathcal{R}_i = \bigoplus_{j=0}^{d-1} u^j (\mathcal{O}_L/\mathfrak{q}_i\mathcal{O}_L)$ is the ring subject to relations $(\ell + \mathfrak{q}_i\mathcal{O}_L) u = u (\theta(\ell) + \mathfrak{q}_i\mathcal{O}_L)$ and $u^d = \gamma + \mathfrak{q}_i$.

When γ is a unit, $\Lambda^{\vee} = \bigoplus_i u^i \mathcal{O}_L^{\vee}$. The above lemma is also valid when each instance of $\mathcal{O}_L$ and Λ is replaced by its respective dual. For the following we will assume γ is a unit. In this case, as a consequence of Wedderburn's theorem, each $\mathcal{R}_i$ is isomorphic to the matrix ring $M_d(\mathbb{F}_q)$.

In order to obtain a search to decision reduction for LWE in suborders of Λ, we need to obtain a decomposition similar to that stated in the above lemma. Here we restrict our proof to a large class of suborders and prime moduli as follows, in order to guarantee such a decomposition.

Let $\mathcal{A} = (L/K, \theta, \zeta_n)$ be a CDA constructed as usual, and let $\mathcal{O}' \subset \mathcal{O}_L$ and $\mathcal{O} = \oplus_{i=0}^{d-1} u^i \mathcal{O}'$ be as above. Denote the conductor ideal of $\mathcal{O}'$ by $\mathfrak{c}$. Then

Proposition 11. *Let $\mathcal{A} = (L/K, \theta, \zeta_n)$, Λ, $\mathcal{O}' \subset \mathcal{O}_L$, and $\mathcal{O} = \oplus_{i=0}^{d-1} u^i \mathcal{O}'$ be as above. Let q be an integer prime, either completely split in K and inert in L/K, or completely split in L. If $\gcd(q\mathcal{O}', \mathfrak{c}) = 1$, then $\mathcal{O}/q\mathcal{O} \cong \Lambda/q\Lambda$.*

Proof. We have

$$\mathcal{O}/q\mathcal{O} = (\oplus_{i=0}^{d-1} u^i \mathcal{O}')/q(\oplus_{i=0}^{d-1} u^i \mathcal{O}') \cong \oplus_{i=0}^{d-1} u^i \mathcal{O}'/q\mathcal{O}'$$
$$\cong \oplus_{i=0}^{d-1} u^i \mathcal{O}_L/q\mathcal{O}_L \cong \Lambda/q\Lambda,$$

where the second isomorphism follows from Lemma 2.   □

This result means that we can use a decomposition of $\mathcal{O}/q\mathcal{O}$ into the direct product of matrix rings, provided that q is coprime to $\mathfrak{c}$, which we assume. The rest of the reduction is then identical to that of [15], and we obtain

Theorem 8. *Let $\mathcal{O} \subset \Lambda$ be a suborder of the natural order of a CDA $\mathcal{A} = (L/K, \theta, \zeta_n)$ as above, $q \in \mathrm{poly}(n)$ completely split in K such that $\gcd(q\mathcal{O}', \mathfrak{c}) = 1$, and $\alpha q \geq \eta_\varepsilon (\Lambda^{\vee})$ for a negligible $\varepsilon = \varepsilon(n)$. Then there is a ppt. reduction from $COLWE_{\mathcal{O}, q, \Sigma_\alpha, G}$ for any pairwise difference set $G \subset \Lambda_q^{\vee}$ to $DCOLWE_{\mathcal{O}, q, \Upsilon_\alpha}$.*

Above, a pairwise difference set G is a set $G = \prod_{i=1}^{n} G_i$ where each G_i is such that $g \neq h \in G_i$ implies $g - h$ is invertible. G is of size $|G| \leq q^{nd}$. We are currently unable to avoid this restriction. For more on this restriction, see [15].

COLWE Hardness from Other Ideals. The reduction from ideal lattice problems to search COLWE above used invertible ideals of suborders. Here we weaken this restriction, using analogous methods to [31]. We begin with:

Lemma 14. *Let $\mathcal{O} \subset \mathcal{A}$ be an order, $\mathcal{Q}$ and $\mathcal{I}$ be two-sided $\mathcal{O}$-ideals, and suppose that $(\mathcal{O} : \mathcal{I})_l (\mathcal{I} : \mathcal{O})_l + \mathcal{Q} = \mathcal{O}$. Then there exists $t \in (\mathcal{I} : \mathcal{O})_l \cap \mathcal{O}_K$ such that $(\mathcal{O} : \mathcal{I})_l t + \mathcal{Q} = \mathcal{O}$, and such a t can be found in polynomial time given $\mathcal{O}, \mathcal{I}$, and the associated $\mathcal{O}$-primes of $\mathcal{Q}$.*

Proof. Let $\{\mathfrak{p}_1, ..., \mathfrak{p}_r\} = \mathrm{ass}_{\mathcal{O}}(\mathcal{Q})$. Let $t \in ((\mathcal{I} : \mathcal{O})_l \setminus \cup_{i=1}^r \mathfrak{p}_i (\mathcal{I} : \mathcal{O})_l) \cap \mathcal{O}_K$. Then

$$(\mathcal{O} : \mathcal{I})_l t + \mathcal{Q} \subset (\mathcal{O} : \mathcal{I})_l (\mathcal{I} : \mathcal{O})_l + \mathcal{Q} = \mathcal{O}.$$

Suppose the containment is strict. Then $(\mathcal{O} : \mathcal{I})_l t + \mathcal{Q} \subset \mathfrak{m}$ for some maximal left ideal $\mathfrak{m}$ of $\mathcal{O}$. Since $\mathcal{Q}$ is two-sided, $\mathcal{Q} \subset \mathfrak{m}$ implies that the associated two-sided prime ideal of $\mathfrak{m}$ is one of the $\mathfrak{p}_i$. We thus have $(\mathcal{O} : \mathcal{I})_l t \subset \mathfrak{m}_i$. Recall that $(\mathcal{O} : \mathcal{I})_l$ is two-sided when $\mathcal{I}$ is two-sided; thus in fact $(\mathcal{O} : \mathcal{I})_l t \subset \mathfrak{p}_i$ also. Then

$$\mathcal{O}t = ((\mathcal{O} : \mathcal{I})_l (\mathcal{I} : \mathcal{O})_l + \mathcal{Q})t \subset (\mathcal{O} : \mathcal{I})_l (\mathcal{I} : \mathcal{O})_l t + \mathcal{Q}t.$$

Since $t \in \mathcal{O}_K$, it commutes with other algebra elements, so we have

$$\mathcal{O}t \subset (\mathcal{O} : \mathcal{I})_l t (\mathcal{I} : \mathcal{O})_l + \mathcal{Q}t$$
$$\subset \mathfrak{p}_i (\mathcal{I} : \mathcal{O})_l + \mathfrak{p}_i (\mathcal{I} : \mathcal{O})_l$$
$$= \mathfrak{p}_i (\mathcal{I} : \mathcal{O})_l,$$

which is a contradiction.

Finally, we show there exists such a t, that is, $((\mathcal{I} : \mathcal{O})_l \setminus \cup_{i=1}^r \mathfrak{p}_i (\mathcal{I} : \mathcal{O})_l) \cap \mathcal{O}_K$ is non-empty. It suffices by the CRT to show that $((\mathcal{I} : \mathcal{O})_l \setminus \mathfrak{p}_i (\mathcal{I} : \mathcal{O})_l) \cap \mathbb{Z}$ is non-empty for any i. Note that $(\mathcal{I} : \mathcal{O})_l = \mathcal{I}$, and recall that any ideal of $\mathcal{O}$ has finite index. Then the smallest non-zero integer contained in $\mathcal{I}$ is $|\mathcal{O}/\mathcal{I}|$, and that in $\mathfrak{p}_i\mathcal{I}$ is $|\mathcal{O}/\mathfrak{p}_i\mathcal{I}|$; but since $\mathfrak{p}_i$ is a proper prime ideal, we have $|\mathcal{O}/\mathcal{I}| < |\mathcal{O}/\mathfrak{p}_i\mathcal{I}|$, so we have $|\mathcal{O}/\mathcal{I}| \in ((\mathcal{I} : \mathcal{O})_l \setminus \mathfrak{p}_i (\mathcal{I} : \mathcal{O})_l) \cap \mathbb{Z}$. □

Lemma 15. *Let $\mathcal{O} \subset \mathcal{A}$ be an order, $\mathcal{Q}$ and $\mathcal{I}$ be $\mathcal{O}$-ideals, $\mathcal{J}$ be a fractional $\mathcal{O}$-ideal, and $t \in (\mathcal{I} : \mathcal{O})_l \cap \mathcal{O}_K$ such that $(\mathcal{O} : \mathcal{I})_l t + \mathcal{Q} = \mathcal{O}$. Then the map $\chi_t : \mathcal{A} \to \mathcal{A}, u \mapsto t \cdot u$ induces an $\mathcal{O}$-module isomorphism from $\mathcal{J}/\mathcal{J}\mathcal{Q}$ to $\mathcal{I}\mathcal{J}/\mathcal{I}\mathcal{J}\mathcal{Q}$.*

Proof. We follow [31, Lemma 2.14]. Consider the function $f : \mathcal{J} \to \mathcal{I}\mathcal{J}$ mod $\mathcal{I}\mathcal{J}\mathcal{Q}$ induced by multiplication by t. It is clearly an $\mathcal{O}$-module homomorphism. The kernel of f contains $\mathcal{J}\mathcal{Q}$, because $t \in (\mathcal{I} : \mathcal{O}) = \mathcal{I}$. We now show $\ker f = \mathcal{J}\mathcal{Q}$.

Suppose $tu \in \mathcal{I}\mathcal{J}\mathcal{Q}$ for some $u \in \mathcal{J}$. Then $(\mathcal{O} : \mathcal{I})_l tu \subset (\mathcal{O} : \mathcal{I})_l \mathcal{I}\mathcal{J}\mathcal{Q} \subset \mathcal{J}\mathcal{Q}$. Because $(\mathcal{O} : \mathcal{I})_l t + \mathcal{Q} = \mathcal{O}$, we find $u \in \mathcal{O}u = ((\mathcal{O} : \mathcal{I})_l t + \mathcal{Q}) u \subset \mathcal{J}\mathcal{Q} + \mathcal{Q}\mathcal{J} \subset \mathcal{J}\mathcal{Q}$, as desired. Thus f mod $\mathcal{J}\mathcal{Q}$ is injective.

We now construct a preimage for any $v \in \mathcal{I}\mathcal{J}$. Since $(\mathcal{O} : \mathcal{I})_l t + \mathcal{Q} = \mathcal{O}$ by assumption, we can find some $c \in (\mathcal{O} : \mathcal{I})_l t$ such that $c = 1$ mod $\mathcal{Q}$. Set $a = cv \in (\mathcal{O} : \mathcal{I})_l t \mathcal{I}\mathcal{J} \subset t\mathcal{J}$. Then $a - v = (c - 1)v \in \mathcal{I}\mathcal{J}\mathcal{Q}$. Now set $w = a \cdot t^{-1} \in \mathcal{J}$, so $\chi_t(w) = a = v$ mod $\mathcal{I}\mathcal{J}\mathcal{Q}$, and w mod $\mathcal{J}\mathcal{Q}$ is the preimage of v mod $\mathcal{I}\mathcal{J}\mathcal{Q}$. □

Lemma 16. *Let $\mathcal{O} \subset \Lambda$ be a non-maximal order, and $\mathcal{I}$ be a two-sided $\mathcal{O}$-ideal such that $\mathcal{O}_l(\mathcal{I}) = \mathcal{O}$. Then there is a ppt. algorithm that on input a prime $q \geq 2$, a $q\mathcal{A}\text{-}BDD_{\mathcal{O},\mathcal{I},\alpha q \cdot \omega(\sqrt{log(nd)})/\sqrt{2}ndr}$ instance $y = x + e$ with $x \in \mathcal{I}^{\vee}$, $r \geq \sqrt{2}q \cdot \eta(\mathcal{I})$, and samples from $D_{\mathcal{I},r}$ outputs samples that are within negligible statistical distance of the COLWE distribution $\Pi_{\mathcal{O},q,s,\Sigma}$ where $s = \chi_t(x \bmod q\mathcal{I}^{\vee}) \in \mathcal{O}_q^{\vee}$, χ_t is as in Lemma 15, and Σ is an error distribution as in Lemma 12.*

Proof. The proof is similar to that of Lemma 12. By Lemma 14, compute a $t \in (\mathcal{I} : \mathcal{O})_l \cap \mathcal{O}_K$ to obtain χ_t. Sample $z \leftarrow D_{\mathcal{I},r}$ and create the COLWE sample

$$(a,b) := \left(\chi_t^{-1}(z \bmod q\mathcal{I}), (z \cdot y)/q + e' \bmod \mathcal{O}_l(\mathcal{I})^{\vee}\right) \in \left(\mathcal{O}_q \times \left(\bigoplus_{i=0}^{d-1} u^i L_{\mathbb{R}}\right)/\mathcal{O}_l(\mathcal{I})^{\vee}\right)$$

where $e' \leftarrow D_{\alpha/\sqrt{2}}$. As before, these samples are within negligible statistical distance of the CLWE distribution and s is uniformly random. We have that $a \in \mathcal{O}_q$ is statistically close to uniform since $r \geq q \cdot \eta(\mathcal{I})$ and χ_t^{-1} is a bijection.

As for b, we show that it has the shape $(a \cdot s)/q + e''$ for an error e'' of the specified distribution, and uniformly random s. Observe that

$$b := (z \cdot y)/q + e' \bmod \mathcal{O}_l(\mathcal{I})^{\vee}$$
$$= (z \cdot x)/q + (z \cdot e)/q + e' \bmod \mathcal{O}_l(\mathcal{I})^{\vee}.$$

We now use χ_t with $\mathcal{J} = \mathcal{I}^{\vee}$, and obtain (by Lemma 5)

$$\chi_t : \mathcal{I}^{\vee}/\mathcal{I}^{\vee}q \to \mathcal{I}\mathcal{I}^{\vee}/\mathcal{I}\mathcal{I}^{\vee}q = (\mathcal{I} : \mathcal{I})_l^{\vee}/(\mathcal{I} : \mathcal{I})_l^{\vee}q = \mathcal{O}_l(\mathcal{I})^{\vee}/\mathcal{O}_l(\mathcal{I})^{\vee}q$$

Thus, since $z = t \cdot a \bmod \mathcal{O}_q^{\vee}$, setting $s = t \cdot x$, if s is uniformly distributed over $\mathcal{I}_q^{\vee}$, it is uniformly distributed over $\mathcal{O}_l(\mathcal{I})_q^{\vee} = \mathcal{O}_q^{\vee}$.

The distribution of the error can be analysed as in [15, Lemma 11].   □

6   Cyclic Learning with Rounding

In this section we extend Learning with Rounding to samples taken from the natural order of a CDA. We begin with the definition of the rounding function.

Definition 26. Let $q > p \in \mathbb{Z}_{\geq 2}$. Set $\mathbb{Z}_p := \mathbb{Z}/p\mathbb{Z}$. Define the function $\lfloor \cdot \rceil_p : \mathbb{Z}_q \to \mathbb{Z}_p$ by $\lfloor x \rceil_p = \lfloor \frac{p}{q} \cdot x \rceil \bmod p$, for all $x \in \mathbb{Z}_q$. We extend this to vectors component-wise, and to the ring case coefficient-wise, i.e. for $a = a_0 + a_1 x + ... + a_{n-1}x^{n-1} \in R_q$, a polynomial ring with coefficients in $\mathbb{Z}_q$, we have

$$\lfloor a \rceil_p = \left\lfloor \frac{p}{q} \cdot a_0 \right\rceil + \left\lfloor \frac{p}{q} \cdot a_1 \right\rceil x + ... + \left\lfloor \frac{p}{q} \cdot a_{n-1} \right\rceil x^{n-1} \in R_p.$$

Learning with Rounding. This function is then used to deterministically generate errors, as opposed to probabilistically sampling vectors or polynomials from an error distribution and adding this noise to a lattice point. We state the standard decision problems in the plain and module cases.

Definition 27. Let $s \in \mathbb{Z}_q^n$. Given uniformly random $a \leftarrow \mathbb{Z}_q^n$, output a $\mathrm{LWR}_{q,p}$ sample $(a, \lfloor \langle a \cdot s \rangle \rceil_p) \in \mathbb{Z}_q^n \times \mathbb{Z}_p$. Then the *decision*-$\mathrm{LWR}_{q,p}$ problem is: given m independent $\mathrm{LWR}_{q,p}$ samples, distinguish them with non-negligible probability from m samples $(a, u) \leftarrow U\left(\mathbb{Z}_q^n \times \mathbb{Z}_p\right)$.

Definition 28 (MLWR). Let $s \in R_q^n$ and $A \leftarrow U\left(R_q^{n \times d}\right)$. Output $(A, \lfloor A^t s \rceil_p) \in R_q^{n \times d} \times R_p^d$. The *decision* MLWR problem is: given m independent MLWR samples, distinguish them with non-negligible probability from m samples $(A, u) \leftarrow U(R_q^{n \times d} \times R_p^d)$.

To ensure that $\lfloor \langle a \cdot s \rangle \rceil_p$ is uniformly distributed, one takes p such that $p \nmid q$.

Cyclic Learning with Rounding. Here we extend the function $\lfloor \cdot \rceil_p$ to the natural order as follows: let $a \in \Lambda_q$, with $a = a_0 + u a_1 + \ldots + u^{d-1} a_{d-1}$, with every $a_i \in \mathcal{O}_L/q\mathcal{O}_L$, so we can write $a_i = a_{i,0} + a_{i,1} x + \ldots + a_{i,nd-1} x^{nd-1}$ with $a_{i,j} \in \mathbb{Z}_q$ for $j = 0, \ldots, nd-1$. Then we apply $\lfloor \cdot \rceil_p$ to a coefficient-wise.

Definition 29. Let $s \in \Lambda_q$ and $q >> p$. A $\mathrm{CLWR}_{q,p,s}$ sample is sampled by taking $a \leftarrow U(\Lambda_q)$ and outputting $(a, \lfloor a \cdot s \rceil_p) \in \Lambda_q \times \Lambda_p$. Then the *decision* CLWR ($\mathrm{DCLWR}_{q,p,s}$) problem is: given m independent $\mathrm{CLWR}_{q,p,s}$ samples, distinguish them with non-negligible probability from m samples $(a, u) \leftarrow U(\Lambda_q \times \Lambda_p)$.

The Hardness of LWR. There have been numerous attempts to obtain reductions from LWE to ensure the hardness of LWR. In [6], LWR was introduced, extended to rings (RLWR) and a proof given bounding the hardness of LWR by LWE. The proof relies on the distribution of the error being bounded, as stated in Definition 30, and on the modulus q being super-polynomial.

In [3], a proof is given for MLWR. Other papers giving reductions for LWR and its variants include [4,7,9], and [21].

The Hardness of CLWR. We adapt work done in [6], which shows that decision-LWR is at least as hard as decision-LWE, to cyclic algebras. This is done by applying the triangle inequality to the probability that an adversary can distinguish between samples obtained from various 'games'. Our proof too needs super-polynomial q. After preliminary definitions, we define five games.

Definition 30 [6, §3.1]. A probability distribution χ over $\mathbb{Z}$ is *B-bounded* if $\Pr_{x \leftarrow \chi}[|x| > B] \leq \mathrm{neg}(n)$. A distribution χ over a ring R is *B*-bounded if the marginal distribution of every coefficient (with respect to a fixed basis) of $x \leftarrow \chi$ is *B*-bounded. A distribution χ over Λ is *B*-bounded if the marginal distribution of every coefficient (with respect to the basis $\{u^i\}$) of $x \leftarrow \chi$ is *B*-bounded.

We extend this to the natural order as follows: a distribution over the natural order is B-bounded if the marginal distribution of every coefficient, with respect to the power basis $\{u^i\}$, is B-bounded, in the sense given above.

Definition 31 [6, §2.1]. The distinguishing advantage of an adversary $\mathcal{A}$ for games H_0, H_1 is $\mathrm{Adv}_{H_0,H_1}(\mathcal{A}) := |\Pr[\mathcal{A} \text{ accepts in } H_0] - \Pr[\mathcal{A} \text{ accepts in } H_1]|$.

The variant of CLWE we use below is called the 'primal' form in [14], which we denote $\mathrm{CLWE}_{q,s,\chi}$ and has a reduction from standard CLWE defined over $\Lambda^\vee$.

Let $f_B = b_0 + ub_1 + \ldots + u^{d-1}b_{d-1} \in \Lambda_q$ be such that each $b_i \in \mathcal{O}_{L_q}$ is B-bounded. Then if, for fixed b, $\lfloor b + f_B \rceil_p \neq \lfloor b \rceil_p$, we call this a *bad* event (denoted BAD below). We now define the following distinguishing games (as in [6]):

Game 0: Choose $s \in \Lambda_q$ and generate a number of CLWR samples upon the request of the attacker. The attacker must distinguish these from the same number of samples taken uniformly at random from $\Lambda_q \times \Lambda_p$.
Game 1: Choose $s \in \Lambda_q$. Upon the request of the attacker, generate $(a, b) = (a, a \cdot s + e) \in \Lambda_q \times \Lambda_q$ as in $\mathrm{CLWE}_{q,s,\chi}$, and output $(a, \lfloor b \rceil_p) \in \Lambda_q \times \Lambda_p$. In Game 1, if we encounter a bad event, we abort the game.
Game 2: Upon the request of the attacker, take $(a, b) \leftarrow U(\Lambda_q \times \Lambda_q)$ and output $(a, \lfloor b \rceil_p) \in \Lambda_q \times \Lambda_p$. If we encounter a bad event, we abort the game.
Game 3: Upon the request of the attacker, choose $(a, b) \leftarrow U(\Lambda_q \times \Lambda_q)$ and output $(a, \lfloor b \rceil_p) \in \Lambda_q \times \Lambda_p$ (note there is no condition on bad events occuring).
Game 4: Upon the request of the attacker, choose $(a, b) \leftarrow U(\Lambda_q \times \Lambda_p)$ and output this to the attacker.

Lemma 17. *Let $(a, b) \leftarrow U(\Lambda_q \times \Lambda_q)$ and denote Game i by G_i. Then $\Pr[BAD \text{ occurs on } b \text{ in } G_2] \leq (2B + 1) \cdot p \cdot nd^2/q$.*
Proof. For the case of plain LWR, that is $b \in \mathbb{Z}_q$, by [6] we have

$$\Pr[\text{BAD occurs on } b \text{ in } G_2] \leq (2B + 1) \cdot p/q. \qquad \square$$

Lemma 18. *Let $K/\mathbb{Q}$ be a cyclotomic field of power-of-2 degree, and L/K a cyclic Galois extension of power-of-2 degree. Then the statistical distance $\Delta\left(U\left(\Lambda_q^n \times \Lambda_p\right), U\left(\Lambda_q^n\right) \times \lfloor U\left(\Lambda_q\right) \rceil_p\right) \leq \mathrm{neg}(n)$ for some q exponential in p, n.*

In our cases, the quantities $|\Lambda_p|, |\Lambda_q|$ will always be powers of p and q respectively. We usually take q split completely in K, in which case $|\Lambda_q| = q^{nd^2}$.

Proof. See Appendix A. $\qquad \square$

Theorem 9. *Let χ be an efficiently sampleable B-bounded distribution over Λ and $q \geq pBd^2 \cdot n^{\omega(1)}$ such that $\Delta = \mathrm{neg}(n)$. For any distribution over $s \in \Lambda_q$, $DCLWR_{q,p,s}$ is at least as hard as $DCLWE_{q,s,\chi}$ for the same distribution over s.*
Proof. By Lemmas 17, 18, the same as [6, Theorem 3.2] mutatis mutandis. $\qquad \square$

Thus one could replace the sampled errors of CLWE with errors introduced by deterministic rounding (for specific parameters) with a certain level of confidence in the security of such a procedure. We note the popularity of rounding-based schemes [5,12], and leave open the development of CLWR-based schemes.

A   Proof of Lemma 18

Consider $P(\lfloor \mathcal{O}_{K_q} \rceil_p = x)$. Coefficients are rounded independently, so $P(\lfloor a \rceil_p = x) = \prod_{i=0}^{n-1} P(\lfloor a_i \rceil_p = x_i)$, for $a \in \mathcal{O}_{K_q}$ with $\mathbb{Z}_q$-coefficients a_i and x x_i. Since $P(\lfloor a_i \rceil_p = x_i) \in \{\frac{y}{q}, \frac{y+1}{q}\}$, $P(\lfloor a \rceil_p = x) \in \{\frac{y^n}{q^n}, \frac{y^{n-1}(y+1)}{q^n}, ..., \frac{y(y+1)^{n-1}}{q^n}, \frac{(y+1)^n}{q^n}\}$. Let $q = py + r$ with r minimal. The statistical distance is then

$$\Delta = \frac{1}{2}\left[r^n \left| \frac{1}{p^n} - \frac{(y+1)^n}{q^n} \right| + \binom{n}{1} r^{n-1}(p-r) \left| \frac{1}{p^n} - \frac{y(y+1)^{n-1}}{q^n} \right| + \dots \right.$$

$$\left. + \binom{n}{n-1} r(p-r)^{n-1} \left| \frac{1}{p^n} - \frac{y^{n-1}(y+1)}{q^n} \right| + (p-r)^n \left| \frac{1}{p^n} - \frac{y^n}{q^n} \right| \right]$$

We pair the first and last terms, and consider inner terms one by one. For the outer terms $\frac{1}{2}\left[r^n \left| \frac{1}{p^n} - \frac{(y+1)^n}{q^n} \right| + (p-r)^n \left| \frac{1}{p^n} - \frac{y^n}{q^n} \right| \right]$ $(*)$, note

$$(p-r)^n \left| \frac{1}{p^n} - \frac{y^n}{q^n} \right| = \left| \frac{(p-r)^n}{p^n} - \frac{(p-r)^n y^n}{q^n} \right| = \left| (1 - \frac{r}{p})^n - \frac{(py - ry)^n}{q^n} \right|$$

$$= \left| (1 - \frac{r}{p})^n - \frac{(q - r(y+1))^n}{q^n} \right| = \left| (1 - \frac{r}{p})^n - \left(1 - \frac{r(y+1)}{q}\right)^n \right|.$$

Since n is a power of two this is a difference of two squares. Factoring it as a product of a sum and difference, the difference is also a difference of two squares. We iterate this to factor out $|\frac{r(y+1)}{q} - \frac{r}{p}|$, which is bounded by $\frac{p}{q}$ by [6]. Formally,

$$\left| (1 - \frac{r}{p})^n - (1 - \frac{r(y+1)}{q})^n \right| = \left| \frac{r(y+1)}{q} - \frac{r}{p} \right| \cdot \left| 1 - \frac{r}{p} + 1 - \frac{r(y+1)}{q} \right|$$

$$\cdot \left| (1 - \frac{r}{p})^2 + (1 - \frac{r(y+1)}{q})^2 \right| \cdot \dots \cdot \left| (1 - \frac{r}{p})^{n/2} + \left(1 - \frac{r(y+1)}{q}\right)^{n/2} \right|.$$

We can do similarly for $r^n \left| \frac{1}{p^n} - \frac{(y+1)^n}{q^n} \right|$; we have, factorising,

$$r^n \left| \frac{1}{p^n} - \frac{(y+1)^n}{q^n} \right| = \left| \frac{r^n}{p^n} - \frac{r^n(y+1)^n}{q^n} \right| = \left| \left(\frac{r}{p}\right)^n - \left(\frac{r(y+1)}{q}\right)^n \right|$$

$$= \left| \frac{r}{p} - \frac{r(y+1)}{q} \right| \cdot \left| \frac{r}{p} + \frac{r(y+1)}{q} \right| \cdot \dots \cdot \left| \left(\frac{r}{p}\right)^{n/2} + \left(\frac{r(y+1)}{q}\right)^{n/2} \right|$$

Combining these, we find that the sum of the outer terms

$$(*) = \frac{1}{2}\left[\left| \frac{r}{p} - \frac{r(y+1)}{q} \right| \cdot \left| \frac{r}{p} + \frac{r(y+1)}{q} \right| \cdot \dots \cdot \left| (\frac{r}{p})^{n/2} + (\frac{r(y+1)}{q})^{n/2} \right| + \right.$$

$$\left. \left| \frac{r(y+1)}{q} - \frac{r}{p} \right| \cdot \left| 1 - \frac{r}{p} + 1 - \frac{r(y+1)}{q} \right| \cdot \dots \cdot \left| (1 - \frac{r}{p})^{n/2} + (1 - \frac{r(y+1)}{q})^{n/2} \right| \right].$$

We remove the previously mentioned factor:

$$(*) = \frac{1}{2} \Big[\Big| \frac{r}{p} - \frac{r(y+1)}{q} \Big| \cdot \Big[\Big| \frac{r}{p} + \frac{r(y+1)}{q} \Big| \cdot \ldots \cdot \Big| (\frac{r}{p})^{n/2} + (\frac{r(y+1)}{q})^{n/2} \Big| $$

$$+ \Big| 1 - \frac{r}{p} + 1 - \frac{r(y+1)}{q} \Big| \cdot \ldots \cdot \Big| (1 - \frac{r}{p})^{n/2} + (1 - \frac{r(y+1)}{q})^{n/2} \Big| \Big] \Big]$$

$$\leq \frac{p}{2q} \Big[\Big| \frac{r}{p} + \frac{r(y+1)}{q} \Big| \cdot \ldots \cdot \Big| (\frac{r}{p})^{n/2} + (\frac{r(y+1)}{q})^{n/2} \Big| $$

$$+ \Big| 1 - \frac{r}{p} + 1 - \frac{r(y+1)}{q} \Big| \cdot \ldots \cdot \Big| (1 - \frac{r}{p})^{n/2} + (1 - \frac{r(y+1)}{q})^{n/2} \Big| \Big].$$

A term inside the large brackets has equal powers of q on the numerator and denominator. Multiplying the bracket by $\frac{p}{q}$, it has as a polynomial in q degree -1, as required. For inner terms of the statistical distance, the ith term is $\frac{1}{2} \binom{n}{i} r^{n-i} (p-r)^i \Big| \frac{1}{p^n} - \frac{y^i(y+1)^{n-i}}{q^n} \Big|$. Since n is even, so is $\binom{n}{i}$. We can write

$$r^{n-i}(p-r)^i \Big| \frac{1}{p^n} - \frac{y^i(y+1)^{n-i}}{q^n} \Big| = \Big| \frac{r^{n-i}(p-r)^i}{p^n} - \frac{r^{n-i}(p-r)^i y^i(y+1)^{n-i}}{q^n} \Big|$$

$$= \Big| \frac{r^{n-i}(p-r)^i q^n - p^n(q - r(y+1))^i(r(y+1))^{n-i}}{p^n q^n} \Big|, \text{ and since}$$

$$p^n(q - r(y+1))^i(r(y+1))^{n-i} = (pq - pr(y+1))^i(pr(y+1))^{n-i}$$

$$= (pq - r(py+p))^i(r(py+p))^{n-i} = (pq - r(q-r+p))^i(r(q-r+p))^{n-i}$$

$$= (pq - rq - r^2 + rp))^i(rq - r^2 + rp))^{n-i} = ((q+r)(p-r))^i(rq + r(p-r))^{n-i}$$

we find that the q^n term has coefficient $(p-r)^i r^{n-i}$. Hence the numerator of $\frac{r^{n-i}(p-r)^i q^n - p^n(q - r(y+1))^i(r(y+1))^{n-i}}{p^n q^n}$ has q-degree $n-1$, and the denominator q-degree n, as required. Putting the above together, we find

$$\Delta = \frac{1}{2} \Big[r^n \Big| \frac{1}{p^n} - \frac{(y+1)^n}{q^n} \Big| + \binom{n}{1} r^{n-1}(p-r) \Big| \frac{1}{p^n} - \frac{y(y+1)^{n-1}}{q^n} \Big| + \ldots$$

$$+ \binom{n}{n-1} r(p-r)^{n-1} \Big| \frac{1}{p^n} - \frac{y^{n-1}(y+1)}{q^n} \Big| + (p-r)^n \Big| \frac{1}{p^n} - \frac{y^n}{q^n} \Big| \Big]$$

$$= \frac{1}{2} \Big[r^n \Big| \frac{1}{p^n} - \frac{(y+1)^n}{q^n} \Big| + (p-r)^n \Big| \frac{1}{p^n} - \frac{y^n}{q^n} \Big| +$$

$$nr^{n-1}(p-r) \Big| \frac{1}{p^n} - \frac{y(y+1)^{n-1}}{q^n} \Big| + \ldots + nr(p-r)^{n-1} \Big| \frac{1}{p^n} - \frac{y^{n-1}(y+1)}{q^n} \Big| \Big]$$

$$\leq \frac{p}{2q} \left[\left| \frac{r}{p} + \frac{r(y+1)}{q} \right| \cdot \ldots \cdot \left| \left(\frac{r}{p} \right)^{n/2} + \left(\frac{r(y+1)}{q} \right)^{n/2} \right| + \right.$$

$$\left. \left| 1 - \frac{r}{p} + 1 - \frac{r(y+1)}{q} \right| \cdot \ldots \cdot \left| \left(1 - \frac{r}{p} \right)^{n/2} + \left(1 - \frac{r(y+1)}{q} \right)^{n/2} \right| \right]$$

$$+ \frac{1}{2} \sum_{i=1}^{n-1} \binom{n}{i} \left| \frac{r^{n-i}(p-r)^i q^n - p^n (q - r(y+1))^i (r(y+1))^{n-i}}{p^n q^n} \right|,$$

and so by the previous analysis there exists some q' exponentially large in p and n such that for all $q \geq q'$, the considered statistical distance is negligible.

We now find $P(\lfloor \Lambda_q \rceil_p = x)$. This is the same as above, since the coefficients are rounded independently: namely, since by assumption $[\Lambda : \mathbb{Q}] = 2^r$ for some r, we have $P(\lfloor a \rceil_p = x) = \prod_{i=0}^{nd^2-1} P(\lfloor a_i \rceil_p = x_i)$ for any $a \in \Lambda_q$ with coefficients a_i. To find the statistical distance, the argument then proceeds identically.

References

1. Albert, A.A.: Structure of Algebras, vol. 24. AMS Colloquium Publications. American Mathematical Society (1939). ISBN: 9780821810248
2. Albrecht, M., Player, R., Scott, S.: On the concrete hardness of learning with errors. J. Math. Cryptol. **9**(3), 169–203 (2015). https://doi.org/10.1515/jmc-2015-0016
3. Alperin-Sheriff, J., Apon, D.: Dimension-preserving reductions from LWE to LWR. Cryptology ePrint Archive, Report 2016/589 (2016). https://eprint.iacr.org/2016/589
4. Alwen, J., Krenn, S., Pietrzak, K., Wichs, D.: Learning with rounding, revisited. In: Canetti, R., Garay, J.A. (eds.) CRYPTO 2013. LNCS, vol. 8042, pp. 57–74. Springer, Heidelberg (2013). https://doi.org/10.1007/978-3-642-40041-4_4
5. Avanzi, R., et al.: CRYSTALS-Kyber. CRYSTALS cryptographic suite for algebraic lattices (2021). https://pq-crystals.org/kyber/data/kyber-specification-round3-20210804.pdf
6. Banerjee, A., Peikert, C., Rosen, A.: Pseudorandom functions and lattices. In: Pointcheval, D., Johansson, T. (eds.) EUROCRYPT 2012. LNCS, vol. 7237, pp. 719–737. Springer, Heidelberg (2012). https://doi.org/10.1007/978-3-642-29011-4_42
7. Bogdanov, A., Guo, S., Masny, D., Richelson, S., Rosen, A.: On the hardness of learning with rounding over small modulus. In: Kushilevitz, E., Malkin, T. (eds.) TCC 2016. LNCS, vol. 9562, pp. 209–224. Springer, Heidelberg (2016). https://doi.org/10.1007/978-3-662-49096-9_9
8. Bolboceanu, M., Brakerski, Z., Perlman, R., Sharma, D.: Order-LWE and the hardness of Ring-LWE with entropic secrets. In: Galbraith, S.D., Moriai, S. (eds.) ASIACRYPT 2019. LNCS, vol. 11922, pp. 91–120. Springer, Cham (2019). https://doi.org/10.1007/978-3-030-34621-8_4
9. Chen, L., Zhang, Z., Zhang, Z.: On the hardness of the computational Ring-LWR problem and its applications. In: Peyrin, T., Galbraith, S. (eds.) ASIACRYPT 2018. LNCS, vol. 11272, pp. 435–464. Springer, Cham (2018). https://doi.org/10.1007/978-3-030-03326-2_15

10. Cheng, Q., Zhang, J., Zhuang, J.: LWE from non-commutative group rings. Des. Codes Crypt. **90**(1), 239–263 (2022). https://doi.org/10.1007/s10623-021-00973-6

11. Conrad, K.: The conductor ideal of an order (2018). https://kconrad.math.uconn.edu/blurbs/gradnumthy/conductor.pdf

12. D'Anvers, J.-P., Karmakar, A., Sinha Roy, S., Vercauteren, F.: Saber: Module-LWR based key exchange, CPA-secure encryption and CCA-secure KEM. In: Joux, A., Nitaj, A., Rachidi, T. (eds.) AFRICACRYPT 2018. LNCS, vol. 10831, pp. 282–305. Springer, Cham (2018). https://doi.org/10.1007/978-3-319-89339-6_16

13. Fesenko, I.B., Vostokov, S.V.: Local Fields and Their Extensions. Translations of Mathematical Monographs. American Mathematical Society (2002). ISBN: 978-0821832592

14. Grover, C.: LWE over cyclic algebras: a novel structure for lattice cryptography. Ph.D. thesis, Imperial College London (2020). https://spiral.imperial.ac.uk/handle/10044/1/97982

15. Grover, C., Mendelsohn, A., Ling, C., Vehkalahti, R.: Non-commutative ring learning with errors from cyclic algebras. J. Cryptol. **35** (2022). https://doi.org/10.1007/s00145-022-09430-6

16. Hasse, H.: Beweis eines Satzes und Wiederlegung einer Vermutung über das allgemeine Normenrestsymbol, German, pp. 64–69 (1931). https://www.digizeitschriften.de/id/252457811_1931|log1

17. Hollanti, C., Lahtonen, J., Lu, H.-F.: Maximal orders in the design of dense space-time lattice codes. IEEE Trans. Inf. Theory **54**(10), 4493–4510 (2008). https://doi.org/10.1109/TIT.2008.928998

18. Janusz, G.: Algebraic Number Fields. American Mathematical Society (1995). ISBN: 9780821872437

19. Kimball, M.: Quaternion algebras in number theory (2017). http://www2.math.ou.edu/~kmartin/quaint

20. Langlois, A., Stehlé, D.: Worst-case to average-case reductions for module lattices. Des. Codes Crypt. **75**(3), 565–599 (2015). https://doi.org/10.1007/s10623-014-9938-4. ISSN: 1573-7586

21. Liu, F.-H., Wang, Z.: Rounding in the rings. In: Micciancio, D., Ristenpart, T. (eds.) CRYPTO 2020. LNCS, vol. 12171, pp. 296–326. Springer, Cham (2020). https://doi.org/10.1007/978-3-030-56880-1_11

22. Lyubashevsky, V., Peikert, C., Regev, O.: On ideal lattices and learning with errors over rings. In: Gilbert, H. (ed.) EUROCRYPT 2010. LNCS, vol. 6110, pp. 1–23. Springer, Heidelberg (2010). https://doi.org/10.1007/978-3-642-13190-5_1

23. Manin, Y.I.: Classical computing, quantum computing, and Shor's factoring algorithm. Séminaire Bourbaki **41**, 375–404 (1998–1999). http://eudml.org/doc/110265

24. Marcus, D.: Number Fields. Universitext. Springer (1977). ISBN: 9783319902326

25. Meli, A.B.: Cyclotomic extensions and quadratic reciprocity (2013). http://math.uchicago.edu/~may/REU2013/REUPapers/Meli.pdf

26. Micciancio, D., Regev, O.: Worst-case to average-case reductions based on Gaussian measures. In: FOCS '04 (2004). https://doi.org/10.1109/FOCS.2004.72

27. Lahtonen, J., Markin, N., McGuire, G.: Construction of multiblock space-time codes from division algebras with roots of unity as nonnorm elements. IEEE Trans. Inf. Theory **54**(11), 5231–5235 (2008)

28. Neukirch, J.: Algebraic Number Theory. Grundlehren der mathematischen Wissenschaften. Springer, Heidelberg (2010). ISBN: 9783642084737

29. Oggier, F., Berhuy, G.: An Introduction to Central Simple Algebras and Their Applications to Wireless Communication. AMS Mathematical Surveys and Monographs. AMS (2013). ISBN: 978-0-8218-4937-8

30. Oggier, F., Sethuraman, B.A.: Quotients of orders in cyclic algebras and space-time codes. Adv. Math. Commun. **7**(4), 441–461 (2013)

31. Peikert, C., Pepin, Z.: Algebraically structured LWE, revisited. In: Hofheinz, D., Rosen, A. (eds.) TCC 2019. LNCS, vol. 11891, pp. 1–23. Springer, Cham (2019). https://doi.org/10.1007/978-3-030-36030-6_1

32. Regev, O.: On lattices, learning with errors, random linear codes, and cryptography. J. ACM **56** (2009). https://doi.org/10.1145/1568318

33. Reiner, I.: Maximal Orders. London Mathematical Society Monographs. Oxford University Press (2003). ISBN: 9780198526735

34. Serre, J.P.: Local Fields. Trans. by M.J. Greenberg. Graduate Texts in Mathematics. Springer, New York (2013). ISBN: 9781475756739

35. Sethuraman, B.A., Rajan, B.S., Shashidhar, V.: Full-diversity, high-rate space-time block codes from division algebras. IEEE Trans. Inf. Theory **49**(10), 2596–2616 (2003)

36. Shor, P.: Algorithms for quantum computation: discrete logarithms and factoring. In: FOCS 1994, pp. 124–134. IEEE Computer Society Press (1994)

37. Stehlé, D., Steinfeld, R., Tanaka, K., Xagawa, K.: Efficient public key encryption based on ideal lattices. In: Matsui, M. (ed.) ASIACRYPT 2009. LNCS, vol. 5912, pp. 617–635. Springer, Heidelberg (2009). https://doi.org/10.1007/978-3-642-10366-7_36

38. Washington, L.C.: Introduction to Cyclotomic Fields. Graduate Texts in Mathematics. Springer, New York (2012). ISBN: 9781461219347

From Worst to Average Case
to Incremental Search Bounds
of the Strong Lucas Test

Semira Einsele[✉] and Gerhard Wunder

Freie Universität Berlin, Taku Str. 9, 14195 Berlin, Germany
{semira.einsele,g.wunder}@fu-berlin.de

Abstract. The strong Lucas test is a widely used probabilistic primality test in cryptographic libraries. When combined with the Miller-Rabin primality test, it forms the Baillie-PSW primality test, known for its absence of false positives, undermining the relevance of a complete understanding of the strong Lucas test.

In primality testing, the worst-case error probability serves as an upper bound on the likelihood of incorrectly identifying a composite as prime. For the strong Lucas test, this bound is 4/15 for odd composites, not products of twin primes. On the other hand, the average-case error probability indicates the probability that a randomly chosen integer is inaccurately classified as prime by the test. This bound is especially important for practical applications, where we test primes that are randomly generated and not generated by an adversary.

The error probability of 4/15 does not directly carry over due to the scarcity of primes, and whether this estimate holds has not yet been established in the literature. This paper addresses this gap by demonstrating that an integer passing t consecutive test rounds, alongside additional standard tests of low computational cost, is indeed prime with a probability greater than $1 - (4/15)^t$ for all $t \geq 1$.

Furthermore, we introduce error bounds for the incremental search algorithm based on the strong Lucas test, as there are no established bounds up to date as well. Rather than independent selection, in this approach, the candidate is chosen uniformly at random, with subsequent candidates determined by incrementally adding 2. This modification reduces the need for random bits and enhances the efficiency of trial division computation further.

Keywords: Primality Test · Strong Lucas Test · Average Case Error Probability · Baillie-PSW Test · Incremental Search Bound

---

The research was supported by the 6G research cluster funded by the Federal Ministry of Education and Research (BMBF) in the programme of "Souverän. Digital. Vernetzt." Joint project 6G-RIC, project identification number: 16KISK020K.

1   Introduction

1.1   Primes and Primality Testing Algorithms in Cryptography

Other than being mathematically interesting, prime numbers are of great importance in cryptography. Many schemes in public key cryptography rely on choosing certain parameters as primes, exemplified by RSA and the Diffie-Hellmann key exchange protocol. These protocols come into play whenever we establish a VPN connection, use secure messaging Apps, or utilize smart cards for contactless payments. The consequences of prime parameter selection mistakes are potentially catastrophic. It is thus essential to have reliable primality testing algorithms that accurately determine whether a given number is composite or prime. These algorithms are integral components in nearly every cryptographic library or mathematical software system and serve two main purposes. First, they are essential for prime generation, a crucial step in many cryptographic protocols. To ensure the freshness of prime parameters, libraries often generate random integers of the desired bit length and test them for primality. Second, primality tests are used in validation and verification functions, such as in cryptographic libraries. These functions evaluate parameter sets, particularly those related to public key parameter sets for RSA or Diffie-Hellman, and perform various tests, including verifying the primality of a prime parameter.

1.2   Probabilistic and Deterministic Primality Tests

Until 2004, determining whether deterministic primality tests, capable of consistently distinguishing between primes and composites falls within the $\mathcal{P}$ complexity class without relying on any mathematical conjectures remained unsolved. This issue was ultimately solved in [1] with the introduction of a polynomial time algorithm. Nonetheless, while theoretically sound, they are impractical for everyday use, especially when confronted with the large inputs typical in cryptography, and probabilistic primality tests still outperform deterministic tests in terms of efficiency and remain the preferred choice in most scenarios. This preference persists despite the trade-off involving reduced accuracy. Furthermore, it has been demonstrated that those employed in practice have an error probability that is practically negligible. These tests operate as randomized Monte Carlo algorithms, ensuring always correct identification of prime numbers while rarely producing false positives. The independent nature of each test round allows for control over the probability of mistakenly classifying a composite as prime by increasing the number of rounds.

In the domain of primality tests, there are two key error probability categories: worst-case and average-case error probabilities. The *worst-case* error probability is the maximum probability that a composite integer will be mistakenly identified as a prime number by the test. Conversely, the *average-case* error probability considers the probability of a random number being misidentified as a prime number by the test. In adversarial scenarios, such as in the Diffie-Hellman key exchange protocol, the parameters could be chosen by an adversary. They might intentionally construct composites with a higher likelihood of being wrongly declared prime by the probabilistic primality test compared to randomly selected numbers. Therefore, it is essential for the primality test to exhibit a small worst-case error probability. However, in numerous other applications, understanding how the test performs in the average-case is more important. Typically, most randomly chosen composites are less likely to be accepted compared to those with the highest-probability of fooling the primality test.

1.3   The Miller-Rabin, Strong Lucas and Bailllie-PSW Test

The Miller-Rabin primality test is frequently utilized in cryptographic libraries and mathematical software due to its relatively straightforward implementation, efficient running time, and well-established error bounds, which have garnered trust within the cryptographic community. The work in [8] provides average-case error estimates, while the works in [13], [11] contribute worst-case error estimates. Detailed mathematical information about the test is available in Appendix A.1 for those interested.

Another notable probabilistic prim ality test is the Lucas test and its more stringent variant, the strong Lucas test. Similar to the Miller-Rabin test, there exist both worst-case [4] and average-case error estimates [9]. Besides being one of the main primality tests implemented in cryptolibraries, it gains importance due to its role in the Baillie-PSW test, which is a specific combination of the Miller-Rabin and Lucas test, see Algorithm 3. So far, no false positives passing this combined test have been identified, and the challenge of constructing a single concrete example remains an unsolved problem. Indeed, Gilchrist [10] computed the number of Baillie-PSW pseudoprimes up to 2^{64} and showed that there are none. Empirical data suggest that it seems very unlikely that integers of cryptographic size would be Baillie-PSW pseudoprimes. However, Pomerance [12] presents a heuristic argument positing the existence of infinitely many Baillie-PSW pseudoprimes. It is worth noting that the parameter selection of the test as implemented in practice is deterministic, hence, the result will remain constant. This can also be seen as an advantage since it eliminates the necessity for randomness. This reasoning undermines the significance of fully understanding the strong Lucas test.

The interested reader can refer to Appendix A.1 for details on the Fermat, Miller-Rabin and Lucas test. Appendix A.2 provides the algorithm for the Baillie-PSW test. Appendix A.3 provides a heuristic argument for their apparent independence. Lastly, Appendix A.4 discusses parameter selection for independence between the (strong) Lucas test and the Fermat/Miller-Rabin test.

1.4   The Algorithm Using the Strong Lucas Test

Let us introduce at following algorithm, which will be the main part of the first sections. It is a primality testing algorithm, based on the strong Lucas test, which will be discussed in Subsect. 2.3.

Algorithm 1: STRONGLUC(t, k)

Input: $t, k, D \in \mathbb{N}, k \geq 2$

Output: First probable prime found

1. Choose an odd k-bit integer n uniformly at random
2. If $\left(\frac{D}{n}\right) \neq -1$, discard n and go to step 1
3. If $(\gcd(D, n) \neq 1)$ or (n is divisible by any of the first 8 odd primes) or (Newton's method finds a square root for $n - \left(\frac{D}{n}\right)$): discard n, go to step 1
4. Else, execute the following loop:
   For $i = 1$ to t:
   – Perform the strong Lucas test to n with randomly chosen bases
   – If n fails any round of the test, discard n and go to step 1
   – Else output n and **stop**

In the context of algorithm STRONGLUC(t, k), let us define the following quantity.

Definition 1. *Let $q_{k,t}$ represent the probability that an integer selected by algorithm* STRONGLUC*(t, k) is composite. Here k represents the bit size of the integer, and t corresponds to the number of independent rounds conducted in the strong Lucas test. The average-case behaviour of the algorithm can then be defined as:*

$$q_{k,t} = \mathbb{P}(\text{STRONGLUC}(t, k) \text{ outputs a composite}). \tag{1}$$

It is noteworthy that OpenSSL already incorporates the practice of dividing by small primes before running the Miller-Rabin test to speed up prime generation. Hence, attaining the error bound of the algorithm, which includes trial division, typically incurs no additional computational costs. Furthermore, the occurrence of twin-prime products can also easily be avoided, as discussed in Subsect. 2.5. Therefore, achieving this error probability is not necessarily associated with an increase in running time.

1.5 Contributions

Arnault provided in [4] worst-case estimates of the Strong Lucas test. Specifically, he showed that the maximal probability of a composite number (which is not a product of twin primes, is relatively prime to D and distinct from 9, being falsely declared prime is at most $4/15$. So any composite passes t independent rounds of the strong Lucas tests with a probability less than or equal to $(4/15)^t$.

While it may seem logical to directly deduce that $q_{k,t} \leq \left(\frac{4}{15}\right)^t$ from this result, such a conclusion is wrong, as demonstrated in Sect. 2. The question of whether this estimate holds has not been answered yet.

The first contribution of this paper is to close this gap by proving that the bound $q_{k,t} \leq \left(\frac{4}{15}\right)^t$ is true for all $k \geq 2$ and $t \geq 1$.

In order to do so, we use a method introduced in [7] established for the Miller-Rabin test and adapt it for the strong Lucas test. For $k \geq 101$, we show that our claim is a trivial consequence of the average-case error results from [9]. In order to prove it for smaller values of k, we need to extend some of the results in [9], enabling us to confirm that $q_{k,t} \leq \left(\frac{4}{15}\right)^t$ for $k \geq 17$ and $t \geq 1$. For the values $2 \leq k \leq 16$, we compute $q_{k,t}$ exactly, which proves our claim.

The second contribution of this paper is the derivation of error bounds for an adapted algorithm known as "incremental search". Rather than uniformly selecting each candidate at random, the initial candidate is chosen using this approach, while all subsequent candidates are generated by incrementally adding 2. This approach provides the benefit of both conserving random bits and enhancing the efficiency of the trial division calculations. While similar bounds exist for the Miller-Rabin test [6], they have not been established for the strong Lucas test. Our work addresses this gap by providing the necessary error bounds.

2   Preliminaries

2.1   Lucas Sequences

Let $D, Q \in \mathbb{Z}$, and $P \in \mathbb{N}$ such that $D = P^2 - 4Q \geq 0$. Let $U_0(P,Q) = 0$, $U_1(P,Q) = 1$, $V_0(P,Q) = 2$ and $V_1(P,Q) = P$. The Lucas sequences $U_n(P,Q)$ and $V_n(P,Q)$ associated with the parameters P, Q are defined recursively for $n \geq 2$ by

$$
\begin{aligned}
U_n(P,Q) &= PU_{n-1}(P,Q) - QU_{n-2}(P,Q), \\
V_n(P,Q) &= PV_{n-1}(P,Q) - QV_{n-2}(P,Q).
\end{aligned}
\tag{2}
$$

2.2   The Lucas Test

For fixed $D \in \mathbb{Z}$ and $n \in \mathbb{N}$, let $\epsilon_D(n)$ denote the Jacobi symbol $\left(\frac{D}{n}\right)$. The following theorem is a more relaxed variant of our main theorem, which will be introduced in the next subsection.

Theorem 1 (Baillie, Wagstaff [5]). *Let P and Q be integers, and $D = P^2 - 4Q$. Let $U_p(P,Q)$ be the Lucas sequence of the first kind. If p is an odd prime such that $(p, QD) = 1$, then the following congruence holds*

$$
U_{p-\epsilon_D(p)} \equiv 0 \bmod p.
\tag{3}
$$

This theorem serves as a basis for the so called *Lucas (primality) test* which checks the congruence (3) for several randomly chosen bases P and Q. Composites p satisfying this congruence are called *Lucas pseudoprimes with parameters P and Q*, short *lpsp(P,Q)* However, similar to Carmichael numbers, which are composites that always pass the primality test based on Fermat's little theorem, a weaker variant of the Miller-Rabin theorem, there are composites, that completely defeat the Lucas test.

Definition 2 (Lucas-Carmichael numbers). *Let D be a fixed integer. A Lucas-Carmichael number is a composite number n, relatively prime to $2D$, such that for all integers P, Q with $\gcd(P, Q) = 1$, $D = P^2 - 4Q$ and $\gcd(n, QD) = 1$, n is a $lpsp(P, Q)$.*

The following theorem further highlights their resemblance to Carmichael numbers.

Theorem 2 (Williams [14]). *Let D be a fixed integer. Then n is a $lpsp(P, Q)$ if and only if n is square-free and $\epsilon_D(p_i) - 1 \mid \epsilon_D(n) - 1$ for every prime $p_i \mid n$.*

In fact, if n is a Lucas-Carmichael number with $D = 1$ or D being a perfect square, then n is a Carmichael number. Hence, any results about the infinitude of Lucas-Carmichael numbers, which is still an open question, would generalize the findings concerning Carmichael numbers [2], a result that itself took 84 years to prove.

2.3 The Strong Lucas Test

Since the Lucas test can never detect Lucas-Carmichael numbers as composites, slight modifications to the test can eliminate this misidentification. The following theorem will serve as the basis of the strong Lucas (primality) test.

Theorem 3 (Baillie, Wagstaff [5]). *Let P and Q be integers, and $D = P^2 - 4Q$. Let p be a prime number not dividing $2QD$. Write $p - \epsilon_D(p) = 2^\kappa q$, where q is odd. Then*

$$\text{either } p \mid U_q \text{ or } p \mid V_{2^i q} \text{ for some } 0 \leq i < \kappa. \tag{4}$$

By checking property (4) for many uniformly at random chosen bases P, Q with $1 \leq P, Q \leq n$, $\gcd(Q, n) = 1$ and $P = D^2 - 4Q$, we obtain a primality test called *the strong Lucas test*.

In Algorithm 1, introduced in Subsect. 1.4, we only considered integers for which $\epsilon_D(n) = -1$. The rationale behind this choice is explained in Appendix A.4.

2.4 Strong Lucas Pseudoprimes

While Theorem 3 is generally not applicable to composites, there exist specific bases P and Q for which the theorem holds. We call a composite number n relatively prime to $2QD$ that satisfies congruence (4) a *strong Lucas pseudoprime* with respect to the parameters P and Q, for short $slpsp(P, Q)$. Any $slpsp(P, Q)$ is also a $lpsp(P, Q)$ for the same parameter pair, but the converse is not necessarily true, see [5]. Therefore, the *strong* Lucas test is a more stringent test for primality.

Definition 3. *Let D and n be fixed integers and let $n - \epsilon(n) = 2^\kappa q$. We define $SL(D, n)$ to denote the number of pairs P, Q with $0 \leq P, Q < n$, $\gcd(Q, n) = 1$, and $P^2 - 4Q \equiv D \bmod n$, such that n satisfies (4). For an integer n, which is not relatively prime to $2D$, we set $SL(D, n) = 0$. Moreover, for n prime, we set $SL(D, n) = n - 1 - \epsilon(n)$.*

If n is composite, then $SL(D,n)$ counts the number of pairs P,Q that make n a $slpsp(P,Q)$.

From now on, let D always denote a random but fixed integer. For n such that $\gcd(D,2n) = 1$, Arnault [4] gave an exact formula on how many pairs P,Q with $0 \le P, Q < n$, $\gcd(Q,n) = 1$, and $P^2 - 4Q \equiv D \bmod n$ exist that make n a $slpsp(P,Q)$, if we know the prime decomposition of n.

Theorem 4 (Arnault [4]). *Let D be an integer and $n = p_1^{r_1} \dots p_s^{r_s}$ be the prime decomposition of an integer $n \ge 2$ relatively prime to $2D$. Put*

$$\begin{cases} n - \epsilon_D(n) = 2^\kappa q \\ p_i - \epsilon_D(p_i) = 2^{k_i} q_i \ \text{for } 1 \le i \le s \end{cases} \quad \text{with } q, q_i \ \text{odd,}$$

ordering the p_i's such that $k_1 \le \dots \le k_s$. The number of pairs P,Q with $0 \le P, Q < n$, $\gcd(Q,n) = 1$, $P^2 - 4Q \equiv D \bmod n$ and such that n is an $slpsp(P,Q)$ is expressed by the formula

$$SL(D,n) = \prod_{i=1}^{s}(\gcd(q,q_i) - 1) + \sum_{j=0}^{k_1-1} 2^{js} \prod_{i=1}^{s} \gcd(q,q_i). \tag{5}$$

2.5 Worst-Case Error Estimates

Arnault also gave worst-case error estimates.

Theorem 5 (Arnault [4]). *For every integer D and composite number n relatively prime to $2D$ and distinct from 9, we have*

$$SL(D,n) \le \frac{4n}{15},$$

except when n is the product of twin-primes. Then we have $SL(D,n) \le n/2$.

For certain types of twin-prime products, half of the bases P,Q declare the integer as a prime. Fortunately, excluding all twin-prime products from consideration does not impose a significant restriction. Whenever $n = p(p+2)$ with $\epsilon_D(n) = -1$ and p prime, we can rewrite $n - \epsilon_D(n) = (p+1)^2$, which is a perfect square. This can be efficiently detected, by for example, implementing Newton's method for square roots as a subroutine before executing the actual strong Lucas test.

2.6 Why Worst-Case Estimates Do Not Imply Average-Case Estimates

Transitioning from the worst-case to the average-case scenario, our algorithm guarantees the exclusion of integers divisible by the first l odd primes and those forming twin-prime products. This perspective enables us to characterize the algorithm as randomly sampling from a set that avoids these specific numbers. This observation leads to the following definition.

Definition 4. *For $k, l \in \mathbb{N}$ with $k \geq 2$ and, let $M_{k,l}$ denote the set of odd k-bit integers that are neither twin-prime products nor divisible by the first l odd primes.*

With this notation, we can express that the process outlined in Algorithm 1 effectively corresponds to a uniform sampling from the set $M_{k,8}$.

One might think that from Theorem 5 it would immediately follow that $q_{k,t} \leq \left(\frac{4}{15}\right)^t$, where $q_{k,t}$ was introduced in Definition 1. However, this reasoning is wrong since it does not take into account the distribution of primes, as the following discussion manifests.

Let X be the event that a number chosen uniformly at random from $M_{k,l}$ is composite, and let E_i denote the event that an integer chosen uniformly at random from $M_{k,l}$ passes the i-th round of the strong Lucas test. Moreover, let Y_t denote the event that this integer passes t consecutive rounds of the strong Lucas test with uniformly chosen bases. Hence, $Y_t = E_1 \cap E_2 \cap \cdots \cap E_t$. Theorem 5 states that $\mathbb{P}[Y_t \mid X] \leq \left(\frac{4}{15}\right)^t$. Critical to the estimation of $q_{k,t}$ is the value of $\mathbb{P}[X \mid Y_t]$, given that in fact $q_{k,t} = \mathbb{P}[X \mid Y_t]$. Naturally, we also have $\mathbb{P}[Y_t] \geq \mathbb{P}[X^c]$. Then, by Bayes' Theorem, we have

$$\mathbb{P}[X \mid Y_t] = \frac{\mathbb{P}[X]\mathbb{P}[Y_t \mid X]}{\mathbb{P}[Y_t]} \leq \frac{\mathbb{P}[Y_t \mid X]}{\mathbb{P}[Y_t]} = \frac{\mathbb{P}[E_1 \cap \cdots \cap E_t \mid X]}{\mathbb{P}[Y_t]}$$

$$= \frac{1}{\mathbb{P}[Y_t]} \prod_{i=1}^{t} \mathbb{P}[E_i \mid X] \leq \frac{1}{\mathbb{P}[Y_t]} \left(\frac{4}{15}\right)^t \leq \frac{1}{\mathbb{P}[X^c]} \left(\frac{4}{15}\right)^t,$$

where X^c denotes the complement of X.

We generally assume that primes in $M_{k,l}$ are scarce, which means that $\mathbb{P}[X^c]$ is small. This would imply that our estimate for $\mathbb{P}[X \mid Y_t]$ may be considerably larger than $\left(\frac{4}{15}\right)^t$ and close to 1. However, intuitively, $q_{k,t}$ is small, and in [9] explicit upper bounds have been established, confirming that in fact it is small. So this approach cannot be used to claim that $q_{k,t} \leq (4/15)^t$ for $k \geq 2$ and $t \geq 1$. In this work, however, we indeed show that this bound holds. We demonstrate this by establishing that the probability of an integer passing a round of the test, denoted as $\mathbb{P}[E_i \mid X]$, is significantly lower than $4/15$.

3 Proof Strategy

In this section, we outline our approach to proving the first main result.

Let us introduce the following quantity, which will be used frequently:

Definition 5. *For $n \in \mathbb{N}$, let*

$$\overline{\alpha}_D(n) = \frac{SL(D, n)}{n - \epsilon_D(n) - 1}$$

be the proportion of number of pairs P, Q that declare n to be a strong Lucas pseudoprime.

Let us first establish some important lemmas.

3.1   Important Lemma and Its Consequences

To prove the first main result, we use the next lemma, adapted from [7] for the strong Lucas test, where we choose r accordingly. For ease of notation, let $\sum'$ denote the sum over composites.

Lemma 1. *Let* $r, t, k \in \mathbb{N}$ *with* $r < t$ *and* $k \geq 2$. *Then*

$$q_{k,t} \leq \left(\frac{4}{15}\right)^{t-r} \frac{q_{k,r}}{1 - q_{k,r}}.$$

Proof. We follow Burthe's [7] proof step by step with adequate adaptations for the strong Lucas test. We include the proof for the sake of completeness. For every $n \in M_{k,l}$, we have $\overline{a}_D(n) \leq 4/15$, since twin-prime products by definition do not belong to $M_{k,l}$. Moreover, we have $\mathbb{P}[X \cap Y_i] = \frac{1}{|M_{k,l}|} \sum_{n \in M_{k,l}} \overline{a}_D(n)^i$. For $r < t$, we have

$$q_{k,t} = \mathbb{P}[X \mid Y_t] = \frac{\mathbb{P}[X \cap Y_t]}{\mathbb{P}[Y_t]} = \frac{\mathbb{P}[X \cap Y_t]}{\mathbb{P}[X \cap Y_{t-1}]} \frac{\mathbb{P}[X \cap Y_{t-1}]}{\mathbb{P}[X \cap Y_{t-2}]} \cdots \frac{\mathbb{P}[X \cap Y_{r+1}]}{\mathbb{P}[X \cap Y_r]} \frac{\mathbb{P}[X \cap Y_r]}{\mathbb{P}[Y_t]}.$$

We bound the fractions as follows:

$$\frac{\mathbb{P}[X \cap Y_i]}{\mathbb{P}[X \cap Y_{i-1}]} = \frac{\sum'_{n \in M_{k,l}} \overline{a}_D(n)^i}{\sum'_{n \in M_{k,l}} \overline{a}_D(n)^{i-1}} \leq \frac{\sum'_{n \in M_{k,l}} \frac{4}{15}\overline{a}_D(n)^{i-1}}{\sum'_{n \in M_{k,l}} \overline{a}_D(n)^{i-1}} = \frac{4}{15}.$$

This implies

$$q_{k,t} \leq \left(\frac{4}{15}\right)^{t-r} \frac{\mathbb{P}[X \cap Y_r]}{\mathbb{P}[Y_r]} \frac{\mathbb{P}[Y_r]}{\mathbb{P}[Y_t]} = \left(\frac{4}{15}\right)^{t-r} q_{k,r} \frac{\mathbb{P}[Y_r]}{\mathbb{P}[Y_t]}.$$

Primes in $M_{k,l}$ always pass the strong Lucas test, thus we have that $\mathbb{P}[X^c \cap Y_t] = \mathbb{P}[X^c] = \mathbb{P}[X^c \cap Y_r]$. Therefore,

$$\frac{\mathbb{P}[Y_r]}{\mathbb{P}[Y_t]} \leq \frac{\mathbb{P}[Y_r]}{\mathbb{P}[X^c \cap Y_t]} = \frac{\mathbb{P}[Y_r]}{\mathbb{P}[X^c \cap Y_r]} = \frac{1}{\mathbb{P}[X^c \mid Y_r]} = \frac{1}{1 - q_{k,r}},$$

which completes the proof. □

Thus, to establish $q_{k,t} \leq \left(\frac{4}{15}\right)^t$, it is sufficient to show that for any given $r, k \in \mathbb{N}$ and $k \geq 2$, we have that

$$q_{k,r} \leq \frac{1}{1 + \left(\frac{15}{4}\right)^r} \quad \text{and} \quad q_{k,r'} \leq \left(\frac{4}{15}\right)^{r'} \text{ for all } r' < r,$$

since, by utilizing Lemma 1, we can then conclude $q_{k,t} \leq \left(\frac{4}{15}\right)^t$ for all $t, k \in \mathbb{N}$ and $k \geq 2$.

Now, let $\pi(x)$ denote the prime counting function up to x, and let p always denote a prime. Using the law of conditional probability, we have

$$q_{k,r} = \mathbb{P}[X \mid Y_r] = \frac{\mathbb{P}[X \cap Y_r]}{\mathbb{P}[Y_r]} = \frac{\displaystyle\sideset{}{'}\sum_{n \in M_{k,l}} \overline{\alpha}_D(n)^r}{\displaystyle\sum_{n \in M_{k,l}} \overline{\alpha}_D(n)^r}$$

$$= \frac{\displaystyle\sideset{}{'}\sum_{n \in M_{k,l}} \overline{\alpha}_D(n)^r}{\displaystyle\sideset{}{'}\sum_{n \in M_{k,l}} \overline{\alpha}_D(n)^r + \sum_{p \in M_{k,l}} 1} = \frac{\displaystyle\sideset{}{'}\sum_{n \in M_{k,l}} \overline{\alpha}_D(n)^r}{\displaystyle\sideset{}{'}\sum_{n \in M_{k,l}} \overline{\alpha}_D(n)^r + \pi(2^k) - \pi(2^{k-1})}.$$

Given that the expression $\frac{x}{x+\pi(2^k)-\pi(2^{k-1})}$ is a monotonically increasing function in x, our goal is to find suitable values N_r and P such that

$$\sideset{}{'}\sum_{n \in M_{k,l}} \overline{\alpha}_D(n)^r \leq N_r \quad \text{and} \quad P \leq \pi(2^k) - \pi(2^{k-1})$$

With these choices, we can express the bound for $q_{k,r}$ as:

$$q_{k,r} \leq \frac{N_r}{N_r + P}. \tag{6}$$

For a fixed $r \in \mathbb{N}$, our aim is to demonstrate that

$$\frac{N_r}{N_r + P} \leq \frac{1}{1 + \left(\frac{15}{4}\right)^r}.$$

3.2   Lower Bound P

The following result serves as our lower bound P:

Proposition 1. *For an integer $k \geq 8$, we have*

$$\pi(2^k) - \pi(2^{k-1}) > (0.71867)\frac{2^k}{k}.$$

Proof. For $k \geq 21$ this is proven in [8]. By running a Python program that computes the actual value of $\pi(2^k) - \pi(2^{k-1})$ for $k \leq 20$, we can show that the proposition is true for all $k \geq 8$. □

Let us now turn to the upper bound N_r.

3.3   Upper Bound N_r

We aim to find an upper bound N_r for $\sum\limits_{n \in M_{k,l}}' \overline{\alpha}_D(n)^r$. First, let us introduce the following set:

Definition 6. *Let*

$$C_{m,D} = \{n \in \mathbb{N} : \gcd(n, 2D) = 1, n \text{ composite and } \alpha_D(n) > 2^{-m}\}.$$

Arnault proved in [3] that $\alpha_D(n) \leq \frac{1}{4}$, so we know that $C_{1,D} = C_{2,D} = \emptyset$. The set $C_{3,D}$ has been classified in [9].

To get our upper bound N_r, we use a number-theoretic function introduced by Arnault. This function serves as a variant of the well-known Euler's totient function $\varphi(n)$, and will play a crucial role in the subsequent analysis.

Definition 7 (Arnault [3]). *Let D be an integer. The following function is defined only on integers relatively prime to $2D$:*

$$\begin{cases} \varphi_D(p^r) = p^{r-1}(p - \epsilon_D(p)) \text{ for any prime } p \nmid 2D \text{ and } r \in \mathbb{N}, \\ \varphi_D(p_1 p_2) = \varphi_D(p_1)\varphi_D(p_2) \text{ if } \gcd(p_1, p_2) = 1. \end{cases}$$

Moreover, for odd $n \in \mathbb{N}$, let

$$\alpha_D(n) = \frac{SL(D, n)}{\varphi_D(n)}.$$

In our analysis, we seek to bound $\overline{\alpha}_D(n)$. However, unlike $\varphi(n)$, $\varphi_D(n)$ is not necessarily bounded by n. Consequently, we cannot straightforwardly establish that $\overline{\alpha}_D(n) \leq \alpha_D(n)$. Nonetheless, the next lemma offers an upper bound and implies that, for sufficiently large l, these two quantities become closely aligned. Before looking into the lemma, let us introduce the following definition:

Definition 8. *For $l \in \mathbb{N}$, let $\tilde{p}_l$ denote the l-th odd prime and $\rho_l = 1 + \frac{1}{\tilde{p}_{l+1}}$.*

Now we can state the lemma, which connects the two functions.

Lemma 2 (Einsele, Paterson [9]). *Let $k, m, l \in \mathbb{N}$ and $n \in C_{m,D} \cap M_{k,l}$ be relatively prime to $2D$. Then,*

$$\overline{\alpha}_D(n) \leq \rho_l^m \alpha_D(n).$$

Lemma 3. *For $k, r, M, l \in \mathbb{N}$ with $3 \leq M \leq 2\sqrt{k-1} - 1$, we have*

$$\sum\limits_{n \in M_{k,l}}' \overline{\alpha}_D(n)^r \leq 2^r |M_{k,l}| \sum\limits_{m=M+1}^{\infty} \left(\frac{\rho_l}{2}\right)^{mr} + 2^r \sum\limits_{m=2}^{M} \left(\frac{\rho_l}{2}\right)^{mr} |C_{m,D} \cap M_{k,l}|.$$

Proof. We use Lemma 2 in the proof of Proposition 20 in [9] and get

$$\sideset{}{'}\sum_{n \in M_{k,l}} \overline{\alpha}_D(n)^r \leq \sum_{m=2}^{\infty} \sum_{n \in M_{k,l} \cap C_{m,D} \setminus C_{m-1,D}} \rho_l^{mr} \alpha_D(n)^r. \tag{7}$$

With $n \in C_{m,D} \cap C_{m-1,D}$ we have $\alpha_D(n) \leq 2^{-(m-1)}$. We use this in (7), split our sum in two parts, bound $|C_{m,D} \cap M_{k,l}| \leq |M_{k,l}|$ for the first sum to prove our claim.   □

Definition 9. *Let $\tilde{M}_{k,l}$ denote the set of k-bit integers that are not divisible by the first odd l primes.*

We have the following bound:

Lemma 4. *For $k \geq 12$, we have*

$$2^{k-2.92} \leq |\tilde{M}_{k,2}| \leq 2^{k-2.9}.$$

Proof. Let A_i be the set of k-bit integers divisible by i. By the inclusion-exclusion principle, we have

$$\left| \bigcup_{i=2,3,5} A_i \right| = |A_2| + |A_3| + |A_5| - |A_2 \cap A_3| - |A_2 \cap A_5| - |A_3 \cap A_3| + |A_2 \cap A_3 \cap A_5|$$

$$= \left\lfloor \frac{2^{k-1}}{2} \right\rfloor + \left\lfloor \frac{2^{k-1}}{3} \right\rfloor + \left\lfloor \frac{2^{k-1}}{5} \right\rfloor - \left\lfloor \frac{2^{k-1}}{6} \right\rfloor - \left\lfloor \frac{2^{k-1}}{10} \right\rfloor - \left\lfloor \frac{2^{k-1}}{15} \right\rfloor + \left\lfloor \frac{2^{k-1}}{30} \right\rfloor$$

$$\geq \frac{2^{k-1}}{2} + \frac{2^{k-1}}{3} - 1 + \frac{2^{k-1}}{5} - 1 - \frac{2^{k-1}}{6} - \frac{2^{k-1}}{10} - \frac{2^{k-1}}{15} + \frac{2^{k-1}}{30} - 1$$

$$= \frac{11}{15} 2^{k-1} - 3. \tag{8}$$

With the same argument, we have that

$$\left| \bigcup_{i=2,3,5} A_i \right| \leq \frac{2^{k-1}}{2} + \frac{2^{k-1}}{3} + \frac{2^{k-1}}{5} - \frac{2^{k-1}}{6} - \frac{2^{k-1}}{10} - \frac{2^{k-1}}{15} + \frac{2^{k-1}}{30} + 3$$

$$= \frac{11}{15} 2^{k-1} + 3 \tag{9}$$

By inequality (8), we have

$$|\tilde{M}_{k,2}| = 2^{k-1} - \left| \bigcup_{i=2,3,5} A_i \right| \leq \frac{4}{15} 2^{k-1} + 3 \leq 2^{k-2.9},$$

for $k \geq 12$. Moreover, we have by inequality (9) that

$$|\tilde{M}_{k,2}| = 2^{k-1} - \left| \bigcup_{i=2,3,5} A_i \right| \geq \frac{4}{15} 2^{k-1} - 3 \geq 2^{k-2.92}$$

for $k \geq 12$.   □

The next lemma is easily established and gives us our first upper bound N_1, as described in inequality (6).

Lemma 5. *For $k, M, l \in \mathbb{N}$ with $3 \le M \le 2\sqrt{k-1} - 1$, we have*

$$\sum_{n \in M_{k,l}}' \overline{\alpha}_D(n) \le 2^{k-1.9-M} \frac{\rho_l^{M+1}}{2 - \rho_l} + 2^{k-2\sqrt{k-1}+1} \rho_l^M M(M-1).$$

Proof. We use the bound from Lemma 3, and set $r = 1$. From the proof of Theorem 13 in [9], we have that $|C_{m,D} \cap M_k| \le \sum_{j=2}^{m} 2^{k+1+m-j-\frac{k-1}{j}}$. Moreover, with $M_{k,l} \subseteq M_k$, we have $|C_{m,D} \cap M_{k,l}| \le |C_{m,D} \cap M_k|$.

We now follow the proof of Proposition 26 in [9]. For the first part of the sum, we obtain $\sum_{m=M+1}^{\infty} \left(\frac{\rho_l}{2}\right)^m = \frac{2^{-M}\rho_l^{M+1}}{2-\rho_l}$ and bound $|M_{k,l}| \le |M_{k,2}|$ using Lemma 4. For the second part of the sum, we use the same argument as in Proposition 26 in [9], which concludes the proof. □

To get a good estimate for N_1, every M that satisfies $3 \le M \le 2\sqrt{k-1} - 1$ is a free parameter and will for each k be chosen such that it tightens the bound.

However, for values $k < 60$, we will see that we need a tighter bound.

To get an even tighter bound, we need to do more analysis. The next result serves as an estimate of the number of odd integers, which are both in $M_{k,l}$ and $C_{m,D}$.

Proposition 2. *If m, k are positive integers with $m + 1 \le 2\sqrt{k-1}$, then*

$$|C_{m,D} \cap M_{k,l}| \le 2^k \sum_{j=2}^{m} \frac{2^{m+1-j} - 1}{2^{\frac{k-1}{j}} - 1}.$$

Proof. From the proof of Theorem 13 in [9], we have that $|C_{m,D} \cap M_k| \le 2^k \sum_{j=2}^{m} \frac{2^{m+1-j}-1}{2^{\frac{k-1}{j}}-1}$. The proposition follows with $M_{k,l} \subseteq M_k$. □

The next result will serve as our second bound N_1 used in inequality (6).

Lemma 6. *For $k, M, l \in \mathbb{N}$ with $3 \le M \le 2\sqrt{k-1} - 1$, we have*

$$\sum_{n \in M_{k,l}}' \overline{\alpha}_D(n) \le 2^{1-M} \frac{\rho_l^{M+1}}{2 - \rho_l} |M_{k,l}| + 2^k \sum_{m=2}^{M} \sum_{j=2}^{m} \left(\frac{\rho_l}{2}\right)^m \frac{2^{m+1-j} - 1}{2^{\frac{k-1}{j}} - 1}.$$

Proof. We use Lemma 3 with $r = 1$. We have $\sum_{m=M+1}^{\infty} \left(\frac{\rho_l}{2}\right)^m = \frac{2^{-M}\rho_l^{M+1}}{2-\rho_l}$ and use Proposition 2 to get the desired result. □

Similarly to Lemma 5, we choose for each k an M that it minimizes our bound.

For small values of k, we can compute the quantity $|M_{k,l}|$ exactly. However, for large values of k, the precise computation is not feasible. Hence, it will be useful to upper bound this quantity. We use the trivial fact that $|M_{k,l}| \le |M_{k,2}|$. In the next section, we shall use these preliminary results for bounding $q_{k,t}$.

4 Intermediate Result

We now establish the main theorem for all $k \geq 42$. However, for smaller values of k, further analysis is required, and this will be discussed in Sects. 5 and 6.

For values of $k \geq 101$, we use the following theorem to establish our claim.

Theorem 6 (Einsele, Paterson [9]). *For $k, l \in \mathbb{N}$ with $k \geq 2$, we have*

$$q_{k,1} < k^2 4^{1.8-\sqrt{k}} \rho_l^{2\sqrt{k-1}-2}.$$

Let us introduce an intermediate result that holds for all $t \geq 1$ and $k \geq 42$.

Theorem 7. *For all $t \geq 1$ and $k \geq 42$ we have*

$$q_{k,t} \leq \left(\frac{4}{15}\right)^t.$$

Proof. We let $r = 1$ in Lemma 1. It suffices to show that $q_{k,1} \leq 4/19$ for all $k \geq 42$ to prove the theorem. Theorem 6 gives us $q_{k,1} \leq \frac{4}{19}$ for each $k \geq 101$, since $k^2 4^{1.8-\sqrt{k}} \rho_l^{2\sqrt{k-1}-2}$ is a strictly decreasing function for $k \geq 10$ and $q_{101,1} \leq \frac{4}{19}$. Thus, we get $q_{k,t} \leq \left(\frac{4}{15}\right)^t$ for $k \geq 101$. By inequality (6), we have $q_{k,1} \leq \frac{N_1}{N_1+P}$, where N_1 is our upper bound for $\sum'_{n \in M_{k,l}} \overline{\alpha}_D(n)$ and P our lower bound for $\pi(2^k) - \pi(2^{k-1})$. Proposition 1 serves as our value for P, and Lemma 5 as our first value for N_1. For each k, we take $3 \leq M \leq 2\sqrt{k-1}-1$ to be the positive integer that minimizes our upper bound for $q_{k,1}$. Table 1 displays the results of our computations, proving that $q_{k,1} < 4/19 < 0.210527$ for $60 \leq k \leq 100$.

Table 1. All $k < 101$ such that $u_k < 4/19$ using M_{opt}, where u_k is the upper bound for $q_{k,1}$ and M_{opt} is the optimal value for M minimizing u_k, using Lemma 5 for the bound for N_1 and $P = (0.71867)2^k/k$.

k	M_{opt}	u_k	k	M_{opt}	u_k	k	M_{opt}	u_k
60	9	0.204541	75	10	0.113743	89	12	0.063918
61	9	0.196467	76	10	0.109708	90	12	0.061105
62	9	0.188917	77	11	0.105817	91	12	0.058467
63	9	0.181868	78	11	0.101064	92	12	0.055994
64	9	0.175296	79	11	0.096609	93	12	0.053676
65	9	0.169176	80	11	0.092435	94	12	0.0515047
66	9	0.163486	81	11	0.088527	95	12	0.0494708
67	9	0.158204	82	11	0.084870	96	12	0.0475661
68	10	0.151309	83	11	0.081449	97	12	0.0457829
69	10	0.144807	84	11	0.078251	98	12	0.044114
70	10	0.138718	85	11	0.075262	99	13	0.043620
71	10	0.133020	86	11	0.072471	100	13	0.040361
72	10	0.127693	87	11	0.069865			
74	10	0.122717	88	12	0.066918			

For the remaining values of k, we use Lemma 6 for determining N_1. We bound $|M_{k,l}|$ using Lemma 4, and for each k we choose $3 \leq M \leq 2\sqrt{k-1} - 1$ to be the positive integer minimizing our upper bound given by inequality (6) for $q_{k,1}$. Table 2 shows the results of the computations, proving $q_{k,1} \leq 4/19$ for $k \geq 42$.

Table 2. All $k < 60$ such that $u_k < 4/19$ using M_{opt}, where u_k is the upper bound for $q_{k,1}$ and M_{opt} is the optimal value for M that minimizes u_k, where we used the bound for N_1 given by Lemma 6 and for $P = (0.71867)2^k/k$.

k	M_{opt}	u_k	k	M_{opt}	u_k	k	M_{opt}	u_k
42	8	0.199683	48	8	0.154860	54	9	0.115831
43	8	0.189917	49	9	0.147791	55	9	0.111229
44	8	0.181164	50	9	0.140038	56	9	0.107117
45	8	0.173352	51	9	0.133018	57	10	0.102671
46	8	0.166410	52	9	0.126677	58	10	0.097171
47	8	0.160268	53	9	0.120964	59	10	0.092159

$\square$

5   Main Result

In this section, we show the main theorem for $k \geq 18$ and $t \geq 1$. In Sect. 6, we compute $q_{k,t}$ exactly using an equation proven by Arnault, proving that the theorem holds for $k \geq 2$ altogether.

Theorem 8. *For all $t \geq 1$ and $k \geq 18$ we have*

$$q_{k,t} \leq \left(\frac{4}{15}\right)^t.$$

Theorem 7 only proves our claim for $k \geq 42$. As the last two terms in the second sum of Lemma 6 dominate our estimate, we divide the set $M_{k,l}$ into two disjoint sets, where the sum terminates earlier for one set, enabling us to lower the value of k. Let $M_{k,l,d_1} \subseteq M_{k,l}$ denote the subset consisting of square-free integers having a prime $p \mid n$ such that $\frac{(p-\epsilon_D(p), n-\epsilon_D(n))}{p-\epsilon_D(p)} \geq \frac{1}{3}$. Let $M_{k,l,d_2} \subseteq M_{k,l}$ denote the subset consisting of all integers for which for every prime $p \mid n$ we have $\frac{(p-\epsilon_D(p), n-\epsilon_D(n))}{p-\epsilon_D(p)} < \frac{1}{3}$, unified with the set of non-square free integers.

Let X_{d_1} denote the event that a number chosen uniformly at random from $M_{k,l}$ is composite and lies in M_{k,l,d_1}, and X_{d_2} that a number chosen uniformly at random from $M_{k,l}$ is composite and lies in M_{k,l,d_2}. Let N_{r,d_1} be the upper bound for $\sum'_{n \in M_{k,l,d_1}} \overline{\alpha}_D(n)^r$ and N_{r,d_2} for $\sum'_{n \in M_{k,l,d_2}} \overline{\alpha}_D(n)^r$ respectively. Then,

$$q_{k,r} = \frac{\mathbb{P}[X \cap Y_r]}{\mathbb{P}[Y_r]} = \frac{\mathbb{P}[(X_{d_1} \cup X_{d_2}) \cap Y_r]}{\mathbb{P}[Y_r]} = \frac{\mathbb{P}[X_{d_1} \cap Y_r] + \mathbb{P}[X_{d_2} \cap Y_r]}{\mathbb{P}[Y_r]}$$

$$= \frac{\displaystyle\sum_{n \in M_{k,l,d_1}}' \overline{\alpha}_D(n)^r + \sum_{n \in M_{k,l,d_2}}' \overline{\alpha}_D(n)^r}{\sum_{n \in M_{k,l}} \overline{\alpha}_D(n)^r}$$

$$= \frac{\displaystyle\sum_{n\in M_{k,l,d_1}}' \overline{\alpha}_D(n)^r + \sum_{n\in M_{k,l,d_2}}' \overline{\alpha}_D(n)^r}{\displaystyle\sum_{n\in M_{k,l,d_1}}' \overline{\alpha}_D(n)^r + \sum_{n\in M_{k,l,d_2}}' \overline{\alpha}_D(n)^r + \pi(2^k) - \pi(2^{k-1})}$$

$$= \frac{N_{r,d_1} + N_{r,d_1}}{N_{r,d_1} + N_{r,d_2} + \pi(2^k) - \pi(2^{k-1})}.$$

For ease of notation, let us define the following quantities:

Definition 10. *Let $\omega(n)$ denote the number of distinct prime factors of n and let $\Omega(n)$ denote the number of prime factors of n counted with multiplicity. Thus, $\omega(n) = s$ and $\Omega(n) = \sum_{i=1}^{s} r_i$.*

We use the following lemmas for proving the Theorems 9 and 10:

Lemma 7. *Let $n = p_1^{r_1} \ldots p_s^{r_s} > 1$ be odd. Then*

$$\alpha_D(n) \le 2^{1-\Omega(n)} \prod_{i=1}^{s} \left(\frac{2}{p_i}\right)^{r_i-1} \frac{\gcd(p_i - \epsilon_D(p_i), n - \epsilon_D(n))}{p_i - \epsilon_D(p_i)}.$$

Proof. This follows from Lemma 10 from [9], where we use the fact that $2^{1-s} = 2^{1-\Omega(n)} \prod_{i=1}^{s} 2^{r_i-1}$.

$\square$

Lemma 8. *Let $n = p_1^{r_1} \ldots p_s^{r_s} \in M_{k,l,d_2}$, where either $\frac{\gcd(p_i-\epsilon_D(p_i),n-\epsilon_D(n))}{p_i-\epsilon_D(p_i)} < \frac{1}{3}$ for all i, or $r_i \ge 2$ for some i. Then,*

$$\alpha_D(n) \le 2^{-\Omega(n)-1}.$$

Proof. Let $n \in M_{k,l,d_2}$. Let us first look at the case where n is square-free. Since $n \in M_{k,l,d_2}$, we must have that $\frac{\gcd(p_i-\epsilon_D(p_i),n-\epsilon_D(n))}{p_i-\epsilon_D(p_i)} < \frac{1}{3}$, and hence $\frac{\gcd(p_i-\epsilon_D(p_i),n-\epsilon_D(n))}{p_i-\epsilon_D(p_i)} \le \frac{1}{4}$. Lemma 7 directly yields $\alpha_D(n) < \frac{2^{1-\Omega(n)}}{4} = 2^{-\Omega(n)-1}$. Now let us look at the case where n is not square-free, meaning that $r_i \ge 2$ for some i. Since $p_1 \ge \tilde{p}_{l+1}$, Lemma 7 directly yields that $\alpha_D(n) \le \frac{2^{1-\Omega(n)}}{\tilde{p}_{l+1}} \le 2^{-\Omega(n)-1}$.

$\square$

The following two lemmas will be used in proving the Theorems 9 and 10:

Lemma 9 (Einsele, Paterson [9]). *If $t \in \mathbb{R}$ with $t \ge 1$, then*

$$\sum_{n=\lfloor t \rfloor+1}^{\infty} \frac{1}{n(n-1)} = \frac{1}{\lfloor t \rfloor} < \frac{2}{t}.$$

Lemma 10 (Damgård et al. [8]). *If t is a real number with $t \ge 1$, then*

$$\sum_{n=\lfloor t \rfloor+1}^{\infty} \frac{1}{n^2} < \frac{\pi^2 - 6}{3t}.$$

By treating the two disjoint sets M_{k,l,d_1} and M_{k,l,d_2} differently in our analysis, we get a tighter estimates than Proposition 2.

Theorem 9. *If m, k are positive integers with $m + 1 \leq 2\sqrt{k-1}$, then*

$$|C_{m,D} \cap M_{k,l,d_1}| \leq 2^k \sum_{j=2}^{m} \frac{3}{\prod_{i=1}^{j-1} \tilde{p}_{l+i}} \frac{1}{2^{\frac{k-1}{j}} + 1}.$$

Proof. For $n \in C_{m,D} \cap M_{k,l}$, we have by Lemma 7 that $2^m > 1/\alpha_D(n) > 2^{\Omega(n)-1}$, and thus $m + 1 > \Omega(n)$. $\Omega(n) \in \mathbb{N}$ implies $\Omega(n) \leq m$. Let $N_D(m, k, j) = \{n \in C_{m,D} \cap M_{k,l,d_1} \mid \Omega(n) = j\}$. Hence,

$$|C_{m,D} \cap M_{k,l,d_1}| = \sum_{j=2}^{m} |N_D(m, k, j)|. \tag{10}$$

Let $n \in N_D(m, k, j)$ with $2 \leq j \leq m$, and let p be the largest prime factor of n. Now, $2^{k-1} < n \leq p^j$ implies that $p > 2^{(k-1)/j}$. Given p and d, where p is a prime with the property that $p > 2^{(k-1)/j}$ and d is such that $d \mid p - \epsilon_D(p)$, we want to get an upper bound for the number of $n \in N_D(m, k, j)$ with the largest prime factor p such that $d_D(p, n) = d$. Let $S_{D,k,d,p} = \{n \in M_{k,l,d_1} : p \mid n, d = \frac{p-\epsilon_D(p)}{(p-\epsilon_D(p), n-\epsilon_D(n))}, n \text{ composite}\}$ for $d = 1, 2, 3$. Since $n \in M_{k,l,d_1}$, we know that n is a product of distinct primes $n = p_1 \ldots p_{j-1}p$, with $p_i \geq \tilde{p}_{l+i}$ for all $i = 1, \ldots, j-1$. Thus, $p = \frac{n}{p_1 \cdots p_{j-1}} \leq \frac{n}{\prod_{i=1}^{j-1} \tilde{p}_{l+i}} < \frac{2^k}{\prod_{i=1}^{j-1} \tilde{p}_{l+i}}$. The size of $S_{D,k,d,p}$ is at most the number of solutions of the system

$$n \equiv 0 \mod p, \quad n \equiv \epsilon_D(n) \mod \frac{p - \epsilon_D(p)}{d}, \quad p < \frac{2^k}{\prod_{i=1}^{j-1} \tilde{p}_{l+i}},$$

where $d = 1, 2, 3$. Via the Chinese Remainder Theorem this, set has fewer than $\frac{2^k d}{\prod_{i=1}^{j-1} \tilde{p}_{l+i}} \frac{1}{p(p-\epsilon_D(p))}$ elements. With p and n odd, $(p - \epsilon_D(p))/d(p - \epsilon_D(p), n - \epsilon_D(n))$ must be even. So,

$$|N_D(m, k, j)| \leq \sum_{\substack{d=1,2,3 \\ p-\epsilon_D(p) \in 2\mathbb{Z}}} \sum_{p>2^{(k-1)/j}} \frac{2^k d}{\prod_{i=1}^{j-1} \tilde{p}_{l+i}} \frac{1}{p(p-\epsilon_D(p))}$$

$$= \frac{2^k}{\prod_{i=1}^{j-1} \tilde{p}_{l+i}} \sum_{d=1,2,3} \sum_{2u>2^{(k-1)/j}-\epsilon_D(p)} \frac{d}{(2u+\epsilon_D(p))2u}.$$

Now let us first look at the case $\epsilon_D(p) = 1$. By Lemma 10, we have

$$\frac{1}{4d} \sum_{u > \frac{2^{\frac{k-1}{j}} - \epsilon_D(p)}{2d}} \frac{1}{(u + \frac{\epsilon_D(p)}{2d})u} \leq \frac{1}{4d} \sum_{u > \frac{2^{\frac{k-1}{j}} - 1}{2d}} \frac{1}{u^2} \leq \frac{1}{4d} \frac{\pi^2 - 6}{3 \frac{2^{\frac{k-1}{j}} - 1}{2d}} = \frac{\pi^2 - 6}{6} \frac{1}{2^{\frac{k-1}{j}} - 1}.$$

Now let us look at the case $\epsilon_D(p) = -1$. By Lemma 9 we have

$$\frac{1}{4d} \sum_{u > 2^{\frac{k-1}{j}} - \epsilon_D(p)} \frac{1}{(u + \frac{\epsilon_D(p)}{2d})u} = \frac{1}{4d} \sum_{u > 2^{\frac{k-1}{j}} + 1} \frac{1}{(u-1)u} \leq \frac{1}{4d} \frac{2}{\frac{2^{\frac{k-1}{j}}+1}{2d}} = \frac{1}{2^{\frac{k-1}{j}} + 1}.$$

For $k, j \in \mathbb{N}$ with $j \leq \frac{k-1}{\log_2(-\frac{\pi^2}{\pi^2 - 12})}$, we have that $\frac{\pi^2 - 6}{6} \frac{1}{2^{\frac{k-1}{j}} - 1} \leq \frac{1}{2^{\frac{k-1}{j}} + 1}$.
With $j \leq m - 2$ and $m + 1 \leq 2\sqrt{k-1}$, this is naturally satisfied. Hence, we get

$$|N_D(m,k,j)| \leq \frac{2^k}{\prod_{i=1}^{j-1} d} \sum_{=1,2,3} \frac{1}{2^{\frac{k-1}{j}} + 1} \leq \frac{2^k}{\prod_{i=1}^{j-1}} \frac{3}{2^{\frac{k-1}{j}} + 1}.$$

Using this estimate in (10) proves the theorem.   □

Theorem 10. *If m, k are positive integers with $m + 1 \leq 2\sqrt{k-1}$, then*

$$|C_{m,D} \cap M_{k,l,d_2}| \leq 2^k \sum_{j=2}^{m-2} \frac{2^{m+1-j} - 4}{2^{\frac{k-1}{j}} + 1}.$$

Proof. By the same argument as in the proof of Theorem 9, we get by applying Lemma 8 that $\Omega(n) \leq m - 2$ for $n \in C_{m,D} \cap M_{k,l,d_2}$. Now let $N_D(m,k,j) = \{n \in C_{m,D} \cap M_{k,l} \mid \Omega(n) = j\}$. We see that

$$|C_{m,D} \cap M_{k,l}| = \sum_{j=2}^{m-2} |N_D(m,k,j)|. \tag{11}$$

Let $n \in N_D(m,k,j)$ with $2 \leq j \leq m - 2$, and let p be the largest prime factor of n. Now $2^{k-1} < n \leq p^j$ implies that $p > 2^{(k-1)/j}$. Let $d_D(p,n) = (p - \epsilon_D(n))/(p - \epsilon_D(n), n - \epsilon_D(n))$. Lemma 7 implies that $2^m > 1/\alpha_D(n) \geq 2^{\Omega(n)-1} d_D(p,n) = 2^{j-1} d_D(p,n)$, so we must have $d_D(p,n) < 2^{m+1-j}$. That $n \in M_{k,l,d_2}$ implies $d_D(p,n) > 3$. Given p, d, where p is a prime with the property that $p > 2^{(k-1)/j}$ and d is such that $d \mid p - \epsilon_D(p)$ and $d < 2^{m+1-j}$, we want to get an upper bound for the number of $n \in N_D(m,k,j)$ with largest prime factor p such that $d_D(p,n) = d$. Let $S_{D,k,d,p} = \{n \in M_{k,l} : p \mid n, d = \frac{p-\epsilon_D(p)}{(p-\epsilon_D(p), n-\epsilon_D(n))}, n \text{ composite}\}$. The size of the set $S_{D,k,d,p}$ is at most the number of solutions of the system

$$n \equiv 0 \bmod p, \quad n \equiv \epsilon_D(n) \bmod \frac{p - \epsilon_D(p)}{d}, \quad p < n < 2^k.$$

Via the Chinese Remainder Theorem, this is less than $\frac{2^k d}{p(p-\epsilon_D(p))}$. If $S_{D,k,d,p} \neq \emptyset$, then there exists an $n \in S_{D,k,d,p}$ with $(n - \epsilon_D(n), p - \epsilon_D(p)) = (p - \epsilon_D(p))/d$. Again $(p - \epsilon_D(p))/d = (p - \epsilon_D(p), n - \epsilon_D(n))$ must be even. Hence,

$$|N_D(m,k,j)| \leq \sum_{\substack{p > 2^{\frac{k-1}{j}}}} \sum_{\substack{d \mid p - \epsilon_D(p) \\ 3 < d < 2^{m+1-j} \\ \frac{p-\epsilon_D(p)}{d} \in 2\mathbb{Z}}} \frac{2^k d}{p(p - \epsilon_D(p))} = 2^k \sum_{\substack{3 < d < 2^{m+1-j}}} \sum_{\substack{p > 2^{\frac{k-1}{j}} \\ d \mid p - \epsilon_D(p) \\ \frac{p-\epsilon_D(p)}{d} \in 2\mathbb{Z}}} \frac{d}{p(p - \epsilon_D(p))}.$$

Now, for the inner sum we have,

$$\sum_{\substack{p>2^{\frac{k-1}{j}} \\ d|p-\epsilon_D(p) \\ \frac{p-\epsilon_D(p)}{d}\in 2\mathbb{Z}}} \frac{d}{p(p-\epsilon_D(p))} < \sum_{2ud>2^{\frac{k-1}{j}}-\epsilon_D(p)} \frac{d}{(2ud+\epsilon_D(p))2ud} = \frac{1}{4d}\sum_{u>\frac{2^{\frac{k-1}{j}}-\epsilon_D(p)}{2d}} \frac{1}{(u+\frac{\epsilon_D(p)}{2d})u}.$$

By the same argument as in Theorem 9 we get

$$|N_D(m,k,j)| \leq 2^k \sum_{3<d<2^{m+1-j}} \frac{1}{2^{\frac{k-1}{j}}+1} \leq 2^k \frac{2^{m+1-j}-4}{2^{\frac{k-1}{j}}+1}.$$

Using this estimate in (11) concludes the proof.   □

Theorem 11. *Let* $m,k,l,r \in \mathbb{N}$, $k \geq 2$ *with* $m+1 \leq 2\sqrt{k-1}$. *Then,*

$$\sideset{}{'}\sum_{n\in M_{k,l}} \overline{\alpha}_D(n)^r \leq 2^{r(1-M)}|M_{k,l}|\frac{\rho_l^{(M+1)t}}{2^t-\rho^r} + 2^{k+r}\left(\sum_{m=2}^{M}\sum_{j=2}^{m}\left(\frac{\rho_l}{2}\right)^{mr}\frac{3}{\prod_{i=1}^{j-1}\tilde{p}_{l+i}}\frac{1}{2^{\frac{k-1}{j}}+1}\right.$$
$$\left.+\sum_{m=2}^{M}\sum_{j=2}^{m-2}\left(\frac{\rho_l}{2}\right)^{mr}\frac{2^{m+1-j}-4}{2^{\frac{k-1}{j}}+1}\right).$$

Proof. With Lemma 2 we get

$$\sideset{}{'}\sum_{n\in M_{k,l}} \overline{\alpha}_D(n)^t = \sum_{m=2}^{\infty}\sum_{n\in C_{m,D}\cap M_{k,l}\backslash C_{m-1,D}} \overline{\alpha}_D(n)^r$$

$$\leq \sum_{m=2}^{\infty}\sum_{n\in C_{m,D}\cap M_{k,l}\backslash C_{m-1,D}} \rho_l^{mt}2^{-(m-1)r}$$

$$\leq 2^r\left(|M_{k,l}|\sum_{m=M+1}^{\infty}\left(\frac{\rho_l}{2}\right)^{mr} + \left(\sum_{m=2}^{M}\left(\frac{\rho_l}{2}\right)^{mr}|M_{k,l,d_1}\cap C_{m,D}|\right.\right.$$

$$\left.\left.+\sum_{m=2}^{M}\left(\frac{\rho_l}{2}\right)^{mr}|M_{k,l,d_2}\cap C_{m,D}|\right)\right).$$

With $\sum_{m=M+1}^{\infty}\left(\frac{\rho_l}{2}\right)^{mr} = 2^{-Mr}\frac{\rho_l^{(M+1)r}}{2^r-\rho^r}$ and Theorem 9 and 10 we are done.   □

Proof of Theorem 8. We use inequality (6) with Theorem 11 for N_1 and N_2. For each k, we choose M to be the positive integer $3 \leq M \leq 2\sqrt{k-1}-1$ that minimizes each upper bound in inequality (6) for $q_{k,1}$ and $q_{k,2}$ with $P = (0.71867)\frac{2^k}{k}$. This shows that $q_{k,1} \leq 4/19$ for $k \geq 34$. For $30 \leq k \leq 33$ we have that $q_{k,1} \leq 4/15$ and $q_{k,2} \leq 16/241$. So, $q_{k,t} \leq (4/15)^t$ for $k \geq 30$ (Table 3).

Table 3. Upper bounds $v_{k,1}$ and $v_{k,2}$ for $q_{k,1}$ and $q_{k,2}$, respectively, derived using inequality (6). The values of N_r are determined via Theorem 11, and Lemma 4 is used to bound $|M_{k,l}|$. We use $P = (0.71867)2^k/k$. The optimal choices for M, namely $M_{\text{opt},1}$ and $M_{\text{opt},2}$, minimize $v_{k,1}$ and $v_{k,2}$, respectively. For $34 \le k \le 41$, $v_{k,1} \le \frac{4}{19}$, while for $30 \le k \le 33$, $v_{k,1} \le \frac{4}{15}$ and $v_{k,2} \le \frac{16}{241}$.

k	$M_{opt,1}$	$v_{k,1}$	$M_{opt,2}$	$v_{k,2}$
30	6	0.239294	8	0.000602
31	6	0.235818	8	0.000544
32	7	0.232670	8	0.000360
33	7	0.220337	9	0.000314
34	7	0.209791		
35	7	0.200868		

k	$M_{opt,1}$	$v_{k,1}$
36	7	0.193406
37	7	0.187248
38	7	0.182247
39	7	0.178267
40	7	0.175183
41	8	0.166822

We now compute the exact values of $|M_{k,l}|$ and $\pi(2^k) - \pi(2^{k-1})$ to get improved results in Theorem 11 with $r = 1, 2$. For each k, we choose $3 \le M \le 2\sqrt{k-1} - 1$ that minimizes the upper bound in (6). With this, we have $q_{k,1} \le 4/19$ for $k = 27, 28, 29$, and $q_{k,1} \le 4/15$ and $q_{k,2} \le 16/241$ for $17 \le k \le 26$, which proves that $q_{k,t} \le (4/15)^t$ for $17 \le k \ge 29$, see Table 4.

Table 4. Upper bounds $v_{k,1}$ and $v_{k,2}$ for $q_{k,1}$ and $q_{k,2}$, respectively, derived using inequality (6). The values of N_r are determined via Theorem 11, and the exact values of $|M_{k,l}|$ and $\pi(2^k) - \pi(2^{k-1})$ are computed. The optimal choices for M, namely $M_{\text{opt},1}$ and $M_{\text{opt},2}$ minimize $v_{k,1}$ and $v_{k,2}$, respectively. For $k = 27, 28, 29$, $v_{k,1} \le \frac{4}{19}$, while for $17 \le k \le 26$, $v_{k,1} \le \frac{4}{15}$ and $v_{k,2} \le \frac{16}{241}$.

k	$M_{opt,1}$	$v_{k,1}$	$M_{opt,2}$	$v_{k,2}$
17	4	0.253449	6	0.004786
18	4	0.256262	6	0.004075
19	4	0.260073	6	0.003510
20	5	0.247789	6	0.003088
21	5	0.235446	6	0.002760
22	5	0.226473	7	0.001935
23	5	0.220211	7	0.001650

k	$M_{opt,1}$	$v_{k,1}$	$M_{opt,2}$	$v_{k,2}$
24	5	0.216189	7	0.001424
25	5	0.214003	7	0.001246
26	5	0.213406	8	0.000926
27	6	0.209426		
28	6	0.197899		
29	6	0.188524		

$\square$

6 Exact Values

In this section, we finally prove the main theorem.

Theorem 12. *For $k \ge 2, t \ge 1$ we have $q_{k,t} \le \left(\frac{4}{15}\right)^t$.*

The approach mentioned above does not prove the desired results for $k \leq 16$. We now suppose divisibility by the first two odd primes instead of the first nine, which in our case is a stronger assumption for $q_{k,t}$. We have an exact formula for $SL(D,n)$ given in (5), which we will use. We consider all odd k-bit integers that are not divisible by 3 and 5 and store them in a list. To compute $SL(D,n)$, all of the prime factors of n must be determined computationally, so this can only be computed for small values of k. We store all primes less than the square root of 2^{16} in a list. Moreover, for each $n \in M_{k,2}$, we determine q, k_1, s, and q_i for all $1 \leq i \leq s$. We then calculate $\overline{\alpha}_D(n)$. Additionally, we only sum over the integers with $\gcd(n, 2D) = 1$. For the bound to hold for every D, we take the one that maximizes $\sum'_{n \in M_{k,l}} \overline{\alpha}_D(n)$ (Table 5).

Table 5. The exact values for $q_{k,1}$ for $2 \leq k \leq 16$.

k	$q_{k,1}$	k	$q_{k,1}$	k	$q_{k,1}$	k	$q_{k,1}$
2	0	6	0.009725	10	0.012924	14	0.003987
3	0	7	0.027481	11	0.008977	15	0.001641
4	0	8	0.019684	12	0.006131	16	0.001095
5	0	9	0.016090	13	0.006737		

Together with Theorems 7 and 8, this proves that $q_{k,t} \leq \left(\frac{4}{15}\right)^t$ for $k \geq 2, t \geq 1$.

7   Average Case Behaviour on Incremental Search

So far, we have explored a method for generating primes by selecting a fresh and random k-bit integer and using Algorithm 1.4 for primality testing until a passing candidate is found. However, an often recommended alternative is to choose a random starting point $n_0 \in M_k$, test it for primality, and if it fails, consecutively test $n_0 + 2, n_0 + 4, \ldots$ until one is found that passes all stages of the test. Numerous adaptations are possible, such as other step sizes and various sieving techniques, yet the basic principle remains unchanged. This method, commonly known as *incremental search*, offers several practical advantages. It is more efficient in using random bits and test division by small primes can be conducted much more efficiently compared to the conventional "uniform choice" method. A drawback is a bias in the distribution of the generated primes.

The key advantage lies in the fact that a complete trial division is only necessary for the starting candidate n_0. The remainders $r_p \equiv n_0 \mod p$ are computed for all primes $p < B$ below a given threshold B and stored in a table. As the candidate sequence progresses, the values in the table are efficiently updated by adding 2 to each stored remainder modulo p, so $n_i = n_0 + 2i \equiv r_p + 2i \mod p$, where n_i is the i-th candidate and $i \in \mathbb{N}_0$. The candidate passes the trial division stage if none of the table values are equal to 0.

7.1   The Incremental Search Algorithm

The analysis of the average-case error probability done in this paper and in [9] depends on the assumption that the candidates are independent, hence the method cannot be directly taken over to the incremental search method. The average case error behaviour of the incremental search algorithm of the Miller-Rabin test was studied in [6]. Notably, no analysis of the average-case error behaviour for the strong Lucas test exists. We now give a more precise version of the algorithm.

Algorithm 2: $\text{PRIMEINCLUC}(t, k, s)$

Input: Bit-size $k \in \mathbb{N}$, testing rounds $t \in \mathbb{N}$, maximum number of candidates before returning "fail" $s \in \mathbb{N}$.

Output: First probable prime found or "fail" after $n_0 + 2s$ iterations.

  1. Choose an odd k-bit integer n_0 uniformly at random
  2. $n = n_0$
  3. If n is divisible by 2, 3, and 5: set $n = n + 2$. If $n \geq n_0 + 2s$ **output** "fail" and **stop**. Else, go to step 3.
  4. Else execute the following loop until it stops:
     For $i = 1$ to t:
     (a) Perform the strong Lucas test to n with randomly chosen bases.
     (b) If n fails any round of test, set $n = n + 2$. If $n \geq n_0 + 2s$, **output** "fail" and **stop**. Else, go to step 3.
     (c) Else output n and **stop**

To enhance the algorithm's efficiency, we can incorporate test division by additional small primes before applying the strong Lucas test. Regardless of the number of primes used, the optimized algorithm's error probability remains at most that of PRIMEINCLUC. This is because test division can never reject a prime, only improving our chances of rejecting composites. However, the following analysis does not explore the error probability of the optimized version.

7.2   Error Estimates of the Incremental Search Algorithm

In this section, we focus on the probability that PRIMEINCLUC outputs a composite. Let $y_{k,t,s}$ denote the probability that one execution of the loop (steps 1(a) and (b)) outputs a composite number.

Definition 11. *Let*

$$\overline{C}_{m,D} = \{n \in \mathbb{N} : \gcd(n, 2D) = 1, n \text{ composite}, \overline{\alpha}_D(n) = \frac{SL(D, n)}{n - \epsilon_D(n) - 1} > 2^{-m}\}.$$

Lemma 11. $\overline{C}_{m,D} \subseteq C_{1.2m,D}$.

Proof. Since $7 = \tilde{p}_3$ is the third odd prime, we have by Lemma 2 for every $n \in \tilde{M}_{k,2}$ that

$$\overline{\alpha}_D(n) \le \rho_1^m \alpha_D(n) = \left(\frac{8}{7}\right)^m \alpha_D(n). \tag{12}$$

For $n \in \overline{C}_{m,D}$, we have by inequality (12) that $2^{-m} < \overline{\alpha}_D(n) \le \left(\frac{8}{7}\right)^m \alpha_D(n)$. Hence, $2^{-m} \left(\frac{7}{8}\right)^m < \alpha_D(n)$ and since $2^{-1.2m} < 2^{-m} \left(\frac{7}{8}\right)^m$, we get $2^{-1.2m} < \alpha_D(n)$. Thus, for $n \in \overline{C}_{m,D}$, we have $n \in C_{1.2m,D}$. □

Let us define the set $\mathcal{D}_{m,D} = \{n \in \tilde{M}_{k,2} \mid [n, \ldots, n + 2(s-1)) \cap \overline{C}_{m,D} \ne \emptyset\}$ for $m \ge 3$. A number belonging to $\overline{C}_{m,D}$ can be in at most s distinct intervals of the form $[n, \ldots, n + 2(s-1))$, Therefore, the next lemma easily follows.

Lemma 12. $\mathcal{D}_{m,D} \subset \mathcal{D}_{m+1,D}$ and $|\mathcal{D}_m| \le s|\tilde{M}_{k,2} \cap \overline{C}_{m,D}|.$

The motivation behind defining the sets $\mathcal{D}_{m,D}$ is that, if we luckily select a starting point n_0 for the inner loop that is *not* a member of $\mathcal{D}_{m,D}$, all composites tested before the loop terminates are going to pass with a probability of at most 2^{-m}. This is reflected in the bound $y_{k,s,t}$ as outlined below:

Theorem 13. *Let* $k \in \mathbb{N}, k \ge 2$ *and* $s = c\ln(2^k)$ *for some constant* c. *Then, for any* $3 \le M \le 2\sqrt{k-1} - 1$, *we have*

$$y_{k,t,s} \le 0.5(ck)^2 \sum_{m=3}^{\lceil 1.2M \rceil} \frac{|C_{m,D} \cap M_k|}{|\tilde{M}_{k,2}|} 2^{-t(m-1)} + 0.7ck2^{-tM}.$$

Proof. Let E' denote the event that the inner loop outputs a composite, and let $D_{m,D}$ be associated with the event that the starting point n_0 is in $\mathcal{D}_{m,D}$. Let X^c be defined as the complement of the event X. Then,

$$y_{k,t,s} = \sum_{m=3}^{M} \mathbb{P}[E' \cap (D_{m,D} \setminus D_{m-1,D})] + \mathbb{P}[E' \cap D_{m,D}^c]$$

$$\le \sum_{m=3}^{M} \mathbb{P}[D_{m,D}]\mathbb{P}[E' \mid (D_{m,D} \setminus D_{m-1,D})] + \mathbb{P}[E' \cap D_{m,D}^c].$$

Consider the scenario where a fixed $n_0 \notin \mathcal{D}_{m,D}$ is chosen as the starting point. In this case, no candidate n that we test will belong to $\overline{C}_{m,D} \cap \tilde{M}_{k,2}$, and thus, each candidate will pass all tests with probability at most 2^{-mt}. The probability of outputting a composite is maximized when all numbers in the considered interval are composite. Consequently, in this scenario, we accept one of the candidates with a probability of at most $s2^{-mt}$. Combining this observation with Lemma 12 and using the fact that $\tilde{M}_{k,2} \le M_k$, we obtain

$$y_{k,t,s} \leq s^2 \sum_{m=3}^{M} \frac{|\overline{C}_{m,D} \cap M_k|}{|\tilde{M}_{k,2}|} 2^{-t(m-1)} + s2^{-tM}$$

$$\leq 0.5(ck)^2 \sum_{m=3}^{M} \frac{|\overline{C}_{m,D} \cap M_k|}{|\tilde{M}_{k,2}|} 2^{-t(m-1)} + 0.7ck2^{-tM}. \tag{13}$$

With Lemma 11 and a substitution we get the desired result. □

By the proof of Theorem 13 in [9], we have $|C_{m,D} \cap M_k| \leq 2^{k+1} \sum_{j=2}^{m} 2^{m-j-\frac{k-1}{j}}$. Using Theorem 13 and Lemma 4 in inequality (13), we get

$$y_{k,t,s} \leq 2^{3.42+t}(ck)^2 \sum_{m=3}^{\lceil 1.2M \rceil} 2^{m(1-t)} \sum_{j=2}^{m} 2^{-j-\frac{k-1}{j}} + 0.7ck2^{-tM}.$$

We can now directly derive numerical estimates for $y_{k,t,s}$ for any given value of s. Table 6 shows concrete values for $c = 1, 5$ and 10, where the value of M was chosen such that it minimizes the estimate.

The next proposition gives a rough idea of how bound's behaviour for large k.

Proposition 3. *Given constants c, where $s = c\ln(2^k)$, and t, the function $y_{k,t,s}$ with respect to $k \geq 11$ satisfies*

$$y_{k,t,s} \leq \lambda k^3 2^{-\sqrt{k}}$$

for a constant λ.

Proof. It is sufficient to show the proposition for $t = 1$. Using Lemma 12 from [9] and the proof of Theorem 13 in [9], we have $|C_{m,D} \cap M_k| \leq 2^{k+1} \sum_{j=2}^{m} 2^{m-j-\frac{k-1}{j}}$ and Lemma 4 with the fact that $3 - 2\sqrt{k-1} < -\sqrt{k}$ for $k \geq 11$ we get

$$\frac{|C_{m,D} \cap \tilde{M}_{k,2}|}{|\tilde{M}_{k,2}|} \leq 2^{m+3.92} \sum_{j=2}^{m} 2^{-j-\frac{k-1}{j}} \leq m2^{m+3-2\sqrt{k-1}} < m2^{m-\sqrt{k}}, \tag{14}$$

for $k \geq 18$. Using inequality (14) in Theorem 13 with $M + 1 \leq 2\sqrt{k-1}$, we get

$$y_{k,t,s} \leq 0.5(ck)^2 2^{-\sqrt{k}} \sum_{m=3}^{\lceil 1.2M \rceil} m + 0.7ck2^{-M} = (ck)^2 2^{-\sqrt{k}}(M-1) + 0.7ck2^{-M}$$

$$< 2c^2 k^{2.5} 2^{-\sqrt{k}} + 0.7ck2^{-M} \leq \lambda k^3 2^{-\sqrt{k}},$$

for some constant $\lambda \in \mathbb{R}$. □

In our analysis, we have focused on the probability of a single iteration of the loop producing a composite output. To assess the overall error probability

Table 6. Lower bounds of $-\log_2(y_{k,t,s})$ as a function of k and t, where $s = c\ln(2^k)$ with $c = 1, 5, 10$.

c	$k\backslash t$	1	2	3	4	5	6	7	8	9	10
1	100	0	6	12	17	21	25	28	31	33	35
	200	3	15	24	32	38	43	48	52	56	59
	400	11	30	42	53	62	69	76	83	89	94
	512	15	36	51	62	72	81	83	97	104	110
	1024	31	61	81	98	112	125	137	148	158	167
	2048	54	96	125	149	169	188	205	220	235	249
	4096	89	147	187	221	251	277	302	324	345	365
5	100	0	2	8	12	17	20	23	26	28	31
	200	0	11	20	27	33	38	43	47	51	55
	400	7	25	38	48	57	65	72	78	84	90
	512	11	32	46	58	68	77	85	92	99	106
	1024	26	56	76	93	108	120	132	143	153	163
	2048	50	91	120	144	165	183	200	216	230	244
	4096	84	142	183	217	246	273	297	320	341	361
10	100	0	0	6	10	15	18	21	24	26	29
	200	0	9	18	25	31	36	41	45	49	53
	400	5	23	36	46	55	63	70	76	82	88
	512	9	30	44	56	66	75	83	90	97	104
	1024	24	54	74	91	106	118	130	141	151	161
	2048	48	89	118	142	163	181	198	214	228	242
	4096	82	140	181	215	244	271	295	318	339	359

of the algorithm, we observe that the inner loop always terminates when initiated with a prime as a starting point. According to Lemma 4, this termination occurs with a probability of $(\pi(2^k) - \pi(2^{k-1}))/|\tilde{M}_{k,2}| \geq 5.3/k$. Consequently, there exists an exponentially small, relative to k, probability that the number of iterations exceeds k^2. Furthermore, it is noteworthy that the error probability of our algorithm is not more than that of a procedure running the inner loop up to k^2 times, outputting a composite only in the event of all iterations failing. This observation substantiates an upper bound on the expected running time. Let $Y_{k,t,s}$ denote the probability of PRIMEINCLUC outputting a composite. Thus, we arrive at the inequality:

$$Y_{k,t,s} \leq k^2 y_{k,t,s} + \left(1 - \frac{5.3}{k}\right)^{k^2}.$$

A   Appendix

Let $n = \prod_{i=1}^{s} p_i^{r_i}$ be the prime decomposition of an integer n.

A.1   The Fermat, Miller-Rabin Primality and Lucas Test

The Fermat test is a simple primality tests and exploits the following theorem:

Theorem 14 (Fermat's Little Theorem). *Let p be a prime number. For all a relatively prime to p, we have*

$$a^{p-1} \equiv 1 \bmod p.$$

To test whether p is prime, we can check if a random integer a coprime to p satisfies Fermat's Little Theorem. This is called the *Fermat test*. A *pseudoprime base a*, or *psp(a)*, is a composite number n such that $a^{n-1} \equiv 1 \bmod n$. The next theorem counts the number of bases that make an integer pass the test:

Theorem 15 (Baillie et al. [5]). *Let $n = \prod_{i=1}^{s} p_i^{r_i}$ be a positive integer. Then, the number of bases $a \bmod n$ for which n is a psp(a) is given by*

$$F(n) = \prod_{i=1}^{s} \gcd(n - 1, p_i - 1).$$

Unfortunately, infinitely many integers satisfying the thereom for all a coprime to n exist and are known as *Carmichael numbers* [2]. Consequently, Carmichael numbers can never be identified as composites by the test, making the test impractical for standalone implementation.

Let us look at a more stringent variant of Fermat's Little Theorem.

Theorem 16 (The Miller-Rabin Theorem). *Let $n > 1$ be an integer, and write $n - 1 = 2^{\kappa}q$, for $q, k \in \mathbb{N}$, where q is odd. Then, n is a prime if and only if for every $a \not\equiv 0 \bmod n$ one of the following is satisfied:*

$$a^q \equiv 1 \bmod n$$

or   (15)

there exists an integer $i < \kappa$ with $a^{2^i q} \equiv -1 \bmod n$.

We can extend this to a primality test, known as the *Miller Rabin test*, by testing property (15) for several bases a. If the property holds for some pair n, a, we say n is a *strong pseudoprime base a*, in short *spsp(a)*.

Besides the strong Lucas test, there is also its weaker variant, the so-called *Lucas test*, which relies on the following theorem:

Theorem 17 (Baillie et al. [5]). *Let $U_p(P, Q)$ be a Lucas sequence of the first kind. If p is an odd prime such that $(p, QD) = 1$, then we have that*

$$U_{p-\epsilon_D(p)} \equiv 0 \bmod p.$$ (16)

An integer that satisfies congruence (16) using parameters (P, Q) is called a *Lucas pseudoprime* for (P, Q), short *lpsp(P, Q)*. We obtain the *Lucas test* by repeatedly checking congruence (16) for various pairs (P, Q).

For a fixed integer D, the number of parameter pairs (P, Q) that lead to a Lucas pseudoprime for a composite n are characterized by the following formula:

Theorem 18 (Baillie et al. [5]). *Let D be a fixed positive integer, and let $n = \prod_{i=1}^{s} p_i^{r_i}$ be a positive odd integer with $\gcd(D, n) = 1$. Then, the number of distinct values of P modulo n, for which there is a Q such that $P^2 - 4Q \equiv D \bmod n$ and n is a Lucas pseudoprime for (P, Q) is given by:*

$$L(D, n) = \prod_{i=1}^{s} (\gcd(n - \epsilon_D(n), p_i - \epsilon_D(p_i)) - 1).$$

A.2   The Baillie-PSW Primality Test

The Baillie-PSW primality test combines a single Fermat/ Miller-Rabin test with base 2 with a (strong) Lucas test. Let us formally introduce the algorithm.

Algorithm 3: BAILLIEPSW(n)

Input: Odd integer n to test for primality.
Output: A probable prime using the Baillie-PSW test or "Composite".
  1. If n is divisible by any prime less than some limit, e.g. 1000, **output** "Composite", else continue.
  2. If n is not a (strong) pseudoprime base 2, **output** "Composite", else continue.
  3. Check if n is not a perfect square, use one of the methods to determine (P, Q):
     – *Method A:* Let D be the first element of the sequence $5, -7, 9, -11, 13, \ldots$ with $\epsilon_D(n) = -1$. Let $P = 1$ and $Q = (1 - D)/4$.
     – *Method B:* Let D be the first element of the sequence $5, 9, 13, 17, \ldots$ with $\epsilon_D(n) = -1$. Let $P = \min\{m \in \mathbb{N} \mid m \text{ odd}, m > \sqrt{D}\}$ and $Q = (P^2 - D)/4$.
  4. If n is not a (strong) Lucas pseudoprime for (P, Q), **output** "Composite", else **output** n.

To date, no composites have been identified passing a Baillie-PSW test, leading to the conjecture that this test could, in fact, determinstically test primality. Gilchrist [10] even verified that using Method A of Algorithm 3 no Baillie-PSW pseudoprimes below 2^{64} exist. In the next section we discuss why the Fermat/ Miller-Rabin and (strong) Lucas test with well-chosen parameters might be independent of each other.

A.3   Orthogonality in the Baillie-PSW Test

If we choose the parameters D, P and Q as described in method B of Algorithm 3, then the first 50 Carmichael numbers and several other base-2 Fermat pseudoprimes will never be Lucas pseudoprimes [5]. We now give a heuristic argument for the potential orthogonality of the tests, most of them given by Arnault [3]. However, to understand this, let us first establish the necessary mathematical prerequisites.

Let $\mathcal{O}$ be the set of algebraic integers and $\mathcal{O}_{\mathbb{Q}[\sqrt{D}]}$ the ring of integers of $\mathbb{Q}[\sqrt{D}]$.

Lemma 13 (Arnault [3]). *Let P, Q be integers such that $D = P^2 - 4Q \neq 0$. Let n be an integer relatively prime to $2QD$. For the Lucas sequences $(U_n), (V_n)$ associated with P, Q we have tha relations*

$$U_k = \frac{\alpha^k - \beta^k}{\alpha - \beta}, \; V_k = \alpha^k + \beta^k, \quad \text{for all } k \in \mathbb{N},$$

where α, β are the two roots of the polynomial $X^2 - PX + Q$. We put $\tau = \alpha\beta^{-1} \in \mathcal{O}_{\mathbb{Q}[\sqrt{D}]}$. For $k \in \mathbb{N}$, we have the equivalences

$$n \mid U_k \Leftrightarrow \tau^k = 1,$$
$$n \mid V_k \Leftrightarrow \tau^k = -1.$$

In particular, if n is composite and relatively prime to $2QD$, it is a slpsp(P, Q) if and only if

$$\tau^q \equiv 1 \bmod n \quad \text{or} \quad \exists i \text{ such that } 0 \leq i < \kappa \text{ and } \tau^{2^i q} \equiv -1 \bmod n,$$

where $n - \epsilon_D(n) = 2^\kappa q$ with q odd.

For $\alpha = a + b\sqrt{D} \in \mathbb{Q}[\sqrt{D}]$, we define its *conjugate* $\overline{\alpha} = a - b\sqrt{D}$. The *norm* of α is defined by $N(\alpha) = \alpha\overline{\alpha}$. We denote the multiplicative group of norm 1 elements of $\mathcal{O}_{\mathbb{Q}[\sqrt{D}]}/(n)$ by $(\mathcal{O}_{\mathbb{Q}[\sqrt{D}]}/(n))^{\wedge}$. The following proposition connects P and Q defined by the Lucas sequence with the elements τ of norm 1 in $\mathcal{O}_{\mathbb{Q}[\sqrt{D}]}$:

Proposition 4 (Arnault [3]). *Let D be an integer that is not a perfect square and let $n > 1$ be an odd integer with $\gcd(n, D) = 1$. Then, for every integer P, there exists an integer Q, uniquely determined modulo n, such that $P^2 - 4Q \equiv D \bmod n$. Furthermore, the set of integers P, such that*

$$\begin{cases} 0 \leq P < n, \\ \gcd(P^2 - D, n) = 1 \text{ i.e. } \gcd(Q, n) = 1 \end{cases}$$

is in one-to-one correspondence with the elements $\tau \in (\mathcal{O}_{\mathbb{Q}[\sqrt{D}]}/(n))^{\wedge}$, such that $\tau - 1$ is a unit in $\mathcal{O}_{\mathbb{Q}[\sqrt{D}]}/(n)$. This correspondence is expressed by the formulas:

$$\begin{cases} \tau \equiv (P + \sqrt{D})(P - \sqrt{D})^{-1} \\ P \equiv \sqrt{D}(\tau + 1)(\tau - 1)^{-1} \end{cases} \quad \bmod n\mathcal{O}_{\mathbb{Q}[\sqrt{D}]}.$$

The next lemma will be used in the heuristic argument:

Lemma 14. *Let* $n = p_i m_i$ *be an odd integer. Then the following holds*

$$\gcd(n-1, p_i - 1) = \gcd(n-1, m_i - 1),$$
$$\gcd(n+1, p_i + 1) = \gcd(n+1, m_i - 1).$$

Proof. For all a, b we have that $\gcd(a+b, a) = \gcd(a, b)$ and $\gcd(a-1, a) = 1$. With this, we obtain for the first equality

$$\gcd(n-1, p_i - 1) = \gcd((m_i - 1)p_i + (p_i - 1), p_i - 1) = \gcd((m_i - 1)p_i, p_i - 1)$$
$$= \gcd(m_i - 1, p_i - 1) = \gcd((p_i - 1)m_i, m_i - 1)$$
$$= \gcd((p_i - 1)m_i + (m_i - 1), m_i - 1) = \gcd(n-1, m_i - 1)$$

The second equality can be established in a similar manner.

Let us now give the heuristic argument why composite numbers rarely pass the Baillie-PSW test.

Let P, Q, D and b be integers satisfying $P^2 - 4Q = D$. Consider $n = p_1 \ldots p_r$ with $\gcd(n, QD)$ and $\left(\frac{D}{n}\right) = -1$ such that n is both a $psp(b)$ and $lpsp(P, Q)$. Let τ be the element associated with the pair (P, Q) in Proposition 4. Fermat's Little Theorem implies:

$$\begin{cases} b^{\gcd(n-1, p_i - 1)} \equiv 1 \text{ in } \mathbb{Z}/n\mathbb{Z}, \\ \tau^{\gcd(n+1, p_i - \epsilon_D(p_i))} \equiv 1 \text{ in } \mathcal{O}_{\mathbb{Q}[\sqrt{D}]}/n\mathcal{O}_{\mathbb{Q}[\sqrt{D}]} \end{cases} \quad \text{for every } i.$$

We let

$$\begin{cases} d_i = \gcd(n-1, p_i - 1), \\ d_i' = \gcd(n+1, p_i - \epsilon_D(p_i)). \end{cases}$$

We can express the equivalences as follows:

$$b^{n-1} \equiv 1 \bmod p_i \Leftrightarrow b \text{ is a } (p_i - 1)/d_i\text{th root} \bmod p$$
$$\tau^{n+1} = 1 \text{ in } \mathcal{O}/p_i\mathcal{O} \Leftrightarrow \tau \text{ is a } (p_i - \epsilon_D(p_i))/d_i'\text{th root in } (\mathcal{O}/p_i\mathcal{O})^{\wedge}.$$

Heuristically, these relations have a small chance of being true when the integers d_i and d_i' are small relative to the group order of $(\mathbb{Z}/p_i\mathbb{Z})^{\times}$ and $(\mathcal{O}/p_i\mathcal{O})^{\wedge}$.

For $\epsilon_D(p_i) = 1$, we observe that

$$d_i d_i' = \gcd(n-1, p_i - 1) \gcd(n+1, p_1 - 1) \le 2(p_1 - 1),$$

implying that it is not possible for both gcds to be large.

For $\epsilon_D(p_i) = -1$, we let $n = p_i m_i$. By Lemma 14, we have

$$\begin{cases} \gcd(n-1, p_i - 1) = \gcd(n-1, p_i - 1) = \gcd(n-1, m_i - 1), \\ \gcd(n+1, p_i + 1) = \gcd(n+1, p_i + 1) = \gcd(n+1, m_i - 1). \end{cases}$$

We conclude that

$$d_i d_i' = \gcd(n-1, p_i - 1) \gcd(n+1, p_i + 1) \le 2(m_i - 1),$$

and, following the same argument as above, it is evident that the gcds cannot be large. Consequently, d_i and d_i' rather small, leading to a scarcity of pseudoprimes.

A.4   Rationale for Avoiding $\epsilon_D(n) = 1$

We now discuss why it is best to avoid the case where $\epsilon_D(n) = 1$, as in such instances, the Fermat/ Miller-Rabin test and the (strong) Lucas test are not independent. Let us consider the various scenarios that can arise.

When D is a Perfect Square. If $D \neq 0$ is a perfect square, the strong Lucas test reduces to the Rabin-Miller test. If $\gcd(n, 2D) = 1$, we put $T = \alpha\beta^{-1} \bmod n$, with $\alpha, \beta \in \mathbb{Z}$. Lemma 13 yields the following equivalences for $k \in \mathbb{N}$:

$$n \mid U_k \Leftrightarrow T^k \equiv 1 \bmod n$$
$$n \mid V_k \Leftrightarrow T^k \equiv -1 \bmod n.$$

For every i, the decompositions $n - 1 = n - \epsilon_D(n) = 2^\kappa q$, $p_i - 1 = p_i - \epsilon_D(p_i) = 2^{k_i} q_i$ are the same, making n a $slpsp(P,Q)$ if and only if it is an $spsp(2QD)$. To ensure that D is not a perfect square, we can simply apply Newton's method for square roots.

When D is a Square Modulo n. Now, let D not necessarily be a perfect square, but a square modulo n. The next lemma allows us to construct Lucas pseudoprimes from Fermat pseudoprimes.

Lemma 15. *Let n be an odd integer that is both a $psp(b)$ and $psp(c)$. Then, n is a $lpsp(P,Q)$ for $P \equiv b + c$ and $Q \equiv bc \bmod n$.*

Proof. Let α and β represent the distinct roots of the polynomial $X^2 - PX + Q$. Then $\{\alpha, \beta\} \equiv \{b, c\} \bmod n$, because the quadratic polynomials $X^2 - PX + Q$ and $X^2 - (b + c)X + bc$ have congruent coefficients modulo n. Since $\left(\frac{D}{n}\right) = \left(\frac{P^2 - 4Q}{n}\right) = \left(\frac{(b-c)^2}{n}\right) = 1$, we get $U_{n-1} = \frac{b^{n-1} - c^{n-1}}{b - c} \equiv 0$, so n is a $lpsp(P,Q)$.   □

Let P and Q satisfy $D = r^2 \bmod n$. Solving the simultaneous equations $P = b+c$ and $Q = bc$ modulo n yields $b = \frac{(P-r)(n+1)}{2}$ and $c = \frac{(P+r)(n+1)}{2}$. For instance, if n is a $psp(2)$, it might be a $psp(b)$ and a $psp(c)$, assuming that $\gcd(n, bc) = 1$. This is plausible, considering the possibility of it being a Carmichael number or either b or c being ± 1. Lemma 15 implies that n is a $lpsp(P,Q)$, indicating that the Lucas test will not be independent of the Fermat test.

When the Jacobi Symbol is 1. Following [5], let $n = \prod_{i=1}^s p_i^{r_i}$ and $\epsilon_D(n) = 1$. By Theorem 18, there are $\prod_{i=1}^s (\gcd(n - 1, p_i - \epsilon_D(p_i)) - 1)$ values of P mod n, for which there is a Q making n a $lpsp(P,Q)$. Similarly, by Theorem 15 n is a $psp(a)$ for $\prod_{i=1}^s \gcd(n - 1, p_i - 1)$ values of a mod n. The product $\gcd(n-1, p_i - 1) \cdot \gcd(n - 1, p_i + 1)$ is less than $2(p_1 + 1)$, but for $\epsilon_D(p_i) = +1$, the gcds are equal, allowing both to be large. The computations in [5] support that if n is a $lpsp(P,Q)$ for many values of P with $\epsilon_D(n) = +1$, then it might also be a $psp(a)$ for many a values. For the strong Lucas test, let $n = \prod p_i^{r_i}$ and

$\epsilon_D(n) = 1$. Let $n - 1 = 2^\kappa q$, $p_i - \epsilon_D(p_i) = 2^{k_i} q_i$, and $p_i - 1 = 2^{l_i} s_i$ with q_i, s_i odd. The number of pairs (P, Q) with $0 \leq P, Q < n$, $D = P^2 - 4Q$, $\gcd(Q, n) = 1$, making n a $slpsp(P, Q)$ is, by Theorem 4, equal to $\prod_{i=1}^{s}(\gcd(q, q_i) - 1) + \sum_{j=0}^{k_1-1} 2^{js} \prod_{i=1}^{s} \gcd(q, p_i - \epsilon_D(p_i))$. The number of bases a such that n is a $spsp(a)$ is, by [11] and [13] equal to $(1 + \sum_{j=0}^{k_1-1} 2^{js}) \prod_{i=1}^{s} \gcd(q, p_i - 1)$. Again, the product $\gcd(q, p_i - 1) \cdot \gcd(q, p_i + 1)$ cannot exceed $2 \gcd(q, p_i + 1)$, but for $\epsilon_D(p_i) = +1$, the gcds are the same, allowing both to be large.

References

1. Agrawal, M., Kayal, N., Saxena, N.: PRIMES is in P. Ann. Math. **160**, 781–793 (2004)
2. Alford, W.R., Granville, A., Pomerance, C.: There are infinitely many carmichael numbers. Ann. Math. **139**(3), 703–722 (1994)
3. Arnault, F.: Sur Quelques Tests Probabilistes de Primalité. Ph.D. thesis, Poitiers (1993)
4. Arnault, F.: The Rabin-Monier theorem for Lucas pseudoprimes. Math. Comput. **66**(218), 869–881 (1997)
5. Baillie, R., Wagstaff, S.S.: Lucas pseudoprimes. Math. Comput. **35**(152), 1391–1417 (1980)
6. Brandt, J., Damgård, I.: On generation of probable primes by incremental search. In: Brickell, E.F. (ed.) CRYPTO 1992. LNCS, vol. 740, pp. 358–370. Springer, Heidelberg (1993). https://doi.org/10.1007/3-540-48071-4_26
7. Burthe, R., Jr.: Further investigations with the strong probable prime test. Math. Comput. **65**(213), 373–381 (1996)
8. Damgård, I., Landrock, P., Pomerance, C.: Average case error estimates for the strong probable prime test. Math. Comput. **61**(203), 177–194 (1993)
9. Einsele, S., Paterson, K.: Average case error estimates of the strong Lucas test. Des. Codes Crypt. **92**(5), 1341–1378 (2024)
10. Gilchrist, J.: Pseudoprime Enumeration with Probabilistic Primality Tests. http://gilchrist.great-site.net/jeff/factoring/pseudoprimes.html?i=1
11. Monier, L.: Evaluation and comparison of two efficient probabilistic primality testing algorithms. Theor. Comput. Sci. **12**(1), 97–108 (1980)
12. Pomerance, C.: Are there counter-examples to the baillie-PSW primality test. In: Dopo Le Parole aangeboden aan Dr. AK Lenstra. Citeseer (1984)
13. Rabin, M.O.: Probabilistic algorithm for primality testing. J. Number Theory **12**, 128–138 (1980)
14. Williams, H.C.: On numbers analogous to the Carmichael numbers. Can. Math. Bull. **20**(1), 133–143 (1977)

Algebraic Structures and Public-Key Cryptography

A New Public Key Cryptosystem Based on the Cubic Pell Curve

Michel Seck[1] and Abderrahmane Nitaj[2(✉)]

[1] Ecole Polytechnique de Thies, LTISI, Thies, Senegal
mseck@ept.sn
[2] Normandie Univ, UNICAEN, CNRS, LMNO, 14000 Caen, France
abderrahmane.nitaj@unicaen.fr

Abstract. Since its invention in 1978 by Rivest, Shamir and Adleman, the public key cryptosystem RSA has become a widely popular and a widely useful scheme in cryptography. Its security is related to the difficulty of factoring large integers which are the product of two large prime numbers. For various reasons, several variants of RSA have been proposed, and some have different arithmetics such as elliptic and singular cubic curves. In 2018, Murru and Saettone proposed another variant of RSA based on the cubic Pell curve with a modulus of the form $N = pq$.

In this paper, we present a new public key cryptosystem based on the arithmetic of the cubic Pell curve with a modulus of the form $N = p^r q^s$. Its security is based on the hardness of factoring composite integers, and on Rabin's trapdoor one way function. In the new scheme, the arithmetic operations are performed on a cubic Pell curve which is known only to the sender and the recipient of a plaintext.

Keywords: Public Key Cryptography · Cubic Pell curve · RSA · KMOV · Rabin's trapdoor

1 Introduction

The RSA cryptosystem is one of the earliest and most popular public key encryption schemes. It was proposed by Rivest, Shamir, and Adleman in 1978 [31] after the introduction of the concept of a trapdoor one-way function by Diffie and Hellman in 1976 [11]. In RSA, to generate a public key $\mathcal{PK}$ and a private key $\mathcal{SK}$, the following steps are taken. First, two prime numbers, p and q, are generated, and the modulus N is computed as $N = pq$. An element e in $\mathbb{Z}/N\mathbb{Z}$ is then chosen such that it is coprime with $\phi(N) = (p-1)(q-1)$, where ϕ is the Euler function. Next, the private exponent $d = e^{-1} \pmod{\phi(N)}$ is computed. The public key is $\mathcal{PK} = (N, e)$, and the private key is $\mathcal{SK} = (N, d)$. To encrypt a message $m \in \mathbb{Z}/N\mathbb{Z}$ with the public key $\mathcal{PK} = (N, e)$, the ciphertext $c \equiv m^e \pmod{N}$ is computed. To decrypt the ciphertext c using the private key $\mathcal{SK} = (N, d)$, the original message is obtained as $m \equiv c^d \pmod{N}$.

In 1979 Rabin [29,43] published a public-key encryption scheme similar to RSA whose security is also related to factoring composite integers. It is known that breaking the Rabin scheme is equivalent to factoring N while for RSA this equivalence is not proven. For Rabin encryption scheme, to generate a public key $\mathcal{PK}$ and a private key $\mathcal{SK}$, one first generates two primes p and q such that $p, q \equiv 3 \pmod{4}$, and computes the modulus $N = pq$. The public key is $\mathcal{PK} = N$ and the private key is $\mathcal{SK} = (p, q)$. To encrypt a message m, one computes $C \equiv m^2 \pmod{N}$. For the decryption of C, one proceeds as follows. First, solve the equation $X^2 = C \pmod{p}$ and $X^2 = C \pmod{q}$. By the Chinese Remainder Theorem, this leads to four solutions which include the plaintext.

Certain vulnerabilities of RSA are know for particular choices of the prime factors p, q, the public exponent e, and the private exponent d [2, 4–6, 9, 23, 40–42]. For example, when $p < q < 2p$, and d is too small with respect to N such as $d < \frac{1}{3}N^{0.25}$, one can use Wiener's attack [23, 42] to efficiently recover the secrete d. It is shown that RSA with a low public exponent e, e.g. $e = 3$, is vulnerable to Håstad's broadcast attack [2, 15]. When $d < N^{0.292}$, the RSA modulus N can be factored in polynomial time [6] using Coppersmith's method [9] which is based on lattice reduction techniques, especially the by the LLL algorithm [19].

Some RSA-like cryptosystems have been also proposed over non-singular and singular curves when certain groups or rings can be defined. In 1991, Koyama, Maurer, Okamoto, and Vanstone [17] proposed a variant of RSA based on elliptic curves over $\mathbb{Z}/N\mathbb{Z}$ when $N = pq$ is the product of two primes with $p, q \equiv 2 \pmod{3}$. This variant is known as KMOV. In KMOV, to generate a public key $\mathcal{PK}$ and a private key $\mathcal{SK}$, on first generates two primes p and q such that $p, q \equiv 2 \pmod{3}$, and compute the modulus $N = pq$ as in RSA. Then, one chooses an exponent e that is invertible modulo $\psi(N) = \operatorname{lcm}(p+1, q+1)$ and compute $d = e^{-1} \pmod{\psi(N)}$. The public key is $\mathcal{PK} = (N, e)$, and the private key is $\mathcal{SK} = (N, d)$. To encrypt a message $m = (x_m, y_m) \in (\mathbb{Z}/N\mathbb{Z}) \times (\mathbb{Z}/N\mathbb{Z})$ with the public key $\mathcal{PK}$, one first computes $b \equiv y_m^2 - x_m^3 \pmod{N}$, and then the ciphertext $c = e(x_m, y_m)$ on the elliptic curve $E_b : y^2 = x^3 + b$ over the ring $\mathbb{Z}/N\mathbb{Z}$. To decrypt the ciphertext $c = (x_c, y_c)$ with the private key $\mathcal{SK}$, one computes $b \equiv y_c^2 - x_c^3 \pmod{N}$, and then $m = d(x_c, y_c)$ on the elliptic curve $E_b : y^2 = x^3 + b$. Note that the condition $p, q \equiv 2 \pmod{3}$ ensures that the supersingular elliptic curve $E_b : y^2 = x^3 + b$ has order $p + 1$ modulo p and order $q + 1$ modulo q. This guarantee the correctness of the decryption phase in KMOV.

Another RSA variant over elliptic curves was proposed by Demytko [10] in 1993. Since then, several variants of KMOV and Demytko constructions have been proposed in the last decades with an RSA modulus $N = pq$. One was proposed by Koyama [16] based on the singular cubic curve $y^2 + axy = x^3 \pmod{N}$, a seconde one was proposed in 1995 by Kuwakado, Koyama, and Tsuruoka [18] based on the singular cubic curve $y^2 \equiv x^3 + bx^2 \pmod{N}$, and a third one was proposed in 2018 by Murru and Saettone [22] using the cubic Pell curve $C_r : x^3 + ry^3 + r^2z^3 - 3rxyz \equiv 1 \pmod{N}$.

Several variants of the RSA cryptosystem have been proposed where the modulus is of a different shape. A first variant was proposed by Takagi [39] in 1998 with a modulus of the form $N = p^k q$. In 1998, Okamoto, Uchiyama, and Fujisaki [26] proposed EPOC and ESIGN Algorithms [27], where the modulus is of the form $N = p^2 q$. The same modulus was used by Schmidt-Samoa [36] in 2005 to design a trapdoor one-way permutation. In 2000, Lim, Kim, Yie and

Lee [20] proposed a variant of RSA where the modulus is a multi-prime power integer of the form $N = p^r q^s$ with $r, s \geq 1$. Two more variants based on elliptic curves and Edwards curves were proposed by Boudabra and Nitaj [7,8] with a multi-power modulus $N = p^r q^s$.

Contributions: In this paper, we study the arithmetic of cubic Pell curves $\mathcal{PC}_a(N) : x^3 + ay^3 + a^2 z^3 - 3axyz = 1$ over $\mathbb{Z}/N\mathbb{Z}$ where $N = p^r q^s$ is a multi-prime power integer, and propose a new scheme. We summarize our contributions as follows.

- We present a detailed study of the cubic Pell curves with the equation $\mathcal{PC}_a(N) : x^3 + ay^3 + a^2 z^3 - 3axyz = 1$, specifically regarding the number of solutions modulo p^r, q^s, and $N = p^r q^s$.
- We propose a new scheme using the arithmetic of the cubic Pell curve $\mathcal{PC}_a(N) : x^3 + ay^3 + a^2 z^3 - 3axyz = 1$ over $\mathbb{Z}/N\mathbb{Z}$ where $N = p^r q^s$, and study its security.

The new scheme works as follows.

1. The public parameters in the new scheme are a prime power modulus $N = p^r q^s$, and a public exponent e.
2. To encrypt a message M with the new scheme, one represents it as $(x_M, y_M, 0)$, and then computes $a \equiv \frac{1 - x_M^3}{y_M^3} \pmod{N}$, and $(x_C, y_C, z_C) = e \otimes (x_M, y_M, 0)$ on the cubic Pell curve $\mathcal{PC}_a(N) : x^3 + ay^3 + a^2 z^3 - 3axyz = 1$ over $\mathbb{Z}/N\mathbb{Z}$.
3. To decrypt a ciphertext $(x_C, y_C, z_C) \in \mathcal{PC}_a(N)$, one first find a solution a of the quadratic equation $x_C^3 + ay_C^3 + a^2 z_C^3 - 3ax_C y_C z_C - 1 \equiv 0 \pmod{N}$. Then, one computes $d = e^{-1} \pmod{|\mathcal{PC}_a(N)|}$, and $(x_D, y_D, z_D) = d \otimes (x_C, y_C, z_C)$ on the cubic Pell curve $\mathcal{PC}_a(N) : x^3 + ay^3 + a^2 z^3 - 3axyz = 1$ over $\mathbb{Z}/N\mathbb{Z}$. Then, one of the values a leads to the plaintext $(x_D, y_D, z_D) = (x_M, y_M, 0)$.

   Since the modular equation $x_C^3 + ay_C^3 + a^2 z_C^3 - 3ax_C y_C z_C - 1 \equiv 0 \pmod{N}$ has four solutions a_i, $i = 1, 2, 3, 4$, then there are four decryption exponents d_i, $i = 1, 2, 3, 4$, and four potential plaintexts $(x_i, y_i, z_i) = d_i \otimes (x_C, y_C, z_C)$. One of these plaintexts is the original one $(x_M, y_M, 0)$. For the other plaintexts, we do not know if $z_i = 0$ for one of them. We show that this scenario has negligible probability. We have extensively experimented our scheme. In all cases, the decryption performed correctly and uniquely.

Any attack should start by trying to find the solutions of the quadratic equation $x_C^3 + ay_C^3 + a^2 z_C^3 - 3ax_C y_C z_C - 1 \equiv 0 \pmod{N}$. This is known to be equivalent to factoring as in Rabin's scheme [29]. To our knowledge, the new scheme is the first KMOV-like public key encryption scheme that has this additional property. A proof of concept implementation of our scheme with SimulaMath [35] and SageMath [33] is provided in [34].

We note that the decryption protocol in our scheme has negligible failure.

Paper Organization: The rest of this paper is organized as follows.

In Sect. 2, we review essential concepts related to curves over finite fields and the ring $\mathbb{Z}/N\mathbb{Z}$, focusing on the cubic Pell curves. In Sect. 3, we investigate the properties of the cubic Pell curve over $\mathbb{Z}/N\mathbb{Z}$ where $N = p^r q^s$ is a product of two distinct prime powers. Our public key encryption scheme is presented in Sect. 4. In Sect. 5, we provide a security analysis of our scheme by examining different attacks. We conclude the paper in Sect. 6.

2   Preliminaries

In this section, we begin by revisiting some useful definitions and properties associated with quadratic residues modulo a prime p and cubic residues modulo an integer n. Additionally, we recapitulate key properties concerning the arithmetic of the cubic Pell curve, which serves as a generalization of the traditional Pell curve in the cubic setting. And finally, we recall the Chinese remainder theorem, the Hensel lemma and some properties related to the number of solution of multivariate polynomial functions overs the set of integers modulo prime power.

2.1   Quadratic and Cubic Residue Modulo Prime Powers

Definition 1 (quadratic residue). *Let p be a prime number and r a positive integer. An integer a is a quadratic residue modulo p^r if the equation $x^2 \equiv a \pmod{p^r}$ has at least one solution, otherwise, a is a quadratic non-residue modulo p^r.*

Definition 2 (cubic residue). *Let p be a prime number. An integer a is a cubic residue modulo p if the equation $x^3 \equiv a \pmod{p}$ has at least one solution, otherwise, a is a cubic non-residue modulo p.*

Notice that when $p \equiv 1 \pmod 3$, there are $(p-1)/3$ non-zero cubic residues modulo p, and when $p \not\equiv 1 \pmod 3$, every element in $\mathbb{Z}/p\mathbb{Z}$ is a cubic residue modulo p.

Theorem 1 ([32]). *The equation $x^k \equiv a \pmod p$ has a solution if and only if $a^{(p-1)/d} \equiv 1 \pmod p$, where $d = \gcd(k, p-1)$. If the congruence has a solution, then it has d incongruent solutions modulo p.*

Corollary 1. *Let $p \geq 3$ be a prime number.*

1. *An integer a is a quadratic residue modulo p if and only if $a^{(p-1)/2} \equiv 1 \pmod p$.*
2. *An integer a is a cubic residue modulo p if and only if $a^{(p-1)/\gcd(3,p-1)} \equiv 1 \pmod p$.*

The Euler totient function ϕ is defined by $\phi(p^r) = p^{r-1}(p-1)$ if p is a prime number, and satisfies $\phi(mn) = \phi(m)\phi(n)$ if $\gcd(m, n) = 1$.

Theorem 2 ([32]). *Let n be a positive integer with a primitive root. If k is a positive integer and a is an integer relatively prime to n, then the congruence $x^k \equiv a \pmod n$ has a solution if and only if*

$$a^{\phi(n)/d} \equiv 1 \pmod n$$

where $d = \gcd(k, \phi(n))$. If the congruence has a solution, then it has d incongruent solutions modulo n.

2.2   The Cubic Pell Curve over a Field

Let $\mathbb{F}$ be a field and let $a \in \mathbb{F}$. Define the quotient ring $R_a = \mathbb{F}[t]/(t^3 - a)$. Note that R_a is a field if a is a cube non-residue in $\mathbb{F}$. An element $w \in R_a$ can be written as $w = x + yt + zt^2$ for some $(x, y, z) \in \mathbb{F}^3$. Let $w_1 = x_1 + y_1 t + z_1 t^2$ and $w_2 = x_2 + y_2 t + z_2 t^2$ be two elements of R_a. The product $w_1 \cdot w_2$ is defined by

$$w_1 \cdot w_2 = [x_1 x_2 + a(y_2 z_1 + y_1 z_2)] + [x_2 y_1 + x_1 y_2 + a z_1 z_2]t + [y_1 y_2 + x_2 z_1 + x_1 z_2]t^2.$$

The norm of $w = x + yt + zt^2$ is given by $N_a(w) = x^3 + ay^3 + a^2 z^3 - 3axyz$ (see [1]). Consider the set $\mathcal{U}_a$ of unitary elements defined as

$$\mathcal{U}_a = \left\{ x + yt + zt^2 \in R_a : x^3 + ay^3 + a^2 z^3 - 3axyz = 1 \right\},$$

and consider the cubic Pell curve over $\mathbb{F}$

$$P_a^3 = \left\{ (x, y, z) \in \mathbb{F}^3 : x^3 + ay^3 + a^2 z^3 - 3axyz = 1 \right\}.$$

The natural product on $\mathcal{U}_a$ induces the generalized Brahmagupta product $\oplus$ defined as follows.

$$w_1 \oplus w_2 = (x_1 x_2 + a(y_2 z_1 + y_1 z_2), x_2 y_1 + x_1 y_2 + a z_1 z_2, y_1 y_2 + x_2 z_1 + x_1 z_2),$$

where $w_1 = (x_1, y_1, z_1) \in P_a^3$ and $w_2 = (x_2, y_2, z_2) \in P_a^3$. Notice that $(P_a^3, \oplus)$ is a group with neutral element $(1, 0, 0)$, and the inverse of $w = (x, y, z)$ is given by $w^{-1} = (x^2 - ayz, az^2 - xy, y^2 - xz)$.

2.3   Curves over the Ring $\mathbb{Z}/N\mathbb{Z}$

Definition 3. *Let $F(x_1, x_2, \ldots, x_k)$ be a polynomial in k variables with integer coefficients, and N a positive integer. A solution $(a_1, a_2, \ldots, a_k)$ of the modular equation $F(x_1, x_2, \ldots, x_k) \equiv 0 \pmod{N}$ is said to be singular modulo N if it satisfies*

$$\frac{\partial F}{\partial x_1}(a_1, a_2, \ldots, a_k) = 0, \ \ \ldots, \ \ \frac{\partial F}{\partial x_k}(a_1, a_2, \ldots, a_k) = 0.$$

If the equation $F(x_1, x_2, \ldots, x_k) \equiv 0 \pmod{N}$ has only non-singular solutions, the curve is non-singular. We denote by c_N the number of solutions of the equation $F(x_1, x_2, \ldots, x_k) \equiv 0 \pmod{N}$, and by s_N the number of singular solutions. The number of non-singular or regular solutions is $R_N = c_N - s_N$.

The following result is useful to count the number of solutions of the equation $F(x_1, x_2, \ldots, x_k) \equiv 0 \pmod{N}$.

Theorem 3 ([7]). *Let $F(t_1, \ldots, t_k) \in \mathbb{Z}[t_1, \ldots, t_k]$ be a polynomial. Consider the curve*

$$F(t_1, \ldots, t_k) \equiv 0 \pmod{p^r},$$

Then $R_{p^r} = p^{(k-1)(r-1)} R_p$. Moreover, if the curve $F(t_1, \ldots, t_k)$ is non-singular, then $c_{p^r} = p^{(k-1)(r-1)} c_p$.

Theorem 4 ([7]). *Let p^r and q^s be two prime power integers with $\gcd(p, q) = 1$. Then*

$$c_{p^r \cdot q^s} = c_{p^r} \cdot c_{q^s}$$

2.4   Chinese Remainder Theorem and Hensel Lemma

The Chinese Remainder Theorem is often used to solve systems of equations modulo a composite number with known factorisation.

Theorem 5 (Chinese Remainder Theorem [28]**).** *Let* n_i, $i = 1, 2, \ldots, m$ *be* m *pairwise relatively prime numbers. For any set of integers* a_i, $i = 1, 2, \ldots, m$, *the system of congruences*

$$x \equiv a_i \pmod{n_i}, \quad i = 1, 2, \ldots, m, \tag{1}$$

has exactly one solution modulo $N = \prod_{i=1}^{m} n_i$.

The following algorithm gives the details in the Chinese Remainder Algorithm to determine the unique solution of the Eq. (1) modulo $N = \prod_{i=1}^{m} n_i$.

Algorithm 1. Chinese Remainder Algorithm

Input: m, a_i, n_i for $i = 1, 2, \ldots, m$.
Output: The unique solution X modulo $N = \prod_{i=1}^{m} n_i$ of Equation (1).
1: Compute $N_k = \prod_{i=1, i \neq k}^{m} n_i$ for $k = 1, 2, \ldots, m$.
2: Compute $X_i = N_i^{-1} \pmod{n_i}$ for $i = 1, 2, \ldots, m$.
3: Compute $X = \sum_{i=1}^{m} a_i X_i N_i \pmod{N}$.
4: Return X.

The following result is a simple application of Theorem 5.

Corollary 2. *Let* p^r *and* q^s *be two prime powers with* $p \neq q$, *and* $N = p^r q^s$. *Let* a_p *and* a_q *be integers. The unique solution to the system*

$$x = a_p \pmod{p^r}, \quad x = a_q \pmod{q^s},$$

is given by

$$X = a_p \times q^s \times [(q^s)^{-1} \bmod p^r] + a_q \times p^r \times [(p^r)^{-1} \bmod q^s] \pmod{N}.$$

Hensel's Lemma is useful to find a solution of a polynomial equation modulo a prime power p^r when a solution modulo p is known.

Lemma 1 (Hensel's lemma, [13]**).** *Let* p *be a prime number. Let* $f(x) \in \mathbb{Z}[x]$ *be a polynomial and* $f'(x)$ *its derivative. If there exists an integer* $r_1 \in \mathbb{Z}/p\mathbb{Z}$ *such that* $f(r_1) \equiv 0 \pmod{p}$ *and* $f'(r_1) \not\equiv 0 \pmod{p}$, *then there exists a unique sequence* $(r_n)_{n \geq 1}$ *of integers satisfying for all* $n \geq 1$

- $r_n \equiv r_1 \pmod{p}$,

- $f(r_n) \equiv 0 \pmod{p^n}$

Moreover, for $n \geq 2$, r_n is given by

$$r_n = r_{n-1} - \frac{f(r_{n-1})}{f'(r_{n-1})} \pmod{p^n},$$

A possible application of Hensel's lemma is the following result which concerns the solutions of the quadratic equation $ax^2 + bx + c \equiv 0 \pmod{p}$.

Corollary 3 (Quadratic equation). *Let a, b, c be integers et p an odd prime numbers. Suppose that $\gcd(a, p) = 1$ and $\Delta = b^2 - 4ac \neq 0 \pmod{p}$. If Δ is a quadratic residue modulo p, then, for $n \geq 1$, the equation $ax^2 + bx + c \equiv 0 \pmod{p^n}$ has two roots y_n and z_n, recursively defined by*

$$y_1 = \frac{-b + \sqrt{\Delta}}{2a} \pmod{p},$$

$$z_1 = \frac{-b - \sqrt{\Delta}}{2a} \pmod{p},$$

$$y_{n+1} = y_n - \frac{ay_n^2 + by_n + c}{2ay_n + b} \pmod{p^{n+1}},$$

$$z_{n+1} = z_n - \frac{az_n^2 + bz_n + c}{2az_n + b} \pmod{p^{n+1}}.$$

Notice that if $p \equiv 3 \pmod{4}$ and Δ is a quadratic residue modulo p, then

$$\sqrt{\Delta} \equiv \pm \Delta^{\frac{p+1}{4}} \pmod{p}.$$

In the case of $p \equiv 1 \pmod{4}$, one can use the Tonelli-Shank algorithm [44] to compute the square roots of Δ modulo p.

3 The Cubic Pell Curve over the Ring $\mathbb{Z}/N\mathbb{Z}$ with $N = p^r q^s$

Let $N = p^r q^s$ where p and q are two different prime numbers, and r and s are positive integers. Let $a \in Z/NZ\backslash\{0\}$. In this section, we study the properties of the cubic Pell curve with the equation

$$x^3 + ay^3 + a^2 z^3 - 3axyz \equiv 1 \pmod{N}.$$

The generalized Brahmagupta product $\oplus$ defined in Sect. 2.2 works perfectly for two solutions of the cubic Pell curve modulo N. Moreover, for a positive integer n, we define the scalar multiplication of a solution (x, y, z) by n as follows

$$n \otimes (x, y, z) = (x, y, z) \oplus \cdots \oplus (x, y, z) \quad (n \text{ times}).$$

Proposition 1. *Let $N > 3$ be an integer, and $a \in \mathbb{Z}/N\mathbb{Z}$. The cubic Pell curve*

$$\mathcal{PC}_a(N) : x^3 + ay^3 + a^2z^3 - 3axyz \equiv 1 \pmod{N},$$

is nonsingular.

Proof. Let $F(x, y, z) = x^3 + ay^3 + a^2z^3 - 3axyz - 1$. Consider the system of equations modulo N,

$$F(x, y, z) \equiv 0, \quad \frac{\partial F}{\partial x}(x, y, z) \equiv 0, \quad \frac{\partial F}{\partial y}(x, y, z) \equiv 0, \quad \frac{\partial F}{\partial z}(x, y, z) \equiv 0, \qquad (2)$$

that is

$$\begin{cases} x^3 + ay^3 + a^2z^3 - 3axyz \equiv 1 & \pmod{N} \\ 3x^2 - 3ayz \equiv 0 & \pmod{N}, \\ 3ay^2 - 3axz \equiv 0 & \pmod{N}, \\ 3a^2z^2 - 3axy \equiv 0 & \pmod{N}, \end{cases}$$

This implies that

$$\begin{cases} 3x^3 - 3axyz \equiv 0 & \pmod{N}, \\ 3ay^3 - 3axyz \equiv 0 & \pmod{N}, \\ 3a^2z^3 - 3axyz \equiv 0 & \pmod{N}, \end{cases}$$

Summing the three sides, we get

$$3\left(x^3 + ay^3 + a^2z^3 - 3axyz\right) \equiv 0 \pmod{N}.$$

Since $x^3 + ay^3 + a^2z^3 - 3axyz \equiv 1 \pmod{N}$, then $3 \equiv 0 \pmod{N}$, which is impossible. Hence the system (2) has no solution and the cubic Pell curve is nonsingular. $\square$

For a prime power p^r and $a \in \mathbb{Z}/p^r\mathbb{Z}$, let $\mathcal{PC}_a\left(p^r\right)$ be the set of the solutions of the cubic Pell curve

$$x^3 + ay^3 + a^2z^3 - 3axyz \equiv 1 \pmod{p^r}.$$

The proof of the following result can found in [12].

Lemma 2. *Let p be a prime number with $p \equiv 1 \pmod 3$. Let a be an integer with $\gcd(a, p) = 1$. The cardinality of $\mathcal{PC}_a\left(p\right)$ is*

$$|\mathcal{PC}_a\left(p\right)| = \begin{cases} p^2 + p + 1 & \text{if } a \text{ is a cube non-residue modulo} p, \\ (p - 1)^2 & \text{if } a \text{ is a non-zero cube residue modulo} p. \end{cases}$$

When $r \geq 2$, we have the following result.

Lemma 3. *Let p^r be a prime power with $p \equiv 1 \pmod 3$, and $r \geq 1$. Let $a \in \mathbb{Z}/p^r\mathbb{Z}$ with $\gcd(a, p) = 1$. The cardinality of $\mathcal{PC}_a\left(p^r\right)$ is*

$$|\mathcal{PC}_a\left(p^r\right)| = \begin{cases} p^{2(r-1)}\left(p^2 + p + 1\right) & \text{if } a \text{ is a cubic non residue modulo } p, \\ p^{2(r-1)}(p - 1)^2 & \text{if } a \text{ is a cubic residue modulo } p. \end{cases}$$

Proof. Suppose that p is a prime number such that $p \equiv 1 \pmod 3$ and let a be an integer with $\gcd(a, p) = 1$. By Proposition 1, the curve $x^3 + ay^3 + a^2 z^3 - 3axyz \equiv 1 \pmod{p^r}$ is non-singular. Therefore, by applying Lemma 3 with $k = 3$, we get

$$|\mathcal{PC}_a (p^r)| = p^{(3-1)(r-1)} |\mathcal{PC}_a (p)| = p^{2(r-1)} |\mathcal{PC}_a (p)|.$$

Combining with Lemma 2, we get

$$|\mathcal{PC}_a (p^r)| = \begin{cases} p^{2(r-1)} \left(p^2 + p + 1 \right) & \text{if } a \text{ is a cubic non residue modulo } p, \\ p^{2(r-1)}(p-1)^2 & \text{if } a \text{ is a cubic residue modulo } p. \end{cases}$$

This terminates the proof. $\qquad\qquad\qquad\qquad\qquad\qquad\qquad\qquad\qquad\qquad\qquad$ $\square$

For $N = p^r q^s$ and $a \in \mathbb{Z}/N\mathbb{Z}$, let $\mathcal{PC}_a(N)$ be the set of the solutions of the cubic Pell curve

$$x^3 + ay^3 + a^2 z^3 - 3axyz \equiv 1 \pmod{N}.$$

The following result is an easy consequence of Lemma 3.

Corollary 4. *Let* $N = p^r q^s$ *be a prime power modulus with* $p, q \equiv 1 \pmod 3$. *The number of solutions of the cubic Pell curve* $x^3 + ay^3 + a^2 z^3 - 3axyz \equiv 1 \pmod N$ *is given by*

$$|\mathcal{PC}_a(N)| = |\mathcal{PC}_a (p^r)| \, |\mathcal{PC}_a (q^s)|,$$

where $\mathcal{PC}_a (p^r)$ *and* $\mathcal{PC}_a (p^r)$ *are the sets of the solutions of the cubic Pell equation* $x^3 + ay^3 + a^2 z^3 - 3axyz \equiv 1$ *modulo* p^r *and modulo* q^s *respectively.*

Proof. Let $N = p^r q^s$ with $p, q \equiv 1 \pmod 3$. Let $a \in \mathbb{Z}/N\mathbb{Z}$ with $a \neq 0$. By the Chinese Remainder Theorem, there is a bijection between $\mathbb{Z}/N\mathbb{Z}$ and $\mathbb{Z}/p^r\mathbb{Z} \times \mathbb{Z}/q^s\mathbb{Z}$. This induces a bijection between the solutions (x_N, y_N, z_N) of the cubic Pell curve $x^3 + ay^3 + a^2 z^3 - 3axyz \equiv 1 \pmod N$, and the solutions $((x_{p^r}, y_{p^r}, z_{p^r}), (x_{q^s}, y_{q^s}, z_{q^s}))$ formed by a solution of the cubic Pell curve $x^3 + ay^3 + a^2 z^3 - 3axyz \equiv 1$ modulo p^r and a solution of the same curve modulo p^r. Moreover, this implies that

$$|\mathcal{PC}_a(N)| = |\mathcal{PC}_a (p^r)| \, |\mathcal{PC}_a (q^s)|,$$

where $\mathcal{PC}_a (p^r)$ and $\mathcal{PC}_a (p^r)$ are the sets of the solutions of the cubic Pell equation $x^3 + ay^3 + a^2 z^3 - 3axyz \equiv 1$ modulo p^r and modulo q^s respectively. $\quad \square$

The properties of the cubic Pell curve modulo a prime power modulus $N = p^r q^s$ with $p, q \equiv 1 \pmod 3$ can be summarized as follows.

1. For a prime power modulus $N = p^r q^s$, let $\mathcal{R}^3(N)$ be the set of the cubic residues a modulo p with $(\gcd(a, N) = 1)$. Its cardinality is

$$|\mathcal{R}^3(N)| = \frac{p^{r-1} q^{s-1}(p-1)(q-1)}{9}.$$

2. Define the values

$$
\begin{aligned}
\psi_1(N) &= p^{2(r-1)}q^{2(s-1)}\left(p^2+p+1\right)\left(q^2+q+1\right), \\
\psi_2(N) &= p^{2(r-1)}q^{2(s-1)}(p-1)^2(q-1)^2, \\
\psi_3(N) &= p^{2(r-1)}q^{2(s-1)}\left(p^2+p+1\right)(q-1)^2, \\
\psi_4(N) &= p^{2(r-1)}q^{2(s-1)}(p-1)^2\left(q^2+q+1\right).
\end{aligned}
\tag{3}
$$

For $a \in \mathbb{Z}/N\mathbb{Z}$ with $\gcd(a, N) = 1$, the number of solutions of the equation $x^3 + ay^3 + a^2z^3 - 3axyz \equiv 1 \pmod{N}$ is then

$$
|\mathcal{PC}_a(N)| = \begin{cases}
\psi_1(N) & \text{if } a \notin \mathcal{R}^3(p) \text{ and } a \notin \mathcal{R}^3(q), \\
\psi_2(N) & \text{if } a \in \mathcal{R}^3(p) \text{ and } a \in \mathcal{R}^3(q), \\
\psi_3(N) & \text{if } a \notin \mathcal{R}^3(p) \text{ and } a \in \mathcal{R}^3(q), \\
\psi_4(N) & \text{if } a \in \mathcal{R}^3(p) \text{ and } a \notin \mathcal{R}^3(q).
\end{cases}
$$

where $\mathcal{R}^3(p)$ is the set of the cubic residues modulo p, and $\mathcal{R}^3(q)$ is the set of the cubic residues modulo q.

3. Let $\mathcal{PC}_a(N)$ be the set of the solutions of the cubic Pell curve $x^3 + ay^3 + a^2z^3 - 3axyz \equiv 1 \pmod{N}$ in $(\mathbb{Z}/N\mathbb{Z})^3$. Then $(\mathcal{PC}_a(N), \oplus)$ is an abelian group with order $|\mathcal{PC}_a(N)|$.

4. The neutral element of $\mathcal{PC}_a(N)$ is $(1, 0, 0)$.

5. The inverse of a solution $(x, y, z) \in \mathcal{PC}_a(N)$ is $\left(x^2 - ayz, az^2 - xy, y^2 - xz\right)$ $\pmod{N}$.

6. The sum of two solutions $(x_1, y_1, z_1), (x_2, y_2, z_2) \in \mathcal{PC}_a(N)$ is (x_3, y_3, z_3) with

$$
(x_3, y_3, z_3) = (x_1x_2 + a(y_2z_1 + y_1z_2), x_2y_1 + x_1y_2 + az_1z_2, y_1y_2 + x_2z_1 + x_1z_2).
$$

7. The scalar product of a solution $(x, y, z) \in \mathcal{PC}_a(N)$ by an integer n is

$$
n \otimes (x, y, z) = (x, y, z) \oplus \cdots \oplus (x, y, z) \quad (n \text{ times}).
$$

8. For any positive integer k, and any solution $(x, y, z) \in \mathcal{PC}_a(N)$,

$$
(1 + k|\mathcal{PC}_a(N)|) \otimes (x, y, z) = (x, y, z).
$$

In $\mathcal{PC}_a(N)$, the addition $\oplus$, the doubling, and the scalar multiplication by an integer are summarized in the following algorithms.

Algorithm 2. Addition in $\mathcal{PC}_a(N)$

Input: $N = p^rq^s$, $a \in \mathbb{Z}/N\mathbb{Z}\backslash\{0\}$, and $(x_1, y_1, z_1), (x_2, y_2, z_2) \in \mathcal{PC}_a(N)$.
Output: $(x_3, y_3, z_3) = (x_1, y_1, z_1) \oplus (x_2, y_2, z_2) \in \mathcal{PC}_a(N)$.
  1: $x_3 \equiv x_1x_2 + a(y_2z_1 + y_1z_2) \pmod{N}$.
  2: $y_3 \equiv x_2y_1 + x_1y_2 + az_1z_2 \pmod{N}$.
  3: $z_3 \equiv y_1y_2 + x_2z_1 + x_1z_2 \pmod{N}$.
  4: Return (x_3, y_3, z_3).

Algorithm 3. Doubling in $\mathcal{PC}_a(N)$

Input: $N = p^r q^s$, $a \in \mathbb{Z}/N\mathbb{Z}\backslash\{0\}$, and $(x_1, y_1, z_1) \in \mathcal{PC}_a(N)$.
Output: $(x_3, y_3, z_3) = 2 \otimes (x_1, y_1, z_1) \in \mathcal{PC}_a(N)$.
1: $x_3 \equiv x_1^2 + 2ay_1z_1 \pmod{N}$.
2: $y_3 \equiv 2x_1y_1 + az_1^2 \pmod{N}$.
3: $z_3 \equiv y_1^2 + 2x_1z_1 \pmod{N}$.
4: Return (x_3, y_3, z_3).

Algorithm 4. Left-to-right scalar multiplication in $\mathcal{PC}_a(N)$

Input: $N = p^r q^s$, $a \in \mathbb{Z}/N\mathbb{Z}\backslash\{0\}$, $(x_1, y_1, z_1) \in \mathcal{PC}_a(N)$, and an integer $n \geq 2$.
Output: $(x_2, y_2, z_2) = n \otimes (x_1, y_1, z_1) \in \mathcal{PC}_a(N)$.
1: Expand n in base 2, that is $n = (n_{k-1}n_{k-2} \ldots n_1n_0)_2$.
2: $(x_2, y_2, z_2) = (1, 0, 0)$.
3: **For** i from $k - 1$ **downto** 0 **do**
4:     $(x_2, y_2, z_2) = 2 \otimes (x_2, y_2, z_2)$.
5:     **If** $n_i = 1$ **then**
6:         $(x_2, y_2, z_2) = (x_2, y_2, z_2) \oplus (x_1, y_1, z_1)$.
7:     **End If**
8: **End For**
9: Return (x_2, y_2, z_2).

---

4   Our Construction

In this section, we present a new scheme based on the cubic Pell curve. It is a variant of both RSA and KMOV. We also provide a numerical example for our scheme.

4.1   The New Public Key Encryption Scheme

In the following algorithms, we give the algorithms of the new public key encryption scheme, namely, the key generation, the encryption, and the decryption algorithm.

Algorithm 5. Key Generation

Input: A security parameter λ, and two small positive integers r and s.
Output: A public key $\mathcal{PK}$ and a private key $\mathcal{SK}$.
1: Choose a prime number p of λ bit size with $p \equiv 1 \pmod{3}$.
2: Choose a prime number q of λ bit size with $q \equiv 1 \pmod{3}$.
3: Compute $N = p^r q^s$.
4: For $i = 1, 2, 3, 4$, compute $\psi_i(N)$ using (3).
5: Choose an integer $e \in \mathbb{Z}/N\mathbb{Z}$ such that

$$\gcd\left(e, pq\left(p^2 + p + 1\right)\left(q^2 + q + 1\right)(p-1)(q-1)\right) = 1.$$

6: For $i = 1, 2, 3, 4$, compute $d_i \equiv e^{-1} \pmod{\psi_i(N)}$.
7: The public key is $\mathcal{PK} = (N, e)$.
8: The private key is $\mathcal{SK} = (p, q, N, d_1, d_2, d_3, d_4, r, s)$.
9: Return the keypair $(\mathcal{PK}, \mathcal{SK})$.

Algorithm 6. Encryption Process

Input: A message $M = (x_M, y_M) \in (\mathbb{Z}/N\mathbb{Z}) \times (\mathbb{Z}/N\mathbb{Z})$ and a public key $\mathcal{PK} = (N, e)$.
Output: The ciphertext of M.
1: Represent the message M as $M = (x_M, y_M, 0)$.
2: Compute $a \equiv \frac{1 - x_M^3}{y_M^3} \pmod{N}$.   ▷ $(x_M, y_M, 0) \in \mathcal{PC}_a(N)$
3: Compute $(x_C, y_C, z_C) = e \otimes (x_M, y_M, 0)$ on the cubic Pell curve with the equation

$$\mathcal{PC}_a(N) : x^3 + ay^3 + a^2 z^3 - 3axyz \equiv 1 \pmod{N}.$$

4: Return the ciphertext (x_C, y_C, z_C).

Algorithm 7. Decryption Process

Input: A ciphertext (x_C, y_C, z_C), and a private key $\mathcal{SK} = (p, q, N, d_1, d_2, d_3, d_4, r, s)$.
Output: The decryption of (x_C, y_C, z_C).
1: Find the solutions $x = a_{p,1}$ and $x = a_{p,2}$ of the equation

$$x_C^3 + xy_C^3 + x^2 z_C^3 - 3xx_C y_C z_C \equiv 1 \pmod{p^r}.$$

2: Find the solutions $y = a_{q,1}$ and $y = a_{q,2}$ of the equation

$$x_C^3 + yy_C^3 + y^2 z_C^3 - 3yx_C y_C z_C \equiv 1 \pmod{q^s}.$$

3: Using the Chinese Remainder Theorem, compute $a_i \in \mathbb{Z}/N\mathbb{Z}$, $i = 1, 2, 3, 4$ such that

$$a_1 \equiv a_{p,1} \pmod{p^r}, a_1 \equiv a_{q,1} \pmod{q^s},$$
$$a_2 \equiv a_{p,1} \pmod{p^r}, a_2 \equiv a_{q,2} \pmod{q^s},$$
$$a_3 \equiv a_{p,2} \pmod{p^r}, a_3 \equiv a_{q,1} \pmod{q^s},$$
$$a_4 \equiv a_{p,2} \pmod{p^r}, a_4 \equiv a_{q,2} \pmod{q^s}.$$

4: **For** $i = 1, 2, 3, 4$ **do**
5:     Set

$$D_i = \begin{cases} d_1 & \text{if } a_i \notin \mathcal{R}^3(p) \text{ and } a_i \notin \mathcal{R}^3(q), \\ d_2 & \text{if } a_i \in \mathcal{R}^3(p) \text{ and } a_i \in \mathcal{R}^3(q), \\ d_3 & \text{if } a_i \notin \mathcal{R}^3(p) \text{ and } a_i \in \mathcal{R}^3(q), \\ d_4 & \text{if } a_i \in \mathcal{R}^3(p) \text{ and } a_i \notin \mathcal{R}^3(q), \end{cases}$$

where $\mathcal{R}^3(p)$ and $\mathcal{R}^3(q)$ are the sets of the cubic residues in $\mathbb{Z}/p\mathbb{Z}$ and $\mathbb{Z}/q\mathbb{Z}$ respectively.
6:     Compute $M_i = (x_i, y_i, z_i) = D_i \otimes (x_C, y_C, z_C)$ on the cubic Pell curve

$$\mathcal{PC}_{a_i}(N) : x^3 + a_i y^3 + a_i^2 z^3 - 3a_i xyz \equiv 1 \pmod{N}.$$

7: **End For**
8: Return the plaintext (x_i, y_i, z_i) for which $z_i = 0$.   ▷ M=$(x_i, y_i, 0)$ is the original message.

Notice that in Algorithm 7, Step 1 to Step 3 are devoted to the computation of the parameter a of the cubic Pell curve used in the encryption process given only the ciphertext (x_C, y_C, z_C) and the private parameters p, q, r, s. Since these steps require the computation of square roots modulo p and modulo q, one can choose p and q so that $p, q \equiv 7 \pmod{12}$, which implies $p, q \equiv 1 \pmod 3$ and $p, q \equiv 3 \pmod 4$. This allows to compute the square roots of a quadratic residue $\Delta \bmod p$ as $\pm \Delta^{(p+1)/4} \pmod p$.

4.2   The Failure of the Decryption Algorithm

In several schemes such as LWE [30], RLWE [21], Ramstake [38], New Hope [37], and several variants of CRYSTALS-Kyber [3], the decryption protocol is probabilistic with a negligible probability of failure. Despite their possible failure, some of the former schemes are used in many cryptographic applications such as electronic voting, electronic auction, and digital signatures.

We do not know if our new scheme presents a possibility of decryption failure. We have extensively experienced it, and the decryption protocol was always successful and unique. The following result shows that the probability of a possible failure is negligible.

Lemma 4. *Let $N = p^r q^s$ be a prime power modulus, $x_M, y_M \in \mathbb{Z}/N\mathbb{Z}$, and $(x_C, y_C, z_C) = e(x_M, y_M, 0)$ be the ciphertext computed with the cubic Pell curve $\mathcal{PC}_a(N) : x^3 + ay^3 + a^2 z^3 - 3axyz \equiv 1 \pmod N$ with $a \equiv \frac{1 - x_M^3}{y_M^3} \pmod N$. The failure probability of the decryption of the scheme lies in the interval $\left(\frac{1}{16N}, \frac{16}{N} \right)$, and is negligible.*

Proof. Let $(x_M, y_M, 0)$ be a plaintext, and (x_C, y_C, z_C) be the corresponding ciphertext over the cubic Pell curve $x^3 + ay^3 + a^2 z^3 - 3axyz \equiv 1 \pmod N$ where $a \equiv \frac{1 - x_M^3}{y_M^3}$. Suppose that another solution a_0 of the quadratic equation $x_C^3 + a_0 y_C^3 + a_0^2 z_C^3 - 3a_0 x_C y_C z_C \equiv 1 \pmod N$ is such that $d(x_C, y_C, z_C) = (x_1, y_1, 0)$ for one decryption exponent $d \in \{d_1, d_2, d_3, d_4\}$, on the cubic Pell curve $\mathcal{PC}_{a_0}(N) : x^3 + a_0 y^3 + a_0^2 z^3 - 3a_0 xyz \equiv 1 \pmod N$, and $(x_M, y_M, 0) \neq (x_1, y_1, 0)$. Then, $(x_1, y_1, 0) \in \mathcal{PC}_{a_0}^0(N)$ where $\mathcal{PC}_{a_0}^0(N)$ is the set of the solutions of the cubic equation

$$\mathcal{PC}_{a_0}^0(N) : x^3 + a_0 y^3 \equiv 1 \pmod N.$$

This scenario happens with probability

$$\text{Prob}(z = 0) = \frac{|\mathcal{PC}_{a_0}^0(N)|}{|\mathcal{PC}_{a_0}(N)|}. \tag{4}$$

The curve $\mathcal{PC}_{a_0}^0(N)$ is a specific case of the cubic Pell equation. Hence, if $\mathcal{PC}_{a_0}^0(p)$ is the number of the solutions of $x^3 + ay^3 \equiv 1 \pmod p$, then, by Theorem 3, the number of the solutions of the equation $x^3 + ay^3 \equiv 1 \pmod{p^r}$ is

$$\left| \mathcal{PC}_{a_0}^0(p^r) \right| = p^{r-1} \left| \mathcal{PC}_{a_0}^0(p) \right|.$$

Using the Chinese Remainder Theorem, it follows that the number of solutions of the equation $x^3 + ay^3 \equiv 1 \pmod{N}$ is

$$\left|\mathcal{PC}_{a_0}^0(N)\right| = p^{r-1}p^{s-1}\left|\mathcal{PC}_{a_0}^0(p)\right|\left|\mathcal{PC}_{a_0}^0(q)\right|.$$

The curve $x^3 + ay^3 \equiv 1 \pmod{p}$ is a specific form of the Selmer curve. It is birationally equivalent to the elliptic curve

$$E_p : v^2 \equiv u^3 - 432a^2 \pmod{p},$$

under the transformations

$$u = -\frac{12a_1 y}{x-1}, \quad v = 36a_1\frac{x+1}{x-1},$$

$$x = \frac{v + 36a_1}{v - 36a_1}, \quad y = -\frac{6u}{v - 36a_1}.$$

By Hasse Theorem [13], the order of E_p satisfies

$$\left(\sqrt{p} - 1\right)^2 \leq |E_p| \leq \left(\sqrt{p} + 1\right)^2.$$

Since $|E_p| = \left|\mathcal{PC}_{a_0}^0(p)\right|$, then using $\left(\sqrt{p} - 1\right)^2 > \frac{1}{2}p$ and $\left(\sqrt{p} + 1\right)^2 < 2p$, we get

$$\frac{1}{2}p < \left|\mathcal{PC}_{a_0}^0(p)\right| < 2p.$$

By Theorem 3 and the Chinese Remainder Theorem, this implies that $\left|\mathcal{PC}_{a_0}^0(N)\right|$ satisfies

$$\frac{1}{4}p^{r-1}p^{s-1}pq \leq \left|\mathcal{PC}_{a_0}^0(N)\right| \leq 4\frac{1}{4}p^{r-1}p^{s-1}pq,$$

that is

$$\frac{1}{4}N \leq \left|\mathcal{PC}_{a_0}^0(N)\right| \leq 4N. \qquad (5)$$

On the other hand, the orders $\psi_i(N)$, $i = 1, 2, 3, 4$ as defined in 3 satisfy

$$p^{r-1}p^{s-1}(p-1)^2(q-1)^2 \leq \psi_i(N) \leq p^{r-1}p^{s-1}\left(p^2 + p + 1\right)\left(q^2 + q + 1\right).$$

Since $|\mathcal{PC}_{a_0}(N)| \in \{\psi_1(N), \psi_2(N), \psi_3(N), \psi_4(N)\}$, and since $(p-1)^2 > \frac{1}{2}p^2$, and $p^2 + p + 1 < 2p^2$ for $p > 4$, we get

$$\frac{1}{4}N^2 \leq |\mathcal{PC}_{a_0}(N)| \leq 4N^2.$$

Combining this with (5), the probability (4) satisfies

$$\frac{1}{16N} < \text{Prob}(z = 0) < \frac{16}{N}.$$

This shows that the decryption failure is negligible. $\qquad\square$

4.3   A Numerical Example for the Scheme

Let us consider the following small example.

1. **Key Generation**
   - Let $p = 922039$, $q = 760531$, $r = 1$ and $s = 3$. Then

$$N = 922039 \times 760531^3 = 405601968528411801552349.$$

   - Let $e = 190681261905711342654691$. The public key is

$$(N, e) = (405601968528411801552349, 190681261905711342654691).$$

   - The private exponents are

$$
\begin{aligned}
d_1 &= e^{-1} \quad (\text{mod } p^{2(r-1)} q^{2(s-1)} \left(p^2 + p + 1\right) \left(q^2 + q + 1\right)), \\
&= 118972772223283451014251175069491011419223088520, \\
d_2 &= e^{-1} \quad (\text{mod } p^{2(r-1)} q^{2(s-1)} \left(p - 1\right)^2 \left(q - 1\right)^2), \\
&= 526736078136313181690638864666078459519302224 11, \\
d_3 &= e^{-1} \quad (\text{mod } p^{2(r-1)} q^{2(s-1)} \left(p^2 + p + 1\right) \left(q - 1\right)^2), \\
&= 110562086970292565355181851346394599567010668711, \\
d_4 &= e^{-1} \quad (\text{mod } p^{2(r-1)} q^{2(s-1)} \left(p - 1\right)^2 \left(q^2 + q + 1\right)), \\
&= 155064179962520723245280314053380086273645670395.
\end{aligned}
$$

   - The private key is $(p, q, N, d_1, d_2, d_3, d_4, r, s)$.
2. **The plaintext**
   Consider the plaintext $(x_M, y_M, 0)$ with

$$
\begin{aligned}
x_M &= 94727413669590175405397, \\
y_M &= 400429216716868987768230.
\end{aligned}
$$

3. **Encryption :**
   - First we compute

$$a \equiv \frac{1 - x_M^3}{y_M^3} \quad (\text{mod } N) = 402129345655132093067351.$$

   - The cubic Pell curve is then

$$\mathcal{PC}_a(N) : x^3 + ay^3 + a^2 z^3 - 3axyz \equiv 1 \quad (\text{mod } N).$$

   - We compute $(x_C, y_C, z_C) = e \otimes (x_M, y_M, 0)$ on the cubic Pell curve $\mathcal{PC}_a(N)$ using the Algorithm 4. We get the ciphertext (x_C, y_C, z_C) with

$$
\begin{aligned}
x_C &= 296657492079316956423913, \\
y_C &= 336170831341196089366817, \\
z_C &= 351828474470867029080629.
\end{aligned}
$$

4. **Decryption :**
   - Solving the equation

$$x_C^3 + xy_C^3 + x^2 z_C^3 - 3xx_C y_C z_C \equiv 1 \pmod{p},$$

we get the solutions $x_1 = 124693$, and $x_2 = 29301$.
   - Apply Hensel's lemma with x_1 and x_2 to solve the equation

$$x_C^3 + xy_C^3 + x^2 z_C^3 - 3xx_C y_C z_C \equiv 1 \pmod{p^r}.$$

We get $x = a_{p,1} = 124693$ and $x = a_{p,2} = 29301$.
   - Solve the equation

$$x_C^3 + yy_C^3 + y^2 z_C^3 - 3yx_C y_C z_C \equiv 1 \pmod{q}.$$

We get the solutions $y_1 = 272543$, and $y_2 = 758118$.
   - Apply Hensel's lemma with y_1 and y_2 to solve the equation

$$x_C^3 + yy_C^3 + y^2 z_C^3 - 3yx_C y_C z_C \equiv 1 \pmod{q^s}.$$

We get $y = a_{q,1} = 362045505517707447$ and $y = a_{q,2} = 2288749681600$
44609.
   - Using the Chinese theorem with $a_{p,1}$ and $a_{q,1}$, we get

$$a_1 = 40212934565513209306735 1 \pmod{N}.$$

   - Using the Chinese theorem with $a_{p,2}$ and $a_{q,2}$, we get

$$a_2 = 261500816821281874691178 \pmod{N}.$$

   - Using the Chinese theorem with $a_{p,1}$ and $a_{q,2}$, we get

$$a_3 = 170916396245462831245876 \pmod{N}.$$

   - Using the Chinese theorem with $a_{p,2}$ and $a_{q,1}$, we get

$$a_4 = 87111797702539334960304 \pmod{N}.$$

   - We can check that a_1 is a cubic residue modulo p and a cubic non-residue modulo q. So $D_1 = d_4$. We then compute $(x_1, y_1, z_1) = d_4 \otimes (x_C, y_C, z_C)$ on the cubic Pell curve $\mathcal{PC}_{a_1}(N)$ using Algorithm 4. We get

$$(x_1, y_1, z_1) = (94727413669590175405397, 400429216716868987768230, 0),$$

which is the original plaintext.
   - We can check that a_2 is a cubic non-residue modulo p and a cubic non-residue modulo q. So $D_2 = d_1$. We then compute $(x_2, y_2, z_2) = d_1 \otimes (x_C, y_C, z_C)$ on the cubic Pell curve $\mathcal{PC}_{a_1}(N)$ using Algorithm 4. We get a solution (x_2, y_2, z_2) with

$$x_2 = 315084178973498538996923,$$
$$y_2 = 334849906408238591863534,$$
$$z_2 = 119465479892270850302989.$$

which is not the original plaintext.

- We can check that a_3 is a cubic residue modulo p and cubic non-residue modulo q. So $D_3 = d_4$. We then compute $(x_1, y_2, z_2) = d_4 \otimes (x_C, y_C, z_C)$ on the cubic Pell curve $\mathcal{PC}_{a_1}(N)$ using Algorithm 4. We get a solution (x_3, y_3, z_3) with

$$x_3 = 34878291015633084269526,$$
$$y_3 = 33424152918940642367808,$$
$$z_3 = 14770289280192790597357.$$

which is not the original plaintext.
- We can check that a_4 is a cubic non-residue modulo p and a cubic non-residue modulo q. So $D_4 = d_1$. We then compute $(x_4, y_4, z_4) = d_1 \otimes (x_C, y_C, z_C)$ on the cubic Pell curve $\mathcal{PC}_{a_1}(N)$ using Algorithm 4. We get a solution (x_4, y_4, z_4) with

$$x_4 = 61028682486757871707051,$$
$$y_4 = 40103759393570115595368,$$
$$z_4 = 37736455561875474588176.$$

which is not the original plaintext.

We notice that the decryption performs perfectly, and is unique.

5 Security Analysis

In this section, we study the security of the new scheme as described in Sect. 4.

5.1 Resistance Against Finding the Cubic Pell Curve

In the new scheme, the value of a in the Pell curve $x^3 + ay^3 + a^2z^3 - 3axyz \equiv 1 \pmod{N}$ is not public. Indeed, the public parameters are N and a solution $(x_C, y_C, z_C) \in \mathcal{PC}_a(N)$. To compute the parameter a, one should solve the modular equation

$$z_C^3 a^2 + \left(y_C^3 - 3x_C y_C z_C\right) a + x_C^3 - 1 \equiv 0 \pmod{N},$$

which is quadratic in a. The discriminant of the equation is

$$\Delta = \left(y_C^3 - 3x_C y_C z_C\right)^2 - 4z_C^3 \left(x_C^3 - 1\right).$$

Then, finding a is equivalent to solving the quadratic equation

$$x^2 \equiv \Delta \pmod{N},$$

where the factorization of N is unknown. This is known as the SQRT-MOD-N problem, and is equivalent to the integer factoring problem [13].

5.2   Resistance Against the Small Private Exponent Attacks

It is known that using a small private exponent is insecure in several schemes such as RSA [6,9,42], KMOV [24] and others [25]. The main known techniques are based on the continued fraction algorithm [42] or on Coppersmith's method [6,9]. The attacks based on the continued fraction algorithm use the following well known result of Legendre (see Theorem 184 of [14]).

Theorem 6 (Legendre). *Let ξ be a positive number. Let a and b be integers such that $\gcd(a, b) = 1$ and*

$$\left| \xi - \frac{a}{b} \right| < \frac{1}{2b^2}.$$

Then $\frac{a}{b}$ is one of the convergents of the continued fraction expansion of ξ.

In most cases, the RSA moduli and their variants are the product of large prime numbers of the same bit size. In our schemes, we also suppose that the prime numbers p and q in the modulus $N = p^r q^s$ are of the same bit size, and ordered so that $q < p < 2q$. The following result gives effective bounds for p and q.

Proposition 2. *Let $N = p^r q^s$ be a prime power modulus with $q < p < 2q$. Then*

$$2^{\frac{-1}{r+s}} N^{\frac{1}{r+s}} < q < N^{\frac{1}{r+s}} < p < 2^{\frac{s}{r+s}} N^{\frac{1}{r+s}}.$$

Proof. Suppose that $q < p < 2q$. Then $q^r < p^r < 2^r q^r$ and $q^s < p^s < 2^s q^s$. Multiplying the former inequalities, we get

$$q^{r+s} < N < 2^{r+s} q^{r+s},$$

which implies that $q < N^{\frac{1}{r+s}}$ and $2^{\frac{-1}{r+s}} N^{\frac{1}{r+s}} < q$.

Multiplying $q^s < p^s < 2^s q^s$ by p^r, we get

$$N < p^{r+s} < 2^s N,$$

and $N^{\frac{1}{r+s}} < p < 2^{\frac{s}{r+s}} N^{\frac{1}{r+s}}$. Summarizing all inequalities, we get

$$2^{\frac{-1}{r+s}} N^{\frac{1}{r+s}} < q < N^{\frac{1}{r+s}} < p < 2^{\frac{s}{r+s}} N^{\frac{1}{r+s}}.$$

This terminates the proof.   □

For $i = 1, 2, 3, 4$, and $\psi(N) \in \{\psi_1(N), \psi_2(N), \psi_3(N), \psi_4(N)\}$ as given in (3), the following result gives an approximation of $\psi_i(N)$ in terms of N.

Proposition 3. *Let $N = p^r q^s$ be a prime power modulus with $q < p < 2q$. Let $\psi(N) \in \{\psi_1(N), \psi_2(N), \psi_3(N), \psi_4(N)\}$ as given in (3). Then N^2 is an approximation of $\psi(N)$ with*

$$\left| \psi(N) - N^2 \right| < 8N^{2 - \frac{1}{r+s}}.$$

Proof. Let $\psi(N) \in \{\psi_1(N), \psi_2(N), \psi_3(N), \psi_4(N)\}$. Then

$$p^{2(r-1)}q^{2(s-1)}(p-1)^2(q-1)^2 \leq \psi(N) \leq p^{2(r-1)}q^{2(s-1)}\left(p^2+p+1\right)\left(q^2+p+1\right).$$

This can be rewritten as

$$N^2\left(1 - \frac{2}{p} + \frac{1}{p^2}\right)\left(1 - \frac{2}{q} + \frac{1}{q^2}\right) \leq \psi(N) \leq N^2\left(1 + \frac{1}{p} + \frac{1}{p^2}\right)\left(1 + \frac{1}{q} + \frac{1}{q^2}\right)$$

By Proposition 2, we have $N^{\frac{1}{r+s}} < p$. Then

$$1 - \frac{2}{p} + \frac{1}{p^2} > 1 - \frac{2}{p} > 1 - 2N^{\frac{-1}{r+s}}.$$

Similarly, by Proposition 2, we have $2^{\frac{-1}{r+s}}N^{\frac{1}{r+s}} < q$. Then

$$1 - \frac{2}{q} + \frac{1}{q^2} > 1 - \frac{2}{q} > 1 - 2^{1+\frac{1}{r+s}}N^{\frac{-1}{r+s}}.$$

Using the former inequalities, we get

$$\psi(N) > N^2\left(1 - 2N^{\frac{-1}{r+s}}\right)\left(1 - 2^{1+\frac{1}{r+s}}N^{\frac{-1}{r+s}}\right)$$

$$= N^2\left(1 - 2^{1+\frac{1}{r+s}}N^{\frac{-1}{r+s}} - 2N^{\frac{-1}{r+s}} + 2^{2+\frac{1}{r+s}}N^{\frac{-2}{r+s}}\right)$$

$$> N^2\left(1 - 8N^{\frac{-1}{r+s}}\right).$$

Using this, we get

$$\psi(N) - N^2 > -8N^{2-\frac{1}{r+s}}. \qquad (6)$$

Also, by Proposition 2, we have $N^{\frac{1}{r+s}} < p$. Then

$$1 + \frac{1}{p} + \frac{1}{p^2} < 1 + \frac{2}{p} < 1 + 2N^{\frac{-1}{r+s}}.$$

Similarly, by Proposition 2, we have $2^{\frac{-1}{r+s}}N^{\frac{1}{r+s}} < q$. Then

$$1 + \frac{1}{q} + \frac{1}{q^2} < 1 + \frac{2}{q} < 1 + 2^{\frac{r+s+1}{r+s}}N^{\frac{-1}{r+s}}.$$

Plugging this in $\psi(N)$, we get

$$\psi(N) < N^2\left(1 + 2N^{\frac{-1}{r+s}}\right)\left(1 + 2^{\frac{r+s+1}{r+s}}N^{\frac{-1}{r+s}}\right)$$

$$= N^2\left(1 + 2^{\frac{r+s+1}{r+s}}N^{\frac{-1}{r+s}} + 2N^{\frac{-1}{r+s}} + 2^{1+\frac{r+s+1}{r+s}}N^{\frac{-2}{r+s}}\right)$$

$$< N^2\left(1 + 8N^{\frac{-1}{r+s}}\right).$$

Using this, we get

$$\psi(N) - N^2 < 8N^{2-\frac{1}{r+s}}. \qquad (7)$$

Combining (6) and (7), we get

$$\left|\psi(N) - N^2\right| < 8N^{2-\frac{1}{r+s}}.$$

This completes the proof. $\qquad\qquad\qquad\qquad\qquad\qquad\qquad\qquad\qquad\qquad\square$

The following result which is based on the continued fraction algorithm shows that using a small private exponent d is vulnerable.

Proposition 4. *Let $N = p^r q^s$ be a prime power modulus with $q < p < 2q$. Let e be a public exponent, and $d \equiv e^{-1} \pmod{\psi(N)}$ where $\psi(N)$ is one of the orders $\psi_1(N)$, $\psi_2(N)$, $\psi_3(N)$, or $\psi_4(N)$. If $e < \psi(N)$, and*

$$d < \frac{\sqrt{2}}{4} N^{\frac{1}{2(r+s)}},$$

then one can find d and factor N in polynomial time.

Proof. The relation $d \equiv e^{-1} \pmod{\psi(N)}$ can be rewritten as $ed - k\psi(N) = 1$ with a positive integer k. This can be rewritten as

$$\left|\frac{e}{\psi(N)} - \frac{k}{d}\right| = \frac{1}{d\psi(N)}.$$

Suppose $e < \psi(N)$. Using Proposition 3, we have $\left|\psi(N) - N^2\right| < 8N^{2-\frac{1}{r+s}}$. Then

$$\left|\frac{e}{N^2} - \frac{k}{d}\right| < \left|\frac{e}{N^2} - \frac{e}{\psi(N)}\right| + \left|\frac{e}{\psi(N)} - \frac{k}{d}\right|$$

$$< \psi(N)\left|\frac{\psi(N) - N^2}{N^2\psi(N)}\right| + \frac{1}{d\psi(N)}$$

$$< \frac{8N^{2-\frac{1}{r+s}}}{N^2} + \frac{1}{d\psi(N)}$$

$$< \frac{8}{N^{\frac{1}{r+s}}} + \frac{1}{N^2 d}.$$

Suppose that $d < \frac{\sqrt{2}}{8} N^{\frac{1}{2(r+s)}}$. Then $d < \frac{1}{4}N^2$, and

$$\left|\frac{e}{N^2} - \frac{k}{d}\right| < \frac{1}{4d^2} + \frac{1}{4d^2} = \frac{1}{2d^2}.$$

By Legendre's Theorem 6, this implies that $\frac{k}{d}$ is a convergent of the continued fraction expansion of $\frac{e}{N^2}$, which can be found in polynomial time. Using k and d in the equation $ed - k\psi(N) = 1$, we get

$$\psi(N) = \frac{ed - 1}{k}.$$

Using $\psi(N)$, we get

$$g = \gcd\left(N^2, \psi(N)\right) = p^{2(r-1)}q^{2(s-1)}.$$

In turn, this gives

$$pq = \frac{N}{\sqrt{g}},$$

and, combining with $N = p^r q^s$, we finally get

$$p = \left(\frac{N^{s-1}}{g^{\frac{s}{2}}}\right)^{\frac{1}{s-r}}, q = \left(\frac{N^{r-1}}{g^{\frac{r}{2}}}\right)^{\frac{1}{r-s}}.$$

This completes the proof.   □

6   Conclusion

In this paper, we proposed a new public key cryptosystem based on the cubic
Pell curve modulo a prime power modulus of the form $N = p^r q^s$ to perform
encryption and decryption. We studied its security and showed that it is based on
two computationally hard problems, namely, the integer factorization problem,
and the Rabin trapdoor. The advantage of the new scheme is that the arithmetic
operations have to be performed on a cubic Pell curve which is known only to
the sender and the recipient.

References

1. Barbeau, E.J.: Pell Equation, Chapter 7: The Cubic Analogue of Pell Equation.
   Springer, New York (2003)
2. Boneh, D.: Twenty years of attacks on the RSA cryptosystem. Notices Amer. Math.
   Soc. **46**(2), 203–213 (1999)
3. Bos J., et al.: CRYSTALS-Kyber: a CCA-secure module-latticebased KEM. In:
   2018 IEEE European Symposium on Security and Privacy (EuroSP), pp. 353–367.
   IEEE (2018)
4. Boneh, D., Durfee, G., Frankel, Y.: An attack on RSA given a small fraction of the
   private key bits. In: Ohta, K., Pei, D. (eds.) ASIACRYPT 1998. LNCS, vol. 1514,
   pp. 25–34. Springer, Heidelberg (1998). https://doi.org/10.1007/3-540-49649-1_3
5. Boneh, D., Durfee, G., Howgrave-Graham, N.: Factoring $N = p^r q$ for large r.
   CRYPTO **1999**, 326–337 (1999)
6. Boneh, D., Durfee, G.: Cryptanalysis of RSA with private key d less than $N^{0.292}$.
   In: Advances in Cryptology-Eurocrypt 1999, LNCs, vol. 1592, pp. 1–11. Springer,
   Cham (1999)
7. Boudabra, M., Nitaj, A.: A new generalization of the KMOV cryptosystem. J.
   Appl. Math. Comput. **57**(1–2), 229–245 (2017)
8. Boudabra, M., Nitaj, A.: A new public key cryptosystem based on Edwards curves.
   J. Appl. Math. Comput. **61**, 431–450 (2019)
9. Coppersmith, D.: Small solutions to polynomial equations, and low exponent RSA
   vulnerabilities. J. Cryptol. **10**(4), 233–260 (1997)

10. Demytko, N.: A new elliptic curve based analogue of RSA. In: Helleseth, T. (ed.), EUROCRYPT 1993, LNCS, vol. 765, pp. 40–49. Springer, Cham (1994)
11. Diffie, W., Hellman, M.: New directions in cryptography. Institute of Electrical and Electronics Engineers, vol. IT-22. Trans. Inf. Theory, no. 6, 1976, pp. 644–654 (1976)
12. Dutto, S., Murru, N.: On the cubic Pell equation over finite fields, arXiv:2203.05290 (2022)
13. Galbraith, S.: Mathematics of Public Key Cryptography. Cambridge University Press, Cambridge (2012)
14. Hardy, G.H., Wright, E.M.: An Introduction to Theory of Numbers, 5th edn. The Clarendon Press, Oxford University Press, New York (1979)
15. Håstad, J.: On using RSA with low exponent in a public key network. In: Advances in Cryptology-CRYPTO'85 Proceedings 5, pp. 403-408. Springer, Heidelberg (1986)
16. Koyama, K.: Fast RSA type scheme based on singular cubic curve $y^2 + axy = x^3 \pmod{n}$. In: Proceedings of Eurocrypt'95, LNCS, vol. 921, pp. 329–339. Springer, Cham (1995)
17. Koyama, K., Maurer, U.M., Okamoto, T., Vanstone, S. A.: New public-key schemes based on elliptic curves over the ring $\mathbb{Z}_n$. In: Proceedings of CRYPTO 1991, LNCS, vol. 576, pp. 252–266. pp. 252–266 (1991)
18. Kuwakado H., Koyama K., Tsuruoka, Y.: A new RSA-type scheme based on singular cubic curves $y^2 \equiv x^3 + bx^2 \pmod{n}$. IEICE Trans. Fundamentals, E78-A, 27–33 (1995)
19. Lenstra, A.K., Lenstra, H.W., Lovász, L.: Factoring polynomials with rational coefficients. Math. Ann. **261**, 513–534 (1982)
20. Lim, S., Kim, S., Yie, I., Lee, H.: A generalized Takagi-cryptosystem with a modulus of the form $p^r q^s$, in: Indocrypt, Springer, pp. 283–294 (2000)
21. V. Lyubashevsky, V. , C. Peikert, C., and O. Regev, O.: On ideal lattices and learning with errors over rings. J. ACM (JACM), **60**(6), 1–35 (2013)
22. Murru N., Saettone F.M.: A novel RSA-like cryptosystem based on a generalization of the Rédei rational functions. In: Kaczorowski, J., Pieprzyk, J., Pomykala, J. (eds.) Number-Theoretic Methods in Cryptology. NuTMiC 2017. LNCS, vol. 10737. Springer, Cham (2018)
23. Nitaj, A.: Another generalization of Wiener's attack on RSA. In: Vaudenay, S. (ed.) Africacrypt 2008. LNCS, vol. 5023, pp. 174–190. Springer, Heidelberg (2008). https://doi.org/10.1007/978-3-540-68164-9_12
24. Nitaj, A.: A new attack on the KMOV cryptosystem. Bull. Korean Math. Soc. **2014**(51), 1347–1356 (2014)
25. Nitaj, A., Ariffin, M.R.B.K., Adenan, N.N.H., Abu, N.A.: Classical attacks on a variant of the RSA cryptosystem. In: Longa, P., Rafols, C. (eds.) Progress in Cryptology, LATINCRYPT 2021. LNCS, vol. 12912 (2021)
26. Okamoto, T., Uchiyama, U.: A new public-key cryptosystem as secure as factoring. In: Advances in Cryptology - EUROCRYPT 1998, LNCS. vol. 1403, pp. 308–318. Springer, Berlin (1998)
27. Okamoto, T., Uchiyama, U., Fujisaki, E.: EPOC: efficient probabilistic public-key encryption (1998)
28. Ding, C., Pei, D., Salomaa, A.: Chinese Remainder Theorem: Applications in Computing, Coding. World Scientific, Cryptography (1996)
29. Rabin, M. O.: Digitalized signatures and public key functions as intractable as factorisation. MIT/LCS/TR-212 MIT Laboratory for Computer Science (1979)

30. Regev, O.: On lattices, learning with errors, random linear codes, and cryptography, ser. STOC'05. New York, NY, USA: Association for Computing Machinery, 2005, pp. 84–93 (2005)
31. Rivest, R., Shamir, A., Adleman, L.: A Method for Obtaining digital signatures and public-key cryptosystems. Commun. ACM **21**(2), 120–126 (1978)
32. Rosen, K.H.: Elementary Number Theory and Its Applications, 3rd edn., pp. 285–302. Addison-Wesley, Reading (1993)
33. SageMath, the Sage Mathematics Software System (Version 10.1), The Sage Developers (2023). https://www.sagemath.org
34. Seck, M.: Proof of concept implementation of the proposed encryption scheme. Available on GitHub at https://github.com/mseckept/schemesecknitaj.git
35. SimulaMath, A Software for learning, teaching and research in mathematics (Version 1.1), The SimulaMath Developers (2023). https://simulamath.org
36. Schmidt-Samoa, K.: A new Rabin-type trapdoor permutation equivalent to factoring, Electronic Notes in Theoretical Computer Science, Elsevier, vol.157, no. 3, pp. 79–94 (2006). https://eprint.iacr.org/2005/278.pdf
37. P. Schwabe, R., et al.: Technical report, National Institute of Standards and Technology (2017). https://csrc.nist.gov/projects/post-quantum-cryptography/round-1-submissions
38. A. Szepieniec. Ramstake. Technical report, National Institute of Standards and Technology (2017). https://csrc.nist.gov/projects/post-quantumcryptography/round-1-submissions
39. Takagi, T.: Fast RSA-type cryptosystem modulo $p^k q$. In: Crypto 1998, LNCS, vol. 1462, pp. 318–326 (1998)
40. Teseleanu, G., Cotan, P.: Small private key attack against a family of RSA-like Cryptosystems, Cryptology ePrint Archive, Paper 2023/1356 (2023). https://eprint.iacr.org/2023/1356
41. de Weger, B.: Cryptanalysis of RSA with small prime difference. Appl. Algebra Eng. Commun. Comput. **13**(1), 17–28 (2002)
42. Wiener, M.: Cryptanalysis of short RSA secret exponents. IEEE Trans. Inf. Theory **36**, 553–558 (1990)
43. Williams, H. C.: An M3 public-key encryption scheme. In Conference on the Theory and Application of Cryptographic Techniques, pp. 358–368. Springer, Heidelberg (1985)
44. Zeugmann, T.: Taking Discrete Roots in the Field $\mathbb{Z}_p$ and in the Ring $\mathbb{Z}_{p^e}$, Division of Computer Science Report Series A (2019)

The Case of Small Prime Numbers Versus the Okamoto-Uchiyama Cryptosystem

George Teşeleanu[1,2]([⊠])(iD)

[1] Advanced Technologies Institute, 10 Dinu Vintilă, Bucharest, Romania
tgeorge@dcti.ro
[2] Simion Stoilow Institute of Mathematics of the Romanian Academy,
21 Calea Grivitei, Bucharest, Romania

Abstract. In this paper we study the effect of using small prime numbers within the Okamoto-Uchiyama public key encryption scheme. We introduce two novel versions and prove their security. Then we show how to choose the system's parameters such that the security results hold. Moreover, we provide a practical comparison between the cryptographic algorithms we introduced and the original Okamoto-Uchiyama cryptosystem.

Keywords: public key encryption · p-subgroup · provable security

1 Introduction

The Okamoto-Uchiyama cryptosystem was introduced in [19] and the authors proved that inverting the encryption function is equivalent to factoring. Unlike most factoring based schemes, the Okamoto-Uchiyama encryption is based on factoring numbers of the form p^2q instead of pq, where p and q are prime numbers. Another important security results is that the scheme is semantically secure if the p-subgroup assumption holds. This assumption is comparable to the quadratic residue and high degree residue assumptions [19]. We underline that the Okamoto-Uchiyama cryptosystem is partially homomorphic with respect to message addition and supports ciphertext randomization.

Shortly after the publication of [19], Coron, Naccache and Paillier introduced in [9] a variant of the Okamoto-Uchiyama cryptosystem that reduces the complexity of decryption. The authors claim, without proof, that the scheme retains the same security as the original scheme. We revisit their claims and we argue that is not obvious why the scheme is as secure as factoring. Therefore, in our opinion, the equivalence of Coron *et al.*'s variant to factoring remains an open problem. We also show that the semantic security of this variant is linked to a special kind of p-subgroup problem.

Another variant was introduced in [8], which also aims at reducing decryption complexity. They achieve this by choosing a special type of generator, and therefore instead of doing two exponentiations and a modular inversion, Choi, Choi

A. Dąbrowski et al. (Eds.): NuTMiC 2024, LNCS 14966, pp. 262–282, 2025.
https://doi.org/10.1007/978-3-031-82380-0_9

and Won manage to decrypt using only an exponentiation. Similar to Coron *et al.*, Choi *et al.* claim that their variant is equivalent to factoring. Their claim is refuted in [23] and it remains an open problem to prove that Choi *et al.*'s encryption scheme is not invertible. Note that the Okamoto-Uchiyama cryptosystem allows us to precompute an exponentiation and the corresponding inverse, and thus reduce the complexity of decryption to only one exponentiation. Therefore, we can obtain a decryption time similar to Choi *et al.* at the cost of memorizing an element modulo n and without sacrificing equivalence to factoring.

Since one of the most important features of the Okamoto-Uchiyama cryptosystem is its equivalence to factoring, we aim at finding a way to decrease decryption times, while retaining the cryptosystem's one-way equivalence to some form of factoring. Therefore, we introduce two variants of the scheme: an unbalanced version and a multiprime one. In the first version we show how to decrease the size of p while keeping the system secure and implicitly decreasing the complexity of decryption. The only cryptosystem related to this version is called the unbalanced RSA [24]. Just like our version, the scope is to reduce p, while keeping the system secure. Compared to the unbalanced RSA, the one-way property of the unbalanced Okamoto-Uchiyama is equivalent to factoring.

In the multiprime version, we increase the number of factors while keeping the size of the modulus constant and we manage to prove that inverting encryption is equivalent to computing the square-free factor of n. In the literature, we can find two related cryptosystems: the multiprime RSA [4,22] and the multiprime Joye-Libert [16,26]. The philosophy behind the multiprime RSA is the same as ours, while in the case of the multiprime Joye-Libert, the authors use multiple primes, but they increase the size of the modulus. Note that none of these systems is equivalent to factoring, while our variant is equivalent to partially factoring the modulus. According to [15], the multiprime RSA was implemented in Wireless Transport Layer Security protocol in order to decrease decryption times by use of parallelism. The same trick can be used to speed up the multiprime Okamoto-Uchiyama cryptosystem.

In the final section of our paper, we analyze the complexity of the two novel variants. Then we compare decryption times for all versions of the Okamoto-Uchiyama scheme that are equivalent to some form of factoring. If parallelization is possible, then the multiprime variant is to be preferred since it has a larger message space. Otherwise, for messages under a certain threshold, the unbalanced version has faster decryption times and once the threshold is crossed, the multiprime variant surpasses the efficiency of the unbalanced one.

For completeness we also provide an unbalanced and a multiprime version of the Coron-Naccache-Paillier cryptosystem. Then we prove their semantic security and finally we analyse their performance. We do not present a similar treatment for the Choi-Choi-Won cryptosystem, since an optimization equivalent with theirs can be achieved by simpler means, as states above.

Structure of the Paper. In Sect. 2 we introduce notations, definitions and lemmas used throughout the paper. The original Okamoto-Uchiyama scheme is described in Sect. 3. In Sects. 4 and 5 we present two novel versions of the Okamoto-

Uchiyama scheme. The Coron-Naccache-Paillier variant is tackled in Sect. 6. A performance analysis of some of the Okamoto-Uchiyama variants is provided in Sect. 7. We conclude in Sect. 8. A multiprime version of the Coron-Naccache-Paillier variant is given in Appendix A.

2 Preliminaries

Notations. Throughout the paper, λ denotes a security parameter. By $|n|$ we denote the size of n in bits. We use the notation $x \xleftarrow{\$} X$ when selecting a random element x from a sample space X. We denote by $x \leftarrow y$ the assignment of the value y to the variable x. The probability that event E happens is denoted by $Pr[E]$. Probabilistic polynomial-time algorithms are referred to as PPT algorithms. The set of integers $\{0, \ldots, a - 1\}$ is further denoted by $[0, a)$. For shorthand, we denote the set $[0, a + 1)$ by $[0, a]$. Multidimensional vectors $v = (v_0, \ldots, v_{s-1})$ are represented as $v = \{v_i\}_{i \in [0,s)}$.

2.1 Computational Complexity

In order to determine the computational complexity of our proposed schemes, we use the complexities of the mathematical operations listed in Table 1. These complexities are in accordance with the algorithms presented in [10,17]. We do not use the explicit complexity of multiplication, but instead we refer to it as $M(\cdot)$ for clarity. When presenting the complexity of performing an exponentiation we assume that the exponent has k bits. Also, in the case of the Chinese remainder theorem (CRT) we consider that the resulting modulus has μ bits and that we have r moduli.

Table 1. Computational complexity for μ-bit numbers

Operation	Complexity
Multiplication	$M(\mu) = \mathcal{O}(\mu \log \mu \log \log \mu)$
Exponentiation	$\mathcal{O}(kM(\mu))$
Modular inverse	$\mathcal{O}(\mu M(\mu))$
CRT	$\mathcal{O}(\log r M(\mu))$

2.2 Number Theoretic Prerequisites

We further present some definitions and lemmas from [19] that are needed for describing the public key encryption schemes presented in this paper.

Definition 1. *Let p be an odd prime. We define the p-Sylow subgroup of $\mathbb{Z}_{p^2}^*$ as*

$$\Gamma = \{x \in \mathbb{Z}_{p^2}^* \mid x \equiv 1 \bmod p\}.$$

Remark that $\mathbb{Z}_{p^2}^*$ is a cyclic group of order $p(p-1)$. Therefore, we obtain the following consequence.

Lemma 1. *Let p be an odd prime. Then the p-Sylow subgroup of $\mathbb{Z}_{p^2}^*$ has cardinality p.*

Lemma 2. *Let p be an odd prime, $g \in \mathbb{Z}_{p^2}^*$ and Γ the p-Sylow subgroup of $\mathbb{Z}_{p^2}^*$. If $g^{p-1} \bmod p^2$ has order p, then $g^{p-1} \in \Gamma$.*

Definition 2. *Let p be an odd prime and Γ the p-Sylow subgroup of $\mathbb{Z}_{p^2}^*$. We define the logarithmic function $L(\cdot)$ on Γ as*

$$L(x) = \frac{x-1}{p}, \text{ for any } x \in \Gamma.$$

Lemma 3. *Let $L(\cdot)$ be the logarithmic function on Γ. Then for any $x, y \in \Gamma$ we have $L(xy) \equiv L(x) + L(y) \bmod p$.*

Lemma 4. *Let $L(\cdot)$ be the logarithmic function on Γ. Also, let $x \in \Gamma$ such that $L(x) \neq 0$ and $y \equiv x^m \bmod p^2$, where $m \in \mathbb{Z}_p$. Then the following holds*

$$m \equiv \frac{L(y)}{L(x)} \equiv \frac{y-1}{x-1} \bmod p.$$

2.3 Public Key Encryption

A *public key encryption* (PKE) scheme usually consists of three PPT algorithms: *Setup*, *Encrypt* and *Decrypt*. The *Setup* algorithm takes as input a security parameter and outputs the public key as well as the matching secret key. *Encrypt* takes as input the public key and a message and outputs the corresponding ciphertext. The *Decrypt* algorithm takes as input the secret key and a ciphertext and outputs either a valid message or an invalidity symbol (if the decryption failed).

Definition 3 (Indistinguishability under Chosen Plaintext Attacks - IND-CPA). *The security model against chosen plaintext attacks for a PKE scheme is captured in the following game:*

Setup(λ): The challenger C generates the public key, sends it to adversary A and keeps the matching secret key to himself.

Query: Adversary A sends to C two equal length messages m_0, m_1. The challenger flips a coin $b \in \{0, 1\}$ and encrypts m_b. The resulting ciphertext c is sent to the adversary.

Guess: In this phase, the adversary outputs a guess $b' \in \{0, 1\}$. He wins the game if $b' = b$.

The advantage of an adversary A attacking a PKE scheme is defined as

$$ADV_A^{IND\text{-}CPA}(\lambda) = |Pr[b = b'] - 1/2|$$

where the probability is computed over the random bits used by C and A. A PKE scheme is IND-CPA secure, if for any PPT adversary A the advantage $ADV_A^{IND\text{-}CPA}(\lambda)$ is negligible.

3 The Okamoto-Uchiyama PKE Scheme

The Okamoto-Uchiyama scheme was introduced in [19] and the authors prove that inverting the encryption function is as hard as factoring. The scheme was also proven IND-CPA secure in the standard model under the p-subgroup assumption[1]. We shortly describe the algorithms of the Okamoto-Uchiyama cryptosystem.

Setup(λ): Generate two distinct large prime numbers p, q such that $|p| = |q| = \lambda$ and compute $n = p^2 q$. Randomly select $g \in \mathbb{Z}_n$ such that $g_p \equiv g^{p-1} \bmod p^2$ has order p in $\mathbb{Z}_{p^2}^*$. Let $h \equiv g^n \bmod n$. Output the public key $pk = (n, g, h, \lambda)$ and the corresponding secret key $sk = (p, q)$.

Encrypt(pk, m): To encrypt a message $m \in [0, 2^{\lambda-1})$ we choose $r \xleftarrow{\$} \mathbb{Z}_n$ and compute $c \equiv g^m h^r \bmod n$. Output the ciphertext c.

Decrypt(sk, c): Compute $c_p \equiv c^{p-1} \bmod p^2$, $g_p \equiv g^{p-1} \bmod p^2$ and recover m from the relation

$$m \equiv \frac{L(c_p)}{L(g_p)} \bmod p.$$

4 The Unbalanced Okamoto-Uchiyama PKE Scheme

In the unbalanced Okamoto-Uchiyama scheme we reduce the size of p (denoted λ_p), while keeping the size of n constant (denoted λ_n). This modification only impacts the description of the *Setup* and *Encrypt* algorithms, which we briefly describe below. Therefore, we have $\lambda_n = 2\lambda_p + \lambda_q$, where $\lambda_q = |q|$ and $\lambda_p \leq \lambda_q$. Note that when $\lambda_p = \lambda_q$ we obtain the Okamoto-Uchiyama cryptosystem, which we further refer to as the balanced Okamoto-Uchiyama scheme.

Setup(λ_p, λ_q): Generate two distinct large prime numbers p, q such that $|p| = \lambda_p$ and $|q| = \lambda_q$. Let $n = p^2 q$. Randomly select $g \in \mathbb{Z}_n$ such that $g_p \equiv g^{p-1} \bmod p^2$ has order p in $\mathbb{Z}_{p^2}^*$. Let $h \equiv g^n \bmod n$. Output the public key $pk = (n, g, h, \lambda_p)$ and the corresponding secret key $sk = (p, q)$.

Encrypt(pk, m): To encrypt a message $m \in [0, 2^{\lambda_p-1})$ we choose $r \xleftarrow{\$} \mathbb{Z}_n$ and compute $c \equiv g^m h^r \bmod n$. Output the ciphertext c.

Remark 1. Modifying the size of p does not impact the security proofs from [19]. Therefore, as long as factoring is hard, the unbalanced version is secure. We discuss how to choose λ_p such that factoring remains difficult in Sect. 7.

---

[1] To the author's knowledge, the only method for breaking this assumption is to know the factorisation of n.

5   The Multiprime Okamoto-Uchiyama PKE Scheme

5.1   Description

We further describe the multiprime Okamoto-Uchiyama encryption scheme. In this case we split up n into multiple primes. Therefore, we have $\lambda_n = 2t\lambda_p + \lambda_q$, where λ_q is the size of the square free prime factor q and $\lambda_p \leq \lambda_q$. Note that the case $t = 1$ and $\lambda_p = \lambda_q$ corresponds to the original cryptosystem [19]. Also, if we set $t = 1$, we obtain the unbalanced version.

$Setup(\lambda_p, \lambda_q)$: Generate $t+1$ distinct large prime numbers $p_1, \ldots, p_t, q$ such that $|p_1| = \ldots = |p_t| = \lambda_p$ and $|q| = \lambda_q$. Let $n = p_1^2 \ldots p_t^2 q$. Randomly select $g \in \mathbb{Z}_n$ such that for any $i \in [1, t]$ the element $g_{p_i} \equiv g^{p_i-1} \bmod p_i^2$ has order p_i in $\mathbb{Z}_{p_i^2}^*$. Let $h \equiv g^n \bmod n$. Output the public key $pk = (n, g, h, \lambda_p, t)$ and the corresponding secret key $sk = (P, q)$, where $P = \{p_i\}_{i \in [1,t]}$.

$Encrypt(pk, m)$: To encrypt a message $m \in [0, 2^{\lambda_p t - 1})$, we choose $r \xleftarrow{\$} \mathbb{Z}_n$ and compute $c \equiv g^m h^r \bmod n$. Output the ciphertext c.

$Decrypt(sk, c)$: For each $i \in [1, t]$ compute $c_{p_i} \equiv c^{p_i-1} \bmod p_i^2$, $g_{p_i} \equiv g^{p_i-1} \bmod p_i^2$ and recover m_i from the relation

$$m_i \equiv \frac{L(c_{p_i})}{L(g_{p_i})} \bmod p_i.$$

Let $n' = p_1 \ldots p_t$. Using the CRT, compute the unique $m \in \mathbb{Z}_{n'}$ such that $m \equiv m_i \bmod p_i$.

Correctness. The recovery of m is possible due to the following relation

$$c_{p_i} \equiv c^{p_i-1} \equiv (g^m h^r)^{p_i-1} \equiv g^{m(p_i-1)} (g^{rn/p_i})^{p_i(p_i-1)} \equiv g^{m(p_i-1)} \bmod p_i^2,$$

which according to Lemmas 2 and 4 implies

$$m \equiv \frac{L(c_{p_i})}{L(g_{p_i})} \bmod p_i.$$

Optimizations. In the *Setup* phase, we have to compute a special type of g. An efficient way to perform this step is to first randomly select $g_i \xleftarrow{\$} \mathbb{Z}_{p_i^2}^*$ such that $g_i^{p_i-1} \bmod p_i^2$ has order p_i for $i \in [1, t]$ and then choose a random element $g_{t+1} \xleftarrow{\$} \mathbb{Z}_q^*$. Afterwards use the CRT to compute an element $g \in \mathbb{Z}_n^*$ such that $g \equiv g_i \bmod p_i^2$ for all i and that $g \equiv g_{t+1} \bmod q$.

   Another possible optimisation is to memorize the values used during the *Decrypt* algorithm that are known beforehand. More precisely, during the *Setup* phase we can precompute the values $L(g_{p_i})^{-1} \bmod p_i$, p_i^2 and n', where $i \in [1, t]$. These values can enhance the secret key, and therefore can be used to speed up the decryption process at the cost of using more memory.

5.2   Security Analysis

In this section we first introduce a novel security assumption called the square-free factor problem and prove that inverting the encryption function of our proposal is as hard as breaking this assumption. Note that, when $t = 1$, the SFF assumption is equivalent with factoring the modulus. Then we generalise the p-subgroup problem stated in [19] and prove the IND-CPA security of our proposal. Note that for simplicity we assume, without losing generality, that $\lambda_p = \lambda_q = \lambda$ when conducting the security analysis.

Definition 4 (Square-Free Factor - SFF). *Choose $t + 1$ distinct large prime numbers $p_1, \ldots, p_t, q \geq 2^\lambda$. Let A be a PPT algorithm that returns an integer. We define the advantage*

$$ADV_A^{SFF}(\lambda) = Pr[A(n) = q \mid n = p_1^2 \ldots p_t^2 q].$$

The Square-Free Factor assumption states that for any PPT algorithm A the advantage $ADV_A^{SFF}(\lambda)$ is negligible.

Theorem 1. *Inverting the encryption algorithm of the multiprime Okamoto-Uchiyama PKE is intractable if the SFF assumption is intractable.*

Proof. Let's assume that there exists a PPT adversary A that given a ciphertext c recovers m with non-negligible probability. We further construct another PPT algorithm B that given as input n can find the prime factor q with non-negligible probability.

The first step of B is to generate the elements g and h that are needed to run A. Therefore, B randomly chooses $g \xleftarrow{\$} \mathbb{Z}_n^*$ and computes $h \equiv g^n \bmod n$. Note that $g^{p_i - 1} \bmod p_i^2$ has order p_i with probability $(p_i - 1)/p_i$. Therefore, g and h have the same distribution as in the original scheme with an overwhelming probability $(p_1 - 1) \ldots (p_t - 1)/(p_1 \ldots p_t)$.

In the next step B constructs a "ciphertext" c'. More precisely, B randomly generates $z' \xleftarrow{\$} \mathbb{Z}_n$ and computes $c' \equiv g^{z'} \bmod n$. We further show the relation between the distribution of "ciphertexts" c' and the distribution of real ciphertexts c.

Let $c \equiv g^{m+nr} \bmod n$ be a real ciphertext. We denote by $z = m + nr$. Let the order of g modulo $p_1, \ldots, p_t$ be $p_1 p_1', \ldots, p_t p_t'$, respectively, while the order modulo q is denoted as q'. Let $\ell = \operatorname{lcm}(p_1', \ldots, p_t', q')$.

Then the distribution of c is given by the distribution of $z = (z_1, z_2)$, where we have $z_1 \equiv z \bmod p_1 \ldots p_t$ and $z_2 \equiv z \bmod \ell$. Similarly, we define $z' = (z_1', z_2')$ as $z_1' \equiv z' \bmod p_1 \ldots p_t$ and $z_2' \equiv z' \bmod \ell$. Remark that z_1 is randomly distributed in $[0, 2^{\lambda t - 1})$, while z_1' is randomly distributed in $\mathbb{Z}_{p_1 \ldots p_t}$, which is roughly the size of $[0, 2^{\lambda t})$. Therefore, the probability of obtaining a given z_1 is at most twice than that of z_1'. Hence, a non-negligible fraction of z_1 is also non-negligible in z_1'.

Note that $\gcd(n, \ell) = 1$, since $\gcd(p_i, \ell) = 1$ and $\gcd(q, \ell) = 1$. Therefore, when z_1 and z_1' are fixed, we have that the distributions of z_2 and z_2' are statistically close.

Given the values (g, h, c') generated by B, adversary A will output a message m with non-negligible probability. Bear in mind that m should satisfy $m < 2^{\lambda t - 1}$ and $z' \equiv m \bmod p_1 \ldots p_t$. The probability of z' to be greater or equal than $2^{\lambda t - 1}$ is

$$\frac{n - 2^{\lambda t - 1}}{n} \simeq \frac{2^{\lambda(2t+1)} - 2^{\lambda t - 1}}{2^{\lambda(2t+1)}} = \frac{2^{\lambda t - 1}(2^{\lambda t + \lambda + 1} - 1)}{2^{\lambda t - 1}2^{\lambda t + \lambda + 1}} \simeq 1,$$

which is non-negligible. Therefore, with overwhelming probability we have $z' \not\equiv \bar{m} \bmod n$, and thus $\gcd(z' - m, n) \neq 1$.

We claim that $\gcd(z' - m, n) = p_1 \ldots p_t$ in almost all the cases. Let $i \in [1, t]$. The property $z' - m \equiv 0 \bmod p_i$ is equivalent with $z' - m = \alpha p_i$, where $\alpha > 0$. If $\alpha p_i \equiv 0 \bmod p_i^2$ then we have that $\alpha \equiv 0 \bmod p_i$. Therefore, the probability of having $z' - m \not\equiv 0 \bmod p_i^2$ is $(p_i - 1)/p_i$. In the case of q we have $z' - m \not\equiv 0 \bmod q$ with probability $(q - 1)/q$. Hence, we obtain that $\gcd(z' - m, n) = p_1 \ldots p_t$ with overwhelming probability $(p_1 - 1) \ldots (p_t - 1)(q - 1)/(p_1 \ldots p_t q)$. Therefore, B computes $q = n/(p_1 \ldots p_t)^2$ with non-negligible probability. □

Definition 5 (p-Subgroups - PS). *Choose $t + 1$ distinct large prime numbers $p_1, \ldots, p_t, q \geq 2^\lambda$ and compute $n = p_1^2 \ldots p_t^2 q$. Let Γ_i be the p_i-Sylow subgroup of $\mathbb{Z}_{p_i^2}^*$. We define the set $\Delta_i = \{x \in \mathbb{Z}_n^* \mid x^{p_i - 1} \in \Gamma_i\}$. We denote by $\Delta = \Delta_1 \cap \ldots \cap \Delta_t$ and $\bar{\Delta} = \mathbb{Z}_n^* \setminus \Delta$. Let A be a PPT algorithm that returns 1 on input (x, n) if $x \in \Delta$. We define the advantage*

$$ADV_A^{PS}(\lambda) = \left| Pr[A(x, n) = 1 | x \xleftarrow{\$} \Delta] - Pr[A(x, n) = 1 | x \xleftarrow{\$} \bar{\Delta}] \right|.$$

The p-Subgroups assumption states that for any PPT algorithm A, the advantage $ADV_A^{PS}(\lambda)$ is negligible.

Theorem 2. *The multiprime Okamoto-Uchiyama PKE is* IND-CPA *secure if and only if the* PS *assumption is intractable.*

Proof. We further denote by $E(m)$ the encryption of a message m. Note that breaking the PS assumption is equivalent with distinguishing between $E(0)$ and $E(1)$. To see that we first compute

$$E(0)^{p_i - 1} \equiv (g^0 h^{r_0})^{p_i - 1} \equiv g^{r_0 n (p_i - 1)} \equiv 1 \bmod p_i^2,$$

and

$$E(1)^{p_i - 1} \equiv (g^1 h^{r_1})^{p_i - 1} \equiv g^{p_i - 1} g^{r_1 n (p_i - 1)} \equiv g^{p_i - 1} \bmod p_i^2.$$

This implies that the order of $E(0)^{p_i - 1} \bmod p_i^2$ divides $p_i - 1$ and the order of $E(1)^{p_i - 1} \bmod p_i^2$ is p_i since $g^{p_i - 1} \bmod p_i^2$ has order p_i. Therefore, $E(0)^{p_i - 1} \in \bar{\Delta}$ and $E(1)^{p_i - 1} \in \Delta$, for all $i \in [1, t]$.

Now, let's assume that there exists a PPT adversary A that can distinguish between $E(0)$ and $E(1)$. We further construct another PPT algorithm B that can break the IND-CPA security of our proposal.

Once B receives a ciphertext c, he computes the value $x \equiv cg^{-m_0} \bmod n$ and chooses $r \xleftarrow{\$} \mathbb{Z}_n$. Let $\bar{g} \equiv g^{(m_1-m_0)+rn} \bmod n$, $\bar{h} \equiv \bar{g}^n \bmod n$ and $\ell = \operatorname{lcm}(p_1 - 1, \ldots, p_t - 1, q - 1)$. We denote by $\bar{E}(m)$ the encryption of m using $\bar{g}$ and $\bar{h}$. If B receives the encryption of m_0, we have

$$x \equiv h^{r_0} \equiv \bar{h}^{r_0/(m_1-m_0+rn)} \bmod n,$$

and

$$x \equiv g^{m_1-m_0} h^{r_1} \equiv \bar{g}\bar{h}^{(r_1-r)/(m_1-m_0+rn)} \bmod n,$$

otherwise. Note that since $\gcd(n, \ell) = 1$, $m_1 - m_0 + rn$ is randomly distributed in $\mathbb{Z}_\ell$. Also, $m_1 - m_0 + rn$ is invertible modulo ℓ with non-negligible probability. Therefore, x is either $\bar{E}(0)$ or $\bar{E}(1)$ with overwhelming probability.

We further prove that with a non-negligible probability the order of $\bar{g}_{p_i} \equiv \bar{g}^{p_i-1} \bmod p_i^2$ is p_i for all $i \in [1, t]$. Remark that $\gcd(m_1 - m_0, p_1 \ldots p_t) = 1$ with probability $(p_1 - 1) \ldots (p_t - 1)/(p_1 \ldots p_t)$. Now we assume that there exist $k_i < p_i$ such that $\bar{g}_{p_i}^{k_i} \equiv 1 \bmod p_i$. Then $k_i(m_1 - m_0) \equiv 0 \bmod p_i$ for all $i \in [1, t]$. This implies that $k_1 \ldots k_t(m_1 - m_0) \equiv 0 \bmod p_1 \ldots p_t$. Since $m_1 - m_0$ is coprime with the modulus, we obtain that $k_1 \ldots k_t \equiv 0 \bmod p_1 \ldots p_t$, which leads to a contradiction. Therefore, with an overwhelming probability we have $\bar{g} \in \Delta$.

On input $(\bar{g}, \bar{h}, x)$, adversary A outputs the correct bit b with non-negligible probability. Algorithm B simply relays b and according to the arguments presented above, it guesses correctly whether c is the encryption of m_0 or m_1.

We further prove the converse statement. Thus, let's assume that there exists a PPT adversary A that guesses correctly if a ciphertext encrypts either m_0 or m_1. We assume, without loss of generality, that $m_0 < m_1$. We will construct a PPT machine B which can distinguish between $E(0)$ and $E(1)$.

Given a ciphertext c algorithm B computes $x \equiv c^{m_1-m_0} g^{m_0+rn} \bmod n$, where $r \xleftarrow{\$} \mathbb{Z}_n$. If B receives an encryption of 0, we have

$$x \equiv (h^{r_0})^{m_1-m_0} g^{m_0+nr} \equiv g^{m_0} h^{r_0(m_1-m_0)+r} \bmod n$$

and

$$x \equiv (gh^{r_1})^{m_1-m_0} g^{m_0+nr} \equiv g^{m_1} h^{r_1(m_1-m_0)+r} \bmod n.$$

Therefore, x is either $E(m_0)$ or $E(m_1)$.

On input (g, h, x) adversary A outputs a bit b. Therefore, if algorithm B outputs b, then with non-negligible probability it guesses correctly whether c is the encryption of 0 or 1. □

6   The Coron-Naccache-Paillier PKE Scheme

To decrease the complexity of the decryption process, the authors of [9] introduce a slightly modified version of the Okamoto-Uchiyama encryption scheme. We further describe the Coron-Naccache-Paillier optimisation.

Setup(λ, λ_u): Generate a large prime number u such that $|u| = \lambda_u$. Also, generate two distinct large prime numbers p, q such that $|p| = |q| = \lambda$ and $p - 1 = uv$. Let $n = p^2 q$. Randomly select $g \in \mathbb{Z}_n$ such that $g_p \equiv g^{p-1} \bmod p^2$ has order p in $\mathbb{Z}_{p^2}^*$. Compute $G \equiv g^v \in \mathbb{Z}_n^*$ and $h \equiv G^n \bmod n$. Output the public key $pk = (n, G, H, \lambda)$ and the corresponding secret key $sk = (p, q)$.

Encrypt(pk, m): To encrypt a message $m \in [0, 2^{\lambda_u - 1})$, we choose $r \xleftarrow{\$} \mathbb{Z}_n$ and compute $c \equiv G^m H^r \bmod n$. Output the ciphertext c.

Decrypt(sk, c): Compute $c_p \equiv c^u \bmod p^2$, $g_p \equiv G^u \bmod p^2$ and recover m from the relation

$$m \equiv \frac{L(c_p)}{L(g_p)} \bmod p.$$

Remark 2. In the original paper [9], the authors encrypt messages of size $\lambda - 1$, but that leads to an incorrect decryption since the order of G is $u < 2^{\lambda-1}$, and thus a wrap-around of the message is possible. This problem is fixed in our description.

Remark 3. In [9], the authors claim without proof that their optimisation is equivalent with factoring n. They only state that "equivalence to factoring is easily derived from the original security proof included in [19]". When we tried to follow the same line of reasoning as in Theorem 1, the following problem occurred. When choosing G randomly[2] from $\mathbb{Z}_n^*$, the probability of $G^u \bmod p^2$ having order p is

$$\frac{u}{p} \simeq \frac{2^{\lambda_u}}{2^\lambda} = \frac{2^{\lambda_u}}{2^{\lambda-\lambda_u}},$$

which is negligible[3]. Therefore, the proof breaks down because G^u will almost never have the correct order, namely p, leading to the simulation failing to generate the correct distributed G with non-negligible probability. Hence, we consider that the optimisation is not equivalent to factoring, until proven otherwise.

Remark 4. In the original paper [9], the IND-CPA security of the scheme is claimed without proof to be equivalent with the PS assumption as introduced in [19]. In reality, the claim is partially true. More precisely, the IND-CPA security can be linked to the following version of the PS assumption[4].

---

[2] In [19, Theorem 6], this step is equivalent to choosing $g \xleftarrow{\$} \mathbb{Z}_n^*$.

[3] Compared with the non-negligible value of $(p-1)/p$ as in [19, Theorem 6].

[4] see Appendix A for the proof.

Definition 6 (Strong Prime p-Subgroups - SP-PS). *Choose $t + 1$ distinct large prime numbers $p_1, \ldots, p_t, q \geq 2^\lambda$ such that $p_i - 1$ has a large prime factor denoted u_i for all $i \in [1, t]$. Let Γ_i be the p_i-Sylow subgroup of $\mathbb{Z}^*_{p_i^2}$. We define the set $\Omega_i = \{x \in \mathbb{Z}_n \mid x^{u_i} \in \Gamma_i\}$. We denote by $\Omega = \Omega_1 \cap \ldots \cap \Omega_t$ and $\bar{\Omega} = \mathbb{Z}^*_n \setminus \Omega$. Let A be a PPT algorithm that returns 1 on input (x, n) if $x \in \Omega$. We define the advantage*

$$ADV_A^{SP\text{-}PS}(\lambda) = \left| Pr[A(x, n) = 1 | x \xleftarrow{\$} \Omega] - Pr[A(x, n) = 1 | x \xleftarrow{\$} \bar{\Omega}] \right|.$$

The Strong Prime p-Subgroups assumption states that for any PPT algorithm A, the advantage $ADV_A^{PS}(\lambda)$ is negligible.

7   Implementation and Performance Analysis

7.1   Parameter Selection

The fastest currently known algorithm for factoring composite numbers is the Number Field Sieve (NFS) [14]. The expected running time of the NFS depends on the size of the modulus n and not on the size of its factors. More precisely, the expected running time is approximately

$$L[n] = e^{1.923(\log n)^{1/3}(\log \log n)^{2/3}}.$$

In [13,14], the authors extrapolate the running time needed to factor a modulus of size λ_n from the computational effort required to factor a 512-bit modulus. Hence, a λ_n-bit modulus offers a security equivalent to a block cipher of d-bit security if

$$L[2^{\lambda_n}] \simeq 50 \cdot 2^{d-56} \cdot L[2^{512}]. \tag{1}$$

Since we start from a secure Okamoto-Uchiyama PKE and we want to optimize decryption by decreasing the size of some of the factors of the modulus, while keeping the size of the modulus constant, the NFS cannot be expected to factor n. Unfortunately, this strategy can make the resulting PKEs vulnerable to the Elliptic Curve Method (ECM) [11], if we lower the size of the factors below a certain threshold. Compared to the NFS, the ECM has the running time determined by the size of the smallest factor. Thus, if p is the smallest factor, then the running time of the ECM is

$$E[n, p] = (\log_2 n)^2 e^{\sqrt{2 \log p \log \log p}}.$$

Similarly to the NFS, Lenstra [12] extrapolates the equivalent security provided by a module of size λ_n with the smallest prime of size λ_p to be

$$E[2^{\lambda_n}, 2^{\lambda_p}] \geq 80 \cdot 2^{d-56} \cdot E[2^{768}, 2^{167}]. \tag{2}$$

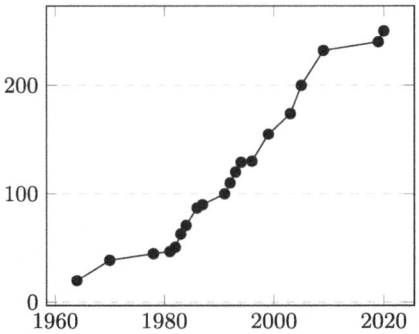

Fig. 1. Size of general number factored versus year

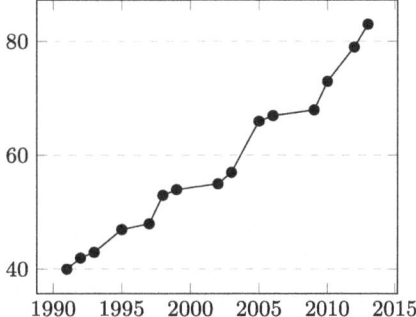

Fig. 2. Size of general number factored using ECM versus year

From Eqs. (1) and (2) we can deduce the following equivalency

$$E[2^{\lambda_n}, 2^{\lambda_p}] \geq 80 \cdot 2^{\log_2(L[2^{\lambda_n}]/(50 \cdot L[2^{512}]))} \cdot E[2^{768}, 2^{167}]. \tag{3}$$

A different model for predicting the security against the NFS and the ECM is provided in [7]. Compared to Lenstra's model, Brent uses known historical factoring records to predict the year a modulus of a given size will be factored. Using the least-squares fit, Brent obtains the following equation for the NFS

$$D_n^{1/3} = \frac{Y - 1928.6}{13.24} \text{ or equivalently } Y = 13.24 \cdot D_n^{1/3} + 1928.6 \tag{4}$$

and for the ECM

$$D_p^{1/2} = \frac{Y - 1932.3}{9.3} \text{ or equivalently } Y = 9.3 \cdot D_p^{1/2} + 1932.3, \tag{5}$$

where D_n is the number of digits of the factored modulus and D_p is the number of digits of the largest prime factor found using the ECM.

Using regression analysis we update Brent's equations using data points from [2,18,25] for the NFS and from [6,27] for the ECM. These data points are presented in Figs. 1 and 2.

Therefore, the updated equation for the NFS is

$$D_n^{1/3} = \frac{Y - 1926}{13.97} \text{ or equivalently } Y = 13.97 \cdot D_n^{1/3} + 1926 \tag{6}$$

and for the ECM

$$D_p^{1/2} = \frac{Y - 1939}{8.207} \text{ or equivalently } Y = 8.207 \cdot D_p^{1/2} + 1939. \tag{7}$$

Equations (4) to (7) are presented in Figs. 3 and 4. Note that the black dots represent the acquired data points. We can see that in the case of the NFS the estimates are close, while in the case of the ECM the new estimate is more

Table 2. Equivalent key sizes

Modulus key size	3072	7680	15360
Lenstra model	800(3)	1617(4)	2761(5)
Regression model	749(4)	1457(5)	2385(6)

pessimistic from a security point of view. Using the updated estimates (Eqs. (6) and (7)) we obtain the following equivalency

$$D_p^{1/2} = \frac{13.97 \cdot D_n^{1/3} - 13}{8.207}.$$

(8)

According to NIST [5], the recommended key sizes for composite modules are $\lambda_n = 3072/7680/15360$. We preferred to use NIST recommendations instead of the ones from [13,14] since these key sizes are the ones used by the industry and the key sizes from [13,14] are criticized as being to conservative [25]. Therefore, using Eqs. (3) and (8) we obtain the equivalent size of the smallest prime. The results are presented in Table 2. Note that in parenthesis we provide the maximum number of prime factors that n can have. Based on these equivalences, we obtain the parameters for the Okamoto-Uchiyama schemes that offer protection against the NFS and the ECM (see Table 3). We can see that the only key sizes that support a multiprime version are 7680 in the regression model and 15360 in both models.

7.2 Complexity

Using the complexities provided in Table 1, we computed the asymptotic run times of the decryption algorithm for each Okamoto-Uchiyama variant. We also

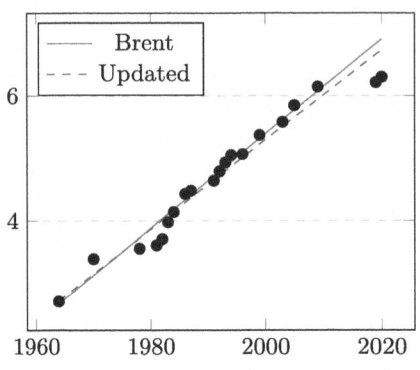

Fig. 3. $D^{1/3}$ versus year Y

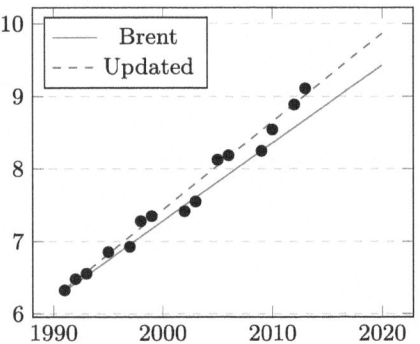

Fig. 4. $D^{1/2}$ versus year Y for ECM

Table 3. Okamoto-Uchiyama parameters' size

$|n|$	t	$|p|$	$|q|$	Model	Type
3072	1	1024	1024	–	Balanced
	1	800	1472	Lenstra	Unbalanced
	1	749	1574	Regression	Unbalanced
7680	1	2560	2560	–	Balanced
	1	1617	4446	Lenstra	Unbalanced
	1	1457	4766	Regression	Unbalanced
	2	1457	1852	Regression	Multiprime
15360	1	5120	5120	–	Balanced
	1	2761	9838	Lenstra	Unbalanced
	1	2385	10590	Regression	Unbalanced
	2	2761	4316	Lenstra	Multiprime
	2	2385	5820	Regression	Multiprime

determined the size of the message space for each variant. The results are provided in Table 4. Note that by parallel multiprime we mean the multiprime version in which we use two separate threads to compute m_1 and m_2.

Table 4. Performance analysis

Scheme	Decryption Complexity	$|m|$
Balanced	$\mathcal{O}(2\lambda M(2\lambda) + \lambda M(\lambda))$	$\lambda - 1$
Unbalanced	$\mathcal{O}(2\lambda_p M(2\lambda_p) + \lambda_p M(\lambda_p))$	$\lambda_p - 1$
Multiprime	$\mathcal{O}(2t\lambda_p M(2\lambda_p) + t\lambda_p M(\lambda_p) + \log t M(t\lambda_p))$	$\lambda_p t - 1$
Parallel Multiprime	$\mathcal{O}(2\lambda_p M(2\lambda_p) + \lambda_p M(\lambda_p) + \log t M(t\lambda_p))$	$\lambda_p t - 1$

The comparison of the computational complexity of the four variants is presented in Figs. 5 and 6. Note that we only provide the comparison for $\lambda_n = 7680/15360$ since these are the only module sizes that support all variants. In the case of $\lambda_n = 3072$, we can easily deduce that the unbalanced version (UnB) will run faster than the balanced one (Bal). Note that in Fig. 6 the two dots for each multiprime version (Mp) correspond to the two equivalence models: Lenstra - upper dot and Regression - lower dot.

From the two plots we can see that the parallel multiprime version (PMp) has a running time comparable to the unbalance version, and since the multiprime version has a larger message space, it should be preferred. Nevertheless, if only one thread is available and messages are below a certain threshold (denoted by dotted lines), then the unbalanced version is preferred, otherwise the multiprime version should be used.

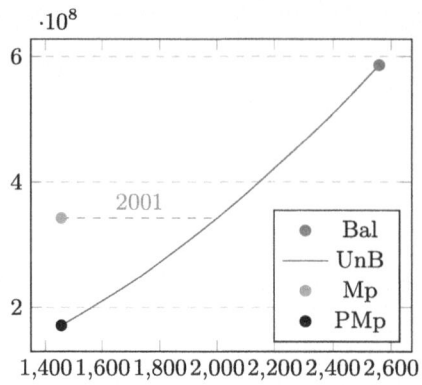
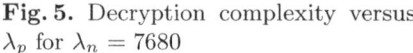
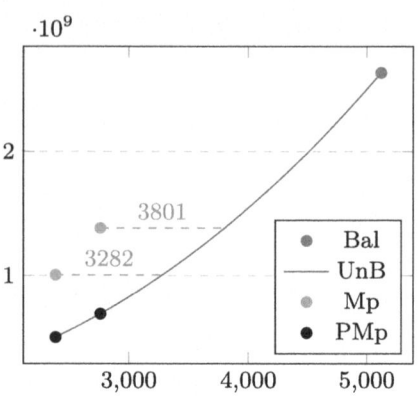

Fig. 5. Decryption complexity versus λ_p for $\lambda_n = 7680$

Fig. 6. Decryption complexity versus λ_p for $\lambda_n = 15360$

7.3   Implementation Details

We further provide the reader with benchmarks for the four Okamoto-Uchiyama PKE schemes. Note that besides the parameters from Table 3, we also used the parameters from Table 5 (see Figs. 5 and 6).

Table 5. Additional Okamoto-Uchiyama parameters' size

$|n|$	t	$|p|$	$|q|$	Model	Type
7680	1	2001	3678	Regression	Unbalanced
15360	1	3801	7758	Lenstra	Unbalanced
	1	3282	8796	Regression	Unbalanced

We ran each of the three sub-algorithms on a CPU Intel i7-4790 4.00 GHz and used GCC to compile it (with the O3 flag activated for optimization). Note that for all computations we used the GMP library [3]. To calculate the running times we used the *omp_get_wtime()* function [1]. For the parallel multiprime variant we used the OMP library [1] to parallelize decryption. To obtain the average running time in seconds we chose to encrypt 100 128/192/256-bit messages. Therefore, we wanted to simulate a key distribution scenario.

The results are provided in Table 6. Note that with blue we marked the additional parameters and the decryption times for the parallel multiprime version are given in the second row of $t = 2$. Also, the optimized version of the decryption algorithm is denoted by *Decrypt* (opt).

In Table 6 we omitted encryption times, since they are similar to each other for a given λ_n. More precisely, for 3072 we obtain 0.009679, for 7680 we have 0.096855 and for 15360 we have 0.550525. We can see from Table 6 that the conclusions from Sect. 7.2 hold. We can also see that the multiprime version has the shortest time to generate the parameters, while the unbalanced version the longest time. Nevertheless, generating parameters is a one-time operation.

Table 6. Message size and running times

$|n|$	t	$|p|$	$|m|$	Setup	Decrypt	Decrypt (opt)
	1	1024	1023	0.058187	0.003073	0.001537
3072	1	800	799	0.120665	0.001535	0.000772
	1	749	748	0.121719	0.001231	0.000614
	1	2560	2559	1.696050	0.034409	0.017205
	1	1617	1616	5.828290	0.010711	0.005353
7680	1	1457	1456	6.644670	0.007872	0.003932
	1	2001	2000	3.158020	0.019550	0.009751
	2	1457	2913	0.476329	0.015782	0.007897
					0.008538	0.004511
	1	5120	5119	17.04490	0.203512	0.101740
	1	2761	2760	70.11920	0.040647	0.020362
	1	3801	3800	37.98000	0.089592	0.044745
	1	2385	2384	103.1440	0.031670	0.015889
15360	1	3282	3281	44.25260	0.063980	0.032011
	2	2761	5521	7.584640	0.081344	0.040684
					0.041635	0.021137
	2	2385	4769	15.86700	0.063377	0.031778
					0.032505	0.016557

8 Conclusions

In this work we introduced two novel versions of the Okamoto-Uchiyama cryptosystem. The first one, called the unbalanced Okamoto-Uchiyama PKE, lowers the size of p in order to decrease decryption time at the cost of shrinking the message space. The second one, called the multiprime Okamoto-Uchiyama PKE, increases the number of factors and achieves a decryption time comparable with the unbalanced version if multiple threads are available. An advantage of the multiprime variant is that it has a larger message space than the unbalanced version and, sometimes, even larger than the original PKE. Therefore, if parallel threads are available we recommend the multiprime version due to its larger message space, otherwise, if short messages are sent we recommend the unbalanced variant.

We also argue why we did not compare our variants to the Coron-Naccache-Paillier PKE. Since equivalence to factoring is not proven in the original paper and there are doubts about why simply modifying the proof of the Okamoto-Uchiyama PKE does not work, we chose to discuss the Coron-Naccache-Paillier PKE separately from the Okamoto-Uchiyama variants. For completeness, in Appendix A we provide an unbalanced and a multiprime variant of the Coron-Naccache-Paillier PKE, study their security and provide performance benchmarks. As in the case of the Okamoto-Uchiyama PKE, we recommend the unbalance version for short messages when decryption parallelization is not possible, and the multiprime one otherwise.

Future Work. We leave for future work the adaptation to unbalanced and multi-prime mode, as well as the performance comparison with other factoring related cryptosystems such as Paillier [20] or RSA [4,22,24].

A   The Multipleprime Coron-Naccache-Paillier PKE Scheme

A.1   Preliminaries

Before presenting the multiprime generalization, we first introduce the generalized CRT as stated in [21].

Theorem 3 (Generalized Chinese Remainder Theorem). *Let m_1, m_2, ..., m_t be positive integers. For a set of integers $a_1, a_2, \ldots, a_t$ the system of congruences*

$$x \equiv a_i \pmod{m_i}, \text{ for } i \in [1, t]$$

has solutions if and only if

$$a_i \equiv a_j \pmod{\gcd(m_i, m_j)}, \text{ for } i \neq j, \ i, j \in [1, t]. \tag{9}$$

If Eq. (9) holds, then the solution will be unique modulo $\operatorname{lcm}(m_1, m_2, \ldots, m_t)$.

A.2   Description

For completeness, we further describe the multiple prime generalisation of the Coron-Naccache-Paillier optimisation. Bear in mind that when $t = 1$ and $\lambda_p = \lambda_q$ we obtain the optimisation presented in [9] and when $t = 1$ we obtain the unbalanced version of the Coron-Naccache-Paillier scheme. The reader can easily see that we can also use the optimisation techniques presented in Sect. 5 to speed-up this proposal.

$Setup(\lambda_p, \lambda_q, \lambda_u)$: Generate t distinct large prime numbers $u_1, \ldots, u_t$ such that $|u_1| = \ldots = |u_t| = \lambda_u$. Also, generate $t + 1$ distinct large prime numbers $p_1, \ldots, p_t, q$ such that $|p_1| = \ldots = |p_t| = \lambda_p$, $|q| = \lambda_q$ and $p_i - 1 = u_i v_i$, for $i \in [1, t]$. Let $n = p_1^2 \cdot \ldots \cdot p_t^2 q$. Randomly select $g \in \mathbb{Z}_n$ such that for any $i \in [1, t]$ the element $g_{p_i} \equiv g^{p_i - 1} \bmod p_i^2$ has order p_i in $\mathbb{Z}_{p_i^2}^*$. Let $\ell = \operatorname{lcm}(p_1 u_1 v_1, \ldots, p_t u_t v_t)$. Using the generalized CRT compute the unique $v \in \mathbb{Z}_\ell$ such that $v \equiv v_i \bmod p_i u_i v_i$. Compute $G \equiv g^v \in \mathbb{Z}_n^*$ and $H \equiv G^n \bmod n$. Output the public key $pk = (n, G, H, \lambda, t)$ and the corresponding secret key $sk = (P, q)$, where $P = \{p_i\}_{i \in [1, t]}$.

$Encrypt(pk, m)$: To encrypt a message $m \in [0, 2^{\lambda_u t - 1})$, we choose $r \xleftarrow{\$} \mathbb{Z}_n$ and compute $c \equiv G^m H^r \bmod n$. Output the ciphertext c.

Decrypt(sk, c): For each $i \in [1, t]$ compute $c_{p_i} \equiv c^{u_i} \bmod p_i^2$, $g_{p_i} \equiv G^{u_i} \bmod p_i^2$ and recover m_i from the relation

$$m_i \equiv \frac{L(c_{p_i})}{L(g_{p_i})} \bmod p_i.$$

Let $n' = p_1 \cdot \ldots \cdot p_t$. Using the CRT compute the unique $m \in \mathbb{Z}_{n'}$ such that $m \equiv m_i \bmod p_i$.

Correctness. The first thing we need to show is that in the *Setup* phase the conditions of Theorem 3 are satisfied. We note that

$$\gcd(p_i u_i v_i, p_j u_j v_j) = \gcd(v_i, v_j) \mid v_i - v_j,$$

and therefore v exists.

To recover m we only need to prove that we end up with the same relations as in the case of the Okamoto-Uchiyama decryption algorithm. Therefore, we have

$$c_{p_i} \equiv c^{u_i} \equiv (G^m H^r)^{u_i} \equiv (g^m h^r)^{v u_i} \equiv (g^m h^r)^{p_i - 1} \bmod p_i^2,$$

since $v \equiv v_i \bmod p_i(p_i - 1)$.

A.3 Security Analysis

Theorem 4. *The multiple prime Coron-Naccache-Paillier PKE is* IND-CPA *secure if and only if the* SP-PS *assumption is intractable.*

Proof (sketch). We further denote by $E(m)$ the encryption of a message m. Note that breaking the SP-PS assumption is equivalent with distinguishing between $E(0)$ and $E(1)$. To see that we first compute

$$E(0)^{u_i} \equiv (G^0 H^{r_0})^{u_i} \equiv G^{r_0 n u_i} \equiv g^{r_0 n(p_i - 1)} \equiv 1 \bmod p_i^2,$$

and

$$E(1)^{p_i - 1} \equiv (G^1 H^{r_1})^{u_i} \equiv G^{u_i} G^{r_1 n u_i} \equiv g^{p_i - 1} g^{r_1 n(p_i - 1)} \equiv g^{p_i - 1} \bmod p_i^2.$$

Table 7. Performance analysis

Scheme	Decryption Complexity
Balanced	$\mathcal{O}(2\lambda_u M(2\lambda) + \lambda M(\lambda))$
Unbalanced	$\mathcal{O}(2\lambda_u M(2\lambda_p) + \lambda_p M(\lambda_p))$
Multiprime	$\mathcal{O}(2t\lambda_u M(2\lambda_p) + t\lambda_u M(\lambda_p) + \log t M(t\lambda_p))$
Parallel Multiprime	$\mathcal{O}(2\lambda_u M(2\lambda_p) + \lambda_u M(\lambda_p) + \log t M(t\lambda_p))$

This implies that the order of $E(0)^{u_i} \bmod p_i^2$ divides u_i and the order of $E(1)^{u_i} \bmod p_i^2$ is p_i since $g^{p_i-1} \bmod p_i^2$ has order p_i. Therefore, $E(0)^{u_i} \in \bar{\Omega}$ and $E(1)^{u_i} \in \Omega$.

Now, let's assume that there exists a PPT adversary A that can distinguish between $E(0)$ and $E(1)$. We further construct another PPT algorithm B that can break the IND-CPA security of our proposal.

Once B receives a ciphertext C, he computes the value $X \equiv CG^{-m_0} \bmod n$ and chooses $r \xleftarrow{\$} \mathbb{Z}_n$. Let $\bar{G} \equiv G^{(m_1-m_0)+rn} \bmod n$, $\bar{H} \equiv \bar{G}^n \bmod n$ and $\ell = \operatorname{lcm}(p_1 - 1, \ldots, p_t - 1, q - 1)$. We denote by $\bar{E}(m)$ the encryption of m using $\bar{G}$ and $\bar{H}$. If B receives the encryption of m_0, we have $X = \bar{E}(0)$ and $X = \bar{E}(1)$, otherwise.

To prove that $\bar{G}^{u_i} \bmod p_i^2$ has order p_i for all $i \in [1,t]$, we first observe that $\bar{G}^{u_i} \equiv \bar{g}^{p_i-1}$, where $\bar{g} \equiv g^{(m_1-m_0)+rn} \bmod n$. Using the same arguments as in the proof of Theorem 2 we obtain that $\bar{G}^{u_i} \in \Gamma_i$, and thus $\bar{G}^{u_i} \in \Omega_i$.

On input $(\bar{G}, \bar{H}, X)$, adversary A outputs the correct bit b with non-negligible probability. Algorithm B simply relays b and according to the arguments presented above, it guesses correctly whether c is the encryption of m_0 or m_1.

We further prove the converse statement. Thus, let's assume that there exists a PPT adversary A that guesses correctly if a ciphertext encrypts either m_0 or m_1. We assume without loss of generality that $m_0 < m_1$. We will construct a PPT machine B which can distinguish between $E(0)$ and $E(1)$.

Given a ciphertext c, algorithm B constructs an element $X \equiv C^{m_1-m_0} G^{m_0+nr}$, where $r \xleftarrow{\$} \mathbb{Z}_n$. If B receives an encryption of 0 we have $X = E(m_0)$ and $X = E(m_1)$, otherwise.

On input (G, H, X), adversary A outputs a bit b. Therefore, if algorithm B outputs b, then with non-negligible probability it guesses correctly whether C is the encryption of 0 or 1. □

A.4 Complexity and Implementation Details

Similarly to Table 4, we computed the complexity of the decryption algorithm for the Coron-Naccache-Paillier variants. The results can be seen in Table 7. Note that the message size for the balanced and unbalanced variants is $\lambda_u - 1$ and for the multiprime versions is $\lambda_u t - 1$. Compared to the Okamoto-Uchiyama, the Coron-Naccache-Paillier variants have faster decryption times, but at the cost of a smaller message space and the loss of equivalence with factoring.

According to [9,14], λ_u should be chosen such that the subgroup generated by G is large enough that it protects the PKE against the baby-step-giant-step method. According to [5], we should choose $\lambda_u = 128/192/256$. However, we want to simulate a key distribution scenario, and thus we have to choose $\lambda_u = 129/193/257$. This implies that the message space is $128/192/256$ for the (un)balanced version and $257/385/513$ for the multiprime ones. We further provide the benchmarks for the Coron-Naccache-Paillier versions in Table 8.

Table 8. Running times

$|n|$	t	$|p|$	Setup	Decrypt	Decrypt (opt)
3072	1	1024	0.063806	0.000436	0.000215
	1	800	0.121260	0.000278	0.000139
	1	749	0.147123	0.000240	0.000119
7680	1	2560	1.481400	0.002870	0.001431
	1	1617	4.954550	0.001399	0.000697
	1	1457	8.247510	0.001149	0.000573
	2	1457	0.570016	0.002308	0.001162
				0.001692	0.000916
15360	1	5120	16.67890	0.011248	0.005606
	1	2761	76.39660	0.004128	0.002063
	1	2385	112.3960	0.003715	0.001856
	2	2761	7.892090	0.008284	0.004155
				0.004840	0.002749
	2	2385	11.83500	0.007450	0.003733
				0.004308	0.002429

Based on Tables 7 and 8, we can see that the parallel multiprime version has a running time comparable to the unbalance version, and since the multiprime version has double the message space, it should be preferred. Nevertheless, if only one thread is available and we only want to distribute symmetric keys, then the unbalanced version is preferable, otherwise the multiprime version should be used.

References

1. OpenMP. https://www.openmp.org/
2. RSA Factoring Challenge. https://en.wikipedia.org/wiki/RSA_Factoring_Challenge
3. The GNU Multiple Precision Arithmetic Library. https://gmplib.org/
4. Cryptography Using Compaq Multiprime Technology in a Parallel Processing Environment (2000). https://www.compaq.com
5. Barker, E.: NIST SP800-57 recommendation for key management, part 1: general. Technical report, NIST (2020)
6. Brent, R.P.: Large factors found by ECM. https://maths-people.anu.edu.au/~brent/ftp/champs.txt
7. Brent, R.P.: Some parallel algorithms for integer factorisation. In: Euro-Par 1999. LNCS, vol. 1685, pp. 1–22. Springer (1999)
8. Choi, D.H., Choi, S., Won, D.: Improvement of probabilistic public key cryptosystems using discrete logarithm. In: ICISC 2001. LNCS, vol. 2288, pp. 72–80. Springer (2001)

9. Coron, J., Naccache, D., Paillier, P.: Accelerating Okamoto-Uchiyama public-key cryptosystem. Electron. Lett. **35**(4), 291–292 (1999)

10. Crandall, R., Pomerance, C.: Prime Numbers: A Computational Perspective. Number Theory and Discrete Mathematics. Springer (2005)

11. Lenstra, H.W., Jr.: Factoring integers with elliptic curves. Ann. Math. **126**(3), 649–673 (1987)

12. Lenstra, A.K.: Unbelievable security. Matching AES security using public key systems. In: ASIACRYPT 2001. LNCS, vol. 2248, pp. 67–86. Springer (2001)

13. Lenstra, A.K., Verheul, E.R.: Selecting cryptographic key sizes. In: PKC 2000. LNCS, vol. 1751, pp. 446–465. Springer (2000)

14. Lenstra, A.K., Verheul, E.R.: Selecting cryptographic key sizes. J. Cryptol. **14**(4), 255–293 (2001)

15. Lim, S., Kim, S., Yie, I., Lee, H.: A generalized Takagi-cryptosystem with a modulus of the form $p^r q^s$. In: INDOCRYPT 2000. LNCS, vol. 1977, pp. 283–294. Springer (2000)

16. Maimuţ, D., Teşeleanu, G.: A new generalisation of the Goldwasser-Micali cryptosystem based on the gap 2^k-residuosity assumption. In: SecITC 2020. LNCS, vol. 12596, pp. 24–40. Springer (2020)

17. Menezes, A., van Oorschot, P.C., Vanstone, S.A.: Handbook of Applied Cryptography. CRC Press (1996)

18. Odlyzko, A.M.: The future of integer factorization. RSA Laboratories' Cryptobytes **1**(2), 5–12 (1995)

19. Okamoto, T., Uchiyama, S.: A new public-key cryptosystem as secure as factoring. In: EUROCRYPT 1998. LNCS, vol. 1403, pp. 308–318. Springer (1998)

20. Paillier, P.: Public-key cryptosystems based on composite degree residuosity classes. In: EUROCRYPT 1999. LNCS, vol. 1592, pp. 223–238. Springer (1999)

21. Pei, D., Salomaa, A., Ding, C.: Chinese Remainder Theorem: Applications in Computing, Coding, Cryptography. World Scientific Publishing (1996)

22. Rivest, R.L., Shamir, A., Adleman, L.M.: Cryptographic communications system and method. US Patent 4,405,829 (1983)

23. Sakurai, K., Takagi, T.: On the security of a modified Paillier public-key primitive. In: ACISP 2002. LNCS, vol. 2384, pp. 436–448. Springer (2002)

24. Shamir, A.: RSA for paranoids. RSA Laboratories' Cryptobytes **1**(3), 1–4 (1995)

25. Silverman, R.D.: A cost-based security analysis of symmetric and asymmetric key lengths. RSA Laboratories' Bulletin 13 (2000)

26. Teşeleanu, G.: The case of small prime numbers versus the Joye-Libert cryptosystem. Mathematics **10**(9), 1577 (2022)

27. Zimmermann, P.: 50 largest factors found by ECM. https://members.loria.fr/PZimmermann/records/top50.html

Compartment-Based and Hierarchical Threshold Delegated Verifiable Accountable Subgroup Multi-signatures

Ahmet Ramazan Ağırtaş[1,2]($\boxtimes$) (iD) and Oğuz Yayla[1] (iD)

[1] Institute of Applied Mathematics, Middle East Technical University,
06800 Çankaya Ankara, Turkey
`oguz@metu.edu.tr`
[2] Nethermind, London, UK
`ahmet@nethermind.io`

Abstract. In this paper, we study the compartment-based and hierarchical delegation of signing power of the verifiable accountable subgroup multi-signature (vASM). ASM is a multi-signature in which the participants are accountable for the resulting signature, and the number of participants is not fixed. After Micali et al.'s and Boneh et al.'s ASM schemes, the verifiable-ASM (vASM) scheme with a verifiable group setup and more efficient verification phase was proposed recently. The verifiable group setup in vASM verifies the participants at the group setup phase. In this work, we show that the vASM scheme can also be considered as a proxy signature in which an authorized user (original signer, designator) delegates her signing rights to a single (or a group of) unauthorized user(s) (proxy signer). Namely, we propose four new constructions with the properties and functionalities of an ideal proxy signature and a compartment-based/hierarchical structure. In the first construction, we apply the vASM scheme recursively; in the second one, we use Shamir's secret sharing (SSS) scheme; in the third construction, we use SSS again but in a nested fashion. In the last one, we use the hierarchical threshold secret sharing (HTSS) scheme for delegation. Then, we show the affiliation of our constructions to proxy signatures and compare our constructions with each other in terms of efficiency and security. Finally we compare the vASM scheme with the existing pairing-based proxy signature schemes.

Keywords: accountable subgroup multi-signatures · proxy signatures · threshold secret sharing · delegation

1 Introduction

Signatory authorities carry out their transactions by assigning a deputy to use this power of signing authority for periods when they cannot be present at their organization. While they can appoint a single deputy among their subordinates, there may also be cases in which they authorize different proxies on different issues. In the literature, several notions define solutions for delegating signing capability to unauthorized users by authorized ones in an organization. *Proxy signature*, which allows an unauthorized user (proxy signer) to sign on behalf of an authorized user (original signer), was first defined by Mambo et al. [27] in 1996. After its first proposal, proxy signatures have been studied in [1,5,18–20,26,29,36,41,43]. Moreover the proxy multi-signatures [42], the multi-proxy signatures [35], and the multi-proxy multi-signatures [34] were proposed according to the number of proxies and original signers in the constructions. Among these variants, *proxy multi-signature* is a notion related to the delegation of signing authority, and it was first proposed by Yi et al. [42] in 2000. It is a multi-signature scheme in which a designated proxy signer generates a signature on a common message on behalf of several original signers. Hwang and Shi [35] proposed the concept of the multi-proxy signature scheme in which an original signer can authorize a group of proxy signers as a proxy agent, and a valid multi-proxy signature can be generated only with the participation of all proxy signers. Another proxy signature concept is *multi-proxy multi-signatures* proposed by Hwang and Chen [34] in 2004. In a multi-proxy multi-signature scheme, only the cooperation of a group of original signers designates a group of proxy signers. Then, only the cooperation of all members of the proxy group can generate a valid signature on behalf of the group of original signers. Several proxy signatures also exist, such as threshold, blind, and ring variants [16,22,24,44]. Most of the proposals in the literature either show that the previous ones were insecure or propose a modified one because of the lack of a standard security model. The security properties of an ideal proxy signature are defined in [27] and extended in [19]. These properties are as follows:

- *Verifiability:* A proxy signature should convince any verifier that the original signer agrees with it.
- *Strong unforgeability:* Except for the designated proxy signer, no one can create a valid proxy signature.
- *Strong identifiability:* From a proxy signature, any verifier should identify the identity of the proxy signer.
- *Strong undeniability:* A proxy signature should have non-repudiation property.
- *Prevention of misuse:* Ensuring that a proxy signing key can be used only proxy signing process, no other purposes.

Boldyreva et al. [5] stated that the above security properties were defined informally and needed to be formalized. For this reason, they first defined a formal security model for the proxy signature schemes and defined the functionalities a proxy signature should have. In this work, we show that the verifiable accountable subgroup multi-signature (vASM) scheme [2] can also be used as a proxy signature scheme. The vASM scheme also supports one or more proxies and original signers. In addition, we also show that the signing power of vASM authority can be delegated via appropriate threshold secret sharing schemes [13,17,32,37,38].

Consider an organization with a complex topology consisting of many compartments; each has many sub-compartments and multi-level hierarchical structures. In this case, the issue of who will deputize for whom would emerge as a challenging problem. For instance, a company may have multiple branches, each with its own manager. Managers might occasionally be unavailable due to reasons such as holidays. In such cases, the manager would need to designate one or more deputies from among their subordinates within the branch. We handle delegation of signing rights of the manager regardless of the graph theoretic structure of her branch. In this paper, we propose four constructions using the vASM scheme and several threshold secret sharing schemes to present alternative solutions to the problem of delegation of signing authority problem. We propose that one can have the functionalities of a proxy signature scheme via accountable subgroup multi-signature schemes [2, 6, 28], in particular via the vASM scheme [2], or a combination of the vASM scheme with some proper threshold secret sharing schemes [13, 17, 32, 37, 38]. Assume a scenario that there exists an organization with lots of compartments, and each has an authorized signer and many unauthorized users, just like managers and their subordinates in a hierarchical organizational structure. The authorized signers have signing rights on behalf of their compartments (also on behalf of the entire organization). They want to assign proxies among unauthorized users in their compartments. To solve this assignment problem, firstly, we apply the vASM scheme recursively. The authorized signer (original signer) and the unauthorized users (proxy signer candidates) jointly participate in a vASM group setup. At the end of this setup phase, each unauthorized user has a *compartment membership key* which can be used to sign on behalf of the authorized user and the entire organization. Then we combine the methods from threshold signatures [4, 10, 14, 15] and vASM scheme to construct solutions for the delegation that supports one or more original/proxy signers and provides accountability at the same time. As a second method, we consider that the authorized users share their membership key via Shamir's secret sharing (SSS) scheme [32] with unauthorized users in their compartment. Then unauthorized users can sign with the sum of their shares and secret keys. Thirdly, we assume a trusted user exists in each compartment and apply Shamir's SSS in a nested fashion. The authorized users first share their membership key via a 2-out-of-2 Shamir's SSS, give one of the shares to the trusted user in their compartment, and then recursively apply another SSS to the second share. At the end of this nested SSS protocol, each unauthorized user has a compartment membership key to sign a proxy signature on behalf of the authorized signer. Finally, we consider a hierarchical structure and think the trusted user(s) are in the first level of the compartment. In this case, the authorized signers share their membership keys via a hierarchical threshold secret sharing (HTSS) scheme [17, 37, 38]. Each unauthorized user can sign on behalf of the authorized signers in an accountable way. After describing the constructions, we see that they all have the main functionalities of a proxy signature according to [5].

The outline of the paper is as follows. In Sect. 2 we give preliminary information, including definitions of proxy signature schemes, bilinear pairings, computational co-DHP/ψ-co-DHP, Shamir's secret sharing scheme (SSS) [32], Feldman's verifiable secret sharing (VSS) protocol [13], hierarchical threshold secret sharing scheme (HTSS) [17,37,38], security discussion of threshold secret sharing schemes and definition of the verifiable-ASM (vASM) scheme [2]. Then we give our proposed constructions in Sect. 3. In the same section, we also provide remarks about the security of the constructions. In Sect. 4, we compare our constructions in terms of the number of operations required in the phases of our proposed constructions. Then we compare the vASM scheme with the existing proxy signatures, proxy multi-signatures, multi-proxy signatures and multi-proxy multi-signatures in terms of their efficiency.

2 Preliminary

In [5] the authors give a detailed definition of a proxy signature and define the first formal security model for this notion. Below we give the formal definitions of the digital signature scheme and proxy signature scheme according to [5].

Definition 1 (Digital signature scheme [5]).
A digital signature scheme $\mathcal{DS} = (\mathcal{G}, \mathcal{K}, \mathcal{S}, \mathcal{V})$ is specified by four algorithms which are defined below:

- $\mathcal{G}(1^\lambda)$ *takes 1^λ, where λ is the security parameter as input, and outputs the public system parameters par including security parameter, hash functions, cyclic groups, generators, etc.*
- $\mathcal{K}(par)$ *takes system parameters par as input, and outputs a secret-public key pair (sk, pk).*
- $\mathcal{S}(par, M, sk)$ *takes system parameters par, message M, and secret key sk as inputs, and returns a signature σ.*
- $\mathcal{V}(par, M, pk, \sigma)$ *takes system parameters par, message $M \in \{0,1\}^*$, public key pk, a signature σ as inputs, and outputs accept or reject.*

Definition 2 (Proxy signature scheme [5]).
A proxy signature scheme is a tuple $\mathcal{PS} = (\mathcal{G}, \mathcal{K}, \mathcal{S}, \mathcal{V}, (\mathcal{D}, \mathcal{P}), \mathcal{PS}, \mathcal{PV}, \mathcal{ID})$, where $\mathcal{DS} = (\mathcal{G}, \mathcal{K}, \mathcal{S}, \mathcal{V})$ are as defined in Definition 1, and the other components are defined below:

- $(\mathcal{D}, \mathcal{P})$ *is an interactive proxy-designation protocol composed of a pair of algorithms, i.e., $\mathcal{D}$ and $\mathcal{P}$. Let i, j be the designator and the proxy, respectively.*
  - $\mathcal{D}(pk_i, sk_i, j, pk_j, \omega)$ *takes both sides' public keys, the designator's secret key, the proxy's identity, and the message space ω as inputs, and gives no local outputs.*
  - $\mathcal{P}(pk_j, sk_j, pk_i)$ *takes public keys of both sides and the secret key of the proxy and outputs proxy signing key skp after the interaction with $\mathcal{D}$.*

- $\mathcal{PS}(skp, M)$ takes proxy signing key skp and the message M as inputs, and outputs a proxy signature $p\sigma$.
- $\mathcal{PV}(pk, M, p\sigma)$ takes public key pk, a message M, and a proxy signature $p\sigma$ as inputs, and outputs accept or reject.
- $\mathcal{ID}(p\sigma)$ takes a proxy signature $p\sigma$ and outputs an identity of a signer i or $\perp$.

We will compare the properties of our constructions with the functionalities given in Definition 2 in Sect. 4.

Definition 3. *Let $\mathbb{G}_1, \mathbb{G}_2$ be cyclic additive groups of prime order q. Let $\mathbb{G}_T$ be another cyclic group that is multiplicative and of the same order. A pairing is a map $e : \mathbb{G}_1 \times \mathbb{G}_2 \to \mathbb{G}_T$ which satisfies the bilinearity and non-degeneracy properties:*

- *Bilinearity: $e(A^\alpha, B^\beta) = e(A, B)^{\alpha\beta}$ for all $\alpha, \beta \in \mathbb{Z}$, $A \in \mathbb{G}_1$ and $B \in \mathbb{G}_2$.*
- *Non-degeneracy: $e \neq 1$.*

Note that, as common in the literature, we use multiplicative notation for groups $\mathbb{G}_1$ and $\mathbb{G}_2$. The definitions of underlying hard problems of the vASM scheme, i.e. computational co-DHP, and computational ψ-co-DHP are given below.

Definition 4 (Computational co-Diffie-Hellman Problem [8]). *For groups $\mathbb{G}_1 = \langle g_1 \rangle$ and $\mathbb{G}_2 = \langle g_2 \rangle$ of prime order q, define $Adv_{\mathbb{G}_1, \mathbb{G}_2}^{co\text{-}CDH}$ of an adversary $\mathcal{A}$ as*

$$Pr\left[y = g_1^{\alpha\beta} : (\alpha, \beta) \xleftarrow{\$} \mathbb{Z}_q^2, y \leftarrow \mathcal{A}(g_1^\alpha, g_1^\beta, g_2^\beta) \right],$$

where the probability is taken over the random choices of $\mathcal{A}$ and the random selection of (α, β). $\mathcal{A}$ (τ, ϵ)-breaks the co-CDH problem if it runs in time at most τ and has $Adv_{\mathbb{G}_1, \mathbb{G}_2}^{co\text{-}CDH} \geq \epsilon$. co-CDH is (τ, ϵ)-hard if no such adversary exists.

Definition 5 (Computational ψ-co-Diffie-Hellman Problem [6]). *For groups $\mathbb{G}_1 = \langle g_1 \rangle$ and $\mathbb{G}_2 = \langle g_2 \rangle$ of prime order q, let $\mathcal{O}^\psi(.)$ be an oracle that on input $g_2^x \in \mathbb{G}_2$ returns $g_1^x \in \mathbb{G}_1$. Define $Adv_{\mathbb{G}_1, \mathbb{G}_2}^{\psi\text{-}co\text{-}CDH}$ of an adversary $\mathcal{A}$ as*

$$Pr\left[y = g_1^{\alpha\beta} : (\alpha, \beta) \xleftarrow{\$} \mathbb{Z}_q^2, y \leftarrow \mathcal{A}^{\mathcal{O}^\psi(.)}(g_1^\alpha, g_1^\beta, g_2^\beta) \right],$$

where the probability is taken over the random choices of $\mathcal{A}$ and the random selection of (α, β). $\mathcal{A}$ (τ, ϵ)-breaks the ψ-co-CDH problem if it runs in time at most τ and has $Adv_{\mathbb{G}_1, \mathbb{G}_2}^{\psi\text{-}co\text{-}CDH} \geq \epsilon$. ψ-co-CDH is (τ, ϵ)-hard if no such adversary exists.

2.1   Shamir's Secret Sharing (SSS) Scheme

Shamir's secret sharing (SSS) scheme [32] is a protocol that is used for sharing a secret among some predetermined players. Assume that we have n players. Let q be prime. The dealer delivers the shares as follows:

- Chooses a polynomial over $\mathbb{Z}_q$ of degree $t - 1 < q$,

$$f(x) = \alpha_{t-1}x^{t-1} + \ldots + \alpha_1 x + \alpha_0$$

  with $\alpha_k \in \mathbb{Z}_q$ for $k = 0, \ldots, t - 1$, where α_0 is the secret to be shared.
- Sends $f(i)$ to the i-th player for $i = 1, 2, \ldots, n$.

If at least t or more players perform Lagrange interpolation with their shares, they can uniquely determine the secret polynomial $f(x)$ and $f(0)$ will yield the secret.

2.2   Feldman's Verifiable Secret Sharing (VSS) Scheme

Feldman's verifiable secret sharing (VSS) scheme [13] is a protocol that is used for sharing a secret in a verifiable fashion, where Shamir's secret sharing scheme [32] is directly used to share and reconstruct the secret. In addition to Shamir's scheme, the shares can be checked for consistency in Feldman's scheme. To this end, the dealer computes commitments with the coefficients of the secret polynomial so that users can verify that they receive consistent shares from the dealer.

Assume that we have n players. Let $\mathbb{G}$ be a cyclic group of prime order q, and g be its generator. The dealer shares a secret as follows:

- Chooses a polynomial over $\mathbb{Z}_q$ of degree $t - 1 < q$,

$$f(x) = \alpha_{t-1}x^{t-1} + \ldots + \alpha_1 x + \alpha_0$$

  with $\alpha_k \in \mathbb{Z}_q$ for $k = 0, \ldots, t - 1$, where α_0 is the secret to be shared.
- Computes a set of commitments $\mathrm{COM} = \{C_k : C_k = g^{\alpha_k}, k = 0, 1, \ldots, t - 1\}$.
- Sends $f(i)$ and COM to the i-th player for $i = 1, 2, \ldots, n$.

After receiving a share and the set of commitments, the i-th player checks

$$g^{f(i)} \stackrel{?}{=} \prod_{k=0}^{t-1} C_k^{i^k}. \tag{1}$$

The received share is consistent only if (1) is satisfied. The reconstruction of the secret is identical to Shamir's secret sharing scheme. The users can uniquely determine the secret polynomial by applying the Lagrange interpolation if and only if threshold t is satisfied.

Definition 6 (Lagrange Interpolation). *Given t points (x_i, y_i) for distinct x_i's and $i = 1, \ldots, t$, the unique polynomial of degree $t - 1$ which satisfies all the points is the linear combination of Lagrange basis polynomials, and given by the equation*

$$P(x) = \sum_{i=1}^{t-1} y_i \ell_i(x), \tag{2}$$

where the Lagrange basis polynomial $\ell_i(x)$ is given by the equation

$$\ell_i(x) = \prod_{\substack{0 \le k \le t-1 \\ i \ne k}} \frac{x - x_k}{x_i - x_k}. \tag{3}$$

Now we define another notion called the Lagrange coefficient, which we will use in the signature aggregation and verification phases of our constructions in Sect. 3.2 and 3.3.

Definition 7 (Lagrange Coefficient). *Given a set of t points (x_i, y_i) for distinct x_i's and $i = 1, \ldots, t$, the Lagrange coefficient λ_i is the evaluation of the the Lagrange basis polynomial $\ell_i(x)$ at 0, i.e.*

$$\lambda_i = \ell_i(0) = \prod_{\substack{0 \le k \le t-1 \\ i \ne k}} \frac{-x_k}{x_i - x_k}. \tag{4}$$

2.3 Hierarchical Threshold Secret Sharing Scheme

Hierarchical threshold secret sharing schemes in [17,37,38] were proposed for sharing secrets in a partitioned structure of users. Assume that we have a set $\mathcal{G}$ of n players, which is composed of m disjoint subsets, $\mathcal{G} = \bigcup_{i=1}^{m} \mathcal{G}_i$ where $\mathcal{G}_i \cap \mathcal{G}_j = \emptyset$ for $i \ne j$. Let $0 < k_1 < \ldots < k_m$ be a sequence of integers. The access structure Γ of the hierarchical threshold secret sharing scheme in [17] is

$$\Gamma = \{\mathcal{V} \subset \mathcal{G} : |\mathcal{V} \cap (\bigcup_{j=1}^{i} \mathcal{G}_j)| \ge k_i | \forall i \in \{1, 2, \ldots, m\}\}$$

Like in Shamir's SSS, the dealer delivers the shares as follows:

– Chooses a random polynomial of degree k_m

$$f(x) = \sum_{i=0}^{k_m} \alpha_i x^i$$

such that $\alpha_0 = s$.
– Sends a share $(x_u, f^{(k_{i-1}+1)}(x_u))$ to the user $u \in \mathcal{G}_i$, where $f^{(k_{i-1}+1)}$ is the $(k_{i-1} + 1)$-th derivative of f, $k_0 = -1$ and $x_u \in \mathbb{Z}_q$ is a part of the share corresponding to the user u.

The secret reconstruction is performed by Birkhoff interpolation defined below only if the level-specific thresholds are satisfied.

Definition 8 (Birkhoff Interpolation). *Let the triplet $\langle X, E, C \rangle$ be as follows:*

- $X = \{x_1, \ldots, x_k\}$ *be a given set of points in* $\mathbb{R}$, *where* $x_1 < x_2 < \ldots < x_k$,
- $E = (e_{i,j})$ *for* $i = 0, \ldots, k$ *and* $j = 0, \ldots, \ell$ *be a matrix with binary entries,* $I(E) = \{(i, j) : e_{i,j} = 1\}$, $d = |I(E)|$, *and*
- $C = \{c_{i,j} : (i, j) \in I(E)\}$ *be a set of* d *real values (we assume hereafter that the right-most column in* E *is nonzero).*

Then, the Birkhoff interpolation problem that corresponds to the triplet $\langle X, E, C \rangle$ *is the problem of finding a polynomial* $P(x) \in \mathbb{R}_{d-1}[x]$ *that satisfies the* d *equalities*

$$P^{(j)}(x_i) = c_{i,j}, (i, j) \in I(E). \tag{5}$$

The matrix E is called the interpolation matrix.

We summarize the Birkhoff interpolation method as described in [12]. Let $\phi = \{g_0, g_1, \ldots, g_{d-1}\}$ be a system of linearly independent, $d - 1$ times continuously differentiable real-valued, functions and $I'(E) = \{\alpha_i : i = 1, \ldots, d\}$ be a vector that is obtained by lexicographically ordering of entries of $I(E)$. Furthermore, let $\alpha_i(1)$ and $\alpha_i(2)$ denote the first and second elements of the pair $\alpha_i \in I'(E)$. Finally, let $C' = \{c'_i : i = 1, \ldots, d\}$ be another vector obtained by lexicographically ordering entries of C (according to the indexes of elements in C).

According to the above definition and clarifications, the Birkhoff interpolation problem is solved by the below equation:

$$P(x) = \sum_{j=0}^{d-1} \frac{det(A(E, X, \phi_j))}{det(A(E, X, \phi))} g_j(x), \tag{6}$$

where

$$A(E, X, \phi_j) = (\theta_{ij})_{d \times d}, \tag{7}$$

$\theta_{ij} = g_{j-1}^{(\alpha_i(2))}(x_{\alpha_i(1)})$ for $i, j = 1, \ldots, d$, and $A(E, X, \phi)$ can be computed by replacing $(j + 1)$-th column of matrix (7) with C'. An explicit example of the application of Birkhoff interpolation using (6) can be found in [12].

Although the Birkhoff interpolation problems can be solved by (6), we cannot directly use this method. In our constructions, we use Birkhoff interpolation for signature aggregation and verification of the aggregated signature. In order to compute (6), any combiner and verifier need to know the sufficient number of shares. However, our constructions require only the shareholders to know their shares, and no one should learn the shares of others. To this end, we use the modified version of (6), which is also given in [12]:

$$P(x) = \sum_{i=0}^{d-1} c'_{i+1}\left(\sum_{j=0}^{d-1}(-1)^{(i+j)}\frac{det(A_i(E,X,\phi_j))}{det(A(E,X,\phi))}g_j(x)\right) \qquad (8)$$

Now we can define the Birkhoff coefficient, which we will use in the signature aggregation and verification phases of our last construction in Sect. 3.4.

Definition 9. *Let the triplet $\langle X, E, C \rangle$ be in Definition 8 then the Birkhoff coefficient β_i is the evaluation of the polynomial*

$$P_i(x) = \sum_{j=0}^{d-1}(-1)^{(i+j)}\frac{det(A_i(E,X,\phi_j))}{det(A(E,X,\phi))}g_j(x),$$

at 0, i.e. $\beta_i = P_i(0)$.

2.4   Security of the Secret Sharing Schemes

Shamir's secret sharing (SSS) scheme [32] is known to be information-theoretically secure, i.e. secure against even computationally unbounded adversaries. The hierarchical threshold secret sharing scheme (HTSS) [17,37,38] is a generalization of Shamir's SSS and is also information-theoretically secure. However, suppose an adversary can behave like a shareholder and interact actively with real shareholders. In that case, he can obtain the secret as described in [39] by Tompa and Woll. The attack works as follows. Consider an adversary behaving actively. He chooses a random *false* share and performs an interaction for reconstruction with $t-1$ honest shareholders. As a result, he will not obtain the secret because there are $t-1$ true shares where there must be at least t. However, he obtains a value, let us say s'. Then he interacts with another group of $t-1$ honest shareholders and gets another value $\hat{s}$. He can compute the secret without having a genuine share using the values s' and $\hat{s}$ and corresponding Lagrange coefficients. To avoid this attack, one should force the shareholders to behave passively. For example, using Feldman's verifiable secret sharing (VSS) scheme [13] would be a solution. Before the reconstruction phase, one checks whether the shares to be interpolated are consistent with the shared secret. This will avoid the active adversary attack defined in [39]. On the other hand, Feldman's VSS has its own security risks.

As we stated before, Feldman's VSS scheme uses SSS, and so it has the same security arguments about sharing and reconstruction phases. However, in the committing phase, it is not information-theoretically secure anymore. The commitment set contains $C_0 = g^s$, where g is the generator for the cyclic group, and s is the secret to be shared. This commitment may leak information about the secret s. The security of the commitments depends on the Discrete Logarithm Problem (DLP), defined over cyclic groups. In some cyclic groups, even with a large order, DLP may not be as hard as it is supposed to be. Therefore the space that we are working in should be chosen carefully. In this paper, all the schemes that we propose are pairing-based constructions. In the literature, there are many secure and efficient pairing-friendly curves that we can choose.

2.5   vASM Scheme

Accountable subgroup multi-signature (ASM) schemes were studied by [2,6,28]. For instance, the Schnorr-based ASM scheme was proposed in 2001 by Micali et al. [28], and the BLS-based ASM scheme was given in 2018 by Boneh et al. [6]. Then, the verifiable-ASM (vASM) scheme was proposed recently in [2] with a different group setup method and more efficient verification. It is also a BLS-based scheme, and its security depends on the hardness of computational co-DHP and computational ψ-co-DHP problems which are given in Definitions 4 and 5. The security proof of vASM scheme in the random oracle model can be found in [2]. In the vASM scheme, each user generates her secret and public key pair independently. Then all users jointly perform a group setup in which they participate in a verifiable secret sharing (VSS) protocol [13]. At the end of this procedure, each user obtains a membership key and a membership public key, which satisfy a common public commitment generated in the group setup phase. Then, any subgroup of users signs a common message M and sends her individual signatures to a designated combiner. This combiner could be either one of the signers or a specifically assigned party.

Let $\mathbb{G}_1, \mathbb{G}_2$ be cyclic additive groups of prime order q. Let $\mathbb{G}_T$ be another cyclic group that is multiplicative and of order q. Let e be an efficient bilinear pairing, defined over the groups $\mathbb{G}_1, \mathbb{G}_2$, and $\mathbb{G}_T$ as in Definition 3. Recall that we use multiplicative notation for groups $\mathbb{G}_1$ and $\mathbb{G}_2$. Let H be a hash function such that H : $\{0,1\}^* \to \mathbb{G}_1$. Finally, assume that we have a group $\mathcal{G}$ of n potential signers, and the subgroup $\mathcal{S} \subseteq \mathcal{G}$ is the set of τ signers among those n potential ones. Below we give the steps of the vASM scheme given in [2].

1. **Key Generation:** Each user $i \in \mathcal{G}$ picks a secret key $sk_i \xleftarrow{\$} \mathbb{Z}_q$, and computes the public key $pk_i \leftarrow g_2^{sk_i}$, where g_2 is a generator of $\mathbb{G}_2$.
2. **Group Setup:** Each user $i \in \mathcal{G}$ proceeds as follows:
   - Chooses a polynomial $f_i(x) = \alpha_{n-1}^{(i)} x^{n-1} + \ldots + \alpha_1^{(i)} x + \alpha_0^{(i)} \in \mathbb{Z}_q[x]$, where $\alpha_0^{(i)} = sk_i$ and $\alpha_k^{(i)}$'s are all nonzero and distinct, for $k = 1, \ldots, n-1$.
   - Computes the set of commitments $\text{COM}_i := \{C_k^{(i)} = g_2^{\alpha_k^{(i)}} | k = 0, \ldots, n-1\}$.
   - Sends $(f_i(j), \text{COM}_i)$ to j-th user in $\mathcal{G}$, for $j = 1, \ldots, n$.
   - After receiving $(f_j(i), \text{COM}_j)$,
     • computes the membership key $mk_i = \sum_{j \in \mathcal{G}} f_j(i)$.
     • computes $\text{COM} := \{C_k = \prod_{j \in \mathcal{G}} C_k^{(j)} | k = 0, \ldots, n-1\}$.
   - Checks:
     (a) $C_0 \stackrel{?}{=} \prod_{i \in \mathcal{G}} pk_i$
     (b) $g_2^{mk_i} \stackrel{?}{=} \prod_{k=0}^{n-1} C_k^{i^k}$
   - If either (a) or (b) fails, then she aborts. Else, she defines MPK = $\{mpk_i = g_2^{mk_i}\}_{i \in \mathcal{G}}$, and makes MPK and COM public.

3. **Signature Generation:** A signer $i \in \mathcal{G}$ computes his/her individual signature $s_i = \mathrm{H}(M)^{mk_i}$ on the message m and sends s_i to the combiner.

4. **Signature Aggregation:** After receiving the individual signatures of the signers, the combiner first forms the set of signers $\mathcal{S} \subseteq \mathcal{G}$. Then, she computes the aggregated subgroup multi-signature $\sigma = \prod\limits_{i \in \mathcal{S}} s_i$.

5. **Verification:** Anyone, who is given $\{par, \mathrm{MPK}, \mathrm{COM}, \mathcal{S}, m, \sigma\}$, can verify the signature σ by checking

$$e\big(\mathrm{H}(M), \prod_{i \in \mathcal{S}} mpk_i\big) \overset{?}{=} e(\sigma, g_2). \tag{9}$$

Correctness of the vASM scheme follows from the following equation array.

$$
\begin{aligned}
e\big(\mathrm{H}(M), \prod_{i \in \mathcal{S}} mpk_i\big) &= e\big(\mathrm{H}(M), \prod_{i \in \mathcal{S}} g_2^{mk_i}\big) \\
&= e\big(\mathrm{H}(M), g_2^{\sum\limits_{i \in \mathcal{S}} mk_i}\big) \\
&= e\big(\mathrm{H}(M)^{\sum\limits_{i \in \mathcal{S}} mk_i}, g_2\big) \\
&= e\big(\prod_{i \in \mathcal{S}} \mathrm{H}(M)^{mk_i}, g_2\big) \\
&= e\big(\prod_{i \in \mathcal{S}} s_i, g_2\big) \\
&= e(\sigma, g_2)
\end{aligned}
$$

Remark 1. Since all users share their secret keys, the first commitments in their individual commitment set have to be equal to their public key. Therefore, the first check in the group setup phase shows that users share their secret keys, not rogue ones [7].

Remark 2. The second check is the standard consistency check of Feldman's VSS scheme as described in Sect. 2.2, whose purpose is to check whether the shares received from other users are consistent with the shared secrets.

Detailed information about the vASM scheme, including security proof, efficiency comparison, and remarks, can be found in [2]. In addition, lattice-based vASM scheme is proposed in [3].

3   Compartment-Based and Hierarchical Threshold Delegation of vASM Authority

In general, the delegation of signing capability can be achieved by giving a power of attorney. For example, consider an organization with a sophisticated and complicated nature whose structure expands in vertical and horizontal directions. In such a case, signing authorities may need to assign multiple proxies simultaneously. In this section, we propose four constructions. In the first one, we apply the vASM scheme recursively. In the second construction, we use Shamir's secret sharing scheme (SSS) [32] directly among the users of each compartment. In the third one, we assume the existence of at least one trusted user in each compartment, and we share the vASM authority by a secret sharing scheme in a nested fashion. In the last one, we use the hierarchical threshold secret sharing scheme [17,37] to delegate the vASM signing authority of an authorized users to the unauthorized users in the same compartment, which is partitioned hierarchically.

Consider a group of users $\mathcal{G} = \bigcup_{i=1}^{m} \mathcal{U}_i$ which is a union of distinct compartments $\mathcal{U}_i$ for $i = 1, \ldots, m$. Without loss of generality, we assume that only one authorized user (or original signer) exists in each compartment $\mathcal{U}_i$, that is AU_i. Assume that each $AU_i \in \mathcal{U}_i$ participates in a vASM group setup, obtains her membership key mk_i, and wants to delegate his vASM signing authority to some unauthorized users (or proxy signers) in her compartment, i.e. $u_{ij} \in \mathcal{U}_i$ for $j = 1, \ldots, k_i$, where $k_i \in \mathbb{Z}$ is the number of authorized user and proxy candidates (unauthorized users) in the i-th compartment. After delegation stage, unauthorized users have their compartment membership keys cmk_i. Now, they can generate a signature by using cmk_i, and send it to a combiner. Recall that this combiner could be either one of the signers or a specifically assigned party as in the definition of vASM [2]. Note that we use italic capital letters to denote the set of users, e.g., $\mathcal{U}_i, \mathcal{G}, \mathcal{S}$, etc. In the remaining part, we use the notation above. Let the functions e and H be as in Sect. 2.5.

3.1   Recursive vASM as a Proxy Signature

The output of a vASM group setup is a membership key and a membership public key for each participant, which are used for signing and verification, respectively. Consider an $AU_i \in \mathcal{U}_i$ has her membership key mk_i that she calculated with the other authorized users $AU_j \in \mathcal{U}_j$, for $j = 1, \ldots, m$, in a group setup as described in Sect. 2.5. Moreover, they also publish a set of membership public keys MPK, and a global commitment set COM at the end of that group setup. Assume that AU_i wants to delegate the signing power of her membership key to a certain number of unauthorized users in her compartment. Let I_i be the union of the indices of AU_i and the set of unauthorized users in i-th compartments that AU_i wants to designate as proxies. To that end, AU_i and the unauthorized users $u_{ij} \in I_i$ jointly participate in another vASM group setup, which we call as *compartment setup*. Each unauthorized participant $u_{ij} \in I_i$ joins the compartment setup protocol with his secret key sk_{ij} for $j = 1, \ldots, k_i$. However, the authorized user AU_i participates in the compartment setup with her membership key mk_i. At the end of this compartment setup, each unauthorized user $u_{ij} \in I_i$ obtains a compartment membership key cmk_{ij} for $j = 1, \ldots, k_i$, along with compartment public key set CPK := $\{cpk_j : j = 1, \ldots, k_i\}$, a proxy commitment set

$\mathrm{PCS} := \{PC_j : j = 1, \ldots, k_i\}$ that contains the commitments of the VSS protocol they participate in, see Fig. 1. Here the compartment membership keys can be seen as proxy signing keys, and the compartment public keys in CPK can be seen as the proxy verification keys. Below we give the steps of this construction.

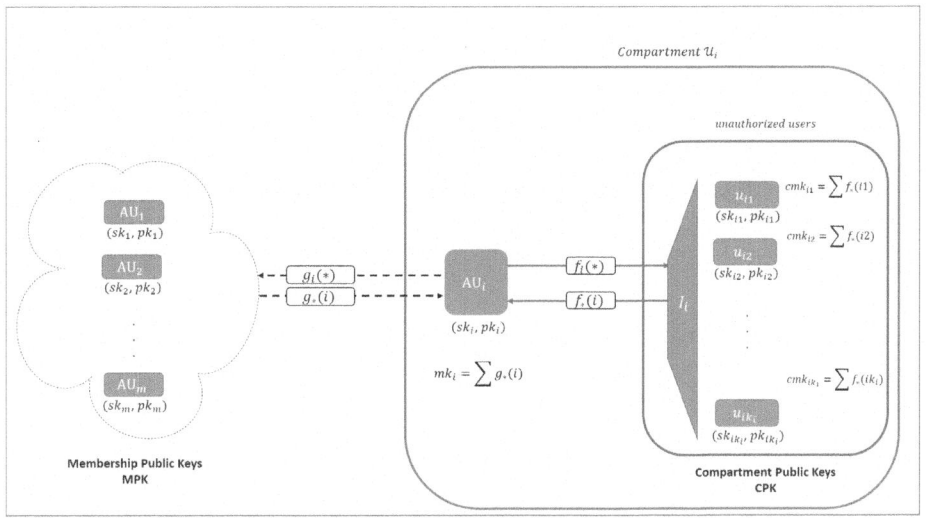

Fig. 1. Membership key generation of recursive vASM

1. **Key Generation:**
   - The authorized user AU_i has her membership key mk_i, membership public key mpk_i, and corresponding commitment set COM by performing a group setup with other authorized users.
   - Each unauthorized user $u_{ij} \in I_i$ has their secret and public key pairs sk_{ij}, pk_{ij} as described in Sect. 2.5.

2. **Compartment Setup:** Both AU_i and $u_{ij} \in I_i$ proceed as follows, see also Fig. 1:
   - Choose a polynomial $f_j(x) = \alpha_{k_i-1}^{(j)} x^{k_i-1} + \ldots + \alpha_1^{(j)} x + \alpha_0^{(j)} \in \mathbb{Z}_q[x]$, where $\alpha_0^{(j)} = sk_{ij}$ (or mk_i for AU_i) and $\alpha_w^{(j)}$'s are all nonzero and distinct, for $w = 1, \ldots, k_i - 1$.
   - Compute the set of commitments $\mathrm{PCS}_j := \{PC_w^{(j)} = g_2^{\alpha_w^{(j)}} | w = 0, \ldots, k_i - 1\}$.
   - Send $(f_j(z), \mathrm{PCS}_j)$ to z-th user in I_i, for $z = 1, \ldots, k_i$.
   - After receiving $(f_z(j), \mathrm{PCS}_z)$,
     - computes the membership key $cmk_{ij} = \sum\limits_{z \in I_i} f_z(j)$, and
     - computes $\mathrm{PCS}_i := \{PC_w = \prod\limits_{z \in I_i} PC_w^{(z)} | w = 0, \ldots, k_i\}$.

– Checks:
  (a) $PC_0 \overset{?}{=} mpk_i \cdot \prod_{j \in I_i} pk_{ij}$ ($mpk_i \in \text{MPK}$ is the membership public key of the authorized user AU_i)

  (b) $g_2^{cmk_{ij}} \overset{?}{=} \prod_{w=0}^{k_i-1} (PC_w)^{j^w}$

– If either (a) or (b) fails, then they abort. Else, define $\text{CPK}_i = \{cpk_{ij} = g_2^{cmk_{ij}}\}_{j \in \mathcal{U}_i}$, and make CPK_i and PCS_i public.

3. **Signature Generation:** A designated proxy signer (among the unauthorized users) $u_{ij} \in I_i$ computes his individual signature $s_{ij} = \text{H}(M)^{cmk_{ij}}$ on the message M and sends s_{ij} to the designated combiner.

4. **Signature Aggregation:** After receiving the individual signatures of the proxy signers, the designated combiner first forms the subgroup of proxy signers $\mathcal{S}_i \subseteq I_i$. Then, she computes the aggregated subgroup multi-signature $\sigma_i = \prod_{j \in \mathcal{S}_i} s_{ij}$.

5. **Verification:** Anyone, who is given $\{par, \text{CPK}_i, \mathcal{S}_i, M, \sigma_i\}$, can verify the signature σ_i by checking

$$e\big(\text{H}(M), \prod_{j \in \mathcal{S}_i} cpk_{ij}\big) \overset{?}{=} e(\sigma_i, g_2). \tag{10}$$

Verification satisfies correctness as given below:

$$e(\sigma_i, g_2) = e(\prod_{j \in \mathcal{S}_i} s_{ij}, g_2)$$
$$= e(\text{H}(M)^{\sum_{j \in \mathcal{S}_i} cmk_{ij}}, g_2)$$
$$= e(\text{H}(M), g_2^{\sum_{j \in \mathcal{S}_i} cmk_{ij}})$$
$$= e(\text{H}(M), \prod_{j \in \mathcal{S}_i} cpk_{ij}).$$

Security. Note that this is indeed a vASM signature scheme. The only difference from the known vASM is the secret that is shared in the compartment setup phase. AU_i shares his membership key mk_i from the previous group setup with the other authorized users of the other compartments (see Fig. 1) while the unauthorized users share their secret keys. Hence, the security of the construction follows from the security of the vASM scheme.

For the compartment setup phase, we should clarify the purpose of consistency checks. There are two consistency checks in the compartment setup. The first check is performed to ensure that the participants know their shared secret. Since the first proxy commitment $PC_0^{(j)} = g_2^{\alpha_0^{(j)}}$, and $\alpha_0^{(j)} = sk_j$(or mk_j), the aggregation of the first commitments $PC_0 = mpk_i \cdot \prod_{j \in I_i} pk_{ij}$ is nothing but the aggregation of the public keys pk_{ij} of the corresponding shared secret keys sk_{ij}(and mpk_i for mk_i). The second check is a standard consistency check for the VSS used in group setup, i.e., to guarantee that the received shares are consistent with the shared secrets. On the other hand, threshold solutions require a trusted combiner and honest majority assumption because any sufficient number (threshold) of malicious users can forge a valid signature. However, using the vASM scheme, each signer is responsible only for his signature. Even if all the unauthorized users (proxy) come together, they cannot forge the membership key of the authorized user.

Remark 3. Any authorized user can assign a proxy via a vASM signature scheme. If she wants to delegate her own or organizational signing power, she participates in the compartment setup with her secret or membership key, respectively.

Remark 4. One can also consider sharing the membership key mk_i of AU_i via a verifiable secret sharing scheme instead of an interactive compartment setup. However, in this case, one should be aware of the need for honest majority assumption.

In the following, we propose three more constructions in which we consider the authorized user AU_i to share her membership key mk_i with different threshold SSS solutions with a trusted user and/or honest majority assumption.

3.2 Shamir's Secret Sharing Scheme Based Delegation

In this construction of the delegation, each $AU_i \in \mathcal{U}_i$ delegates her signing authority by sharing her membership key mk_i to a subset of unauthorized users $u_{ij} \in \mathcal{U}_i$ via (t_i, k_i)-Shamir's SSS for $j = 1, \ldots, k_i$, where t_i is the threshold and k_i is the number of users in compartment $\mathcal{U}_i$ as shown in Fig. 2.

Then, the unauthorized users $u_{ij} \in \mathcal{G}$ for $i = 1, 2, \ldots, m$ and $j = 1, 2, \ldots, k_i$ sign as follows.

1. **Key Generation:** Each user $u_{ij} \in \mathcal{G}$ picks uniformly at random a secret key $sk_{ij} \xleftarrow{\$} \mathbb{Z}_q$, and computes the public key $pk_{ij} \leftarrow g_2^{sk_{ij}}$, where g_2 is a generator of $\mathbb{G}_2$.
2. **Compartment Setup:** Each AU_i shares her membership key mk_i via a (t_i, k_i)-Shamir's SSS [32] to a subset of unauthorized users $u_{ij} \in \mathcal{U}_i$ as shown in Fig. 2. At the end of this secret sharing procedure, each user $u_{ij} \in \mathcal{U}_i$ obtains a compartment membership key $cmk_{ij} = f_i(u_{ij})$.
3. **Signature Generation:** Each user $u_{ij} \in \mathcal{U}_i$ computes his individual signature on the message M as $s_{ij} = H(M)^{sk_{ij}+cmk_{ij}}$, and sends it to the combiner.
4. **Signature Aggregation:** After receiving the individual signatures from the users, the combiner
   (a) forms the subgroup of signers $\mathcal{S} := \{\mathcal{S}_1, \ldots, \mathcal{S}_m\}$, where $\mathcal{S}_i$ is the subgroup of signers from compartment $\mathcal{U}_i$, for $i = 1, \ldots, m$.
   (b) computes $\sigma = \prod_{i=1}^{m} \prod_{j \in \mathcal{S}_i} s_{ij}^{\lambda_{ij}}$, where λ_{ij} is the appropriate Lagrange coefficient as described in Definition 7.

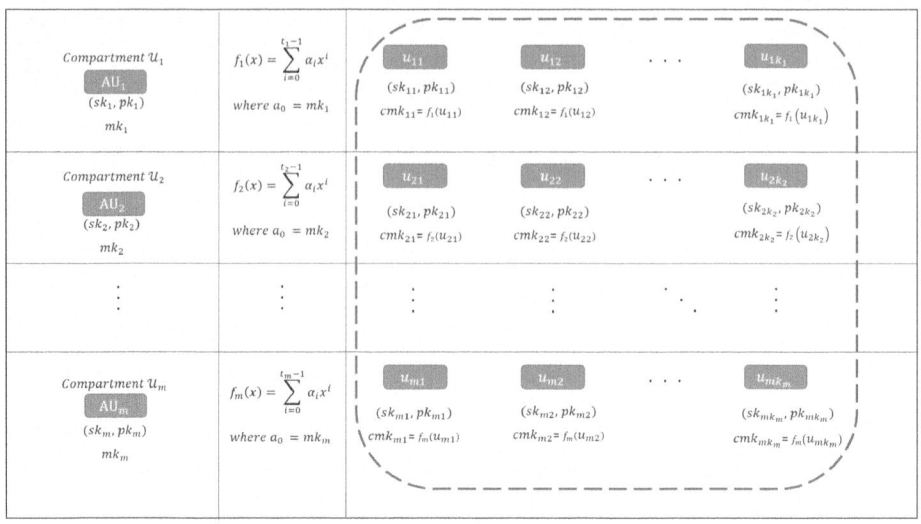

Fig. 2. Shamir's Secret Sharing Scheme Based Delegation

5. **Verification:** If $|\mathcal{S}_i| \geq t_i$ for $i = 1, \ldots, m$, any verifier given $(par, \text{MPK}, \mathcal{PK}, \mathcal{S}, M, \sigma)$ can verify the signature by checking the below equation

$$e\left(\mathrm{H}(M), \prod_{i=1}^{m} mpk_i \prod_{j \in \mathcal{S}_i} pk_{ij}^{\lambda_{ij}}\right) \overset{?}{=} e(\sigma, g_2).$$

The verification equation satisfies correctness as given below:

$$e(\sigma, g_2) = e(\prod_{i=1}^{m} \prod_{j \in \mathcal{S}_i} s_{ij}^{\lambda_{ij}}, g_2)$$

$$= e(\mathrm{H}(M)^{\sum_{i=1}^{m} \sum_{j \in \mathcal{S}_i} \lambda_{ij} sk_{ij} + \lambda_{ij} cmk_{ij}}, g_2)$$

$$= e(\mathrm{H}(M), g_2^{\sum_{i=1}^{m} \sum_{j \in \mathcal{S}_i} \lambda_{ij} sk_{ij} + \lambda_{ij} cmk_{ij}})$$

$$= e(\mathrm{H}(M), g_2^{\sum_{i=1}^{m} \sum_{j \in \mathcal{S}_i} \lambda_{ij} cmk_{ij}} g_2^{\sum_{i=1}^{m} \sum_{j \in \mathcal{S}_i} \lambda_{ij} sk_{ij}})$$

$$= e(\mathrm{H}(M), \prod_{i=1}^{m} mpk_i \prod_{j \in \mathcal{S}_i} pk_{ij}^{\lambda_{ij}}).$$

Security. In this scenario, each user u_{ij} signs a common message M with the sum of his secret and compartment membership keys, i.e. $sk_{ij} + cmk_{ij}$. The resulting individual signature of user u_{ij} is $s_{ij} = \mathrm{H}(M)^{sk_{ij}+cmk_{ij}}$ which can be seen as a BLS signature [8] on message M with the key $sk_{ij} + cmk_{ij}$. Since sk_{ij} is sampled randomly and the cmk_{ij} is computed via an information-theoretically secure SSS scheme, the signing key $sk_{ij} + cmk_{ij}$ will also be random-looking. Assume that the membership key cmk_{ij} is somehow obtained by an adversary. Even in this case, if the secret key sk_{ij} is secure, the individual signature s_{ij} of the user u_{ij} cannot be forged. On the other hand, if a sufficient number of unauthorized users are corrupted, then they can reconstruct the membership key mk_i of AU_i. Therefore, we assume for this scenario that the number of corrupted users is less than threshold t. Assigning a trusted user is a solution to this problem. In the following constructions, unauthorized users cannot obtain the membership key of AU_i without compromising with the trusted user(s).

3.3   Trusted User Based Delegation

In this construction, each $AU_i \in \mathcal{U}_i$ delegates her signing authority by sharing her membership key mk_i to a certain number of unauthorized users $u_{ij} \in \mathcal{U}_i$, including at least one trusted user. For simplicity, we assume that exactly one trusted user exists in each compartment. Note that one can also consider a group of trusted users. $AU_i \in \mathcal{U}_i$ uses Shamir's SSS in a nested way as described in Fig. 3. Without loss of generality, we assume that $u_{i1} \in \mathcal{U}_i$ is the trusted user in the i-th compartment.

$AU_i \in \mathcal{U}_i$ chooses two polynomials and distributes the shares as follows:

1. Chooses $f_{i1}(x) = a_i x + mk_i$ for a random secret $a_i \in \mathbb{Z}_q$.
2. Sends $f_{i1}(1)$ to the trusted user u_{i1}.
3. Chooses $f_{i2}(x) = a_{t_i-2}^{(i)} x^{t_i-2} + \ldots + a_1^{(i)} x_1 + f_{i1}(2)$, where $a_k^{(i)} \in \mathbb{Z}_q$ for $k = 1, \ldots, t_i - 2$.
4. Sends $f_{i2}(u_{ij})$ to the user u_{ij} for $j = 2, \ldots, k_i$.

With the first polynomial evaluation, AU_i shares her membership key mk_i via $(2,2)$-Shamir's SSS and sends one of the shares to the trusted user. For the second share, she applies one more $(t_i - 1, k_i - 1)$-Shamir's SSS, where t_i is the threshold, and k_i is the number of users in compartment $\mathcal{U}_i$.

The unauthorized users $u_{ij} \in \mathcal{G}$ for $i = 1, 2, \ldots, m$ and $j = 1, 2, \ldots, k_i$ sign as follows.

1. **Key Generation:** Each user $u_{ij} \in \mathcal{G}$ picks uniformly at random a secret key $sk_{ij} \xleftarrow{\$} \mathbb{Z}_q$, and computes her public key $pk_{ij} \leftarrow g_2^{sk_{ij}}$, where g_2 is a generator of $\mathbb{G}_2$.
2. **Compartment Setup:** Each AU_i shares her membership key mk_i via Shamir's SSS [32] (first $(2,2)$-Shamir's SSS, then $(t_i - 1, k_i - 1)$-Shamir's SSS) as shown in Fig. 3. At the end of this nested secret sharing procedure, each user in $\mathcal{U}_i$ obtains a compartment membership key cmk_{ij}.
3. **Signature Generation:** Each user $u_{ij} \in \mathcal{U}_i$ computes his individual signature on the message M as $s_{ij} = H(M)^{sk_{ij}+cmk_{ij}}$, and sends it to the combiner.
4. **Signature Aggregation:** After receiving the individual signatures from the users, the combiner

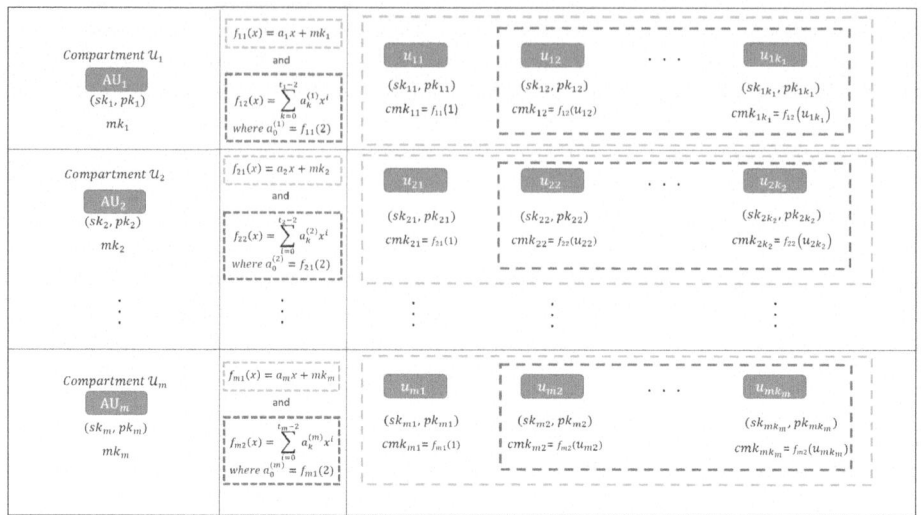

Fig. 3. Trusted User Based Delegation

(a) forms the subgroup of signers $\mathcal{S} := \{\mathcal{S}_1, \ldots, \mathcal{S}_m\}$, where $\mathcal{S}_i$ is the subgroup of signers from the compartment $\mathcal{U}_i$, for $i = 1, \ldots, m$.

(b) computes $\sigma = \prod_{i=1}^{m} s_{i1}^{\gamma_{i1}} \left(\prod_{\substack{j \in \mathcal{S}_i \\ j \neq 1}} s_{ij}^{\lambda_{ij}} \right)^{\gamma_{i2}}$, where γ_{i1}, γ_{i2} and λ_{ij} are the appro-

priate Lagrange coefficients as described in Definition 7, for the first and the second secret sharing scheme, respectively.

5. **Verification:** If the thresholds are satisfied and the trusted users participate in the signing, then any verifier given $(par, \mathrm{MPK}, \mathcal{PK}, \mathcal{S}, M, \sigma)$ can verify the signature by checking the below equation

$$e\left(\mathrm{H}(M), \prod_{i=1}^{m} mpk_i \cdot pk_{i1}^{\gamma_{i1}} \left(\prod_{\substack{j \in \mathcal{S}_i \\ j \neq 1}} pk_{ij}^{\lambda_{ij}} \right)^{\gamma_{i2}} \right) \overset{?}{=} e(\sigma, g_2)$$

The verification equation satisfies correctness as given below:

$$e(\sigma, g_2) = e\left(\prod_{i=1}^{m} s_{i1}^{\gamma_{i1}} \left(\prod_{\substack{j \in \mathcal{S}_i \\ j \neq 1}} s_{ij}^{\lambda_{ij}} \right)^{\gamma_{i2}}, g_2 \right)$$

$$= e\left(\mathrm{H}(M)^{\sum_{i=1}^{m} \gamma_{i1} sk_{i1} + \gamma_{i1} cmk_{i1} + \sum_{i=1}^{m} \gamma_{i2} \left(\sum_{\substack{j \in \mathcal{S}_i \\ j \neq 1}} \lambda_{ij} sk_{ij} + \lambda_{ij} cmk_{ij} \right)}, g_2 \right)$$

$$= e\left(\mathrm{H}(M), g_2^{\sum\limits_{i=1}^{m} \gamma_{i1} sk_{i1}+\gamma_{i1} cmk_{i1}+\sum\limits_{i=1}^{m}\gamma_{i2}(\sum\limits_{\substack{j\in\mathcal{S}_i\\j\neq1}}\lambda_{ij}sk_{ij}+\lambda_{ij}cmk_{ij})}\right)$$

$$= e\left(\mathrm{H}(M), g_2^{\sum\limits_{i=1}^{m} \gamma_{i1} sk_{i1}+\sum\limits_{i=1}^{m}\gamma_{i2}(\sum\limits_{\substack{j\in\mathcal{S}_i\\j\neq1}}\lambda_{ij}sk_{ij})+\sum\limits_{i=1}^{m}\gamma_{i1}cmk_{i1}+\sum\limits_{i=1}^{m}\gamma_{i2}(\sum\limits_{\substack{j\in\mathcal{S}_i\\j\neq1}}\lambda_{ij}cmk_{ij})}\right)$$

$$= e\left(\mathrm{H}(M), g_2^{\sum\limits_{i=1}^{m} \gamma_{i1} sk_{i1}+\sum\limits_{i=1}^{m}\gamma_{i2}(\sum\limits_{\substack{j\in\mathcal{S}_i\\j\neq1}}\lambda_{ij}sk_{ij})+\sum\limits_{i=1}^{m}mk_i}\right)$$

$$= e\left(\mathrm{H}(M), \prod_{i=1}^{m} mpk_i \cdot pk_{i1}^{\gamma_{i1}}\left(\prod_{\substack{j\in\mathcal{S}_i\\j\neq1}} pk_{ij}^{\lambda_{ij}}\right)^{\gamma_{i2}}\right)$$

Security. Security of this construction similarly follows from the security discussion of the previous construction. In this scenario, like in the previous one, each user u_{ij} signs a message M with the sum of his secret and membership keys, i.e. $sk_{ij} + cmk_{ij}$. The resulting individual signature is $s_{ij} = \mathrm{H}(M)^{sk_{ij}+cmk_{ij}}$ which can also be seen as a BLS signature [8] on the message M under the key $sk_{ij} + cmk_{ij}$. Because sk_{ij} is sampled uniformly at random, and cmk_{ij} is computed via the SSS scheme, the signing key $sk_{ij}+cmk_{ij}$ will be uniformly random. Moreover, computing membership keys are done in a nested way. Namely, the membership key mk_i of AU_i is partitioned into two parts. The first one is given to the trusted unauthorized user, and the other one is shared again among the other untrusted unauthorized users. Assume for a moment that the untrusted unauthorized users are corrupted, and they gather their membership keys cmk_{ij}. Even in this case, since the trusted user has the other half, corrupted users cannot forge the membership key mk_i of AU_i.

We can assign a group of trusted users instead of a single one, and instead of using SSS recursively, we can use a hierarchical threshold secret sharing scheme for the same purpose.

3.4   Hierarchical Threshold Secret Sharing Scheme Based Delegation

Consider an organization with m compartments $\mathcal{G} = \bigcup\limits_{i=1}^{m} \mathcal{U}_i$ and each compartment $\mathcal{U}_i$ has a hierarchical structure with r levels as shown in Fig. 4. Assume that each $\mathrm{AU}_i \in \mathcal{U}_i$ delegates her signing capability to the unauthorized users via the hierarchical threshold secret sharing (HTSS) scheme proposed in [17] as defined in Sect. 2.3.

The unauthorized users $u_{ij} \in \mathcal{G}$ for $i = 1, 2, \ldots, m$ and $j = 1, 2, \ldots, k_i$ sign as follows.

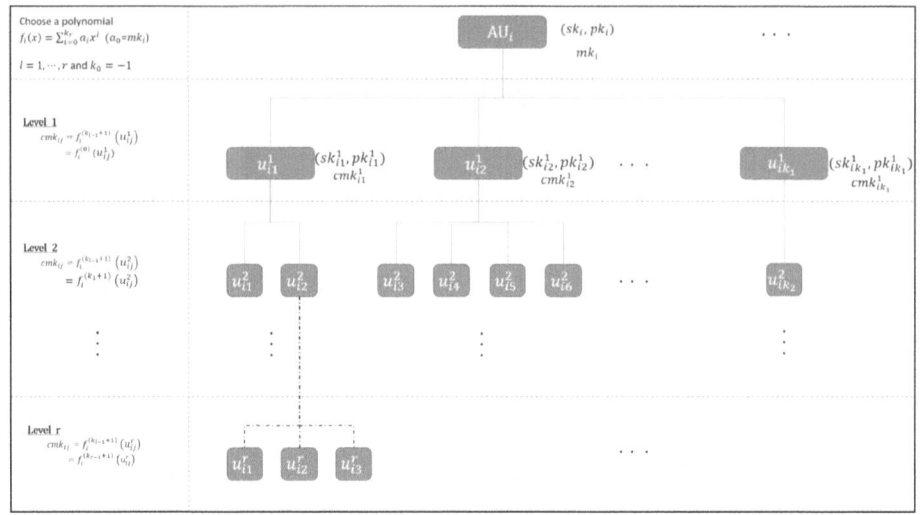

Fig. 4. Hierarchical Threshold Secret Sharing Scheme Based Delegation

1. **Key Generation:** Each user $u_{ij} \in \mathcal{G}$ picks uniformly at random a secret key $sk_{ij} \xleftarrow{\$} \mathbb{Z}_q$, and computes her public key $pk_{ij} \leftarrow g_2^{sk_{ij}}$, where g_2 is a generator of $\mathbb{G}_2$.

2. **Compartment Setup:** Each AU_i shares her membership key mk_i via a HTSS scheme [17] as described in Fig. 4. She sends compartment membership keys $cmk_{ij} = f_i^{(k_{l-1}+1)}(u_{ij})$ to each user $u_{ij} \in \mathcal{U}_i$, for $j = 1, \ldots, k_i$, and where $l = 1, \ldots, r$ is the level of the user u_{ij} belongs to.

3. **Signature Generation:** Each user u_{ij} computes his individual signature on the message M as $s_{ij} = H(M)^{sk_{ij}+cmk_{ij}}$, and sends it to the combiner.

4. **Signature aggregation:** After receiving the individual signatures from the users, the combiner
   (a) forms the subgroup of signers $\mathcal{S} := \{\mathcal{S}_1, \ldots, \mathcal{S}_m\}$, where $\mathcal{S}_i$ is the subgroup of signers from the compartment $\mathcal{U}_i$, for $i = 1, \ldots, m$.
   (b) computes $\sigma = \prod_{i=1}^{m} \prod_{j \in \mathcal{S}_i} s_{ij}^{\beta_{ij}}$, where β_{ij} is the appropriate Birkhoff coefficients as described in Definition 9.

5. **Verification:** If the level-specific thresholds are satisfied, then any verifier given $(par, \text{MPK}, \mathcal{PK}, \mathcal{S}, M, \sigma)$ can verify the signature by checking the below equation

$$e\left(H(M), \prod_{i=1}^{m} mpk_i \prod_{j \in \mathcal{S}_i} pk_{ij}^{\beta_{ij}}\right) \stackrel{?}{=} e(\sigma, g_2)$$

The verification equation satisfies correctness as given below:

$$e(\sigma, g_2) = e\left(\prod_{i=1}^{m} \prod_{j \in \mathcal{S}_i} s_{ij}^{\beta_{ij}}, g_2\right)$$

$$= e\left(\mathrm{H}(M)^{\sum\limits_{i=1}^{m} \sum\limits_{j \in \mathcal{S}_i} \beta_{ij} sk_{ij} + \beta_{ij} cmk_{ij}}, g_2\right)$$

$$= e\left(\mathrm{H}(M), g_2^{\sum\limits_{i=1}^{m} \sum\limits_{j \in \mathcal{S}_i} \beta_{ij} sk_{ij} + \beta_{ij} cmk_{ij}}\right)$$

$$= e\left(\mathrm{H}(M), g_2^{\sum\limits_{i=1}^{m} \sum\limits_{j \in \mathcal{S}_i} \beta_{ij} cmk_{ij}} g_2^{\sum\limits_{i=1}^{m} \sum\limits_{j \in \mathcal{S}_i} \beta_{ij} sk_{ij}}\right)$$

$$= e\left(\mathrm{H}(M), \prod_{i=1}^{m} mpk_i \prod_{j \in \mathcal{S}_i} pk_{ij}^{\beta_{ij}}\right).$$

Security. Like in the previous ones, each user u_{ij} signs a common message M with a combination of his secret and compartment membership keys, i.e. $sk_{ij} + cmk_{ij}$. Then the signature $s_{ij} = \mathrm{H}(M)^{sk_{ij} + cmk_{ij}}$ can be also seen as a BLS signature [8] on the message M with the key $sk_{ij} + cmk_{ij}$. Because sk_{ij} is picked uniformly at random and the cmk_{ij} is computed in an HTSS scheme, the signing key $sk_{ij} + cmk_{ij}$ will also be random. Moreover, AU$_i$ chooses a secret polynomial such that the constant term is her membership key mk_i and shares it among unauthorized users (subordinates) according to their levels. The unauthorized users in the first level get their shares like in Shamir's SSS. Other users in the following levels get their shares from the polynomials' evaluations, which are some certain order derivatives of the first level polynomial. Since an order $d > 0$ derivative kills the constant term, the users in the lower levels cannot obtain the mk_i without cooperating with the first level (i.e. trusted) users. Therefore, while distributing the shares, AU$_i$ assumes that her trusted subordinates are in the first level.

Remark 5. In the original vASM scheme, the messages are signed by using only the membership keys. However, in our constructions, we assume that messages are signed using the sum of secret keys and compartment membership keys. In this way, we discard the natural anonymity of threshold secret sharing schemes; and ensure accountability.

Remark 6. Notice that the methods we give above, except Sect. 3.1, provide conditional security. If enough shareholders are corrupted, they can easily forge the membership key of the authorized user in their compartment. Since it is an accountable subgroup multi-signature, they can sign on behalf of the entire organization. Therefore one should be aware of the need for a trusted user/combiner in each construction except the one in Sect. 3.1.

4 Comparison

4.1 Comparison of the Proposed Constructions

We compare our proposed constructions with each other and give the number of operations in each phase of the proposed schemes in Table 1. The number of main operations, such that group operations and bilinear pairings are the same for Shamir's SSS-based, Trusted user-based, and HTSS-based constructions. The main difference between these three constructions emerges in computing Lagrange and Birkhoff coefficients. For Shamir's SSS-based and trusted user-based delegation, Lagrange coefficients should be computed in the signature aggregation and in the verification phase by the designated combiner and the verifier, respectively. For HTSS-based delegation, Birkhoff coefficients should also be computed for the same phases. One needs to perform simple integer addition and multiplication to compute the Lagrange coefficients, whereas additional matrix operations should be conducted to compute the Birkhoff coefficients. Finally,

Table 1. Comparison of the methods given in Sect. 3.

Phases	Recursive vASM (Sect. 3.1)	Shamir's SSS based (Sect. 3.2)	Trusted user based (Sect. 3.3)	HTSS based (Sect. 3.4)
Key Generation	1 $\text{Exp}_{\mathbb{G}_2}$	1 $\text{Exp}_{\mathbb{G}_2}$	1 $\text{Exp}_{\mathbb{G}_2}$	1 $\text{Exp}_{\mathbb{G}_2}$
Compartment Setup	$m + 2k_i$ $\text{Exp}_{\mathbb{G}_2}$ $m + k_i(k_i + 1) - 2$ $\text{Mul}_{\mathbb{G}_2}$	(t_i, k_i)-SSS	$(2,2)$-SSS $(t_i - 1, k_i - 1)$-SSS	HTSS with m levels
Signature Generation	1 Hash 1 $\text{Exp}_{\mathbb{G}_1}$	1 Hash 1 $\text{Exp}_{\mathbb{G}_1}$	1 Hash 1 $\text{Exp}_{\mathbb{G}_1}$	1 Hash 1 $\text{Exp}_{\mathbb{G}_1}$
Signature Aggregation	$k_i - 1$ $\text{Mul}_{\mathbb{G}_1}$	l $\text{Exp}_{\mathbb{G}_1}$ $l - 1$ $\text{Mul}_{\mathbb{G}_1}$ l Lagrange Coef.	l $\text{Exp}_{\mathbb{G}_1}$ $l - 1$ $\text{Mul}_{\mathbb{G}_1}$ $l + 1$ Lagrange Coef.	l $\text{Exp}_{\mathbb{G}_1}$ $l - 1$ $\text{Mul}_{\mathbb{G}_1}$ l Birkhoff Coef.
Verification	1 Hash $k_i - 1$ $\text{Mul}_{\mathbb{G}_2}$ 2 pairings	1 Hash $m + l$ $\text{Exp}_{\mathbb{G}_2}$ $m + l - 1$ $\text{Mul}_{\mathbb{G}_2}$ 2 pairings l Lagrange Coef.	1 Hash $m + l$ $\text{Exp}_{\mathbb{G}_2}$ $m + l - 1$ $\text{Mul}_{\mathbb{G}_2}$ 2 pairings $l + 1$ Lagrange Coef.	1 Hash $m + l$ $\text{Exp}_{\mathbb{G}_2}$ $m + l - 1$ $\text{Mul}_{\mathbb{G}_2}$ 2 pairings l Birkhoff Coef.

$\text{Exp}_{\mathbb{G}_i}$: Exponentiation in group $\mathbb{G}_i, i \in \{1, 2\}$
$\text{Mul}_{\mathbb{G}_i}$: Multiplication in group $\mathbb{G}_i, i \in \{1, 2\}$
k_i : Number of signers in $\mathcal{U}_i$, i.e. i-th compartment
m : Number of compartments in the organization $\mathcal{G}$
l : Number of all signers in the organization $\mathcal{G}$, i.e. $l = m \cdot k_i$

since they all use threshold schemes, they all are secure when an adversary can corrupt up to t signers, where t is the threshold of the construction.

The number of operations required by the recursive vASM construction is less than the others. Only the compartment setup phase requires more operation than the others constructions. Since it is a one-time setup, it can be omitted. Moreover, the recursive vASM construction has a more robust assumption regarding the number of users an adversary can corrupt. Since the group setup is performed via an interactive (n, n)-VSS scheme, only one honest participant (possibly the authorized user or original signer) suffices to ensure the system's security.

4.2   Comparison with the Existing Proxy Signature Schemes

In literature, many proxy signature schemes and variants of this notion have similar but not the same definitions. There are mainly three parties in all kinds of proxy signatures, i.e., the original signer, the proxy signer, and the verifier. Here is the high level description of existing proxy signatures. An original signer (or a set of original signers) generates and signs a warrant that contains detailed information about the delegation, such as proxy IDs, validity time, etc. Then the original signer sends this warrant signature to the proxy signer (or a set of proxy signers). The proxy signer signs on behalf of the original signer using this warrant signature and the corresponding public keys of the two sides. However, the verifier's job is the same as in standard signature schemes, with the difference that the warrant signature needs to be verified along with the signature on a message.

Boldyreva et al. formally defined the proxy signatures in [5] for the first time, and defined what functionalities a proxy signature scheme should provide to the users. We compare the functionalities of the formal definition of proxy signature that is given in Sect. 2 with our constructions.

1. **Proxy-designation protocol $(\mathcal{D}, \mathcal{P})$ functionality**: Recall that $(\mathcal{D}, \mathcal{P})$ is an interactive proxy-designation protocol composed of a pair of algorithms, i.e., $\mathcal{D}$ and $\mathcal{P}$ (See Definition 2). After this interaction, proxy signer has its proxy signing key, which is used for proxy signing $\mathcal{PS}$. Similarly, in the recursive vASM construction, instead of the protocols $\mathcal{D}$ and $\mathcal{P}$, both designator and the proxy use the same compartment setup in which the designator (original signer) and the proxy signer jointly participate in an interactive Feldman's VSS. In the end, proxy signer gets her compartment membership key cmk_{ij} as an analog of proxy signing keys skp. In the other constructions, we use secret sharing schemes in the compartment setup phases, and the outputs of these phases are the compartment membership keys used for proxy signing.
2. **Signing and verification functionalities:** Analogy between signing and verification functionalities are straightforward.
3. **Identification functionality:** All of our constructions have accountability properties because of the underlying signature scheme, i.e. vASM. The verifiers should know the identities of the proxy signers to verify the proxy signature properly.

Table 2. Comparison with the existing proxy signature schemes.

Schemes	Proxy key generation		Signature generation	Verification	Signature size
	Original signer	Proxy signer			
Boldyreva et al.[(*)] [5]	1 Exp (mod p) 1 $H_{\mathbb{Z}_q}$	2 Exp (mod p) 3 $H_{\mathbb{Z}_q}$ 1 Mul (mod q)	1 Exp (mod p) 3 $H_{\mathbb{Z}_q}$	3 Exp (mod p) 1 $H_{\mathbb{Z}_q}$ 3 Mul (mod q)	$3\|\mathbb{Z}_p\|+\|\mathbb{Z}_q\|+\|M_w\|$
Lee et al. [21]	1 $\mathcal{P}$ 1 $Exp_{\mathbb{G}}$ 1 $H_{\mathbb{G}}$	1 $\mathcal{P}$ 1 $H_{\mathbb{G}}$	2 $\mathcal{P}$ 2 $Exp_{\mathbb{G}}$ 1 $H_{\mathbb{Z}_q}$ 1 $Exp_{\mathbb{G}_T}$	1 $\mathcal{P}$ 1 $Exp_{\mathbb{G}}$ 1 $H_{\mathbb{Z}_q}$	$\|\mathbb{G}\| + \|\mathbb{G}_T\|$
Shim [33]	1 $Exp_{\mathbb{G}}$ 1 $H_{\mathbb{G}}$	2 $\mathcal{P}$ 1 $Exp_{\mathbb{G}}$ 1 $Mul_{\mathbb{G}}$ 2 $H_{\mathbb{G}}$	1 $\mathcal{P}$ 1 $Exp_{\mathbb{G}}$ 1 $Mul_{\mathbb{G}}$ 1 $H_{\mathbb{G}}$	2 $\mathcal{P}$ 2 $Exp_{\mathbb{G}}$ 1 $Mul_{\mathbb{G}}$ 3 $H_{\mathbb{G}}$ 1 $Mul_{\mathbb{G}_T}$	$\|\mathbb{G}_T\| + \|M_w\|$
Zhang et al. [45]	3 $Exp_{\mathbb{G}}$ 2 $Mul_{\mathbb{G}}$ 2 $H_{\mathbb{G}}$	4 $\mathcal{P}$ 1 $Exp_{\mathbb{G}}$ 2 $Mul_{\mathbb{G}}$ 4 $H_{\mathbb{G}}$ 2 $Mul_{\mathbb{G}_T}$	2 $Exp_{\mathbb{G}}$ 1 $Mul_{\mathbb{G}}$ 2 $H_{\mathbb{G}}$	5 $\mathcal{P}$ 2 $Mul_{\mathbb{G}}$ 5 $H_{\mathbb{G}}$ 3 $Mul_{\mathbb{G}_T}$	$3\|\mathbb{G}\| + \|M_w\|$
Seo et al. [31]	2 $Exp_{\mathbb{G}}$ 1 $Mul_{\mathbb{G}}$ 1 $H_{\mathbb{G}}$ 1 $H_{\mathbb{Z}_q}$	3 $\mathcal{P}$ 1 $Exp_{\mathbb{G}}$ 2 $Mul_{\mathbb{G}}$ 1 $H_{\mathbb{G}}$ 1 $Mul_{\mathbb{G}_T}$	2 $Exp_{\mathbb{G}}$ 1 $Mul_{\mathbb{G}}$ 1 $H_{\mathbb{G}}$ 1 $H_{\mathbb{Z}_q}$	4 $\mathcal{P}$ 2 $Exp_{\mathbb{G}}$ 3 $Mul_{\mathbb{G}}$ 4 $H_{\mathbb{G}}$ 2 $H_{\mathbb{Z}_q}$ 2 $Mul_{\mathbb{G}_T}$	$3\|\mathbb{G}\| + \|M_w\|$
Verma & Singh [40]	1 $Exp_{\mathbb{G}}$ 1 $H_{\mathbb{G}}$	2 $\mathcal{P}$ 1 $H_{\mathbb{G}}$	1 $Exp_{\mathbb{G}}$ 1 $H_{\mathbb{Z}_q}$	2 $\mathcal{P}$ 1 $Exp_{\mathbb{G}}$ 1 $Mul_{\mathbb{G}}$ 1 $H_{\mathbb{G}}$ 1 $H_{\mathbb{Z}_q}$	$\|\mathbb{G}\| + \|M_w\|$
vASM [2]	4 $Exp_{\mathbb{G}_2}$ 4 $Mul_{\mathbb{G}_2}$		1 $H_{\mathbb{G}_1}$ 1 $Exp_{\mathbb{G}_1}$	2 $\mathcal{P}$ 1 $Mul_{\mathbb{G}_2}$ 1 $H_{\mathbb{G}_1}$	$\|\mathbb{G}_1\|$

Note that since the group operations determine the efficiency numbers we ignore the integer operations.

(*) Notice that all the schemes except [5] are pairing-based schemes.

$\mathcal{P}$: Bilinear pairing operation.

Exp_A: Exponentiation in group A, where $A \in \{\mathbb{G}, \mathbb{G}_1, \mathbb{G}_2, \mathbb{G}_T\}$.

Mul_A: Multiplication in group A, where $A \in \{\mathbb{G}, \mathbb{G}_1, \mathbb{G}_2, \mathbb{G}_T\}$.

H_A: Hash onto the set A, where $A \in \{\mathbb{G}, \mathbb{G}_1, \mathbb{Z}_q\}$.

$\|M_w\|$: Size of the warrant in bits.

$\|A\|$: Size of the elements of the set A, where $A \in \{G, G_1\}$

Although the vASM scheme is an accountable subgroup multi-signature scheme, it provides the functionalities of a proxy signature with a flexible number of original and proxy signers. It can serve as several types of proxy signatures according to the number of original and proxy signers (n and l, respectively) as follows:

– **vASM signature scheme can serve as a proxy signature ($n = l = 1$).** To that end the original signer and the proxy signer generate their secret and public keys, participate in a group setup and obtain their membership

Table 3. Comparison with the existing proxy multi-signature schemes.

Schemes	Proxy key generation		Signature generation	Verification	Signature size
	Original signer	Proxy signer			
Li & Chen [23]	3 $Exp_\mathbb{G}$ $n\ Mul_\mathbb{G}$ 1 $H_{\mathbb{Z}_q}$	$3n\ \mathcal{P}$ 1 $Exp_\mathbb{G}$ $n\ Mul_\mathbb{G}$ 1 $H_{\mathbb{Z}_q}$ $n\ Exp_{\mathbb{G}_T}$ $n\ Mul_{\mathbb{G}_T}$	1 $\mathcal{P}$ 2 $Exp_\mathbb{G}$ 1 $Mul_\mathbb{G}$ 1 $H_{\mathbb{Z}_q}$ 1 $Exp_{\mathbb{G}_T}$	3 $\mathcal{P}$ $n\ Mul_\mathbb{G}$ 2 $H_{\mathbb{Z}_q}$ 2 $Exp_{\mathbb{G}_T}$ 2 $Mul_{\mathbb{G}_T}$	$2|\mathbb{G}| + |\mathbb{Z}_q| + |M_w|$
Du & Wen [11]	3 $Exp_\mathbb{G}$ $(n+1)\ Mul_\mathbb{G}$ 2 $H_\mathbb{G}$	$4n\ \mathcal{P}$ $(2n-1)\ Mul_\mathbb{G}$ $2n\ H_\mathbb{G}$ $2n\ Mul_{\mathbb{G}_T}$	3 $Exp_\mathbb{G}$ 3 $Mul_\mathbb{G}$ 2 $H_\mathbb{G}$	6 $\mathcal{P}$ $(2n-1)\ Mul_\mathbb{G}$ 4 $H_\mathbb{G}$ 4 $Mul_{\mathbb{G}_T}$	$3|\mathbb{G}| + |M_w|$
vASM [2]	$2n\ Exp_{\mathbb{G}_2}$ $(n^2 + n - 2)\ Mul_{\mathbb{G}_2}$		1 $H_{\mathbb{G}_1}$ 1 $Exp_{\mathbb{G}_1}$	2 $\mathcal{P}$ $(n-1)\ Mul_{\mathbb{G}_2}$ 1 $H_{\mathbb{G}_1}$	$|\mathbb{G}_1|$

Note that since the group operations determine the efficiency numbers we ignore the integer operations.
$\mathcal{P}$: Bilinear pairing operation.
Exp_A: Exponentiation in group A, where $A \in \{\mathbb{G}, \mathbb{G}_1, \mathbb{G}_2, \mathbb{G}_T\}$.
Mul_A: Multiplication in group A, where $A \in \{\mathbb{G}, \mathbb{G}_1, \mathbb{G}_2, \mathbb{G}_T\}$.
H_A: Hash onto the set A, where $A \in \{\mathbb{G}, \mathbb{G}_1, \mathbb{Z}_q\}$.
n : Number of original signers.
$|M_w|$: Size of the warrant in bits.
$|A|$: Size of the elements of the set A, where $A \in \{G, G_1\}$

keys. Then the proxy signer-using his membership key- can sign any message on behalf of the original signer in an accountable way. We compare vASM scheme with existing proxy signatures in Table 2. We choose the best -up to our knowledge- pairing-based proxy signature schemes for comparison. It can be seen from the table that the proxy key generation (group setup) of vASM scheme requires less operations than the existing ones which require at least one or more pairings and a number of elliptic curve additions, scalar multiplications and hash computations. In signature generation and the verification phases, vASM requires less operations than most of the schemes but not all. In terms of the signature size, vASM is better than all other schemes. The size of a vASM signature is just one group element whereas all others have more than one. Note that the actual efficiency depends on the system parameters (e.g., elliptic curve groups, hash functions, etc.) that are used to instantiate the schemes. Moreover, we also add the Schnorr-based scheme proposed by Boldyreva et al. [5] to the table. Although it is not a pairing-based one we add it in our table to make the readers able to compare the number of operations.

- **vASM signature scheme can serve as a proxy multi-signature ($n > 1$ and $l = 1$).** n original signers and the proxy signer jointly participate in the group setup of vASM scheme. This way, each original signer authorizes the proxy signer to sign on behalf of himself. We give an efficiency comparison

Table 4. Comparison with the existing multi-proxy signature schemes.

Scheme	Proxy key generation		Signature generation	Aggregation	Verification	Signature size
	Original signer	Proxy signer				
Li & Chen [23]	3 Exp_G 1 Mul_G 1 H_{Z_q}	3 P 2 Exp_G 3 Mul_G 1 H_{Z_q} 1 Exp_{G_T} 1 Mul_{G_T}	3 Exp_G l Mul_G 1 H_{Z_q}	3l P $(l-1)$ Mul_G l H_{Z_q} l Exp_{G_T} l Mul_{G_T}	3 P l Exp_G $(3l-1)$ Mul_G $(l+1)$ H_G 2 H_{Z_q} 1 Exp_{G_T}	3$|G|+|M_w|$
Cao & Cao [9]	2 Exp_G 1 Mul_G 1 H_G	3 P 1 Exp_G 1 Mul_G 2 H_G 1 H_{Z_q}	2 Exp_G l Mul_G 1 H_G	5 P $(l-1)$ Mul_G 2 H_G 1 H_{Z_q} 1 Exp_{G_T} 3 Mul_{G_T}	5 P 3 Exp_G l Mul_G 2 H_G 1 H_{Z_q} 3 Mul_{G_T}	3$|G|+|M_w|$
Liu et al. [25]	7 Exp_G 2$|M_w|$ Mul_G	2 P $|M_w|$ Mul_G 2 Mul_{G_T}	3 P 5 Exp_G $(2|M_w|+7)$ Mul_G 3 Mul_{G_T}	3l P $(4l-2)$ Mul_G 3 Mul_{G_T}	3 P 2$|M_w|$ Mul_G $(l+2)$ Mul_{G_T}	3$|G|+|M_w|$
vASM [2]	2l Exp_{G_2} (l^2+l-2) Mul_{G_2}		1 H_{G_1} 1 Exp_{G_1}	$(l-1)$ Mul_{G_1}	2 P $(l-1)$ Mul_{G_2} 1 H_{G_1}	$|G_1|$

Note that since the group operations determine the efficiency numbers we ignore the integer operations.

$\mathcal{P}$: Bilinear pairing operation.

Exp_A: Exponentiation in group A, where $A \in \{\mathbb{G}, \mathbb{G}_1, \mathbb{G}_2, \mathbb{G}_T\}$.

Mul_A: Multiplication in group A, where $A \in \{\mathbb{G}, \mathbb{G}_1, \mathbb{G}_2, \mathbb{G}_T\}$.

H_A: Hash onto the set A, where $A \in \{\mathbb{G}, \mathbb{G}_1, \mathbb{Z}_q\}$.

l : Number of proxy signers

$|M_w|$: Size of the warrant in bits.

$|A|$: Size of the elements of the set A, where $A \in \{G, G_1\}$

of vASM and the existing pairing-based proxy multi-signature schemes in Table 3. It can be easily seen that vASM scheme is much more efficient than the existing pairing-based proxy multi-signature schemes in terms of all comparison parameters. vASM does not require pairing operations for proxy key generation phases, whereas the others require many pairings. vASM has also better computational efficiency than the others in terms of signature generation and verification phases. The signature size of vASM is only one group element, while others result in more.

- **vASM signature scheme can serve as a multi-proxy signature ($n = 1$ and $l > 1$).** If an original signer participates in a group setup procedure with l proxy signers, he authorizes the proxy signers by sharing his own secret key in the group setup (for details see Sect. 2.5). In Table 4, we compare the vASM scheme with existing pairing-based multi-proxy signature schemes. In general, vASM scheme requires less number of operations that are hard to compute (e.g. bilinear pairings). But the other schemes may have better computational efficiency as the number of proxies increase. For example, for a very large l, the cost of proxy key generation (group setup) of vASM requires more time than

Table 5. Comparison with the existing multi-proxy multi-signature schemes.

Scheme	Proxy key generation			Signature generation	Aggregation	Verification	Signature size
	Original signer	Proxy signer	Chairman				
Li & Chen [23]	$3\,Exp_G$ $(n+l)\,Mul_G$ $1\,H_{\mathbb{Z}_q}$	$3\,Exp_G$ $(n+l)\,Mul_G$ $1\,H_{\mathbb{Z}_q}$	$3(n+l)\,\mathcal{P}$ $(n + l - 1)\,Mul_G$ $1\,H_{\mathbb{Z}_q}$ $(n+l)\,Exp_{G_T}$ $(n+l)\,Mul_{G_T}$	$3\,Exp_G$ $l\,Mul_G$ $1\,H_{\mathbb{Z}_q}$	$3l\,\mathcal{P}$ $(l-1)\,Mul_G$ $l\,H_{\mathbb{Z}_q}$ $l\,Exp_{G_T}$ $l\,Mul_{G_T}$	$6\,\mathcal{P}$ $(n + 2l - 2)\,Mul_G$ $2\,H_{\mathbb{Z}_q}$ $2\,Exp_{G_T}$ $2\,Mul_{G_T}$	$4|G| + |M_w|$
Sahu & Padhye [30]	$2\,Exp_G$ $(n-1)\,Mul_G$ $1\,H_{\mathbb{Z}_q}$	$2n\,\mathcal{P}$ $(n+l)\,Exp_G$ $(n+l-1)\,Mul_G$ $(n+1)\,H_{\mathbb{Z}_q}$	NA	$1\,\mathcal{P}$ $2\,Exp_G$ $1\,Mul_G$ $1\,H_{\mathbb{Z}_q}$ $1\,Exp_{G_T}$ $(l-1)\,Mul_{G_T}$	$2\,\mathcal{P}$ $l\,Exp_G$ $(nl + l - 1)\,Mul_G$ $l\,H_{\mathbb{Z}_q}$ $l\,Exp_{G_T}$ $l\,Mul_{G_T}$	$2\,\mathcal{P}$ $2\,Exp_G$ $(nl-1)\,Mul_G$ $2\,H_{\mathbb{Z}_q}$ $1\,Exp_{G_T}$ $1\,Mul_{G_T}$	$3|G| + |M_w|$
vASM [2]	$2(n+l)\,Exp_{G_2}$ $((n+l)^2 + (n+l) - 2)\,Mul_{G_2}$			$1\,H_{G_1}$ $1\,Exp_{G_1}$	$(l-1)\,Mul_{G_1}$	$2\,\mathcal{P}$ $(n + l - 1)\,Mul_{G_2}$ $1\,H_{G_1}$	$|G_1|$

Note that since the group operations determine the efficiency numbers we ignore the integer operations.

$\mathcal{P}$: Bilinear pairing operation.

Exp_A: Exponentiation in group A, where $A \in \{\mathbb{G}, \mathbb{G}_1, \mathbb{G}_2, \mathbb{G}_T\}$.

Mul_A: Multiplication in group A, where $A \in \{\mathbb{G}, \mathbb{G}_1, \mathbb{G}_2, \mathbb{G}_T\}$.

H_A: Hash onto the set A, where $A \in \{\mathbb{G}, \mathbb{G}_1, \mathbb{Z}_q\}$.

n : Number of original signers.

l : Number of proxy signers

$|M_w|$: Size of the warrant in bits. $|A|$: Size of the elements of the set A, where $A \in \{G, G_1\}$

a few pairings which are required by the other schemes in the same phase. But for other comparison parameters, i.e. signature generation, aggregation, verification and signature size, vASM has better efficiency numbers than the other existing pairing-based multi-proxy signature schemes.

- **vASM signature scheme can serve as a multi-proxy multi-signature ($n > 1$ and $l > 1$).** To this end, n original signers and l proxy signers jointly perform a group setup. As a result of this group setup, each proxy signer has a membership key, which can be used as a proxy key. Any number of proxy signers can sign on behalf of the original signers using these membership keys. We compare the vASM scheme with the existing pairing-based multi-proxy multi-signature schemes in Table 5. vASM scheme has slightly better efficiency numbers than the other schemes.

Remark 7. Note that vASM scheme provides only a general delegation of signing rights. Since there is no warrant-like information the delegation is valid until it is canceled (by invalidating the commitment set COM.).

Remark 8. One important property of vASM is that it has aggregation property. For N distinct vASM signatures, one can aggregate all signatures into one which is a single $\mathbb{G}_1$ element. In this case, verification of the aggregated vASM signature requires $N+1$ pairing operations instead of $3N+1$. Further information about aggregation of vASM signatures can be found in [2].

5   Conclusion

In this work, we propose four constructions of compartment-based and hierarchical delegation of vASM signing authority for different organizational scenarios. We show that applying the vASM scheme recursively and combining the functionalities of threshold secret sharing schemes with the vASM scheme can solve an organizational problem of delegating the signing power of authorized users to single/multiple proxies in an accountable fashion. We present the comparison of our constructions with the existing proxy signatures in the literature in terms of their properties and functionalities. Our constructions provide us with all the functionalities of the proxy signatures that are defined in [5]. We also compared our constructions with each other according to the number of computations required and the security assumptions. Shamir's SSS-based and trusted user-based delegations require nearly the same number of computations, whereas HTSS-based delegation requires more operations than the others because of the determinant operations performed in signature aggregation and verification phases. On the other hand, all three constructions are insecure in case of an adversary that can corrupt $t+1$ signers, where t is the threshold of the secret sharing schemes used in the constructions. However, our proposed recursive vASM construction provides a much more efficient solution for delegating the vASM signing authority. Moreover, it is more secure than the other constructions based on the threshold secret sharing schemes. Existence of at least one honest participant during the one-time compartment setup is enough to ensure the security of recursive vASM construction. Since the authorized signer (original signer/designator/delegator) also participates in the compartment setup, even though all the unauthorized users are corrupted, they cannot reconstruct the membership key of the authorized user. In contrast, our constructions using threshold secret sharing schemes require at least one trusted unauthorized user. Finally we compare the vASM scheme with the existing pairing-based proxy signature variants, i.e. proxy signatures, proxy multi-signatures, multi-proxy signatures, multi-proxy multi-signatures, in terms of efficiency. vASM scheme can be used as a flexible and practical proxy signature scheme because of its simple structure. On the other hand, it provides a general delegation, i.e., it doesn't have a warrant that gives detailed information about the delegation, such as validity time, delegation context, etc.

Acknowledgments. The authors express their gratitude to the anonymous reviewers for their detailed comments, which have significantly improved the paper.

References

1. Alomair, B., Sampigethaya, K., Poovendran, R.: Efficient generic forward-secure signatures and proxy signatures. In: Public Key Infrastructure: 5th European PKI Workshop: Theory and Practice, EuroPKI 2008 Trondheim, Norway, 16–17 June 2008 Proceedings 5, pp. 166–181. Springer (2008)

2. Ağırtaş, A.R., Yayla, O.: Pairing-based accountable subgroup multi-signatures with verifiable group setup. Cryptology ePrint Archive, Report 2022/018 (2022). https://ia.cr/2022/018

3. Ağırtaş, A.R., Yayla, O.: A lattice-based accountable subgroup multi-signature scheme with verifiable group setup. Cryptology ePrint Archive, Paper 2024/014 (2024). https://eprint.iacr.org/2024/014. https://eprint.iacr.org/2024/014

4. Boldyreva, A.: Threshold signatures, multisignatures and blind signatures based on the gap-Diffie-Hellman-group signature scheme. In: Desmedt, Y.G. (ed.) PKC 2003. LNCS, vol. 2567, pp. 31–46. Springer, Heidelberg (2003). https://doi.org/10.1007/3-540-36288-6_3

5. Boldyreva, A., Palacio, A., Warinschi, B.: Secure proxy signature schemes for delegation of signing rights. J. Cryptol. **25**, 57–115 (2012)

6. Boneh, D., Drijvers, M., Neven, G.: Compact multi-signatures for smaller blockchains. In: Peyrin, T., Galbraith, S. (eds.) ASIACRYPT 2018. LNCS, vol. 11273, pp. 435–464. Springer, Cham (2018). https://doi.org/10.1007/978-3-030-03329-3_15

7. Boneh, D., Gentry, C., Lynn, B., Shacham, H.: Aggregate and verifiably encrypted signatures from bilinear maps. In: Biham, E. (ed.) EUROCRYPT 2003. LNCS, vol. 2656, pp. 416–432. Springer, Heidelberg (2003). https://doi.org/10.1007/3-540-39200-9_26

8. Boneh, D., Lynn, B., Shacham, H.: Short signatures from the weil pairing. J. Cryptol. **17**(4), 297–319 (2004)

9. Cao, F., Cao, Z.: A secure identity-based multi-proxy signature scheme. Comput. Electr. Eng. **35**(1), 86–95 (2009)

10. Desmedt, Y., Frankel, Y.: Threshold cryptosystems. In: Brassard, G. (ed.) Advances in Cryptology - CRYPTO'89 Proceedings, pp. 307–315. Springer, New York (1990)

11. Du, H., Wen, Q.: Certificateless proxy multi-signature. Inf. Sci. **276**, 21–30 (2014)

12. Eslami, Z., Pakniat, N., Nojoumian, M.: Ideal social secret sharing using birkhoff interpolation method. Secur. Commun. Netw. **9** (2016)

13. Feldman, P.: A practical scheme for non-interactive verifiable secret sharing. In: 28th Annual Symposium on Foundations of Computer Science (SFCS 1987), pp. 427–438 (1987)

14. Gennaro, R., Jarecki, S., Krawczyk, H., Rabin, T.: Robust threshold DSS signatures. In: Maurer, U. (ed.) Advances in Cryptology – EUROCRYPT 1996, pp. 354–371. Springer, Heidelberg (1996)

15. Harn, L.: Group-oriented (t, n) threshold digital signature scheme and digital multisignature. IEE Proc. Comput. Digit. Tech. **141**, 307–313 (1994)

16. Herranz, J., Sáez, G.: Revisiting fully distributed proxy signature schemes. In: Canteaut, A., Viswanathan, K. (eds.) Progress in Cryptology - INDOCRYPT 2004, pp. 356–370. Springer, Heidelberg (2005)

17. Käsper, E., Nikov, V., Nikova, S.: Strongly multiplicative hierarchical threshold secret sharing. In: Desmedt, Y. (ed.) Information Theoretic Security, pp. 148–168. Springer, Heidelberg (2009)

18. Kim, S., Park, S., Won, D.: Proxy signatures, revisited. In: Han, Y., Okamoto, T., Qing, S. (eds.) ICICS 1997. LNCS, vol. 1334, pp. 223–232. Springer, Heidelberg (1997). https://doi.org/10.1007/BFb0028478
19. Lee, B., Kim, H., Kim, K.: Strong proxy signature and its applications. In: Proceedings of SCIS, pp. 474–486 (2001)
20. Lee, B., Kim, H., Kim, K.: Secure mobile agent using strong non-designated proxy signature. In: Varadharajan, V., Mu, Y. (eds.) ACISP 2001. LNCS, vol. 2119, pp. 474–486. Springer, Heidelberg (2001). https://doi.org/10.1007/3-540-47719-5_37
21. Lee, J.S., Chang, J.H., Lee, D.H.: Forgery attacks on Kang et al.'s identity-based strong designated verifier signature scheme and its improvement with security proof. Comput. Electr. Eng. 36(5), 948-954 (2010)
22. Li, J., Yuen, T.H., Chen, X., Wang, Y.: Proxy ring signature: Formal definitions, efficient construction and new variant. In: 2006 International Conference on Computational Intelligence and Security, vol. 2, pp. 1259–1264 (2006)
23. Li, X., Chen, K.: ID-based multi-proxy signature, proxy multi-signature and multi-proxy multi-signature schemes from bilinear pairings. Appl. Math. Comput. 169(1), 437-450 (2005)
24. Lin, W., Jan, J.K.: Security personal learning tools using a proxy blind signature scheme. In: Proceedings of Journal of Chinese Language and Computing, pp. 273–277 (2000)
25. Liu, Z., Hu, Y., Zhang, X., Ma, H.: Provably secure multi-proxy signature scheme with revocation in the standard model. Comput. Commun. 34(3), 494–501 (2011)
26. Malkin, T., Obana, S., Yung, M.: The hierarchy of key evolving signatures and a characterization of proxy signatures. In: Cachin, C., Camenisch, J.L. (eds.) EUROCRYPT 2004. LNCS, vol. 3027, pp. 306–322. Springer, Heidelberg (2004). https://doi.org/10.1007/978-3-540-24676-3_19
27. Mambo, M., Usuda, K., Okamoto, E.: Proxy signatures for delegating signing operation. In: Proceedings of the 3rd ACM Conference on Computer and Communications Security, CCS 1996, pp. 48–57. Association for Computing Machinery, New York (1996)
28. Micali, S., Ohta, K., Reyzin, L.: Accountable-subgroup multisignatures: extended abstract. In: Proceedings of the 8th ACM Conference on Computer and Communications Security, CCS 2001, pp. 245–254. Association for Computing Machinery, New York (2001)
29. Okamoto, T., Tada, M., Okamoto, E.: Extended proxy signatures for smart cards. In: ISW 1999. LNCS, vol. 1729, pp. 247–258. Springer, Heidelberg (1999). https://doi.org/10.1007/3-540-47790-X_21
30. Sahu, R.A., Padhye, S.: Identity-based multi-proxy multi-signature scheme provably secure in random oracle model. Trans. Emerg. Telecommun. Technol. 26(4), 547–558 (2015)
31. Seo, S., Choi, K., Hwang, J., Kim, S.: Efficient certificateless proxy signature scheme with provable security. Inf. Sci. 188, 322–337 (2012)
32. Shamir, A.: How to share a secret. Commun. ACM 22(11), 612–613 (1979)
33. Shim, K.A.: Short designated verifier proxy signatures. Comput. Electr. Eng. 37(2), 180–186 (2011). https://doi.org/10.1016/j.compeleceng.2011.02.004
34. Shin-Jia, H., Chiu-Chin, C.: New multi-proxy multi-signature schemes. Appl. Math. Comput. 147(1), 57–67 (2004). https://www.sciencedirect.com/science/article/pii/S0096300302006501
35. Hwang, S.J., Shi, C.: A simple multi-proxy signature scheme for electronic commerce. In: Proceedings of the 10th National Conference on Information Security, Hualien Taiwan, ROC, pp. 134–138 (2000)

36. Sun, H.M.: Design of time-stamped proxy signatures with traceable receivers. IEE Proc.-Comput. Digit. Tech. **147**(6), 462–466 (2000)
37. Tassa, T.: Hierarchical threshold secret sharing. In: Naor, M. (ed.) TCC 2004. LNCS, vol. 2951, pp. 473–490. Springer, Heidelberg (2004). https://doi.org/10.1007/978-3-540-24638-1_26
38. Tassa, T., Dyn, N.: Multipartite secret sharing by bivariate interpolation. J. Cryptol. **22**(2), 227–258 (2009)
39. Tompa, M., Woll, H.: How to share a secret with cheaters. J. Cryptol. **1**(3), 133–138 (1989)
40. Verma, G.K., Singh, B.B.: Short certificate-based proxy signature scheme from pairings. Trans. Emerg. Telecommun. Technol. **28**(12), e3214 (2017)
41. Wu, C.K., Varadharajan, V.: Modified Chinese remainder theorem and its application to proxy signatures. In: Proceedings of the 1999 ICPP Workshops on Collaboration and Mobile Computing (CMC 1999). Group Communications (IWGC). Internet '99 (IWI'99). Industrial Applications on Network Computing (INDAP). Multime, pp. 146–151 (1999)
42. Yi, L., Bai, G., Xiao, G.: Proxy multi-signature scheme: a new type of proxy signature scheme. Electron. Lett. **36**, 527 – 528 (2000)
43. Zhang, F., Safavi-Naini, R., Lin, C.Y.: New proxy signature, proxy blind signature and proxy ring signature schemes from bilinear pairing. Cryptology ePrint Archive, Report 2003/104 (2003). https://ia.cr/2003/104
44. Zhang, K.: Threshold proxy signature schemes. In: Okamoto, E., Davida, G., Mambo, M. (eds.) ISW 1997. LNCS, vol. 1396, pp. 282–290. Springer, Heidelberg (1998). https://doi.org/10.1007/BFb0030429
45. Zhang, L., Zhang, F., Wu, Q.: Delegation of signing rights using certificateless proxy signatures. Inf. Sci. **184**(1), 298–309 (2012)

Towards Message Recovery in NTRU Encryption with Auxiliary Data

Marios Adamoudis[1], Konstantinos A. Draziotis[1], and Eirini Poimenidou[1,2($\boxtimes$)]

[1] School of Informatics, Aristotle University of Thessaloniki, Thessaloniki, Greece
{madamou,drazioti,epoimeni}@csd.auth.gr
[2] CERN, Geneva, Switzerland

Abstract. In the present paper, we implement a message recovery attack on the NTRU-HPS cryptosystem using its state-of-the-art parameters. We assume the knowledge of at most 2 bits of each coefficient of an unknown polynomial $u(x)$. Then, using Babai's nearest plane algorithm, we successfully recover the message. Additionally, we discuss the possibility of a side-channel attack method designed to extract the necessary bit information from the cryptographic operations.

Keywords: Public Key Cryptography · NTRU · Closest Vector Problem · LLL algorithm · Babai's Nearest Plane Algorithm

MSC 2020: 94A60 · 11T71 · 11Y16

1 Introduction

The NTRU cryptosystem was developed in 1996 by Hoffstein, Pipher, and Silverman [12]. To encrypt and decrypt data, NTRU makes use of lattice-based cryptography. The two algorithms that make up this system are NTRUSign for digital signatures and NTRUEncrypt for encryption. Notably, NTRU seems immune to quantum attacks, whereas RSA and Diffie-Hellman are vulnerable to Shor's quantum attack [31]. Compared to RSA, NTRU completes private-key operations substantially more quickly. NTRU became a finalist in the 3rd round of the Post-Quantum Cryptography Standardization Project, but NIST will not standardize it [25]. In May 2016, Daniel Bernstein, Chitchanok Chuengsatiansup, Tanja Lange, and Christine van Vredendaal introduced NTRU Prime. As of August 2022, starting from version 9.0, OpenSSH employs NTRU in conjunction with the X25519 Elliptic Curve Diffie-Hellman (ECDH) key exchange as its default configuration [34]. GoldBug Messenger [30] holds the distinction of being the first chat and email client to incorporate the NTRU algorithm under an open-source license. This implementation is rooted in the Spot-On Encryption

The second author was co-funded by SECUR-EU. The SECUR-EU project funded under Grant Agreement 101128029 is supported by the European Cybersecurity Competence Centre.

Suite Kernels. Another implementation is in wolfSSL, which supports NTRU cipher [35].

In the present work, we outline an attack on NTRU inspired by [3,28]. Initially, we proceed by multiplying the encryption equation by a positive integer k and introducing the polynomial $u(x) = -kh(x) \star r(x)$, where $h(x)$ is the public key and $r(x)$ denotes the ephemeral key. Our approach hinges on a lattice-based attack strategy, predicated on the hypothesis that we possess knowledge of the binary length of each coefficient u_i, where $u(x) = u_{N-1}x^{N-1} + \cdots + u_0$. Particularly, when this binary length equals $\text{bits}(q) - 1$, we require knowledge of a second bit of the coefficient u_i. This type of assumption mirrors common practices; for instance, in [14], authors target the DSA digital signature scheme, presuming that a portion of the bits within each associated ephemeral key can be recuperated. Leveraging this assumption, we have successfully augmented a similar attack outlined in [1]. In [22], the authors use "hints" from an oracle to recover the secret key in Kyber. In an older paper (2008) [24], the authors discuss the use of theoretical constructs called "inversion oracles" in the context of analyzing the NTRU encryption scheme.

The lattice structure utilized aligns with the one employed in [28]. However, a pivotal distinction lies in our approach to selecting k. Unlike in the referenced work, where authors opt for a small k to construct a Voronoi First Kind (VFK) lattice, here we allow k to adopt larger values. Moreover, we elucidate the rationale behind our choice of k.

1.1 Roadmap

In Sect. 2, we present the previous work related to NTRU attacks based on lattices as well as some information about side-channel attacks. In Sect. 3, we provide the fundamental lattice theory that is necessary for understanding our attack. In Sect. 4, we present the NTRU cryptosystem. In Sect. 5, we describe in detail our attack. Finally, in Sect. 6, we summarize our results. Furthermore, in Appendix A, we prove a Proposition that provides the length of the shortest vector in a specific lattice.

Our work's corresponding implementation can be found at https://github.com/drazioti/ntru_cvp_conf.

2 Previous Work

The NTRU cryptographic system first drew attention in 1997 when Coppersmith [7] initiated lattice-based attacks against it. Subsequently, Gentry proposed a particularly effective approach, especially advantageous when the parameter N is composite; details can be found in [11]. In [21], May tackled similar challenges using run-lattices. Expanding on May's work, Silverman [32] introduced a method involving the selection of certain coefficients while simultaneously reducing the lattice's dimension to force them to zero. This strategy resembles the one employed by researchers in [10], where decryption failures were exploited to recover the secret key, provided the decryption oracle supported such recovery. Alternative methods have also been explored, such as transforming the NTRU problem into a multivariate quadratic system over a finite field with two elements using Witt vectors, as outlined in [6,33].

Odlyzko's contribution [16] introduced a meet-in-the-middle attack, effectively dividing the search space into two subspaces, thereby significantly reducing time complexity. Complementing this approach, Howgrave's hybrid attack [13] seamlessly integrates lattice reduction with a meet-in-the-middle algorithm. Extensively tested by researchers, this hybrid method stands as a crucial tool for evaluating the security of NTRU.

In their latest research, detailed in the publication [17], the authors introduce a novel strategy for addressing the most recent versions of the NTRU encryption scheme. Their approach entails the utilization of a meticulously designed lattice and the application of the BKZ algorithm in tandem with the lattice sieving algorithm from the G6K library. Their investigation centers on the substantial benefits gained by deviating from the conventional Coppersmith-Shamir lattice towards a basis from a cyclotomic ring. In this way, they managed to decrypt the NTRU-HPS-171 instance within 83 core days using the cyclotomic ring basis, as opposed to the 172 core days required with the Coppersmith-Shamir basis.

To attack the NTRU cryptosystem with a modulus higher than that outlined in the NTRU-Encrypt standard, similar techniques were independently suggested in [4] and [8]. Kirchner [15], illustrated that the time complexity becomes polynomial when q is set to $2^{\Omega(\sqrt{n \log \log n})}$ in the field $\mathbb{Q}(\zeta_{2^n})$.

Finally, Nguyen [26] improved and elucidated the hybrid and meet-in-the-middle attack. While the subfield attack variation introduced in this study surpasses previous methods, it is not better than the hybrid attack.

2.1   Side-Channel Attacks

Lattice-based cryptosystems are resistant to post-quantum computers in theory, but in practice, they are vulnerable to side-channel attacks (SDAs). Side-channel attacks exploit unintended information leakage from a cryptographic system by targeting weaknesses in the physical implementation of the algorithm or its execution environment. These attacks rely on observing measurable physical properties of the cryptographic device or system, such as power consumption [19, 20], electromagnetic radiation, timing information [18], or sound emanations.

In this paper, we require information about the (unknown) polynomial $u(x) = -kh(x) \star r(x) \mod (q, x^N - 1)$, where $h(x)$ is the public key, $r(x)$ is the ephemeral key, and k is a suitable chosen integer. More specifically, we know the binary length ℓ_i of all the coefficients u_i of $u(x)$, and for all u_i such that $\ell_i = \ell$, where $\ell = \text{bits}(q) - 1$, we also know a second bit. We provide two possible methods for acquiring these data using side-channel attacks. The first one would be through a cold boot attack [27], where the attacker gets a noisy version of the polynomial $u(x)$ from the system's memory during a power-up/power-down cycle before $u(x)$ gets cleaned or overwritten by the system. Instead of getting the whole key like the authors of [27] do, we should be able to get candidates only for the binary length, for every u_i. Finally, the actual bits can be determined by verifying the correctness of the decryption operation for a known plaintext.

The second method for finding some bits of u_i is from a scan-based side-channel attack [2], where one can find the corresponding locations of the flip-flops of $u(x)$ [2, Chapter III] in the scan chain and again, instead of recovering the whole key like the authors of the above-mentioned paper do, the attacker should deduce possible candidates for some bits of u_i. The correctness of the bits can be tested, as before, through trial and error. This second method proves more difficult since such an attack would require detailed knowledge of the circuit layout, including the design of the scan chains and the assignment of flip-flops to various data paths.

Side-channel attack techniques demand precise instrumentation and controlled environments to capture and analyze the subtle signals indicative of sensitive information and often require specialized equipment; therefore, in this paper, we will not be conducting the practical implementations of the side-channel attack. For the rest of the paper, we assume that we have the information necessary to proceed with our attack.

3   Preliminaries on Lattices

In this section, we recall some well-known facts about lattices. In the field of cryptology, lattices serve as a fundamental tool for cryptanalysis and as essential building blocks for new cryptographic designs.

3.1   Basic Definitions

Let $\mathbf{b}_1, \mathbf{b}_2, \ldots, \mathbf{b}_n$ be linearly independent vectors of $\mathbb{R}^m$. The set

$$\mathcal{L} = \left\{ \sum_{j=1}^{n} \alpha_j \mathbf{b}_j : \alpha_j \in \mathbb{Z}, 1 \leq j \leq n \right\}$$

is called a *lattice* and the finite vector set $\mathcal{B} = \{\mathbf{b}_1, \ldots, \mathbf{b}_n\}$ is called a basis of the lattice $\mathcal{L}$. All the bases of $\mathcal{L}$ have the same number of elements, i.e. in our case n, which is called *dimension* or *rank* of $\mathcal{L}$. If $n = m$, then the lattice $\mathcal{L}$ is said to have *full rank*. We consider M be the $n \times m$ matrix, having as rows the vectors $\mathbf{b}_1, \ldots, \mathbf{b}_n$. If $\mathcal{L}$ has full rank, then the *volume* of the lattice $\mathcal{L}$ is defined to be the positive number $|\det M|$. The volume, as well as the rank, are independent of the basis $\mathcal{B}$. It is denoted by $vol(\mathcal{L})$ or $\det \mathcal{L}$. Let now $\mathbf{v} \in \mathbb{R}^m$, then $\|\mathbf{v}\|$ denotes the Euclidean norm of $\mathbf{v}$. Additionally, we denote by $\lambda_1(\mathcal{L})$ the least of the lengths of vectors of $\mathcal{L} - \{\mathbf{0}\}$. Finally, if $\mathbf{t} \in \operatorname{span}(\mathbf{b}_1, ..., \mathbf{b}_n)$, then by $dist(\mathcal{L}, \mathbf{t})$, we denote $\min\{\|\mathbf{v} - \mathbf{t}\| : \mathbf{v} \in \mathcal{L}\}$.

3.2   Computation Problems on Lattices

Here we describe the fundamental problems on lattices.

The Shortest Vector Problem (SVP): Given a lattice $\mathcal{L}$ find a non zero vector $\mathbf{b} \in \mathcal{L}$ that minimizes the (Euclidean) norm $\|\mathbf{b}\|$.

The Closest Vector Problem (CVP): Given a lattice $\mathcal{L}$ and a vector $\mathbf{t} \in \mathbb{R}^m$ that is not in $\mathcal{L}$, find a vector $\mathbf{b} \in \mathcal{L}$ that minimizes the distance $\|\mathbf{b} - \mathbf{t}\|$.

The approximate Shortest Vector Problem (apprSVP): Given a lattice $\mathcal{L}$ and a function $f(n)$, find a non-zero vector $\mathbf{b} \in \mathcal{L}$, such that:

$$\|\mathbf{b}\| \leq f(n)/\lambda_1(\mathcal{L}).$$

Each choice of the function $f(n)$ gives a different approximation of the Shortest Vector Problem.

The approximate Closest Vector Problem (apprCVP): Given a lattice $\mathcal{L}$, a vector $\mathbf{t} \in \mathrm{span}(\mathbf{b}_1, \ldots, \mathbf{b}_n)$ and a function $f(n)$, find a vector $\mathbf{b} \in \mathcal{L}$ such that,

$$\|\mathbf{b} - \mathbf{t}\| \leq f(n)dist(\mathcal{L}, \mathbf{t}).$$

Each choice of the function $f(n)$ gives a different approximation of the Closest Vector Problem.

3.3 Lattice Basis Reduction

The security of various cryptosystems is determined by the difficulty of solving apprSVP or apprCVP in different kinds of lattices. The well known LLL algorithm solves SVP rather well in small dimensions but performs poorly in large dimensions. The inability of LLL and other lattice reduction algorithms to effectively solve apprSVP and apprCVP determines the security of lattice-based cryptosystems.

Definition 1. *A basis* $\mathcal{B} = \{\mathbf{b}_1, \ldots, \mathbf{b}_n\}$ *of a lattice* $\mathcal{L}$ *is called LLL-reduced if it satisfies the following conditions:*
1. $|\mu_{i,j}| = \frac{|\mathbf{b}_i \cdot \mathbf{b}_j^*|}{\|\mathbf{b}_j^*\|^2} \leq \frac{1}{2}$ *for every* i, j *with* $1 \leq j < i \leq n$,
2. $\|\mathbf{b}_i^*\|^2 \geq (\frac{3}{4} - \mu_{i,i-1}^2)\|\mathbf{b}_{i-1}^*\|^2$ *for every* i *with* $1 < i \leq n$.

Proposition 1. *Let* $\mathcal{L}$ *be a lattice of rank* n. *For every LLL-reduced basis* $\mathcal{B} = \{\mathbf{b}_1, \ldots, \mathbf{b}_n\}$ *of a lattice* $\mathcal{L}$, *it is*

$$\|\mathbf{b}_1\| \leq 2^{(n-1)/2}\lambda_1(\mathcal{L}).$$

Thus, an LLL-reduced basis solves the approximate SVP to within a factor of $2^{(n-1)/2}$.

For details on the algorithm you can refer to [9].

3.4 Babai's Algorithm

To solve apprCVP, we usually use Babai's algorithm [5] (which has polynomial running time). In fact, combining this algorithm with the LLL algorithm, we can solve apprCVP for some lattice $\mathcal{L} \subset \mathbb{Z}^m$ having $f(n) = 2^{n/2}$ and $n = rank(\mathcal{L})$, in polynomial time. Below, we present the algorithm.

Babai's Nearest plane Algorithm

INPUT: A $n \times m$-matrix M with rows the vectors of a basis

$$\mathcal{B} = \{\mathbf{b}_i\}_{1 \leq i \leq n} \subset \mathbb{Z}^m$$

of a lattice $\mathcal{L}$ and a vector $\mathbf{t} \in \mathbb{R}^m$.

OUTPUT: $\mathbf{x} \in \mathcal{L}$ such that $||\mathbf{x} - \mathbf{t}|| \leq 2^{n/2} dist(\mathcal{L}, \mathbf{t})$.

1. $\{\mathbf{b}_i^*\}_{1 \leq i \leq n} \leftarrow GSO(M)$ # GSO : Gram-Schmidt Orthogonalization
2. $\mathbf{b} \leftarrow \mathbf{t}$
3. for $j = n$ to 1
4. $\quad\quad c_j \leftarrow \left\lceil \frac{\mathbf{b} \cdot \mathbf{b}_j^*}{||\mathbf{b}_j^*||^2} \right\rfloor$ #$\lceil x \rfloor = \lfloor x + 0.5 \rfloor$
5. $\quad\quad \mathbf{b} \leftarrow \mathbf{b} - c_j \mathbf{b}_j$
6. return $\mathbf{t} - \mathbf{b}$.

If the rank of $\mathcal{L}$ is "quite" small, then we can solve the CVP with the deterministic algorithm of Micciancio-Voulgaris [23].

4   Backgound on NTRU

Let the polynomial ring $\mathcal{R} = \mathbb{Z}[x]/\langle D(x) \rangle$ for some $D(x) \in \mathbb{Z}[x]$ and $\langle D(x) \rangle$ be the ideal generated by $D(x)$. We write $\star$ for the multiplication in the ring $\mathcal{R}$. Also, fix a polynomial $h(x) \in \mathbb{Z}[x]$ of degree $N - 1$. We set,

$$B_h = \begin{pmatrix} - h(x) - \\ - x \star h(x) - \\ \vdots \\ - x^{N-1} \star h(x) - \end{pmatrix}, \tag{1}$$

where with $x^i \star h(x)$, we write the vector with coordinates the coefficients of the polynomial $h(x)$, after multiplication in $\mathcal{R}$ with x^i. In expressing the coefficient vector of $h(x) = a_{N-1}x^{N-1} + \cdots + a_0$, we denoted as $\mathbf{h} = (a_0, ..., a_{N-1}) \in \mathbb{Z}^N$. Then, the multiplication $g(x) \star h(x)$ in $\mathcal{R}$ can be represented as the multiplication of the row matrix $[\mathbf{g}]$ and matrix B_h, i.e., $[\mathbf{g}]B_h$.

The set

$$\mathcal{L}_h = \{(f(x), g(x)) \in \mathcal{R}^2 : g(x) = f(x) \star h(x)\}$$

is a lattice, where $h(x)$ has degree $N - 1$. To see this we write,

$$\mathcal{L}_h = \mathbb{Z}^{2N} B_h',$$

where B_h' is the block matrix,

$$\left[\frac{B_h}{qI_N} \right].$$

If we consider the previous lattice, but taking   mod q (for some positive q), we get a (NTRU type) lattice

$$\mathcal{L}_h^q = \{(f(x), g(x)) \in \mathcal{R}^2 : g(x) = f(x) \star h(x) \quad (\bmod q)\},$$

where we also write it as,

$$\mathcal{L}_h^q = \{(\mathbf{f}, \mathbf{g}) \in \mathbb{Z}^{2N} : [\mathbf{g}] = [\mathbf{f}]M_h\}$$

with

$$M_h = \left[\begin{array}{c|c} I_N & B_h \\ \hline \mathbf{0}_N & qI_N \end{array} \right].$$

The lattice $\mathcal{L}_h^q$ has several interesting properties when B_h is a circulant matrix (see the matrix given by (1)). This occurs for instance if $D(x) = x^N - 1$. In this case, if $(\mathbf{a}, \mathbf{b})$ is a vector in the lattice, then performing a cyclic permutation of $\mathbf{a}$ and $\mathbf{b}$ $k-$times will result in another vector in the lattice. On the other hand, if $D(x) = x^p - x - 1$ (the case of NTRU-Prime), then B_h is not circulant.

4.1   NTRU-HPS

Alice selects public parameters (N, p, q), with N and $p = 3$ being prime numbers, and both co-prime to q. Usually N and q are large, and q is a power of 2. We also assume that $D(x) = x^N - 1$.

We set

- $\mathcal{R} = \mathbb{Z}[x]/\langle D(x)\rangle$, $\mathcal{R}/3 = \mathbb{Z}_3[x]/\langle D(x)\rangle$ and $\mathcal{R}/q = \mathbb{Z}_q[x]/\langle D(x)\rangle$.

- $\mathcal{S} = \mathbb{Z}[x]/\langle \Phi_N(x)\rangle$, $\mathcal{S}/3 = \mathbb{Z}_3[x]/\langle \Phi_N(x)\rangle$ and $\mathcal{S}/q = \mathbb{Z}_q[x]/\langle \Phi_N(x)\rangle$, where $\Phi_N(x) = D(x)/\Phi_1(x) = x^{N-1} + x^{N-2} + \cdots + x + 1$.

Moreover, we define the set of ternary polynomials $\mathcal{T}_\alpha$ of degree α, as the set of polynomials with coefficients from the set $\{-1, 0, 1\}$ and degree at most α. With $\mathcal{T}(d_1, d_2) \subset \mathcal{R}$, we denote the polynomials of $\mathcal{R}$ with d_1 entries equal to one, d_2 entries equal to minus one and the remaining entries are zero.

We assume $q \le 16N/3 + 16$ and we define the following sample spaces:
- $\mathcal{M}_m = \mathcal{M}_g = \mathcal{T}_{N-2}(\frac{q}{16} - 1, \frac{q}{16} - 1)$,
- $\mathcal{M}_f = \mathcal{M}_r = \mathcal{T}_{N-2}$.

Alice, for her private key randomly selects $(f(x), g(x))$ such that $f(x) \in \mathcal{M}_f$ and $g(x) \in \mathcal{M}_g$. It is important that $f(x)$ is invertible in both $\mathcal{S}/q$ and $\mathcal{S}/3$.

The inverses in $\mathcal{S}/3$ and $\mathcal{S}/q$ can be efficiently computed using the Euclidean algorithm and Hensel's Lemma, see [12, Proposition 6.45]. Let $F_q(x)$ and $F_3(x)$ represent the inverses of $f(x)$ in $\mathcal{S}/q$ and $\mathcal{S}/3$, respectively.

Alice next computes

$$h(x) = 3F_q(x) \star g(x) \mod q.$$

The polynomial $h(x)$ is Alice's public key.

The problem of distinguishing $h(x)$ from uniform elements in $\mathcal{R}/q$ is called *decision NTRU problem*. While, the problem of finding the private key $(f(x), g(x))$ is referred to as the *search NTRU problem*. Bob's plaintext is a polynomial $m(x) \in \mathcal{R}$, whose coefficients are in the set $\{-1, 0, 1\}$. Thus, the plaintext $m(x)$ is the centerlift of a polynomial in $\mathcal{R}/3$. Bob chooses a random ephemeral key $r(x) \in \mathcal{M}_r$ and computes the ciphertext,

$$c(x) \equiv h(x) \star r(x) + m(x) \mod q. \tag{2}$$

Finally, Bob sends to Alice the ciphertext $c(x) \in \mathcal{R}/q$.

To decrypt, Alice follows the algorithm:

1. $a(x) \leftarrow c(x) \star f(x) \mod (q, \Phi_1 \Phi_N)$
2. $m(x) \leftarrow a(x) \star f_3(x) \mod (3, \Phi_N)$
3. $m'(x) \leftarrow \text{Lift}_3(m(x))$
4. $r(x) \leftarrow (c(x) - m'(x))h_q(x) \mod (q, \Phi_N)$
5. **if** $(r(x), m(x)) \in \mathcal{M}_r \times \mathcal{M}_m$ **then**
6.   **return** $(m(x), r(x), 0)$
7. **else**
8.   **return** $(0, 0, 1)$

5   The Attack

5.1   The General Idea

We use the encryption Eq. (2) to get:

$$c(x) = h(x) \star r(x) + m(x) \mod (q, x^N - 1).$$

Let k be a positive integer which we shall choose later. We multiply the previous equation by k and we set $b(x) = kc(x) \mod (q, x^N - 1)$ and

$$u(x) = -kh(x) \star r(x) \mod (q, x^N - 1).$$

So,

$$km(x) = b(x) + u(x) \mod (q, x^N - 1). \tag{3}$$

Therefore,

$$km(x) = b(x) + u(x) + qv(x), \text{ for some polynomial } v(x).$$

Polynomials $m(x)$ and $u(x)$ are unknown. Let $\mathbf{m} = (m_i), \mathbf{b} = (b_i), \mathbf{u} = (u_i)$, and $\mathbf{v}$ be the vectors corresponding to $m(x), b(x), u(x)$, and $v(x)$, respectively. We set $\mathbf{V}$ to be the unknown vector $(-\mathbf{m}, \mathbf{u})$.

We observe that $(-\mathbf{m}, \mathbf{b} + \mathbf{u})$ is in $\mathcal{L}_k$, where $\mathcal{L}_k$ is the lattice generated by the rows of the matrix M_k:

$$M_k = \left[\begin{array}{c|c} I_N & -kI_N \\ \hline \mathbf{0}_N & qI_N \end{array} \right]. \tag{4}$$

Indeed, if we consider $(-\mathbf{m}, -\mathbf{v}) \in \mathbb{Z}^{2N}$, then:

$$(-\mathbf{m}, -\mathbf{v})M_k = (-\mathbf{m}, -\mathbf{v}) \left[\begin{array}{c|c} I_N & -kI_N \\ \hline \mathbf{0}_N & qI_N \end{array} \right] = (-\mathbf{m}, k\mathbf{m} - q\mathbf{v}) = (-\mathbf{m}, \mathbf{b} + \mathbf{u}).$$

Now, we shall prove that if we can obtain a precise approximation of the unknown vector $\mathbf{V}$, we shall reveal the message. Assume that we can find a vector $\mathbf{E}' = (\mathbf{0}_N, \mathbf{E}) = (\mathbf{0}_N, E_0, ..., E_{N-1}) \in \mathbb{Z}^{2N}$ such that,

$$||\mathbf{V} - \mathbf{E}'|| < \frac{1}{2}\lambda_1, \text{ where } \lambda_1 \text{ is the length of a shortest vector in } \mathcal{L}_k. \tag{5}$$

Note that, neither $\mathbf{V}$ nor $\mathbf{E}'$ is in $\mathcal{L}_k$. We choose the target vector $\mathbf{t}$ through $\mathbf{E}$ as follows:

$$\mathbf{t} = (0, ..., 0, b_0 + E_0, ..., b_{N-1} + E_{N-1}) \in \mathbb{Z}^{2N},$$

and set $\mathbf{w} \leftarrow CVP(\mathcal{L}_k, \mathbf{t})$. We shall prove that $\mathbf{w}$ provides the message $\mathbf{m}$. First, we remark that:

$$||\mathbf{w} - \mathbf{t}|| \le ||(-\mathbf{m}, \mathbf{b} + \mathbf{u}) - \mathbf{t}||, \tag{6}$$

since $(-\mathbf{m}, \mathbf{b} + \mathbf{u}) \in \mathcal{L}_k$. Then we have:

$$||(-\mathbf{m}, \mathbf{b} + \mathbf{u}) - \mathbf{t})|| =$$

$$= ||(-m_0, ..., -m_{N-1}, b_0 + u_0, ..., b_{N-1} + u_{N-1}) - (0, ..., 0, E_0 + b_0, ..., E_{N-1} + b_{N-1})|| =$$

$$= ||(-m_0, ..., -m_{N-1}, u_0 - E_0, ..., u_{N-1} - E_{N-1})|| =$$

$$= ||(-\mathbf{m}, \mathbf{u}) - \mathbf{E}'|| = ||\mathbf{V} - \mathbf{E}'|| < \frac{1}{2}\lambda_1.$$

Finally:

$$||\mathbf{w} - (-\mathbf{m}, \mathbf{b} + \mathbf{u})|| = ||(\mathbf{w} - \mathbf{t}) + (\mathbf{t} - (-\mathbf{m}, \mathbf{b} + \mathbf{u}))||$$

$$\le ||\mathbf{w} - \mathbf{t}|| + ||\mathbf{t} - (-\mathbf{m}, \mathbf{b} + \mathbf{u})|| \le 2||(-\mathbf{m}, \mathbf{b} + \mathbf{u}) - \mathbf{t}|| < \lambda_1.$$

But $\mathbf{w} - (-\mathbf{m}, \mathbf{b} + \mathbf{u}) \in \mathcal{L}_k$, thus $\mathbf{w} = (-\mathbf{m}, \mathbf{b} + \mathbf{u})$. We conclude therefore that the first N-coordinates of $\mathbf{w}$ provide the message $\mathbf{m}$.

Remark 1. If $q > (k+1)\sqrt{k+1}$, then $\lambda_1(\mathcal{L}_k) = \sqrt{1 + k^2}$. See Appendix A. In general $\lambda_1(\mathcal{L}_k) \le \sqrt{1 + k^2}$ since the first vector of the matrix M_k has Euclidean length $\sqrt{1 + k^2}$. In fact, there are k's such that, $\lambda_1(\mathcal{L}_k) < \sqrt{1 + k^2}$.

For instance if $m(x) = m_{N-1}x^{N-1} + m_{N-2}x^{N-2} + \cdots + m_1 x + m_0$ then $\mathbf{m} = (m_0, m_1, ..., m_{N-2}, m_{N-1})$. In this case $m_i \in \{-1, 0, 1\}$ and $m_{N-1} = 0$ since $\mathcal{M}_m = \mathcal{T}_{N-2}(\frac{q}{16} - 1, \frac{q}{16} - 1)$.

5.2   Choosing E and k

Let $u(x) = u_{N-1}x^{N-1} + \cdots + u_1 x + u_0$ be the same as previously.

Assumption A. We assume that for each coefficient u_i we know ℓ_i such that $u_i \in [2^{\ell_i - 1}, 2^{\ell_i})$. I.e. u_i has binary length ℓ_i.

Assumption B. For all u_i such that $\ell_i = \ell$, where $\ell = \text{bits}(q) - 1$, we also know the second highest order bit, i.e. we know the z_i's such that $u_i = 2^{\ell-1} + z_i 2^{\ell-2} + \cdots$, for $i = 0, 1, ..., N - 1$.

The previous two assumptions can be provided by an oracle which outputs the length of the coefficients (u_i) and in the case (u_i) has the maximum length, i.e. $\text{bits}(q) - 1$, we also know the second highest order bit. We remark here that, in NTRU-HPS and their variants, q is a power of 2. For instance in ntruhps2048509 $q = 2048$, i.e. the minimum number with 12 bits. So taking mod q to the polynomials we get at most 11 bits numbers, that's why we have set $\ell = \text{bits}(q) - 1$ and not $\text{bits}(q)$. We consider the following two cases.

Case 1. $\ell_j = \text{bits}(u_j) = \ell$, then we set $E_j = 2^{\ell-1} + 2^{\ell-2} + 2^{\ell-3}$ if the second highest order bit is 1, else we set $E_j = 2^{\ell-1} + 2^{\ell-3}$.

Case 2. $\ell_j = \text{bits}(u_j) < \ell$, then we set $E_j = 2^{\ell_j - 1} + 2^{\ell_j - 2}$.

That is, if $u_j = x_j 2^{\ell-1} + y_j x^{\ell-2} + \cdots$, where $x_j, y_j \in \{0, 1\}$, then we set,

$$E_j = \begin{cases} 2^{\ell-1} + 2^{\ell-2} + 2^{\ell-3}, & \text{if } x_j = 1, y_j = 1 \\ 2^{\ell-1} + 2^{\ell-3}, & \text{if } x_j = 1, \ y_j = 0 \\ 2^{\ell_j - 1} + 2^{\ell_j - 2}, & \text{if } x_j = 0 \end{cases}$$

Equivalently,

$$E_j = \begin{cases} 2^{\ell-1} + y_j 2^{\ell-2} + 2^{\ell-3}, & \text{if } \text{len}_2(u_j) = \ell, \\ 2^{\ell_j - 1} + 2^{\ell_j - 2}, & \text{if } \ell_j = \text{len}_2(u_j) < \ell \end{cases}$$

We get the following Lemma.

Lemma 1. *We have* $|u_j - E_j| \leq 2^{\ell-3} - 1$.

Proof. Lets see for instance the case $x_j = y_j = 1$. Then,

$$|u_j - E_j| = |(z_j - 1)2^{\ell-3} + \cdots|.$$

Since $z_j \in \{0, 1\}$ we get $|u_j - E_j| \leq 2^{\ell-4} + 2^{\ell-5} + \cdots + 2 + 1 = 2^{\ell-3} - 1$. Similar for the other two cases, we have
• $x_j = 1, y_j = 0$

$$|u_j - E_j| = |(2^{\ell-1} + z_j 2^{\ell-3} + \cdots) - (2^{\ell-1} + 2^{\ell-3})| = |(z_j - 1)2^{\ell-3} + \cdots| \leq 2^{\ell-3} - 1.$$

- $x_j = 0$. We remind that ℓ_j is the binary length of u_j.

$$|u_j - E_j| = |(2^{\ell_j - 1} + r_j 2^{\ell_j - 2} + \cdots) - (2^{\ell_j - 1} + 2^{\ell_j - 2})| =$$

$$= |(r_j - 1)2^{\ell_j - 2} + \cdots| \leq 2^{\ell_j - 2} - 1,$$

and since $\ell_j < \ell$ i.e. $\ell_j \leq \ell - 1$ we get $|u_j - E_j| \leq 2^{\ell - 3} - 1$.

To summarize, our selection of $\mathbf{E}$ is based on an oracle that provides the binary length of the coefficients of $u(x) = -kh(x) \star r(x)$ in $\mathcal{R}$.

Let $\mathbf{E} = (E_0, E_1, ..., E_{N-1})$. We apply the following algorithm.

Input: The ciphertext $\mathbf{c}$, a positive integer k, and the previous $\mathbf{E}$.
Output: The message $\mathbf{m}$ or fail.
1: Set b_i the coefficients of $kc(x)$. Further, let the target vector

$$\mathbf{t} = (0, ..., 0, b_0 + E_0, ..., b_{N-1} + E_{N-1}) = (\mathbf{0}_N, \mathbf{b} + \mathbf{E}).$$

2: Call Babai algorithm to the pair $(\mathcal{L}_k, \mathbf{t})$ and let $\mathbf{w}$ be its output.
3: Return the first N−coordinates of $\mathbf{w}$.
In step 3 we get the possible message $\mathbf{m}$, in this case the first N−coordinates of $\mathbf{w}$ is $-\mathbf{m}$.

We continue with the choice of integer k. The value of k defines the lattice $\mathcal{L}_k$. The Babai algorithm is used to approximate the distance $d_1 = d(\mathcal{L}_k, \mathbf{t})$, where $\mathcal{L}_k$ is our lattice with the parameter k. Typically, Babai's algorithm is employed to find an approximation of a closest vector of a lattice given a target vector $\mathbf{t}$. Here, the goal is to choose k such that d_1 is close to an unknown distance $d_2 = d(\mathbf{u}, \mathbf{E})$. The vector $\mathbf{u}$ is unknown, but through experimentation, an estimation of d_2 is obtained. If $d_1 \approx d_2 = \|\mathbf{u} - \mathbf{E}\|$ then the output of Babai, say the vector $\mathbf{w} = (\mathbf{w}_1, \mathbf{w}_2)$ will be such that $\mathbf{w}_1 = -\mathbf{m}$.

We explain the previous selection of k and why $\mathbf{w}_1 = -\mathbf{m}$. Set the auxiliary vector $\mathbf{t}' = (\mathbf{0}_N, \mathbf{b} + \mathbf{u})$. We have chosen $\mathbf{E}$ such that $\mathbf{u} \approx \mathbf{E}$, so $\mathbf{t}' \approx (\mathbf{0}_N, \mathbf{b} + \mathbf{E}) = \mathbf{t}$. Also, there is the lattice point $\mathbf{W} = (-\mathbf{m}, \mathbf{u} + \mathbf{b})$. $\mathbf{W}$ is such that $\mathbf{W} \approx \mathbf{t}'$, since $\mathbf{m} \approx \mathbf{0}_N$. Thus,

$$\mathbf{W} - \mathbf{t} \approx \mathbf{t}' - \mathbf{t} = (\mathbf{0}_N, \mathbf{u} - \mathbf{E}).$$

That is:

$$\|\mathbf{W} - \mathbf{t}\| \approx \|\mathbf{u} - \mathbf{E}\|. \tag{7}$$

Now if there is k such that $d_1 = d(\mathcal{L}_k, \mathbf{t}) \approx \|\mathbf{W} - \mathbf{t}\|$, a CVP oracle will (probably) return $\mathbf{W}$, therefore we can find $\mathbf{m}$. But if $d_1 \approx \|\mathbf{W} - \mathbf{t}\|$, from (7), we get $d_1 \approx \|\mathbf{u} - \mathbf{E}\| = d_2$.

We execute the following experiment. For a plethora of k's we compute $d_1 = d(\mathcal{L}_k, \mathbf{t})$ (using Babai) and $d_2 = \|\mathbf{u} - \mathbf{E}\|$. The latter distance can be computed, since we have already chosen $\mathbf{u}$ and $\mathbf{E}$.

In the provided figure (Fig. 1), the parameter q is set to 2048, and k ranges from 1 to 1000. Certain parameters for the NTRU cryptosystem, as well as the message $m(x)$ and the nonce $r(x)$, are fixed. For each value of k, the target vector $\mathbf{t}$ is computed based on the previous selection of $\mathbf{E}$, and the unknown vector $\mathbf{u}$ is also computed. In the $y-$axis we compute the difference $|d_1 - d_2|$, where d_1 is computed using Babai's algorithm. We remark that for k, say ≈ 550, the previous difference is minimized. So, using such a k, we expect the output of Babai to reveal the message. Similar for $N = 4096$ we pick $k = 1080$.

Now that we have a way to select both $\mathbf{E}$ and k, we can execute our attack. We applied it for the three variants of NTRU-HPS (the code is in the github repository), namely ntruhps2048509, ntruhps2048677, and ntruhps4096821. These are the suggested parameters for the NTRU-HPS when submitted to the NIST competition. In all the experiments, we revealed the unknown message. The attack time was negligible, approximately 1 s.

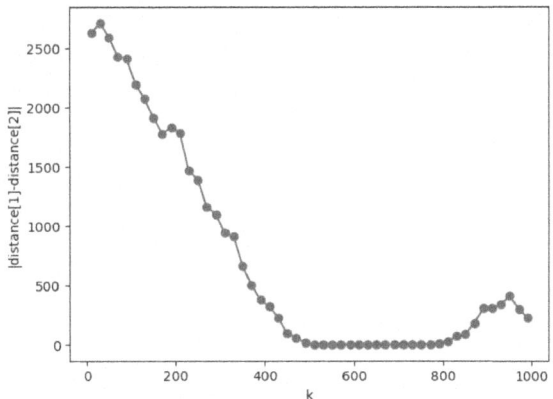

Fig. 1. In this graph we set $q = 2048$, k takes values in the horizontal axis and on the $y-$axis we see the $|\text{distance}(\mathbf{u}, \mathbf{E}) - \text{distance}(\mathcal{L}_k, \mathbf{t})|$. For each k, we generate a new NTRU instance. We remark that Babai's algorithm provides outputs with distances close to distance$(\mathbf{u}, \mathbf{E})$ for $k \in [520, 790]$. Finally, we select k to be 550.

Remark 2. From the encryption equation we get $u(x) = km(x) - b(x)$ and since $m(x)$ is ternary, we get that the polynomial $b(x) + u(x)$ has their coefficients in $\{-k, 0, k\}$. So, the unknown polynomial $u(x)$ has the coefficient of x^i in $\{k - b_i, -b_i, -k - b_i\}$, i.e.

$$\mathbf{u} = (u_1, ..., u_n) \in \{k - b_i, -b_i, -k - b_i\}^N.$$

So, assuming a uniform distribution for the coefficients u_i, we get that the entropy of $\mathbf{u}$ is $\log_2(3) \times N$. This is the same as the entropy of $\mathbf{r}$ or the message $\mathbf{m}$. Hence, there is no inherent advantage in using an oracle for either $\mathbf{u}$ or $\mathbf{m}$. However, our attack strategy leverages the parameter k significantly. The overall setup of our method leads to the exploitation of $\mathbf{u}$ rather than $\mathbf{m}$.

Remark 3. In [3, Example 7] for ntruhps2048509 we get $|u_i - E_i| \leq 36$ and in our attack we get $2^{\ell-3} - 1 = 256$ (here $\ell = \text{bits}(q) - 1 = 11$), which is a significant improvement.

6   Conclusion

Eve, equipped with the public key and a ciphertext, can reveal the message using a straightforward oracle that reveals two specific bits of an unknown polynomial's coefficients. We've effectively executed this attack on NTRU-HPS with the standard parameters suggested in NIST, utilizing a sagemath implementation [29]. The results demonstrated significant efficiency. We anticipate this attack to also be effective against the other two NTRU variants, NTRU-HRSS and NTRU-Prime.

Moreover, an oracle meeting assumptions A and B (see Sect. 5.2) could be devised through a side-channel attack, as detailed in Subsect. 2.1. Also, from Eq. 3, we see that the unknown polynomial $u(x)$ equals to $k(m(x) - c(x))$. Hence, acquiring certain side-channel information about the message $m(x)$ could furnish us with the necessary bits to facilitate our attack.

Acknowledgement. The authors sincerely thank the referees for their valuable suggestions to the initial draft.

A   Appendix

Proposition 2. *Let k, N and q be positive integers with $q \geq (k+1)\sqrt{k^2 + 1}$. We set*

$$M_k = \left[\begin{array}{c|c} I_N & -kI_N \\ \hline 0_N & qI_N \end{array} \right].$$

Let $\mathcal{L}_k$ be the lattice generated by the rows of M_k. Then, $\lambda_1(\mathcal{L}_k) = \sqrt{k^2 + 1}$.

Proof. It is enough to prove that for all non-zero $\mathbf{v} \in \mathcal{L}_k$ we have $\|\mathbf{v}\| \geq \sqrt{k^2 + 1}$. Since the first row of M_k has length $\sqrt{k^2 + 1}$ we are done.

Suppose that there is a vector $\mathbf{v} \in \mathcal{L}_k \setminus \{\mathbf{0}\}$ such that

$$\|\mathbf{v}\| < \sqrt{k^2 + 1}. \tag{8}$$

Let $\mathbf{b}_1, \ldots, \mathbf{b}_{2N}$ be the rows of the matrix M_k. Since $\mathbf{v} \in \mathcal{L}_k$, there are integers $l_1, \ldots, l_{2N}$ such that,

$$\mathbf{v} = l_1 \mathbf{b}_1 + \cdots + l_{2N} \mathbf{b}_{2N} =$$

$$(l_1, \ldots, l_N, -l_1 k + q l_{N+1}, \ldots, -l_N k + q l_{2N})$$

From the inequality (8) we get

$$\begin{cases} |l_1|, |l_2|, \ldots, |l_N| < \sqrt{k^2 + 1} \\ |-l_1 k + q l_{N+1}| < \sqrt{k^2 + 1} \\ \cdots \\ |-l_N k + q l_{2N}| < \sqrt{k^2 + 1} \end{cases} \tag{9}$$

So we can easily see that for $i = 1, \ldots, N$ we get

$$|l_i k| < \sqrt{k^2 + 1}k. \tag{10}$$

Case 1: not all the integers $l_{N+1}, l_{N+2}, \ldots, l_{2N}$ are zero.
Without loss of generality, say l_{N+j} is not zero for some $j \in \{1, \ldots, N\}$. Then from (10) and (9), we get

$$\|\mathbf{v}\| \geq |-l_j k + q l_{N+j}| \geq |l_{N+j}|q - |l_j k| > q - \sqrt{k^2 + 1}k \geq \sqrt{k^2 + 1},$$

which contradicts to inequality (8).

Case 2: Let $l_{N+1} = l_{N+2} = \cdots = l_{2N} = 0$.
In this case

$$\mathbf{v} = (l_1, \ldots, l_N, -l_1 k, \ldots, -l_N k).$$

Then,

$$\|\mathbf{v}\| = \sqrt{l_1^2(1 + k^2) + l_2^2(1 + k^2) + \cdots + l_N^2(1 + k^2)} > \sqrt{k^2 + 1},$$

which contradicts our hypothesis (8). The Proposition follows.

References

1. Adamoudis, M., Draziotis, K.A., Poulakis, D.: Enhancing a DSA attack. In: CAI 2019, pp. 13–25. LNCS, vol. 11545. Springer (2019)
2. Abdel, K., Amr, Y.: A scan-based side channel attack on the NTRUEncrypt cryptosystem. In: 7th International Conference on Availability, Reliability and Security (2012)
3. Adamoudis, M., Draziotis, K.A.: Message recovery attack on NTRU using a lattice independent from the public key (2023, to appear in Advances in Mathematics of Communications (Amer. Inst. of Math. Sciences)). https://doi.org/10.3934/amc.2023040
4. Albrecht, M., Bai, S., Ducas, L.: A subfield lattice attack on overstretched NTRU assumptions. In: Robshaw, M., Katz, J. (eds.) CRYPTO 2016. LNCS, vol. 9814, pp. 153–178. Springer, Heidelberg (2016). https://doi.org/10.1007/978-3-662-53018-4_6
5. Babai, L.: On Lovász lattice reduction and the nearest lattice point problem. Combinatorica **6**(1), 1–13 (1986)
6. Bourgeois, G., Faugère, J.C.: Algebraic attack on NTRU using Witt vectors and Gröbner bases. J. Math. Cryptol. **3**(3), 205–214 (2009)
7. Coppersmith, D., Shamir, A.: Lattice attacks on NTRU. In: Proceedings of the Eurocrypt. LNCS, vol. 1223. Springer (1997)
8. Cheon, J.H., Jeong, J., Lee, C.: An algorithm for NTRU problems and cryptanalysis of the GGH multilinear map without an encoding of zero. Cryptology ePrint Archive, Report 2016/139 (2016)
9. Galbraith, S.: Mathematics of Public key Cryptography, Cambridge University Press (2012)

10. Gama, N., Nguyen, P.Q.: New chosen-ciphertext attacks on NTRU. In: Okamoto, T., Wang, X. (eds.) PKC 2007. LNCS, vol. 4450, pp. 89–106. Springer, Heidelberg (2007). https://doi.org/10.1007/978-3-540-71677-8_7

11. Gentry, C.: Key recovery and message attacks on NTRU-composite. In: Pfitzmann, B. (ed.) EUROCRYPT 2001. LNCS, vol. 2045, pp. 182–194. Springer, Heidelberg (2001). https://doi.org/10.1007/3-540-44987-6_12

12. Hoffstein, J., Pipher, J., Silverman, J.H.: NTRU: a ring-based public key cryptosystem. In: Buhler, J. (eds.) Proceedings of ANTS 1998. LNCS, vol. 1423, pp. 267–288 (1998)

13. Howgrave-Graham, N.: A hybrid lattice-reduction and meet-in-the-middle attack against NTRU. In: Menezes, A. (ed.) CRYPTO 2007. LNCS, vol. 4622, pp. 150–169. Springer, Heidelberg (2007). https://doi.org/10.1007/978-3-540-74143-5_9

14. Howgrave-Graham, N.A., Smart, N.P.: Lattice attacks on digital signature schemes. Des. Codes Cryptogr. **23**, 283–290 (2001)

15. Kirchner, P., Fouque, P.A.: Revisiting lattice attacks on overstretched NTRU parameters. In: Eurocrypt 2017. LNCS, vol. 10210. Springer, Cham (2017). https://doi.org/10.1007/978-3-319-66787-4_12

16. Howgrave-Graham, N., Silverman, J.H., Whyte, W.: Meet-in-the-middle Attack on an NTRU private key, Technical report, NTRU Cryptosystems, July 2006. Report 04. http://www.ntru.com

17. Kirshanova, E., May, A., Nowakowsk, J.: New NTRU Records with. Improved Lattice Bases. eprint: 2023/582

18. Kocher, P.C.: Timing attacks on implementations of Diffie-Hellman, RSA, DSS, and other systems. In: Advances in Cryptology — CRYPTO 1996. Springer, Heidelberg (2001)

19. Kocher, P., Jaffe, J., Jun, B.: Differential power analysis. In: Advances in Cryptology — CRYPTO 1999. Lecture Notes in Computer Science, vol. 1666. Springer, Heidelberg (1999)

20. Mangard, S., Oswald, E., Popp, T.: Power Analysis Attacks, Springer New York (2007). https://link.springer.com/book/10.1007/978-0-387-38162-6

21. May, A.: Cryptanalysis of NTRU (preprint) (1999). http://citeseerx.ist.psu.edu/viewdoc/summary?doi=10.1.1.41.3484

22. May, A., Nowakowski, J.: Too Many Hints – When LLL Breaks LWE (2024). https://eprint.iacr.org/2023/777.pdf

23. Micciancio, D., Voulgaris, P.: A deterministic single exponential time algorithm for most lattice problems based on Voronoi cell computations. In: Proceedings of STOC, pp. 351–358. ACM (2010)

24. Mol, P., Yung, M.: Recovering NTRU secret key from inversion oracles. In: PKC 2008 (2008). https://iacr.org/archive/pkc2008/49390018/49390018.pdf

25. NIST, 3rd round candidate announcement. https://csrc.nist.gov/news/2020/pqc-third-round-candidate-announcement. Accessed 1 Jan 2022

26. Nguyen, P.Q.: Boosting the hybrid attack on NTRU: torus LSH, permuted HNF and boxed sphere. In: Third PQC Standardization Conference (2021)

27. Paterson, K.G., Villanueva-Polanco, R.: Cold boot attacks on NTRU. In: Patra, A., Smart, N.P. (eds.) INDOCRYPT 2017. LNCS, vol. 10698, pp. 107–125. Springer, Cham (2017). https://doi.org/10.1007/978-3-319-71667-1_6

28. Poimenidou, E., Adamoudis, M., Draziotis, K.A., Tsichlas, K.: Message Recovery Attack in NTRU through VFK Lattices. Preprint. https://doi.org/10.48550/arXiv.2311.17022

29. Sage Mathematics Software, The Sage Development Team. http://www.sagemath.org

30. Scott Edwards, GoldBug Crypto Messenger (2018). https://compendio.github.io/goldbug-manual/
31. Shor, P.W.: Algorithms for quantum computation: discrete logarithms and factoring. In: 35th Annual Symposium on Foundations of Computer Science, Santa Fe, New Mexico, USA, 20–22 November 1994, pp. 124–134. IEEE Computer Society (1994)
32. Silverman, J.H.: Dimension-reduced lattices, zero-forced lattices, and the NTRU public key cryptosystem. Technical report 13, Version 1, NTRU Cryptosystems (1999)
33. Silverman, H., Smart, N.P., Vercauteren, F.: An algebraic approach to NTRU ($q = 2n$) via Witt vectors and overdetermined systems of non linear equations. In: Security in Communication Networks – SCN 2004. LNCS, vol. 3352, pp. 278–298. Springer (2005)
34. https://www.openssh.com/txt/release-9.0
35. https://www.wolfssl.com/products/wolfssl/

Author Index

A. Dąbrowski et al. (Eds.): NuTMiC 2024, LNCS 14966, p. 331, 2025.
https://doi.org/10.1007/978-3-031-82380-0

The manufacturer's authorised representative in the EU is Springer
Nature Customer Service Centre GmbH, Europaplatz 3, 69115 Heidelberg,
Germany. If you have any concerns regarding our products, please
contact ProductSafety@springernature.com

Printed and bound by CPI Group (UK) Ltd, Croydon, CR0 4YY

27/04/2026

02097586-0009